Energy Development
for Sustainability

Energy Development for Sustainability

Editors

Wei-Hsin Chen
Alvin B. Culaba
Aristotle T. Ubando
Steven Lim

MDPI • Basel • Beijing • Wuhan • Barcelona • Belgrade • Manchester • Tokyo • Cluj • Tianjin

Editors

Wei-Hsin Chen
National Cheng Kung
University
Taiwan

Alvin B. Culaba
De La Salle University
Philippines

Aristotle T. Ubando
De La Salle University
Philippines

Steven Lim
Universiti Tunku Abdul
Rahman
Malaysia

Editorial Office
MDPI
St. Alban-Anlage 66
4052 Basel, Switzerland

This is a reprint of articles from the Special Issue published online in the open access journal *Sustainability* (ISSN 2071-1050) (available at: https://www.mdpi.com/journal/sustainability/special_issues/EnergyDevelop_Sustainability).

For citation purposes, cite each article independently as indicated on the article page online and as indicated below:

LastName, A.A.; LastName, B.B.; LastName, C.C. Article Title. *Journal Name* **Year**, *Volume Number*, Page Range.

ISBN 978-3-0365-4113-6 (Hbk)
ISBN 978-3-0365-4114-3 (PDF)

Contents

About the Editors

Wei-Hsin Chen

Wei-Hsin Chen is a Distinguished Professor at the Department of Aeronautics and Astronautics, National Cheng Kung University, Taiwan. He was a visiting professor at Princeton University, University of New South Wales, University of Edinburg, University of British Columbia, and University of Lorraine. He was also invited as an Invited Lecturer of the University of Lorraine in 2019 and 2020. His research topics include bioenergy (combustion, torrefaction, pyrolysis, gasification, liquefaction, bioethanol, etc.), hydrogen production and purification, clean energy (clean coal technology, wind power, thermoelectric generation, carbon capture and utilization, etc.), energy system analysis (optimization, evolutionary computation, machine learning, etc.), and atmospheric science (aerosol absorption and PM2.5). He has published around 750 papers with an H-index of 67 (Web of Science). His research has a huge contribution and impact on biomass torrefaction. He is the associate editor and editorial board member of several prestigious international journals, and the guest editors of over 10 international journals. He has published 11 books and chapters concerning energy science and air pollution. He received 2015 and 2018 Outstanding Research Award (Ministry of Science and Technology, Taiwan), 2015 and 2020 Highly Cited Paper Award (Applied Energy, Elsevier), 2016, 2017, 2018, 2019, 2020, and 2021 Highly Cited Researcher (Web of Science), 2017 Outstanding Engineering Professor Award (Chinese Institute of Engineers), 2019 Highly Cited Review Article Award (Bioresource Technology, Elsevier), 2020 Top Download Paper Award (International Journal of Energy Research, Wiley), and 2021 TECO Award, 2022 Top Cited Article Award (International Journal of Energy Research, Wiley).

Alvin B. Culaba

Alvin B. Culaba is an Academician (elected 2008) and co-focal person on energy of the National Academy of Science and Technology Philippines, and spearheads the iSTART Program of DOST designed to help accelerate regional economic development through science, technology, and innovation. He is part of the core team that crafted the country's Science and Technology Foresight 2020–2050 called Pagtanaw2050 and helped formulate the Philippine Development Research Agenda 2020–2025. A multi-awarded scientist, teacher, and administrator, Dr. Culaba served as Philippine Energy Adviser (2007–2010) and an independent board member of the Philippine Electricity Market Corporation (2014–2016). As an expert panel member of the Philippine Joint Congressional Commission on Science, Technology, and Engineering (COMSTE) he assessed the S&T competitiveness of the energy sector and was commissioned to undertake energy policy studies that led to the formulation of the country's energy twin-bill, the Biofuels Act of 2006 (RA9367) and the Renewable Energy Act of 2008 (RA9513). His over 200 scientific publications in Web of Science- and Scopus-indexed journals are amongst the highly cited as evidenced by his Google Scholar h-index of 28, i10 index of 58, and 2867 citations. He serves as reviewer and editorial board member of reputable international journals. He was Past President of the National Research Council of the Philippines (NRCP) and the Philippine-American Academy of Science and Engineering (PAASE). He had previously served as energy consultant for the Asian and Pacific Centre for Transfer of Technology of the United National-Economic and Social Commission for Asia and the Pacific and as member of Technical Working Group of ASEAN Energy Centre-Economic Research Institute for ASEAN and East Asia (ERIA) and the National University of Singapore Energy Studies Institute. He was also a Visiting Professor-Scholar at Florida State University, University of Arizona, Tokyo

Institute of Technology, Advanced Institute of Industrial Technology Japan, National Institute of Technology Akashi College Japan, and National Tsinghua University, among others. Dr. Culaba is a Distinguished Full Professor of Mechanical Engineering, University Fellow, current holder of the Go Peng Kuan Professorial Chair in Mechanical Engineering and Mr. & Mrs. Huang Kong Chim Professorial Chair in Chemical Engineering. He is a former Executive Vice President of De La Salle University Manila (2010–2016). He is the co-founder of the DLSU Center for Engineering and Sustainable Development Center where he is presently the Director, and also the principal founder of the Algae Bio-Innovation Global Hub.

Aristotle T. Ubando

Aristotle T. Ubando is the Chairman, a Full Professor, and a Research Fellow at the Department of Mechanical Engineering of De La Salle University, Philippines. He is the head of the Thermomechanical Analysis Laboratory that provides computational simulation design capability to the Semiconductor Industry in the Philippines. He was a recipient of the Jose M. Reyes Chair in Mechanical Engineering in 2022 and the Ubaldus Alphonsus Bloemen FSC Chair in School for Maintenance Engineering Academic Chair in 2021. He was bestowed the Achievement Award by the National Research Council Philippines in 2020 for his contribution to Engineering research. He was selected as one of the 2017 Asian Scientist 100 for his contribution in the sustainability field. Ubando was the recipient of the Outstanding Young Scientists Award in 2016 by the National Academy of Science and Technology of the Philippines. In 2013, he was granted a dissertation attachment scholarship at the University of Arizona and Texas A&M University by the Fulbright Agriculture program. He has published more than 150 Scopus-indexed journal papers at the national and international level with an H-index of 18 (Scopus). Ubando is currently an RDLeader to the National Research Council of the Philippines which aims to strengthen the research capabilities in the academy and in research and development institutions. His research interest includes finite element analysis, computational fluid dynamics, process integration and optimization, and energy system modelling.

Steven Lim

Dr. Steven Lim is currently an Assistant Professor at Department of Chemical Engineering, Lee Kong Chian Faculty of Engineering and Science, Universiti Tunku Abdul Rahman in Malaysia. He is also the chairperson for the Centre for Photonics and Advanced Materials Research. His professional memberships include Graduate Engineer under Board of Engineers Malaysia, Associate member of IChemE, and Senior Member of Hong Kong Chemical, Biological and Environmental Engineering Society. He has worked on numerous local and international research projects including UTARRF, SATU Joint Research Scheme, FRGS, and Sumitomo Foundation related to palm oil and reutilization of palm oil waste such as oil palm fronds, empty fruit bunch, and palm fatty acid distillate. His research works have been published in several international peer-reviewed journals including *Progress in Energy and Combustion Science, Bioresource Technology, Applied Energy,* and *Chemical Engineering Journal.* Currently, he has published more than 60 journal articles, 5 book chapters, and 9 conference proceedings with an h-index of about 22 and has 1941 citations according to Scopus. He has also been involved in several consultancy projects with the industry on the quality improvement for long fibers produced from empty fruit bunch, processing of palm oil sludge, and advanced material research for rubber and polymeric materials.

Preface to "Energy Development for Sustainability"

Development of any nation must proceed sustainably, as we argue, and energy is at the core of any sustainable development. The United Nations Sustainable Development Agenda ensures that "no one is left behind" which means electricity must be accessible, reliable, and affordable; less wasteful and non-pollutive; and explore the use of more renewable and alternative sources of sustainable energy. This book is written primarily to present new research and innovation in various areas of energy development such as resource-efficient energy systems, energy storage, smart grids, clean fuels, sustainable energy pathways, and evidenced-based policies. The last two years have seen many promising applications of energy technologies that help reduce carbon emissions and address societal impacts of global climate change. Discussions, insights, and analyses in this Special Issue provide essential knowledge and information to energy developers, generators, investors, and policy makers. It is believed that energy will remain a challenging area in ensuring a sustainable future.

Wei-Hsin Chen, Alvin B. Culaba, Aristotle T. Ubando, and Steven Lim
Editors

 *sustainability*

Article

Multi-Objective Optimization of an Integrated Algal and Sludge-Based Bioenergy Park and Wastewater Treatment System

Jayne Lois San Juan [1,2,*], Carlo James Caligan [1], Maria Mikayla Garcia [1], Jericho Mitra [1], Andres Philip Mayol [2,3], Charlle Sy [1,2], Aristotle Ubando [2,3] and Alvin Culaba [2,3]

[1] Industrial Engineering Department, De La Salle University, Manila 0922, Philippines; carlo_caligan@dlsu.edu.ph (C.J.C.); maria_mikayla_garcia@dlsu.edu.ph (M.M.G.); jericho_mitra@dlsu.edu.ph (J.M.); charlle.sy@dlsu.edu.ph (C.S.)
[2] Center for Engineering and Sustainable Development Research, De La Salle University, Manila 0922, Philippines; andres_mayol@dlsu.edu.ph (A.P.M.); aristotle.ubando@dlsu.edu.ph (A.U.); alvin.culaba@dlsu.edu.ph (A.C.)
[3] Mechanical Engineering Department, De La Salle University, Manila 0922, Philippines
*   Correspondence: jayne.sanjuan@dlsu.edu.ph

Received: 20 August 2020; Accepted: 16 September 2020; Published: 21 September 2020

Abstract: Given increasing energy demand and global warming potential, the advancements in bioenergy production have become a key factor in combating these issues. Biorefineries have been effective in converting biomass into energy and valuable products with the added benefits of treating wastewater used as a cultivation medium. Recent developments enable relationships between sewage sludge and microalgae that could lead to higher biomass and energy yields. This study proposes a multi-objective optimization model that would assist stakeholders in designing an integrated system consisting of wastewater treatment systems, an algal-based bioenergy park, and a sludge-based bioenergy park that would decide which processes to use in treating wastewater and sludge while minimizing cost and carbon emissions. The baseline run of the model showed that the three plants were utilized in treating both sludge and water for the optimal answer. Running the model with no storage prioritizes water disposal, while having storage can help produce more energy. Sensitivity analysis was performed on storage costs and demand. Results show that decreasing the demand is directly proportional to the total costs while increasing it can help reduce expected costs through storage and utilizing process capacities. Costs of storage do not cause a huge overall difference in costs and directly follow the change.

Keywords: multi-objective optimization; bioenergy; biomass; microalgae; sludge; wastewater

---

1. Introduction

As global energy consumption continuously increases over time due to economic and population growth, fossil fuel utilization persists to be a prominent contributor to meeting the ever-increasing energy demand. Oil, coal, and natural gas have consistently remained the top 3 energy sources by a significant amount from 1990 to 2017 [1]. Greenhouse gas emissions from the use of fossil fuels pose a threat to the environment, as they are a primary factor in global warming [2]. Moreover, the non-renewable nature of fossil fuels poses a threat to the sustainability of traditional energy practices. As such, interest in finding renewable alternative sources of energy that can compete with fossil fuels in terms of performance while having less environmental impact has become prevalent in recent years [3].

Biofuels, such as biodiesel and bioethanol, are potential alternatives as they are derived from renewable biomass and generate less emissions. Biorefineries have been highly utilized in converting biomass into renewable energy and other bio-products. Microalgal biomass is highly regarded for biofuel production in terms of its performance, product outputs, and environmental benefits. However, challenges regarding its high investment and processing costs make it particularly difficult to adopt commercially [4]. Downstream processes, excluding cultivation, make up 60% of the total production cost [5]. Different alternatives per process exhibits varying costs, environmental damage, and output quality that need to be considered in designing an algal-based biorefinery. Integrating a wastewater treatment system with algal cultivation has gained increasing interest. Wastewater contains the necessary nutrients for algae growth, which makes it a viable cultivation medium and wastewater treatment alternative. This lowers the cultivation medium costs of microalgal-based biofuel production, increasing its economic feasibility [6].

Wastewater treatment is an essential process in society due to the contaminants within wastewater generating significant health and environmental risks when left untreated. Waste generation has been continually increasing with population growth, thus requiring more and more wastewater to be treated [7]. Utilizing the different treatment processes inherently produces sewage sludge that is first treated then disposed of. Conventional treatment and disposal methods however do not take advantage of the energy generating capabilities present within wastewater sludge that can benefit the system economically and environmentally. Oladejo et al. [8] reviewed the different conversion methods in recovering energy from sludge. Tradeoffs such as cost, environmental impact, operational capacity, quality limit, bio-product yield, and energy yield of each conversion route presents the importance of resource management, process utilization, and decision making in sludge-to-energy recovery methods.

Bioenergy parks are known facilities that can convert biomass resources into valuable products such as biofuels and power [9]. This allows exchanges of materials between bioenergy plants in maximizing the use of bioenergy products, by-products, and wastes. Exchanges between microalgae and sludge bioenergy parks have been explored in several studies that present beneficial results. Wang et al. [10] determined that the co-pyrolysis of sewage sludge and microalgae yielded greater bio-oil than individual pyrolysis of sludge. On the other hand, Mahdy et al. [11] co-digested sludge and microalgae, which resulted in a higher methane yield compared to the substrates individually digested. Chen et al. [12] explored the utilization of sludge digestate with wastewater in microalgae cultivation, which reduces the total costs of the integrated system. However, it is important to note that using both sludge and microalgae in a given bioenergy production process will require a certain amount of pretreatment before fully realizing the benefits.

The integration of a wastewater treatment system, algal-based bioenergy park, and sludge-based bioenergy park forms a closed-loop system that considers different alternatives in treating water and managing imminent waste, energy, and environmental problems. The connected system can select different options in accommodating varying inputs of wastewater volume and the potential biomass derived from its treatment. Utilizing the biomass produced for energy production would provide a multitude of benefits, as it could bring in profit when sold in the market and reduce environmental impacts through greenhouse gas emission savings. Mathematical models applied on this network can produce insights on the necessary processes to be used, considering both economic and environmental factors that may be beneficial to actual systems that are similar in nature. The proposed study would be an optimization model on the integrated network between wastewater treatment systems and microalgae and sludge bioenergy park focusing, which could be used by stakeholders in planning out the necessary processes to use on the treatment of wastewater and sludge. The study can also encourage wastewater treatment providers to consider finding alternative processes aside from the traditional treatment of water and sludge that could satisfy proper disposal requirements while providing long-term benefits in terms of cost and emissions.

2. Review of Related Literature

The use of mathematical models has been utilized in designing wastewater treatment systems to find the optimal network of operations [13]. This has been adopted in studies in providing adequate treatment for the influent, while avoiding any unnecessary processes. Tsai et al. [14] developed a decision-making framework in handling the incoming wastewater in the system. With the use of dynamic programming, the study focused on minimizing the economic cost incurred in a wastewater treatment system, while adhering to the constraints set to the final disposal limits for both water and sludge. The study was able to show the technologies that should be adopted in the treatment facility that would ensure cleanliness of the effluent while minimizing economic cost. The study, however, was not able to consider the bioenergy generated in the treatment of sludge, which can further reduce the overall cost of the system. Kim et al. [15] further extended the scope of the study by applying a multi-objective optimization model on wastewater treatment systems that minimizes cost, greenhouse emissions, and pollution from the disposed effluent. Using multi-objective optimization is done when two or more naturally conflicting objectives are considered to find an optimal solution that would not optimize one target at the expense of others. Rather, it would find a solution that would find the balance for all objectives. In comparison with the former study, a fixed pathway for sewage and sludge treatment was established to accommodate the incoming wastewater. The research was able to determine the process configurations of the plant including the flow rates, dissolved oxygen concentration, operating temperature, and decision on whether to use sludge biomass as a biogas.

Ang et al. [16] developed a mathematical model by considering different input configurations for the incoming wastewater, multiple available treatment processes, and different disposal and reuse options. Processes in the wastewater treatment facility were optimized using a non-linear programming model that considers both economic cost and adverse environmental impacts from the quality of effluent discharged. They tested the changes in the behavior in the model by adding other disposal and reuse options compared to the initial scenario wherein one disposal option was available and found that the wastewater would need to undergo different pathways to maintain optimality.

Although proven effective in treating wastewater, traditional wastewater treatment facilities still face its major issue of high greenhouse emissions from its processes [17]. Given this scenario, interests in sustainable bioenergy parks as alternatives in water and sludge treatment are continuously rising. These facilities must adopt long-term design and capacity planning to become a sustainable project that could gradually replace fossil fuel consumption and take up more energy market share [4]. The use of mathematical models has been found to be efficient in the design of bioenergy parks into sustainable systems [18,19]. These studies are mostly concerned with handling biomass using various processing technologies with respect to cost and environmental performance.

As such, a comprehensive review shows that microalgae bioenergy parks are mainly concerned with the selection of cultivation systems, harvesting techniques, and conversion processes of microalgae into bioenergy [20]. A mathematical model developed by Gupta et al. [21] aims to minimize the cost of processes for the microalgae biorefinery while satisfying the demand for biodiesel in the market. The study established a fixed cultivation system, harvesting, and conversion process and was able to find the best design for the microalgae processes including the capacity of the tanks, microalgae growth medium, and growth duration. The study focused on fulfilling the bioenergy demand, although it failed to consider the revenue obtained from selling the products, which could substantially improve the economic cost. It also limited the cultivation system and harvesting option. Hoeltz et al. [22] reviewed the whole microalgae process and found different tradeoffs for cultivation systems and harvesting options in terms of biomass productivity, operating cost, and water quality while Ong et al. [23] considered different biomass conversion processes to energy that fundamentally vary in biodiesel yield and operating cost.

Garcia–Preto et al. [24] designed an algal biorefinery to produce different kinds of biofuels focusing on the economic aspect of using the processes in a multi-period horizon. The model was able to find the best cultivation system and conversion process of microalgae into biodiesel, astaxanthin, and polyhydroxy butyrate with considerations of investment costs, operating costs, land requirements, and revenue for selling the products to ensure economic feasibility.

Culaba et al. [25] developed an optimization model focusing on the biofuel production aspect of microalgae biomass that can help in investment planning and operational and expansion decisions. The study was able to consider both the objective of maximizing net present value and minimizing greenhouse gas emission. They were able to show the connections that the biomass would need to undergo and material streams for the inputs to optimize biofuel production. Solis et al. [26] made a similar study utilizing a life cycle assessment methodology for the algal biorefinery to properly consider the carbon footprint of the processes done. The study was able to show which processes are utilized when either cost or environmental impact was optimized and when both objectives were optimized simultaneously. It was shown that different connections were formed with the consideration of different objectives.

A different approach was used by Caligan et al. [27] by integrating an algal biorefinery plant and wastewater treatment plant through a multi-objective optimization model that considers both cost and environmental impact aspects of the system. Wastewater may be treated by a wastewater treatment plant, algal biorefinery plant, or both plants to reach water quality requirements before disposal. Utilizing the biorefinery plant means selecting the proper cultivation system and harvesting option of microalgae, which will be used as biomass for energy production which could either be sold or used in running the facilities to reduce overall expenses. The study was able to show the connection of the processes in both facilities that can be used to treat the wastewater and handle microalgae biomass. With the consideration of the two objectives, both facilities were utilized in treating wastewater to find a balance between the two objectives.

Sludge treatment systems have also been found in the literature. Mathematical models on wastewater treatment plants have considered handling the accumulated sludge; however, they limit sludge handling to a default treatment process [14,15]. The quality of the digestate is not considered in those studies. Providing multiple alternatives for sludge treatment may be added to account for the unpredictability of the quality of incoming sludge. Sludge-to-energy systems are becoming significant in the field of research due to the high potential of sludge as biomass feedstock for energy production. As such, Cao and Pawlowski [28] introduced anaerobic digestion and pyrolysis processes for sewage sludge treatment in the study. Anaerobic digestion of sewage sludge has been the most common process in sludge treatment since it is cost-efficient and results in biogas production. However, this process is not able to fully utilize the organic matters in sludge compared to in pyrolysis. Pyrolysis, on the other hand, provides higher bioenergy production, although its use would incur higher operating cost. Depending on the quality and quantity of the sludge, different choices may be optimal for its treatment.

Different sludge handling approaches that provide treatment as well as generate energy were presented by Mills et al. [29]. Different sets of anaerobic digestion and pyrolysis processes were included and showed their respective energy flows and product. Lam et al. [30] extended the aforementioned study by developing a life-cycle data envelopment analysis on available technologies used by different facilities. The study included aerobic digestion, anaerobic digestion, and incineration as the main sludge treatment processes. The efficiency of the processes was evaluated based on multiple performance metrics which include volatile solids reduction, energy generation, chemical consumption, energy use and recovery, sludge residue generation, and environmental impact. Both aerobic and anaerobic digestion were able to reduce the volatile solids content of sludge; however, neither aerobic digestion nor anaerobic digestion are not as effective as the incineration process. Incineration was able to show high volatile solids reduction and sludge reduction, though it consumed higher energy and resulted in more adverse impacts to the environment. The aforementioned studies regarding sludge treatment, however, were mostly experimental and do not provide the best options to be used for different scenarios.

Vadenbo et al. [31] developed a mixed-integer linear programming for waste incineration which considered both economic and environmental impact objectives. With the consideration of multiple activities, incineration configurations, materials, and the final product, the model was able to optimize the system configuration by providing the best choices based on the established objective. This study, however, is only focused on the incineration process, which may not always be the best treatment to be used for the sludge produced.

The use of pyrolysis and incineration processes, compared to other stabilization processes, generally requires low moisture content to increase the efficiency of the processes and reduce the generation of non-condensable gases [32,33]. Therefore, the sludge is usually subjected to dewatering and drying processes before being processed in these kinds of treatments.

Though recent works have generally been efficient in planning and designing each of the facilities separately, they were not able to analyze their scope from a system perspective since not all factors and relationships were considered. Some studies on wastewater treatment systems were not able to take into consideration the sludge by-product generated in the processes used. Relevant studies, particularly on sludge were mostly experimental studies focusing on the evaluation of different available technologies and providing possible configurations to be implemented by the facilities. These studies did not intend to optimize the processes and hence did not always meet particular targets and constraints for their outputs. The application of optimization models assures us that the design has reached the optimal solution in terms of the objectives specified while meeting all the targets and constraints.

To date, no optimization model has been done considering an integrated network of wastewater treatment and bioenergy production from microalgae and sludge. The proposed model will be able to determine the best treatment and conversion processes to select that would minimize both economic cost and carbon emission depending on the wastewater demand.

3. Network Definition

An overview of the integrated wastewater treatment system (WTS), sludge bioenergy park (SBP), and microalgae bioenergy park (MBP) network is shown in Figure 1. The scope of the system would begin with the input of wastewater to either the WTS or MBP and end in the reuse and disposal of treated wastewater. The system would accept various wastewater sources that are generally found in wastewater systems including industrial wastewater, municipal wastewater, and stormwater runoffs [34]. The accommodated input would have varying contents in terms of volume and constituent concentrations. The WTS would consist of different treatment options, with each having distinct treatment rates for each quality type and generating different characteristics of sewage sludge. The accumulated sludge will be sent to the SBP for sludge treatment. The network would utilize a mixed input configuration in which the entering input in a treatment process would result in an aggregate quality level for the mixture that will serve as a basis for quality improvement. Once done, the output may either go through succeeding treatment options available or be disposed of afterwards if the quality of the effluent has satisfied the requirements for disposal. It may enter the MBP for additional treatment, if necessary, to comply with the requirements.

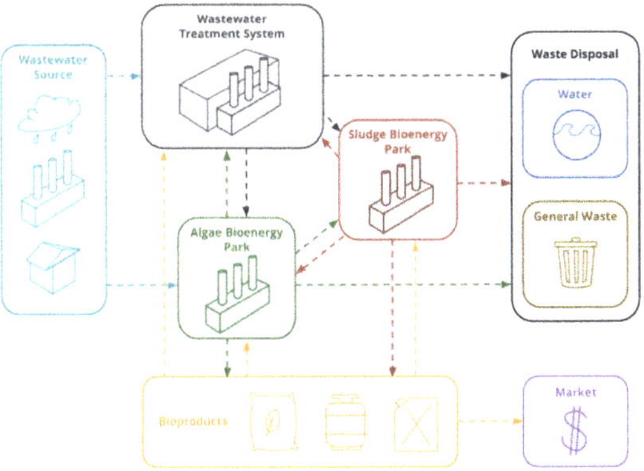

Figure 1. Network Diagram.

The MBP would use wastewater as a medium for growing microalgae in a cultivation system. The microalgae will utilize the contents of the water as nutrients for growth, which in return will serve as a treatment to the water by removing any undesirable contents that are present. It may also be mixed with fresh water to balance the mixture when the contaminant concentrations are deemed undesirable for microalgae growth. Cultivation systems, or the containers used for growing microalgae, are classified as open and closed systems. Digested sludge coming from the sludge bioenergy park can also be sent to the cultivation system in the MBP to increase algal biomass production [35]. The microalgae produced are then collected through different harvesting techniques, which can have varying results in terms of cost and algal biomass yield. The water used as a medium is then disposed of or stored for additional treatment in the following period if necessary. The collected algal biomass then proceeds to the conversion stage where different options are available for generating various bioenergy products.

The SBP is concerned with handling the sludge produced in the WTS, which would initially enter the thickening stage for volume reduction. Thickened sludge may be selected from different stabilization processes available for treatment that may vary in terms of cost, sludge treatment rate, and gas emissions. Using the digestion process for stabilization is also known to produce methane that can be used for energy production [36]. Sludge may also bypass the stabilization stage if advanced treatments are necessary to further remove organic matters and other undesirable contents in sludge, as well as for producing better bioenergy products. Microalgae biomass from the algae bioenergy park can be brought to energy generating processes in the stabilization and advanced treatment stages to be combined with sewage sludge to boost bioenergy production. Treated sludge is then disposed of from the facility and is brought to landfill sites.

Multi-objective optimization has been widely utilized in analyzing and optimizing complex systems including industrial symbiosis and process integration. In particular, the use of mixed integer nonlinear programming (MINLP) models has been applied in optimization studies for designing huge systems such as bioenergy parks and wastewater treatment systems with consideration of economic and environmental aspects [15,16]. This will be used in finding the best connections and material exchanges between facilities. Adopting multi-objective optimization is appropriate for fully capturing both economic costs and environmental impacts simultaneously. This is because focusing on only aspect of a model compromises the other aspects and neglects the tradeoff that exists between the objectives. The goal programming approach has been efficient in previous multi-objective optimization studies in achieving a solution that balances the two objectives [37]. This methodology will be utilized to find the balance between the two objectives considered.

A multi-period system is adopted in the model to recognize the difference in treatment routes to be utilized for sludge and water for dealing with the uncertainty of wastewater demand and quality. The objective components considered by the system would include both costs and environmental impacts. The total economic cost would consist of the operating costs of utilized processes, holding cost of storage use, and disposal cost of the treated water and sludge into output sites. Potential savings and profit from the energy produced by both sludge and microalgae feedstocks are deducted in the cost component. On the other hand, carbon emissions produced from the processes used and disposal in landfill sites are concerned with the environmental considerations in the model. Greenhouse emission savings for the use of the renewable energies generated will be considered.

4. Model Formulation

A mixed integer nonlinear programming model (MINLP) was formulated for the network integrating the three facilities with the goal of finding operational decisions in treating the wastewater demand and handling sludge and algal biomass. This methodology would be able to simultaneously minimize cost and environmental objectives while considering the available treatment processes water and biomass, capacity constraints, disposal options for the treated water and sludge, and market prices of potential bioproducts produced in a bioenergy park.

4.1. Model Nomenclature

Table 1 shows the indices, decision variables, and parameters used in formulating the model.

Table 1. Notations.

	Indices		
i	Input Type	t	Time Period
j	Cultivation System	s	Stabilization Process
k	Wastewater Treatment Process	a	Advanced Treatment Process
e	Algae Conversion Process	b	Bioproduct
m	Quality Type	l	Sludge Disposal Site
n	Harvesting Type	d	Bioproduct to be sold or reused
o	Output site		

	Decision Variables	
U_{ikt}	Volume of water type i to enter WTS process k at time period t	
V_{ijt}	Volume of water type i to enter cultivation system j at time period t	
se_{it}	Volume of water type i to be stored at time period t	
P_{kkt}	Volume of water in WTS from treatment process k to process k at time period t	
stw_{kt}	Volume of water from storage to treatment process k at time period t	
Y_{kot}	Binary, 1; water coming from process k will be disposed of at site o at time period t, 0; otherwise	
wst_{kt}	Volume of water from treatment process k to storage at time period t	
X_{mkt}	Quality type m entering treatment process k at time period t	
ast_{nt}	Volume of water to storage after going through harvesting type n at time period t	
w_{mnt}	Quality type m going through harvesting option n at time period t	
$\text{\DH}_{mit}$	Quality type m of water type i at time period t	
ß_{mt}	Quality type m of water of total water stored at time period t	
sta_{jt}	Volume of water from storage to cultivation system j at time period t	
wd_{kot}	Volume of water from treatment process k to disposal site o at time period t	
C_{jnt}	Volume of water from cultivation type j to harvesting type n at time period t	
fw_t	Volume of freshwater to be used at time period t	
Y_{not}	Binary, 1; water coming from harvesting type n will be disposed at site o at time period t, 0; otherwise	
wd_{not}	Volume of water that went through harvesting type n to disposal site o at time period t	
ε_m	Quality type m of freshwater to be used at time period t	
∂_{mjt}	Quality type m of entering water in cultivation system j at time period t	
bm_{sjt}	Amount of sludge from stabilization process s to cultivation system j at time period t	
P_{mst}	Quality type m of sludge after going through stabilization process s at time period t	
cy_{net}	Amount of algae cells from harvesting type n to conversion process e at time period t	
ra_{nt}	Amount of algae cells from harvesting type n to be sold in the market at time period t	
alg_{nat}	Amount of algae cells from harvesting type n to undergo advanced treatment process a at time period t	
alg_{nst}	Amount of algae cells from harvesting type n to undergo stabilization process s at time period t	
BE_{ebtd}	Amount of bioproducts b produced through conversion process e to become d at time period t	
SS_{st}	Amount of sludge to undergo stabilization process s at time period t	
SS_{at}	Amount of sludge to undergo advanced treatment process a at time period t	
Q_{mt}	Quality type m of sludge produced at time period t	
ds_{st}	Amount of sludge from stabilization process s to be disposed of at time period t	
Ym_{st}	Binary, 1; if microalgae will be mixed with sludge during stabilization process s at time period t	
P_{mat}	Quality type m of sludge after going through advanced treatment process a at time period t	
Ym_{at}	Binary, 1; if microalgae will be mixed with sludge at advanced treatment process a at time period t, 0; otherwise	
y_{slt}	Binary, 1; if sludge from stabilization process s will be disposed of at disposal site l at time period t, 0; otherwise	
y_{alt}	Binary, 1; if sludge from advanced treatment a will be disposed of at disposal site l at time period t, 0; otherwise	
BE_{abtd}	Amount of bioproducts b produced through advanced treatment a to become d at time period t	
BE_{sbtd}	Amount of bioproducts b produced through stabilization process s to become d at time period t	
sd_{alt}	Volume of sludge that went through advanced treatment process a to disposal site l at time period t	
sd_{slt}	Volume of sludge that went through stabilization process s to disposal site l at time period t	

Table 1. *Cont.*

Indices	
Parameters	
dem_{it}	Demand of wastewater type i to be treated at time period t
cap_k	Capacity of water on treatment process k
tr_{mj}	Treatment rate of water going to cultivation system j for quality type m
acr_{mjn}	Amount of algae cells produced from cultivation system j to harvesting option n for quality type m
er_{emb}	Bioenergy product b production rate for biomass in conversion process e for quality type m
vth	Volume reduction rate for sludge entering the thickening process
sr_k	Sludge quantity production rate for water going through treatment process k
er_{abm}	Bioenergy product b production rate for biomass in advanced treatment process a for quality type m
qi_{sm}	Sludge treatment rate for stabilization process s for quality type m
er_{sbm}	Bioenergy product b production rate for biomass in stabilization process s for quality type m
cer_{abm}	Bioenergy product b production rate for combined sludge and algae biomass in advanced treatment process a for quality type m
cer_{sbm}	Bioenergy product b production rate for combined sludge and algae biomass in stabilization process s for quality type m
vr_a	Volume reduction rate for sludge entering advanced treatment process a
vde	Volume reduction rate for sludge entering the dewatering process
D_{ml}	Requirement on sludge quality m in output site l
R_{mo}	Requirement on water quality m in output site o
qi_{am}	Sludge treatment rate for advanced treatment process a for quality type m
N_k	Rate of water lost after going through process k
N_j	Rate of water lost after going through cultivation system j
tr_{mk}	Treatment rate of water going to treatment process k for quality type m
A_{mit}	Water quality type m for water input i entering the system at time t
M	A very large number
cap_j	Capacity of water for cultivation system j
cap_s	Capacity of sludge for stabilization process s
cap_a	Capacity of sludge for advanced treatment a
vr_s	Volume reduction rate for sludge entering stabilization process s
tr_{mn}	Treatment rate of water going through harvesting option n for quality type m
vc_k	Processing cost for water entering treatment process k
vc_j	Processing cost for water entering cultivation system j
vc_n	Processing cost for water to be processed in harvesting option n
vc_e	Processing cost for algae to be processed at conversion process e
vc_s	Processing cost for sludge/biomass to enter stabilization process s
vc_a	Processing cost for sludge/biomass to enter advanced treatment process a
hc	Holding cost for storage
vct	Processing cost for sludge to enter the thickening process
vcd	Processing cost for sludge to enter the dewatering process
dc_o	Disposal cost for water to be discharged at output site o
sdc_l	Disposal cost for sludge to be discharge at output site l
mp_t	Market price for algae at time period t
mp_{btd}	Profit from bioproduct b becoming d at time period t
em_j	Carbon emissions per mass produced in cultivation system j
em_n	Carbon emissions per mass produced in harvesting option n
em_s	Carbon emissions per mass produced in stabilization process s
em_a	Carbon emissions per mass produced in advanced treatment process a
em_o	Carbon emissions per mass produced in disposing water at disposal site o
sem_l	Carbon emissions per mass produced in disposing sludge at disposal site l
ghs_{btd}	Greenhouse emission savings for bioproduct b to become d at time period t

4.2. Constraints

Equation (1) displays the total demand of wastewater to be treated at a certain period. These could be accommodated by either the WTS, MBP, or to be stored to be treated on the following period. This is shown to be equivalent to the wastewater demand for the given period.

$$\sum_k U_{ikt} + \sum_j V_{ijt} + se_{it} = dem_{it} \; \forall \; i,t \tag{1}$$

The following set of equations are concerned with the water being stored during a particular period. Equation (2) shows that the volume of water in the inventory consists of the treated water in the wastewater treatment process, water after going microalgae harvesting, and water planned to be treated for the succeeding periods. The stored water would then be treated in the WTS or used as a cultivation medium for the following period. Ang et al. [16] utilized the method of calculating the total quality of water through the weighted average of the input quality with respect to its volume. This is used to calculate the quality of water stored at the current period and is displayed in Equation (3).

$$\sum_k wst_{kt} + \sum_n ast_{nt} + \sum_I se_{it} = \sum_k stw_{kt} + \sum_j sta_{jt} \; \forall \; t \tag{2}$$

$$\sum_k X_{mkt} (wst_{kt}) + \sum_n w_{mnt} (ast_{nt}) + \sum_I se_{it} (c_5) = \beta_{mt} (\sum_k wst_{kt} + \sum_n ast_{nt} + \sum_I se_{it}) \; \forall m, t \tag{3}$$

The following sets of equations are concerned with the processes done in the wastewater treatment facility. Equation (4) displays the volume quantity of the water entity entering and exiting each treatment process that would be utilized. The constraint provides assurance that wastewater will not return to previous treatment processes. It would move to succeeding treatments in WTS if necessary, be disposed of, or moved to storage. Water coming from the algae bioenergy park (stw_{kt}) can be treated by the wastewater treatment facility in the following period. A percentage of water input (N_k) would be lost due to evaporation and other factors and would therefore be considered for each treatment process. The quality of water entering the treatment process is calculated in Equation (5). This will be equal to the total weighted average of the water entering the process. Equation (6) makes sure that the effluent quality requirements are met before letting water be discharged from the facility.

$$(1 - N_k) [\sum_i U_{ikt} + \sum_{k'} P_{k'kt} + stw_{k(t-1)}] = \sum_{k''} P_{kk''t} + Y_{kot} (wd_{kot}) + wst_{kt} \; \forall k, t$$
$$k' \le k$$
$$k'' \ge k \tag{4}$$

$$\sum_i U_{ikt} (A_{mit}) + \sum_{k'} P_{k'kt} (X_{mk't}) (tr_{mk'}) + stw_{k(t-1)} (\beta_{m(t-1)}) = X_{mkt} (\sum_i U_{ikt} + \sum_{k'} P_{k'kt} + stw_{k(t-1)}) \; \forall \; m,k,t \tag{5}$$

$$X_{mkt} (1 - tr_{mk}) \le R_{mo} + M(1 - Y_{kot}) \; \forall m,k,o,t \tag{6}$$

Equations (7)–(14) are concerned with the processes that are done at the microalgae bioenergy park. The mass balance of the volume of water to enter the cultivation systems available are shown in Equation (7). (N_j) serves as the amount of water lost after the cultivation process. Freshwater is allowed to be combined with the entering wastewater if necessary to improve water quality and reduce the chances of microalgae contamination. The quality of water entering each cultivation system is shown in Equation (8).

$$(1 - N_j) [\sum_I V_{ijt} + sta_{j(t-1)} + fw_t] = \sum_n C_{jnt} \; \forall j,t \tag{7}$$

$$\sum_i V_{ijt} (A_{mit}) + fw_t (\varepsilon_m) + sta_{j(t-1)} (\beta_{m(t-1)}) = \partial_{mjt} (\sum_i V_{ijt} + fw_t + sta_{j(t-1)}) \; \forall j,m,t \tag{8}$$

The volume flow of water to undergo each harvesting option and quality after each process are shown in Equations (9) and (10), respectively. Water used after the harvesting process would either be disposed of or be stored for additional treatment in the succeeding period. Equation (11) ensures that water quality can meet the specified requirements before being discharged from the facility. The microalgae cells produced after going through cultivation and harvesting steps can either be sold, processed further to be converted into a bioproduct, or be co-processed with sludge in the next period, as shown in Equation (12). Equation (13) shows the total amount of algae to be co-processed with sludge in the stabilization and advanced treatment process for the succeeding period. Equation (14) shows the conversion of microalgae cells into bioproducts. The volatile solid content of biomass is associated with the bioenergy production [38]

$$\sum_j C_{jnt} = Y_{not} (wd_{not}) + ast_{nt} \; \forall n,t \tag{9}$$

$$\sum_j C_{jnt} [\varepsilon_{mjt} (1 - tr_{mj})] = w_{mnt} (\sum_j C_{jnt}) \; \forall m,n,t \tag{10}$$

$$w_{mnt} (1 - tr_{mn}) \leq R_{mo} + M (1 - Y_{not}) \; \forall m,n,o,t \tag{11}$$

$$[\sum_j C_{jnt} (\sum_m \varepsilon_{mjt}) + \sum_s \sum_j bm_{sj(t-1)} (P_{mst})] \, acr_{mjn} = \sum_e cy_{net} + ra_{nt} + alg_{nt} \; \forall n,t \tag{12}$$

$$\sum_n alg_{nt} = \sum_s \sum_n alg_{nst} + \sum_a \sum_n alg_{nat} \; \forall t \tag{13}$$

$$\sum_b er_{emb} [\sum_n w_{mnt} (cy_{net})] = \sum_d \sum_b BE_{ebtd} (cy_{net}) \; \forall e,t; \, m = VS. \tag{14}$$

The equations below are concerned with the handling of sludge produced in the system. Equation (15) shows the total amount of sludge produced in the current period that would either go through stabilization or advanced treatment process for treatment. Sludge would undergo the thickening process (vth) first, which would reduce the total volume. Sludge production is dependent on the treatment process of wastewater as well as its quality, particularly the biochemical oxygen demand of the water [39]. Equation (16) tackles the quality of the total sludge produced in the current period.

$$vth \, (\sum_k [sr_k (X_{mkt})][\sum_i \sum_k U_{ikt} + \sum_{k'} \sum_k P_{k'kt} + \sum_k stw_{k(t-1)}]) = \sum_s SS_{st} + \sum_a SS_{at} \; \forall t; \, m = BOD \tag{15}$$

$$\sum_k X_{mkt} [\sum_i U_{ikt} + \sum_{k'} P_{k'kt} + stw_{k(t-1)}] = Q_{mt} (\sum_k [\sum_i U_{ikt} + \sum_{k'} P_{k'kt} + stw_{k(t-1)}]) \; \forall m,t \tag{16}$$

Treatment of sludge and added algae after going through the stabilization process is tackled in Equation (17). Algae produced in the previous period can be added in the stabilization process to enhance the sludge quality, which would ultimately increase the bioenergy production. Stabilized sludge can either be disposed of afterwards or transferred to a microalgae bioenergy park to be added in the cultivation system, as shown in Equation (18). After going through a stabilization process, the sludge would then undergo through the dewatering step, which would further reduce the amount of sludge disposed of. Equation (19) displays the total amount of sludge that came from the stabilization process to be disposed of at a particular period and Equation (20) makes sure that the quality requirement is met before disposal.

$$qi_{sm}[(SS_{st})Q_{mt} + \sum_n alg_{ns(t-1)} (w_{mn(t-1)})] = P_{mst} [(SS_{st}) + \sum_n alg_{ns(t-1)}] \; \forall m,s,t \tag{17}$$

$$\sum_s vr_s(SS_{st}) = (\sum_s \sum_j bm_{sjt} + \sum_s ds_{st}) \; \forall t \tag{18}$$

$$vde \, (ds_{st} + vr_s (alg_{ns(t-1)})) = \sum_l Y_{slt} (sd_{slt}) \; \forall s, t \tag{19}$$

$$P_{mst} \leq D_{ml} + M (1 - Y_{slt}) \; \forall m,l,s, t \tag{20}$$

Equation (21) shows the quality improvement of sludge and microalgae added after going through an advanced treatment process. Unlike the stabilization step, sludge will first go through the dewatering step before the advanced treatment to reduce moisture content, which will consequently reduce the total volume processed. The total amount of sludge disposed after going through an advanced treatment process and meeting quality requirements for disposal are shown in Equations (22) and (23), respectively.

$$qi_{am}[(vde)(SS_{at})Q_{mt} + \sum_n alg_{na(t-1)} (w_{mn(t-1)})] = P_{mat} [(SS_{at})(vde) + \sum_n alg_{na(t-1)}] \; \forall m,a,t \tag{21}$$

$$vde\ (SS_{at} + (alg_{na(t-1)})) = \Sigma_l\ Y_{alt}\ (sd_{alt})\ \forall a,\ t \tag{22}$$

$$P_{mat} \leq D_{ml} + M\ (1 - Y_{alt})\ \forall m,l,a,t \tag{23}$$

Energy production for the stabilization process is shown in Equation (24), while Equation (25) is concerned with the energy production for the advanced treatment process. Different energy rates are considered with sludge processed alone and when additional algae biomass is used. Similar to during energy conversion for algal biomass, the volatile solids contents of the sludge are considered in these sets of equations. Equations (26) and (27) activate the combined energy rate for biomass when microalgae is added for stabilization and advanced treatment processes, respectively.

$$er_{sbm}\ (1 - Ym_{st}) + cer_{sbm}\ (Ym_{st})\ [SS_{st}\ (Q_{mt}) + \Sigma_n\ alg_{ns(t-1)}\ w_{mn(t-1)}] = \Sigma_d\ BE_{sbtd}\ \forall s,t,b;\ m = VS \tag{24}$$

$$er_{abm}\ (1 - Ym_{at}) + cer_{abm}\ (Ym_{at})\ [SS_{at}\ (Q_{mt}) + \Sigma_n\ alg_{na(t-1)}\ w_{mn(t-1)}] = \Sigma_d\ BE_{abtd}\ \forall a,t,b;\ m = VS \tag{25}$$

$$\Sigma_n\ alg_{ns(t-1)} \leq M\ (Ym_{st})\ \forall\ s,t \tag{26}$$

$$\Sigma_n\ alg_{na(t-1)} \leq M\ (Ym_{at})\ \forall\ a,t \tag{27}$$

Equation (28) limits the amount of water entering the treatment process in the WTS. Each cultivation system in the ABP also has a specific capacity that it can handle for a particular period, as presented in Equation (29). The capacity of sludge and algae biomass to enter stabilization and advanced treatment processes is shown in Equations (30) and (31), respectively.

$$\Sigma_i\ U_{ikt} + \Sigma_{k'}\ P_{k'kt} + stw_{k(t-1)} \leq cap_k\ \forall\ k,t \tag{28}$$

$$\Sigma_i\ V_{ijt} + sta_{j(t-1)} + fw_t \leq cap_j\ \forall\ j,t \tag{29}$$

$$SS_{st} + alg_{s(t-1)} \leq cap_s\ \forall\ s,t \tag{30}$$

$$SS_{at} + alg_{a(t-1)} \leq cap_a\ \forall\ a,t \tag{31}$$

4.3. Objective Function

The model has been developed to maximize the performance of both cost and environmental objectives. A balance between the two objectives is achieved by maximizing the least desirable value to avoid the optimization of one objective and sacrificing the other as seen in Equation (32). This methodology has been utilized in multiple studies in striking a balance between two objectives, particularly cost and environmental impact [25,26] As such, the overall efficiency value of the system is the lower efficiency value between the cost and the environmental impact. The efficiency values are calculated by dividing the value of the achieved improvement (difference between worst and actual values) to the potential improvement (difference between worst and best values). These values correspond to the results of the bi-objective optimization being compared to its single objective counterparts. The best values ($Cost_{best}$ and Env_{best}) for cost and environmental impact are obtained through the application of single objective optimization for the respective objective. The worst values ($Cost_{worst}$ and Env_{worst}), on the other hand, are obtained by treating one particular objective as a system variable and optimizing the other conflicting objective. In the case of $Cost_{worst}$, it is the cost value when a single-objective optimization is set for environmental impact and for Env_{worst}, it is set for optimizing cost.

$$Max\ Z = min\ [\{(Cost_{worst} - Cost)/(Cost_{worst} - Cost_{best})\},\ \{(Env_{worst} - Env)/(Env_{worst} - Env_{best})\}] \tag{32}$$

4.3.1. Economic Objective

The total economic objective function shown in Equation (33) includes the operating cost for using different available processes for water, sludge, and microalgae, holding cost for the total amount of water held in storage, the cost of freshwater entering the cultivation system, the disposal cost of water and sludge discharged from the facilities, and the deduction of cost from potential savings from the biomass produced. The operating cost for each process on the wastewater treatment facility (vc_k) is considered on a per liter basis and is incurred depending on the amount of water to be treated. The cost of holding water to be treated in the succeeding period is defined as (hc). The operating costs for microalgae cultivation and harvesting are also calculated on a per liter basis and are defined as (vc_j)and (vc_h), respectively. Using freshwater for growing microalgae would incur additional costs (cc) if this medium is added. Processing of microalgae biomass for biofuel production would incur additional operating costs (vc_e). The sludge would be treated in the sludge bioenergy park through either stabilization or advanced treatment options, which the costs are represented as (vc_s) and (vc_a), respectively. Finally, the disposal cost of water (dc_o) and sludge (sdc_l) are considered depending on the amount released.

The bioproducts and raw microalgae may be sold in the market resulting in profits, while the bioproducts may be used as energy to run the facilities, which is tackled in Equation (34). The microalgae produced may be sold afterwards without further processing, which gives a profit of (mp_t). The profit and utilization of the biodiesel produced by sludge and algae bioenergy parks are represented by (mp_{btd}). These are considered as savings and would lessen the total economic cost incurred by the system.

$$
\begin{aligned}
\text{Cost} = {} & \Sigma_k \, \Sigma_t \, [\Sigma_I \, U_{ikt} + \Sigma_{k'} \, P_{k'kt} + stw_{k(t-1)}] \, (vc_k) + \Sigma_j \, \Sigma_t \, (\Sigma_n \, C_{jnt} + \Sigma_s \, bm_{sjt}) \, (vc_j) + fw_t(cc) + \Sigma_n \, \Sigma_t \\
& (\Sigma_j \, C_{jnt}) \, (vc_n) + \Sigma_e \, \Sigma_t \, (\Sigma_n \, CY_{net} + alg_{et}) \, (vc_e) + \Sigma_t \, (\Sigma_k \, stw_{kt} + \Sigma_j \, stw_{jt}) \, (hc) + \Sigma_s \, \Sigma_t \, (SS_{st}) \, (vc_s) + \\
& \Sigma_a \, \Sigma_t \, (SS_{at})(vc_a) + \Sigma_s \Sigma_t \, alg_{st} \, (vc_s) + \Sigma_a \Sigma_t \, alg_{at}(vc_a) + \Sigma_e \Sigma_t \, (CY_{net} + alg_{et}) \, (vc_e) + \Sigma_t \Sigma_o (\Sigma_k \, wd_{kot} + \\
& \Sigma_n \, wd_{not})(dc_o) + \Sigma_l \, \Sigma_t \, (\Sigma_s \, sd_{slt} + \Sigma_a \, sd_{alt}) \, (sdc_l) - \text{Savings}
\end{aligned}
\tag{33}
$$

$$
\text{Savings} = \Sigma_t \, ra_t \, (mp_t) + \Sigma_d \, \Sigma_b \, \Sigma_t \, (\Sigma_e \, BE_{ebtd} + \Sigma_a \, BE_{abtd} + \Sigma_s \, BE_{sbtd}) \, (mp_{btd}) \tag{34}
$$

4.3.2. Environmental Impact Objective

Carbon emissions obtained from using the processes available in each of the facilities and the disposal of water and sludge are considered in the environmental impact shown in Equation (35). Carbon emissions in the wastewater treatment facility are calculated depending on the amount of water treated on each process (em_k). Cultivation and harvesting processes would also incur emissions depending on the amount of water to be processed and are represented by (em_j) and (em_n), respectively. Carbon emissions in the conversion process of microalgae are calculated depending on the amount of biomass processed for each process (em_e). Emissions from stabilization and advanced treatments of sludge are defined as (em_s) and (em_a), respectively. Lastly, disposal of the treated water and sludge would also add carbon emissions, which are represented as (em_o) and (sem_l), respectively.

The use of bioproducts for energy in running the facilities are considered as greenhouse emission savings shown in Equation (36). (ghs_{btd}) represents the greenhouse emission reduction from the generation of bioenergy products, which will reduce the overall environmental impact caused by the system.

$$
\begin{aligned}
\text{Env} = {} & \Sigma_k \, \Sigma_t \, (\Sigma_I \, U_{ikt} + \Sigma_{k'} \, P_{k'kt} + stw_{k(t-1)}) + \Sigma_j \, \Sigma_t \, ((\Sigma_n \, C_{jnt} + \Sigma_s \, bm_{sjt}) \, (em_j) + \Sigma_n \, \Sigma_t \, \Sigma_j \, (C_{jnt}) \, (em_n) + \\
& \Sigma_s \, \Sigma_t \, (SS_{st} + alg_{st}) \, (em_s) + \Sigma_a \, \Sigma_t \, (SS_{at} + alg_{at}) \, (em_a) + \Sigma_o \, \Sigma_t \, (\Sigma_k \, wd_{kot} + \Sigma_n \, wd_{not})(em_o) + \Sigma_l \, \Sigma_t \\
& (\Sigma_s \, sd_{slt} + \Sigma_a \, sd_{alt}) \, (sem_l) - \text{Greenhouse Emission Savings}
\end{aligned}
\tag{35}
$$

$$
\text{Greenhouse Emission Savings} = \Sigma_d \Sigma_b \Sigma_t \, [\Sigma_e \, BE_{ebtd} + \Sigma_a \, BE_{abtd} + \Sigma_s \, BE_{sbtd}] \, (ghs_{btd}) \tag{36}
$$

5. Model Validation

The model has been implemented in MATLAB and solved using IBM CPLEX, with a solution time of 186.5 min on a Dell XPS 15 with a 2.8 GHz Intel Core i5 processor and 8 GB 2133 MHz LPDDR3 RAM. The case study considered four available wastewater treatment processes in the WTS, two cultivation systems, two harvesting options, two options for stabilization, and two advanced treatment processes. Four quality types were considered for the entering water and sludge and two disposal sites were included for both water and sludge treated in a span of two time periods. The parameter values used were based on various literature sources shown in the Appendix A. Table A1 shows the quality requirements for water and sludge discharge while the cell production rate after passing through the cultivation and harvesting processes are shown in Table A2. Tables A3 and A4 summarize the quality improvement rate of the processes for water and sludge, respectively. Energy production rate for the processes in the bioenergy parks are shown in Table A5. The cost and carbon emission parameters used in the model are summarized in Tables A6 and A7, respectively.

Running the model through individual optimization first would collect the worst possible outcome for the other objective, as shown in Table 2. In order to compute the model results for multi-objective function, the model maximized the efficiency of both costs and environmental impacts (also shown in Table 2), which resulted in an efficiency rate of 73.72%. The individual optimizations have also proved that the tradeoffs considered need to be met by both objectives. The economic objective was able to choose technology and pathways that yield in higher biomass from closed systems and co-processing of microalgae and sludge.

Table 2. Objective Function of Baseline Scenario.

Objective Function	Economic Cost (in Dollars)	Environmental Impact (in kg)
Economic cost only	25,555.36	28,476.57
Environmental impact only	265.33	217.63
Dual optimization	26,323.14	230.17

Based on the results obtained, all three of the individual facilities were utilized by the model to some extent as displayed in Figure 2. In the first period, the wastewater was treated using the algae bioenergy park by going through cultivation system 2 and harvesting option 2. Since the treatment in the ABP is not sufficient for reaching effluent discharge requirements, further treatment needed to be done in the following period. Water was held in storage, which incurs a holding cost. It was then transferred to the WTS in the following period. This pathway implies the importance of the connection of the three separate systems in obtaining the optimal solution. As in this case, the MBP could still be used to treat the stored water; however, the model decided to transfer it to the WTS since only minimal treatment is required to achieve the effluent requirements. Simply using the MBP again to treat the stored water may only lead to unnecessary costs being incurred.

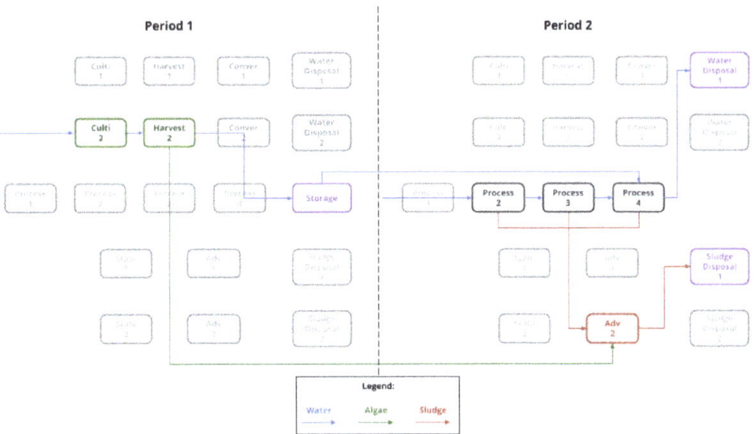

Figure 2. Treatment Network of Baseline Scenario.

The wastewater demand in period 2 was treated solely using the processes available in the wastewater treatment plant before disposal. In treatment process 4 in the WTS, the wastewater was mixed with the stored wastewater from period 1 before receiving treatment and being sent to the disposal site. The sludge produced in this period was treated using advanced treatment option 2 before being disposed of. The algae produced in the first period was also transferred to the sludge plant for co-processing to further increase bioenergy production. Table 3 shows the summary of the decision pathways used in the scenario.

Table 3. Decision Variables of Baseline Scenario

Wastewater Treatment Plant		Algae Bioenergy Park		Sludge Bioenergy Park	
Amount of wastewater treated (L)	10,000	Amount of wastewater treated (L)	9000	Amount of sludge produced (kg)	1381.16
Amount of wastewater disposed (L)	16,781.75	Amount of wastewater used as cultivation medium (L)	9000	Amount of sludge to be used as biomass to algae park (L)	0
Amount of sludge disposed (kg)	583.88	Amount of bioenergy products produced (kg)	171.72	Amount of bioenergy products produced (kg)	84.26

In analyzing the importance of storage types in each major step or process of the system, all types were disregarded in the base model run. Restrictions in the transfer of water and biomass between plants were implemented in this scenario. With this setup, there are no water exchanges between the WTS and MBP, or biomass transfer between the MBP and SBP. As seen in Table 4, the economic and environmental objectives were relatively closer to each other; however, the efficiency of the model was only 50.03%. In comparison to the previous model, it was able to find a more efficient balance between the two conflicting objectives.

Table 4. Objective Function without forms of Storage.

Objective Function	Economic Cost (in Dollars)	Environmental Impact (in kg)
Economic cost only	23,383.78	29,312.21
Environmental impact only	388.88	337.11
Dual optimization	26,347.85	362.99

The decision pathways located in Figure 3 also display some differences as compared to the previous model with different types of storage available. Since transfer of effluent is not allowed, the algae bioenergy park was not utilized since the treatment that the facility provides does not reach the discharge requirements. While this removes the holding cost incurred for retaining water to be treated in the succeeding period, it also restricts the flexibility of processes that wastewater may go through in receiving full treatment. It also prohibits combining sludge and algae for increased biomass production. Hence, in the first period, processes 2 and 4 in the WTS were utilized to treat the wastewater before disposal. The sludge produced was treated using stabilization process 1 and disposed of afterwards. On the other hand, processes 2, 3, and 4 were necessary to treat the wastewater in period 2. The sludge produced was treated using the 2nd advanced treatment option available and was disposed of afterwards. Table 5 shows the summary of the decision variables of the scenario without storage.

Table 5. Decision Variables without Storage.

Wastewater Treatment Plant		Algae Bioenergy Park		Sludge Bioenergy Park	
Amount of wastewater treated	19,000 L	Amount of wastewater treated	0 L	Amount of sludge produced	2210.6032 kg
Amount of wastewater disposed	16,696.25 L	Amount of wastewater used as cultivation medium	0 L	Amount of sludge to be used as biomass to algae park	0 kg
Amount of sludge disposed	983.9764 kg	Amount of bioenergy products produced	0 kg	Amount of bioenergy products produced	104.36 kg

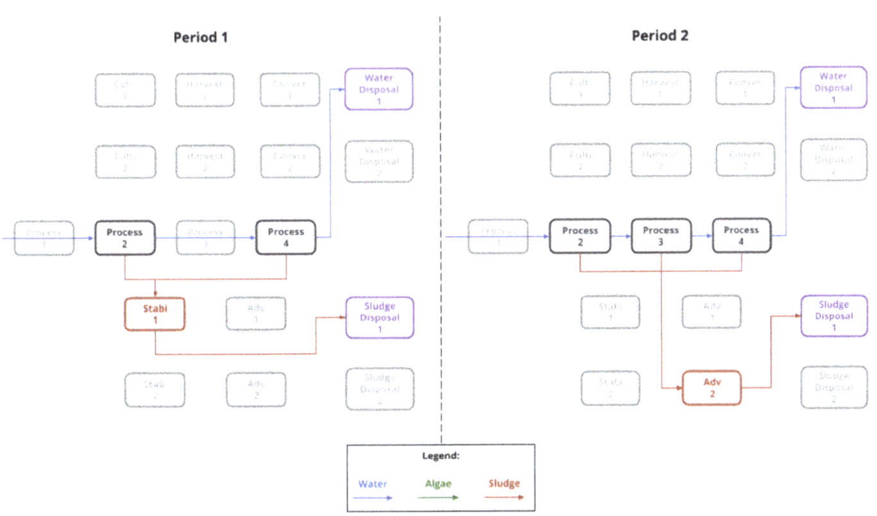

Figure 3. Treatment Network of Scenario with No Storage.

The two scenarios tested above were analyzed below. The cost slightly increased with the scenario with no storage allowed compared to the baseline scenario, as shown in Figure 4. Although no holding cost was incurred for with the restriction of storages, it also had an impact on the potential increase in profit by combining the sludge and algae biomass, which slightly increased the economic component overall. In terms of the environmental objective, higher carbon emissions were obtained when no storage was allowed. Solely relying on the WTS for treating the wastewater demand highly affects the objective compared to having the option to use the microalgae treatment. Emissions of both the baseline and no storage scenarios are shown in Figure 5.

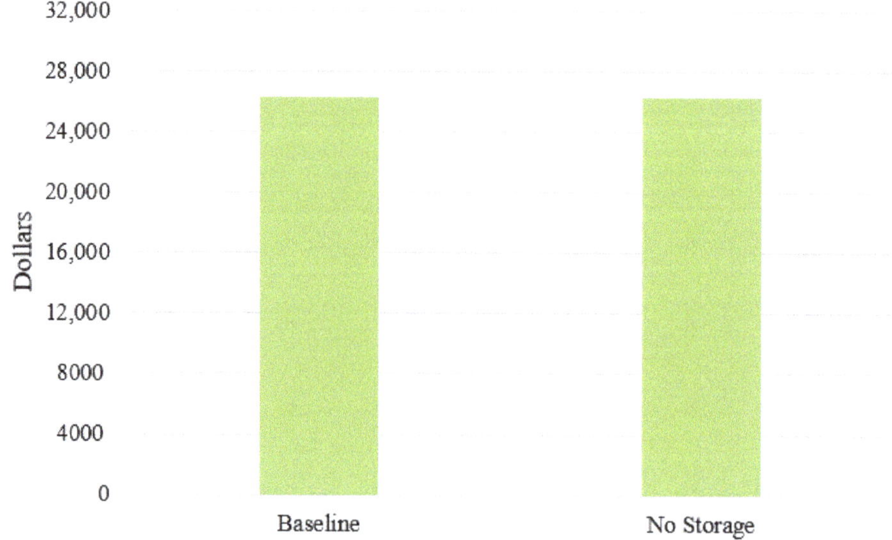

Figure 4. Comparison of Optimal Cost Results.

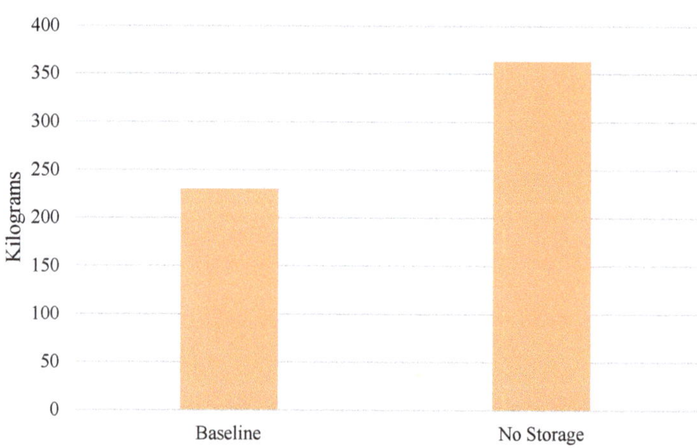

Figure 5. Comparison of Optimal Emissions Results.

6. Scenario Analysis

Due to the nature of uncertainty of available resources such as wastewater and inputs, a scenario analysis was done through sensitivity analysis. The sensitivity analysis varied the demand with increments of 5% starting from −20% up to +20%. As seen on Figure 6, the behavior of the costs relative to the percent increase of the demand. Decreasing the current demand by up to 20% yielded lower costs linearly. This is expected since all facilities involved would have less resources to use and process, with less savings to produce. In the portion of increasing demand, once it was increased by 5%, it had a slight decrease in costs and continued to spike up linearly. The slight decrease came from the cultivation variable choosing to store the excess instead of directly processing it all at once. During the following period, it processed and mixed with the new inputs and chose to produce algae and energy with the more expensive energy conversion options, as these options yield more bioenergy products. This continues as demand increases to 20% as long as capacities for cultivation and processes are large enough to accommodate these inputs, since this would save enough processing costs.

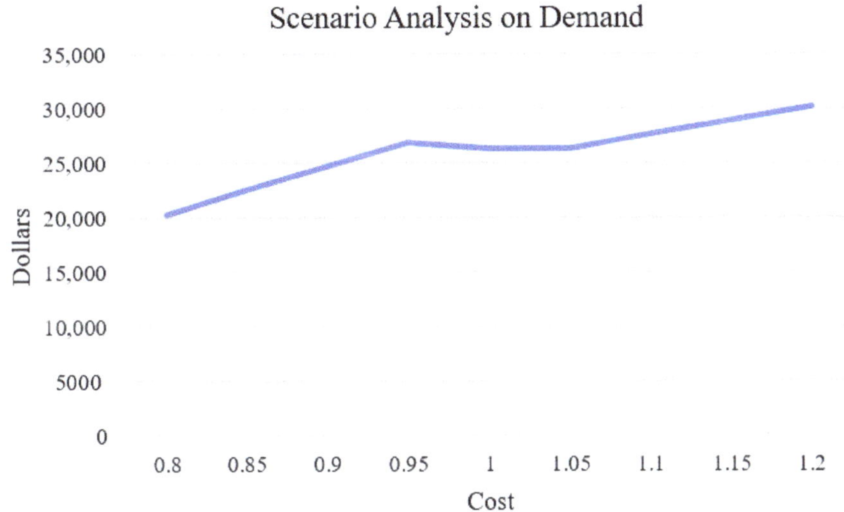

Figure 6. Sensitivity Analysis of Demand.

Due to the nature of inventory costs, the changes in prices were also analyzed over time as shown in Figure 7. The costs were increased and decreased by 20% from the original inventory costs, in increments of 5% each. Unlike demand which changed its behavior midway, inventory costs showed a linear progression over time as costs increased. The results did not change from the preceding time period, which means costs have to be significantly large to change from the original decision variables.

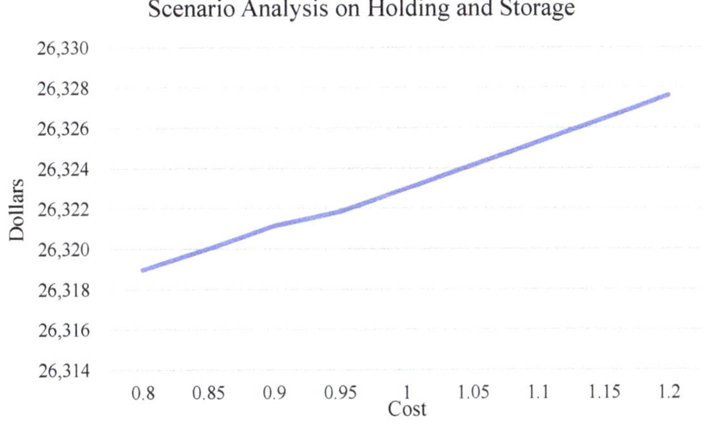

Figure 7. Sensitivity Analysis of Storage Costs.

7. Conclusions

The study has proposed a multi-objective optimization model for integrating wastewater treatment systems and sludge and algae-based bioenergy parks that considers water and sludge quality before discharging them from their respective facilities. The model can be used by stakeholders in finding the best processes to utilize in an integrated system that could treat the wastewater demand and sludge generated while minimizing both economic costs and carbon emissions. It will encourage treatment providers to consider a flexible treatment system rather than sticking to the traditional wastewater and sludge treatment by showing them the benefits in terms of costs and emissions. The system contains alternative treatments for wastewater and sludge to manage the uncertainty in the quality and volume of the incoming water. Solely optimizing either a cost or environmental objective could lead to high value for the other due to the conflicting nature between the two objectives, thus goal programming methodology was utilized to strike a balance between the two components. In running the baseline model with dual objective optimization, the whole system was able to incur a cost of 26,323.14 dollars and carbon emissions of 230.17 kg and was able to utilize the three plants. A scenario wherein no storage was allowed was conducted to see the effect on the objective functions if no water exchanges between facilities for treatment would be considered, or co-processing of algae and sludge biomass. With this scenario, the cost incurred reached 26,347.85 dollars and there were carbon emissions of 362.99 kg. The slight increase in cost when no storage was allowed was attributed to lower profits provided by sludge alone compared to combined microalgae and sludge. Having no storage immensely increased the overall carbon emission, since generally, the emissions of the processes in the wastewater treatment plant are relatively higher compared to those in the algae bioenergy park. Increasing the demand of inputs also causes a change in behavior of the model. The model would utilize the storage and process the input during the following period to fill up the capacity of processes to lessen costs (as seen in Figure 6). Additionally, changes in inventory costs were investigated over time and showed that the results remained the same even if costs were gradually increased, resulting in a linear proportion with time (as seen in Figure 7). Thus, inventory costs would have to increase significantly in order for storage to be unnecessary.

Future works may include the effect of storing microalgae and sludge biomass on biomass quality for future processing in accommodating different energy demands for each period. Moreover, the parameters used in the study are literature-based and may be too optimistic compared to the actual production. As such, the combined energy production for biomass may vary depending on the ratio of microalgae and sludge that needs to be added. Different pathways may also be available for water and sludge treatment, while other processes for microalgae cultivation and energy conversion may be available that could be added in a future study.

Author Contributions: Conceptualization, J.L.S.J.; Data Curation, C.J.C., M.M.G., and J.M.; Formal Analysis, C.J.C., M.M.G., and J.M., Supervision, A.P.M., J.L.S.J., C.S., A.C., and A.U.; Validation, C.J.C., M.M.G., and J.M.; Writing–original draft, C.A.C., M.M.G., and J.M.; Writing–Review, C.J.C., M.M.G., J.M., A.P.M., J.L.S.J., C.S., A.C., and A.U. All authors have read and agreed to the published version of the manuscript.

Funding: This research received no external funding.

Acknowledgments: The authors wish to acknowledge the financial support from the Office of the Vice Chancellor of Research and Innovation of De La Salle University.

Conflicts of Interest: The authors declare no conflict of interest.

Appendix A

Table A1. Quality requirements.

Quality Type	Quality Requirement	Quality Type	Quality Requirement
BOD	70	P	5
N	14	VS (sludge)	90

Table A2. Microalgae Cell Rate Production.

| | Cultivation System | | | | | |
| | Open | | | Closed | | |
	N	**P**	**VS**	**N**	**P**	**VS**
Floating	0.00012	0.00012	0.00012	0.00013	0.00013	000013
Flotation	0.0009	0.0009	0.0009	0.00012	0.00012	0.00012

Table A3. Quality Improvement Rate for Water.

	BOD	**N**	**P**
Wastewater treatment plant			
Pretreatment	0.54	0.54	0.56
Primary treatment	0.7	0.7	0.82
Secondary treatment	0.4	0.4	0.34
Tertiary treatment	0.08	0.08	0.13
Cultivation system			
Closed	0.7	0.28	0.3
Open	0.7	0.28	0.3
Harvesting option			
Floating	0.02	0.01	0.01
Filtration	0.01	0.01	0.01

Table A4. Quality Improvement Rate for Sludge.

Process	**VS**	**Process**	**VS**
Anaerobic digestion	25.1%	Incineration	95%
Aerobic digestion	25.1%	Pyrolysis	69%

Table A5. Energy Production Rate.

| | Microalgae | | | Sludge | | Co-Processing | |
	Biodiesel	**Biogas**		**Biodiesel**	**Biogas**	**Biodiesel**	**Biogas**
Anaerobic digestion	0.42	0.25	Aerobic digestion	0.20	0.20	0.33	0.3628
Gasification	0.42	0.25	Incineration	0.05	0.0299	0..05	0.0299
			Pyrolysis	0.2523	0.1501	0.58	0.216

Table A6. Cost parameter.

Wastewater Treatment Facility	
Pretreatment	0.42
Primary treatment	0.65
Secondary treatment	0.37
Tertiary treatment	0.3
Microalgae Bioenergy Park	
Cultivation system	
Closed	0.8
Open	0.7

Table A6. *Cont.*

Harvesting option	
Floating	0.1
Filtration	0.2
Conversion option	
Anaerobic digestion	2
Gasification	1.9
Sludge Bioenergy Park	
Stabilization process	
Anaerobic digestion	2
Aerobic digestion	2
Advanced treatment process	
Incineration	1.75
Pyrolysis	2.1
Other Costs	
Wastewater disposal	0.01
Sludge disposal	0.43
Freshwater cost	0.5
Holding cost	0.0003

Table A7. Emission Parameter.

Process	Emission
Wastewater treatment processes	0.007
Cultivation system-Open/Closed	0
Harvesting option-Floating	0.0001
Harvesting option-Filtration	0.0002
Conversion option-Anaerobic digestion/Gasification	0.04
Stabilization Option-Anaerobic/Aerobic digestion	0.04
Advanced Treatment-Incineration	0.08
Advanced Treatment-Pyrolysis	0.04
Water disposal	0.0005
Sludge disposal	0.0005

References

1. The International Energy Agency. Electricity Generation by Fuel and Scenario, 2018–2040. 2019. Available online: https://www.iea.org/fuels-and-technologies/electricity (accessed on 30 June 2020).
2. Smith, C.J.; Forster, P.M.; Allen, M.; Fuglestvedt, J.; Millar, R.J.; Rogelj, J.; Zickfeld, K. Current fossil fuel infrastructure does not yet commit us to 1.5 °C warming. *Nat. Commun.* **2019**, *10*, 101. [CrossRef] [PubMed]
3. Butturi, M.A.; Lolli, F.; Sellitto, M.A.; Balugani, E.; Gamberini, R.; Rimini, B. Renewable energy in eco-industrial parks and urban-industrial symbiosis: A literature review and a conceptual synthesis. *Appl. Energy* **2019**, *255*, 113825. [CrossRef]
4. Culaba, A.B.; Lim Ching, P.; San Juan, J.L.G.; Philip Mayol, A.; Sybingco, E.; Ubando, A.T. A Dynamic Sustainability Assessment of Algal Biorefineries for Biofuel Production. In Proceedings of the 2019 IEEE 11th International Conference on Humanoid, Nanotechnology, Information Technology, Communication and Control, Environment, and Management, Laoag, Philippines, 29 November–1 December 2019; pp. 3–6. [CrossRef]
5. Kim, J.; Yoo, G.; Lee, H.; Lim, J.; Kim, K.; Kim, C.W.; Park, M.S.; Yang, J.W. Methods of downstream processing for the production of biodiesel from microalgae. *Biotechnol. Adv.* **2013**, *31*, 862–876. [CrossRef] [PubMed]
6. Cheah, W.Y.; Ling, T.C.; Show, P.L.; Juan, J.C.; Chang, J.S.; Lee, D.J. Cultivation in wastewaters for energy: A microalgae platform. *Appl. Energy* **2016**, *179*, 609–625. [CrossRef]

7. Mateo-Sagasta, J.; Liqa, R.; Thebo, A. Wastewater: Economic asset in an urbanizing world. *Wastewater Econ. Asset Urban. World* **2015**, 1–282. [CrossRef]

8. Oladejo, J.; Shi, K.; Luo, X.; Yang, G.; Wu, T. A review of sludge-to-energy recovery methods. *Energies* **2019**, *12*, 60. [CrossRef]

9. Benjamin, M.F.D.; Ubando, A.T.; Razon, L.F.; Tan, R.R. Analyzing the disruption resilience of bioenergy parks using dynamic inoperability input–output modeling. *Environ. Syst. Decis.* **2015**, *35*, 351–362. [CrossRef]

10. Wang, X.; Zhao, B.; Yang, X. Co-pyrolysis of microalgae and sewage sludge: Biocrude assessment and char yield prediction. *Energy Convers. Manag.* **2016**, *117*, 326–334. [CrossRef]

11. Mahdy, A.; Mendez, L.; Ballesteros, M.; González-Fernández, C. Algaculture integration in conventional wastewater treatment plants: Anaerobic digestion comparison of primary and secondary sludge with microalgae biomass. *Bioresour. Technol.* **2015**, *184*, 236–244. [CrossRef]

12. Di Chen, Y.; Li, S.; Ho, S.H.; Wang, C.; Lin, Y.C.; Nagarajan, D.; Chang, J.S.; Ren, N. Integration of sludge digestion and microalgae cultivation for enhancing bioenergy and biorefinery. *Renew. Sustain. Energy Rev.* **2018**, *96*, 76–90. [CrossRef]

13. Boix, M.; Montastruc, L.; Azzaro-Pantel, C.; Domenech, S. Optimization methods applied to the design of eco-industrial parks: A literature review. *J. Clean. Prod.* **2015**, *87*, 303–317. [CrossRef]

14. Tsai, J.C.C.; Chen, V.C.P.; Beck, M.B.; Chen, J. Stochastic dynamic programming formulation for a wastewater treatment decision-making framework. *Ann. Oper. Res.* **2004**, *132*, 207–221. [CrossRef]

15. Kim, D.; Bowen, J.D.; Ozelkan, E.C. Optimization of wastewater treatment plant operation for greenhouse gas mitigation. *J. Environ. Manag.* **2015**, *163*, 39–48. [CrossRef] [PubMed]

16. Ang, M.; Duyag, J.; Tee, K.; Sy, C. A multiple input type optimization model integrating reuse and disposal options for a wastewater treatment facility. *Chem. Eng. Trans.* **2018**, *70*, 199–204. [CrossRef]

17. Campos, J.L.; Valenzuela-Heredia, D.; Pedrouso, A.; Val Del Río, A.; Belmonte, M.; Mosquera-Corral, A. Greenhouse Gases Emissions from Wastewater Treatment Plants: Minimization, Treatment, and Prevention. *J. Chem.* **2016**, *2016*, 1–12. [CrossRef]

18. Benjamin, M.F.D.; Tan, R.R.; Razon, L.F. A methodology for criticality analysis in symbiotic bioenergy parks. *Energy Procedia* **2014**, *61*, 41–44. [CrossRef]

19. Ng, R.T.L.; Ng, D.K.S.; Tan, R.R. Optimal planning, design and synthesis of symbiotic bioenergy parks. *J. Clean. Prod.* **2015**, *87*, 291–302. [CrossRef]

20. Wu, W.; Chang, J.S. Integrated algal biorefineries from process systems engineering aspects: A review. *Bioresour. Technol.* **2019**, *291*, 121939. [CrossRef]

21. Gupta, S.S.; Bhartiya, S.; Shastri, Y. Model-based optimisation of integrated algae biorefinery. *IFAC Proc. Vol.* **2014**, *47*, 1011–1018. [CrossRef]

22. Hoeltz, M.; Silvana, M.; Moraes, A.; Cassia, R.; Schneider, D.S. Microalgae: Cultivation techniques and wastewater phycoremediation. *J. Environ. Sci. Health Part A Toxic Hazard. Subst. Environ. Eng.* **2015**, *50*, 585–601. [CrossRef]

23. Ong, H.C.; Chen, W.H.; Farooq, A.; Gan, Y.Y.; Lee, K.T.; Ashokkumar, V. Catalytic thermochemical conversion of biomass for biofuel production: A comprehensive review. Renewable and Sustainable. *Energy Rev.* **2019**, *113*, 109266. [CrossRef]

24. García Prieto, C.V.; Ramos, F.D.; Estrada, V.; Villar, M.A.; Diaz, M.S. Optimization of an integrated algae-based biorefinery for the production of biodiesel, astaxanthin and PHB. *Energy* **2017**, *139*, 1159–1172. [CrossRef]

25. Culaba, A.B.; San Juan, J.L.; Ching, P.M.; Mayol, A.P.; Sybingco, E.; Ubando, A. Optimal synthesis of algal biorefineries for Biofuel production based on techno-economic and environmental efficiency. In Proceedings of the 2019 IEEE 11th International Conference on Humanoid, Nanotechnology, Information Technology, Communication and Control, Environment, and Management, Laoag, Philippines, 29 November–1 December 2019. [CrossRef]

26. Solis, C.A.; Mayol, A.P.; San Juan, J.G.; Ubando, A.T.; Culaba, A.B. Multi-objective optimal synthesis of algal biorefineries toward a sustainable circular bioeconomy. *IOP Conf. Ser. Earth Environ. Sci.* **2020**, *463*, 6–12. [CrossRef]

27. Caligan, C.J.A.; Garcia, M.M.S.; Mitra, J.L.; Mayol, A.P.; San Juan, J.L.G.; Culaba, A.B. Multi-objective optimization of water exchanges between a wastewater treatment facility and algal biofuel production plant. *IOP Conf. Ser. Earth Environ. Sci.* **2020**, *463*. [CrossRef]

28. Cao, Y.; Pawłowski, A. Sewage sludge-to-energy approaches based on anaerobic digestion and pyrolysis: Brief overview and energy efficiency assessment. *Renew. Sustain. Energy Rev.* **2012**, *16*, 1657–1665. [CrossRef]

29. Mills, N.; Pearce, P.; Farrow, J.; Thorpe, R.B.; Kirkby, N.F. Environmental & economic life cycle assessment of current & future sewage sludge to energy technologies. *Waste Manag.* **2014**, *34*, 185–195. [CrossRef]

30. Lam, C.M.; Hsu, S.C.; Alvarado, V.; Li, W.M. Integrated life-cycle data envelopment analysis for techno-environmental performance evaluation on sludge-to-energy systems. *Appl. Energy* **2020**, *266*, 114867. [CrossRef]

31. Vadenbo, C.; Hellweg, S.; Guillén-Gosálbez, G. Multi-objective optimization of waste and resource management in industrial networks—Part I: Model description. *Resour. Conserv. Recycl.* **2014**, *89*, 52–63. [CrossRef]

32. Jiao, F.; Zhang, L.; Dong, Z.; Namioka, T.; Yamada, N.; Ninomiya, Y. Study on the species of heavy metals in MSW incineration fly ash and their leaching behavior. *Fuel Process. Technol.* **2016**, *152*, 108–115. [CrossRef]

33. Domínguez, A.; Menéndez, J.A.; Inguanzo, M.; Pís, J.J. Production of bio-fuels by high temperature pyrolysis of sewage sludge using conventional and microwave heating. *Bioresour. Technol.* **2006**, *97*, 1185–1193. [CrossRef]

34. Amoatey, P.; Bani, R. Wastewater Management. In *Waste Water—Evaluation and Management*; InTech: London, UK, 2011.

35. Amenorfenyo, D.K.; Huang, X.; Zhang, Y.; Zeng, Q.; Zhang, N.; Ren, J.; Haung, Q. Microalgae Brewery Wastewater Treatment: Potentials, Benefits and the Challenges. *Int. J. Environ. Res. Public Health* **2019**, *16*, 1910. [CrossRef] [PubMed]

36. Xu, H.; Li, Y.; Hua, D.; Zhao, Y.; Mu, H.; Chen, H.; Chen, G. Enhancing the anaerobic digestion of corn stover by chemical pretreatment with the black liquor from the paper industry. *Bioresour. Technol.* **2020**, *306*. [CrossRef] [PubMed]

37. Rollan, C.D.; Li, R.; San Juan, J.L.; Dizon, L.; Ong, K.B. A planning tool for tree species selection and planting schedule in forestation projects considering environmental and socio-economic benefits. *J. Environ. Manag.* **2018**, *206*, 319–329. [CrossRef] [PubMed]

38. Oosterkamp, W.J. Use of Volatile Solids from Biomass for Energy Production. In *Bioenergy Research: Advances and Applications*; Elsevier: Amsterdam, The Netherlands, 2014; pp. 203–217. [CrossRef]

39. Canziani, R.; Spinosa, L. Sludge from wastewater treatment plants. In *Industrial and Municipal Sludge*; Butterworth-Heinemann: Oxford, UK, 2019. [CrossRef]

 *sustainability*

Article

Biofuel from Microalgae: Sustainable Pathways

Alvin B. Culaba [1,2], Aristotle T. Ubando [1,2,3,*], Phoebe Mae L. Ching [4,5], Wei-Hsin Chen [6,7,8] and Jo-Shu Chang [7,9,10]

[1] Mechanical Engineering Department, De La Salle University, 2401 Taft Avenue, Manila 0922, Philippines; alvin.culaba@dlsu.edu.ph

[2] Center for Engineering and Sustainable Development Research, De La Salle University, 2401 Taft Avenue, Manila 0922, Philippines

[3] Thermomechanical Laboratory, De La Salle University, Laguna Campus, LTI Spine Road, Laguna Blvd, Biñan, Laguna 4024, Philippines

[4] Industrial Engineering Department, De La Salle University, 2401 Taft Avenue, Manila 0922, Philippines; pmlching@connect.ust.hk

[5] Industrial Engineering and Decision Analytics Department, Hong Kong University of Science and Technology, Clear Water Bay, Kowloon, Hong Kong 100025, China

[6] Department of Aeronautics and Astronautics, National Cheng Kung University, Tainan 701, Taiwan; chenwh@mail.ncku.edu.tw

[7] Research Center for Smart Sustainable Circular Economy, Tunghai University, Taichung 407, Taiwan; changjs@mail.ncku.edu.tw

[8] Department of Mechanical Engineering, National Chin-Yi University of Technology, Taichung 411, Taiwan

[9] Department of Chemical and Materials Engineering, College of Engineering, Tunghai University, Taichung 407, Taiwan

[10] Department of Chemical Engineering, National Cheng Kung University, Tainan 701, Taiwan

* Correspondence: aristotle.ubando@dlsu.edu.ph

Received: 5 September 2020; Accepted: 23 September 2020; Published: 28 September 2020

Abstract: As the demand for biofuels increases globally, microalgae offer a viable biomass feedstock to produce biofuel. With abundant sources of biomass in rural communities, these materials could be converted to biodiesel. Efforts are being done in order to pursue commercialization. However, its main usage is for other applications such as pharmaceutical, nutraceutical, and aquaculture, which has a high return of investment. In the last 5 decades of algal research, cultivation to genetically engineered algae have been pursued in order to push algal biofuel commercialization. This will be beneficial to society, especially if coupled with a good government policy of algal biofuels and other by-products. Algal technology is a disruptive but complementary technology that will provide sustainability with regard to the world's current issues. Commercialization of algal fuel is still a bottleneck and a challenge. Having a large production is technical feasible, but it is not economical as of now. Efforts for the cultivation and production of bio-oil are still ongoing and will continue to develop over time. The life cycle assessment methodology allows for a sustainable evaluation of the production of microalgae biomass to biodiesel.

Keywords: algae; biofuel production; environmental policy; life cycle assessment

---

1. Introduction

To sustain our population's growth and advancement, it is critical that renewable sources of energy be found. While it is important for energy to be produced without harmful emissions and long-term damage to the environment, it is equally important for the form of energy to be reproducible over extended periods. Biofuels are a form of renewable energy by the fact that they are generated in shorter cycles, as compared to the geological processes required to generate fossil fuels. This property gives them two major advantages: consistency and scalability. Unlike sunlight and wind, which are subject to the variability of weather and other external forces, the volume of energy produced can be expected based on internal process parameters. The fact that these process parameters can be decided beforehand also means that the volume of production and the choice of technology are within our control.

While the concept of biofuel is promising in itself, the following questions remain: (1) what are the best sources or raw material for biofuels? (2) what are the processes involved in transforming the raw energy source into biofuel? (3) what are the optimum set-ups for implementing these processes? and finally, (4) how can biofuels replace fossil fuels? For all of the aforementioned questions, one of the key difficulties is the variety of options available. In the selection of a raw material for biofuel production, there are countless forms of biomass available. Each raw material source has its own advantages and disadvantages, and requires a different set of processes to be converted into biofuel. Furthermore, each process has its own considerations, such as the choice of equipment and operating procedures. The extensive amount of options available complicated decisions for the industry and academia alike. The research interests of both sectors require them to invest in testing and discovering effective methods of producing biofuels. The industry, in particular, faces difficulty in shifting from fossil fuels to biofuels, and potentially from inferior to superior biofuels when the availability arises.

Yet the efforts regarding biofuel have yielded some promising directions for biofuel production. In particular, microalgae have been proven to be an efficient and productive source of energy. In terms of infrastructure, they offer a higher oil yield than other biomass sources [1,2]. Table 1, containing a comparison between the productivity metrics of microalgae and those of other energy crops, shows the former's superior performance in every metric. In this sense, algae-based biofuel requires less land to produce the same amount of oil as other energy crops. This negates the need for large refineries, that would need to be situated in remote areas, resulting in high distribution costs. In terms of operational cost, algae-based biofuels can be cultivated and produced with wastewater, which was found to significantly reduce production costs when implemented [3]. These biofuels are generally easy to cultivate, not requiring any further processing aside from being kept in a growth medium [2].

Table 1. Comparison of biodiesel feedstock (adapted from [1]).

Plant Source	Seed Oil Content (% Oil by Weight in Biomass)	Biodiesel Productivity (kg Biodiesel/year)	Photosynthetic Efficiency (%)	Oil yield (L/ha)	References
Corn	44	152	0.79	172	
Coconut	50	2367	2.40	2689	[1]
Sugarcane	53	3696	2.24–2.59	-	
Palm Oil	36	4747	3.20	5950	
Canola/rapeseed (*Brassica napus L.*)	41	862	-	974	[2]
Microalgae (30% oil by wt.)	80	51,927	2.00–6.48	58,700	[1]
Microalgae (70% oil by wt.)	70	121,104	2.30–15.0	136,900	

While this addresses the first of the aforementioned questions (i.e., "What are the best sources or raw material for the biofuel?"), there are still the next three to consider. To address the second question ("What are the processes involved in transforming the raw energy source into biofuel?"), there are various methods of algae cultivation and oil production that may be evaluated. Various set-ups have been developed, with each seeking to introduce an improvement or a solution to a problem in algae-based biofuel production. Such problems include the trade-offs between lipid production and growth rate for algae. To address the third question ("What are the optimum set-ups for implementing these processes?"), requires the evaluation of decisions on facility planning, logistics, and co-production. Finally, the issue of adoption (posed by the question "How can biofuels replace fossil fuels?") involves business model design, plans for expansion, and the anticipation of the consequences for society if biofuels were to replace fossil fuels to a large extent.

It is clear that, in order to produce and use algae-based biofuels in a sustainable manner, many factors would need to be considered. This problem requires a holistic perspective, promoting the complete identification of all components involved in production; able to consider all possible outcomes of a decision; and extending this evaluation to the environmental, social, and techno-economic implications of the decision. In identifying sustainable pathways for biofuel production from microalgae, this chapter will follow the approach of a Life Cycle Assessment (LCA). It is an internationally-standardized tool for evaluating the environmental performance of a single product [4]. Through its cradle-to-grave perspective, all the relevant processes are identified and accounted for. Impact assessment converts all material consumption and emission into logical implications.

This chapter is segmented into six parts. In this part, the issue of the sustainability of biofuels was decomposed into key questions, each representing a challenging aspect of biofuel production. In the second part, a collection of solutions from the academic literature, addressing each question, is presented. In the third part, the most viable solution alternatives are further evaluated from an LCA cradle-to-grave perspective, providing insights on the general direction of progress in this area. The challenges of sustaining growth and feasibility are evaluated separately from the other questions, as they follow a different timeline. The fifth part contains the practical implications of the proposed solutions, segmented by the nature of their impact as environmental, technical, economic, and social. Finally, new directions and ideas for the refinement of existing solutions are discussed in the sixth part.

2. Sustainable Technological Solutions

Understandably, one of the most effective ways of improving the sustainability of a product is to utilize contemporary technologies. In this part, the technologies will be divided into two major groups: those involved in algae cultivation and those involved in processing or production. The algae cultivation stage covers the growth and drying of algae. It prepares the raw materials for the state where they may be used for biofuel production. This is primarily a natural state and does not require processing, although the following sections will discuss some methods and set-ups that can influence the outcomes of cultivation. On the other hand, processing covers the phases where dried algae are converted into biofuel through chemical processes. These are primarily technology-driven processes, and the corresponding section will discuss various means of converting dried algae into energy.

2.1. Algae Cultivation Technology

Based on the nature of solutions, the algae cultivation stage can be further dissected into two parts: the choice of algae species to cultivate and the cultivation set-up for the algae. The former concerns the inherent properties of algae. Specific species have been identified as especially apt for lipid production. However, high lipid production is usually indicative of lower growth rates, which is a disadvantage of species with this characteristic. Table 2 shows a summary of studies on specific species of algae and the advantages that were identified with each species.

Table 2. Summary of algae species characteristics.

Algae Species	Advantage	Source
Chlorella, Dunaliella, Chlamydomonas, Scenedesmus, Spirulina	Appropriate for bioethanol production	[5]
Scenedesmus	Higher yield than duckweed, its feedstock has a higher heating value.	[6]
Chlorella	Yields bio-oil with low oxygen content and a comparatively high heating value when subjected to pyrolysis.	[7]
Nannocholoropsis	Yields bio-oil with lower oxygen content when subjected to pyrolism with higher amounts of the catalyst HZSM-5.	[8]
Chlorococcum, Chlorella vulgaris	Can be mass cultivated with ease using current farming technology; shows applicability in ethanol production.	[9,10]
P. ellipsoidea, S. almeriensis	High energy yield without the disadvantage of high enzyme requirements, high capital cost, and being energy intensive, as experienced with other species.	[11]

There have also been attempts to genetically modify existing species to promote desirable qualities of algae species. In particular, there is much interest to enhance lipid production due to its association with energy yield. Microalgae are highly suitable for genetic manipulation, with its amenity for genetic transformation with foreign genes. A summary of successful genetic modifications is given in Table 3.

Table 3. Summary of successful genetic modifications of algae species.

Algae Species	Modification	Impact	Source
Madhuca indica, Balanites aegyaptiaca	De-oiled seedcake	Fast-growing, with lesser dependence on insecticides and fertilizer.	[12]
Clostridium tyrobutyricum	Over-expressions of aldehyde/alcohol dehydrogenase	Enhanced butanol production and tolerance of butanol.	[13]
Escherichia coli	Modification of amino acid biosynthetic pathway	Higher production of 1-butanol and 1-propanol.	[14]

Another way of improving the output from cultivation is through the very process of cultivation. At the moment, three forms of cultivation can be observed: open-culture systems (see Figure 1), closed-culture systems (see Figure 2), and hybrid two-stage cultivation. Open-culture systems are set-ups where algae are allowed to grow in open raceway ponds, subject to the conditions of the environment. This means that minimum human intervention is needed in cultivating the algae, but it also means that some yield potential is lost. This is not a problem with closed-culture systems, where the environment is controlled, yet these advantages may be offset by the higher operating and capital cost required by such facilities. Two-stage hybrid cultivation systems attempt to achieve a balance between the advantages of either system. In this case, algae are grown in enclosures, but allowed to yield in open ponds, less susceptible to yield loss from contaminants. The use of optimal cultivation techniques can simultaneously lower resource consumption (and thereby cost), while promoting a high energy yield for any species.

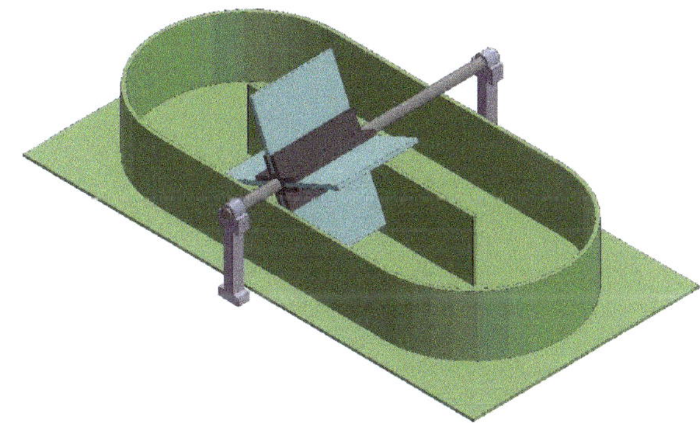

Figure 1. Open-culture set-up illustration.

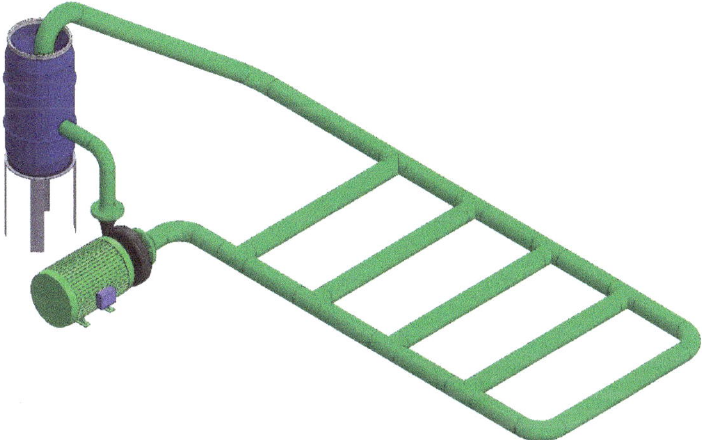

Figure 2. Closed-culture set-up illustration.

New approaches to the cultivation and growth of microalgae were developed and designed. Recently, an indoor hybrid helical-tubular photobioreactor was proposed by Hashemi et al. [15] for beta-carotene production *Dunaliella salina* cells under salt stress condition. A newly developed microgravity-capable membrane raceway photobioreactor for *Chlorella vulgaris* SAG 211-12 was introduced by Helisch et al. [16] for life support in space. On the other hand, Khalekuzzaman et al. [17] developed a hybrid anaerobic baffled reactor and photobioreactor for a simplified method of algal biofuel production. It is important for algal biofuel production to be simple and cost-effective to make it competitive with commercially available biofuels.

2.2. Biofuel Production Technology

There are also various methods that have been developed for biofuel production. Studies that develop or investigate the use of these methods are able to give insight on the optimal parameters for each process. A summary of developments in this area are given in Table 4.

Table 4. Summary of biofuel production processes.

Process	Description	Recent Developments
Biochemical Conversion	Utilization of bacteria and other microorganisms in breaking down the carbohydrate content of algae into biofuels. Includes: Fermentation, anaerobic digestion	Increasing yield of feedstock, finding more efficient bacteria enzymes and more robust microorganisms.
Thermochemical Conversion	Application of high-temperature technologies to convert biomass into alcohol, hydrocarbon fuels, chemicals, and power. Includes: Gasification, pyrolysis, liquefaction	Identification of high-yielding gas mixtures for the conversion process, assessment of various algae species for aptness toward pyrolysis.
Chemical Reaction (Transesterification)	Conversion of oil into biodiesel by utilizing excess methanol.	Identification of seed oils which have a high heating value when converted into biodiesels.
Direct Combustion	Conversion of solid residue from microalgal biomass using heat and energy to produce electricity.	Studying and standardizing microalgal biomass allocation of solid residue.

2.3. Conversion of Microalgae to Biofuel

There are several processes involved in converting microalgae to biofuel. These processes are the following: Cultivation, Harvesting, Drying, and Oil Extraction. Several authors have reviewed these processes over the years. [1] published an article that tackles the conversion of microalgae to biofuel. His discussion concluded that biofuel conversion will be more efficient if the cultivation process is optimized. In addition, he also made several recommendations on what type of cultivation systems to use to enhance microalgae production. Following this study, [18] reviewed the technologies for algal biofuel production, its processing, and extractions. It proposed coupling algal conversion to biofuel with carbon sequestration and wastewater treatment in order to enhance the yield and cut down its production costs. Not just in the macro side of production but also looking deeper inside the algal cell, [19] discussed the possibility of algal production via genetic alteration to produce more algal oil.

2.3.1. Cultivation and Species of Microalgae

Microalgae have various commercial applications in the areas of pharmaceuticals, cosmetics, aquaculture, nutraceuticals, and energy [20–23]. Dating back to 1939–1945, during World War II, microalgae were cultured by the military for food purposes [24,25]. In addition, as a response to environmental issues in the 1950s, algae were used to capture carbon dioxide (CO_2) [26]. Hence, microalgae have paved the way for addressing issues such as climate change, energy demand, and food security.

Microalgae offer higher oil yield per litter per hectare (see Table 5). Based on the table, microalgae have 100× yield of oil to convert to biodiesel. Microalgae have a lot of species discovered and undiscovered. Table 6 shows the different microalgae species and their oil content.

Table 5. Chemical composition of microalgae (adapted from [27]).

Microalgae	Protein	Carbohydrates	Lipid
Scenedesmus obliquus	50–56	10–17	12–14
Scenedesmus quadricauda	47	-	1.9
Scenedesmus dimorphus	8–18	21–52	16–40
Chlamydomonas rheinhardii	48	17	21
Chlorella vulgaris	51–58	12–17	14–22
Chlorella pyrenoidosa	57	26	2
Spirogyra sp.	6–20	33–64	11–21
Dunaliella bioculata	49	4	8
Dunaliella salina	57	32	6
Euglena gracilis	39–61	14–18	14–20
Prymnesium parvum	28–45	25–33	22–39
Tetraselmis maculata	52	15	3

Table 6. Oil content of microalgae (adapted from [1]).

Microalgae	Oil Content (%Dry wt)
Botryococcus braunili	25–75
Chlorella sp.	28–32
Crypthecodinium cohnii	20
Cylindrotheca sp.	16–37
Cylindrotheca sp.	23
Dunaliella primolecta	25–33
Isochrysis sp.	>20
Monallanthus salina	20–35
Nannochloris sp.	31–68
Neochloris oleoabundans	35–54
Nitzschia sp.	45–47
Phaeodactylum tricornutum	20–30
Schizochytrium sp.	50–77
Tetraselmis sueica	15–23

Culturing microalgae for microalgae production was discussed in [1], which analyzed the biorefinery concept, and advances in photo-bioreactor engineering will further reduce the cost of production. This was followed by a research using an empirical and critical analysis to translate laboratory research into a full-scale commercialization application [3]. However, [28] looked at the cost, energy balance, and environmental impacts of algal biofuels and highlighted the uncertainties for microalgae production.

In culturing microalgae there are several parameters that we need to consider, such as light, CO_2, Temperature, pH level, and, most importantly, mixing. Microalgae are photosynthetic cells; they need light to grow. Hence, light intensity is an important aspect of growing microalgae.

There are different ways to illuminate light in the algal culture; one is with sunlight, and by using artificial ponds. These lighting systems also vary with respect to the size of the culture system that one is growing. Hence, the bigger the area that one has to cultivate, such as the open pond cultivation system, the bigger the light needed, such as sunlight. In addition, if we stick to photo-bioreactors such as flat plates, tubular, airlifts, or inclined photo-bioreactors, artificial light will suffice. There are several studies that use light as a light source (flat plate PBR and tubular PBR). Comparing both systems, flat plate PBR has higher light capture than tubular PBR, since flat plate has higher surface area than the tubular PBR [29]. There are several studies dealing with light control, since natural solar light cannot be controlled. The main focus of these studies is on PBR systems using artificial light. To give sufficient light to microalgae, the optimal range of wavelength must be between 600 and 700 nm [26].

Borowitzka [24] discussed the commercialization of the production of microalgae. He concluded that there is a need for an indoor or closed system in algal cultivation, so that there will be a controlled environment which can be more efficient in terms of yield. Hydrothermal liquefaction technology shows promising results in the production of algal biofuels, as it enables the processing of wet microalgae biomass to produce biofuels. Recent works on hydrothermal liquefaction for algal biofuels are discussed as follows. Devi and Parthiban [30] proposed the application of hydrothermal liquefaction on microalgae Nostoc ellipsosporum was cultivated in municipal wastewater to generate high bio-oil yield. Dandamudi et al. [31] applied hydrothermal liquefaction on Cyanidioschyzon merolae and Salicornia bigelovii Torr. for the production of high-valued biofuel intermediates. Arun et al. [32] used hydrothermal liquefaction on Scenedesmus obliquus to produce biohydrogen. The employment of hydrothermal liquefaction on microalgae biomass to biofuel production provides positive benefits in terms of cost-effectiveness, value-addition, and environmental impact of the algal biofuels.

2.3.2. Harvesting

Harvesting of algae means the detachment of algal cells to the growth medium of the cultivation system. Methods of harvesting rely on the type of algae being harvested, depending on the density, size, or even the description of the final product of the produced algae [3,33,34]. Different harvesting methods are used in algae technology. There are mechanical, chemical, biological, or even electrical means of harvesting the algae.

The harvesting process is one of the important processes in algal production in terms of cost, accounting for 30% of the overall production cost for biofuel [35]. The harvesting techniques listed are used on the application of biofuel, human and animal food, high valued products, and water quality restoration. These harvesting techniques passed all the six important criteria—biomass quantity and quality, cost, processing time, species specific, and toxicity. Coagulation/flocculation, centrifugation, and filtration are amongst the harvesting techniques that can be feasible in biofuel production [36].

2.3.3. Drying and Dewatering

Drying is one of the main bottlenecks in algae to biofuel production, as it accounts for around 84.9% of the total energy consumed in the whole production line [37]. There are different methods used in drying microalgae for biofuel production, such as solar dying, convective drying, rotary drying, spray drying, crossflow drying, vacuum drying, and flashing drying. Convective drying was experimented on algae with variation of temperature and wind velocity to determine the lipid yield of algae [38]. Prakash et al. [39] used a solar device to dry microalgae, specifically the spirulina and scenedesmus species. Solar drying is assumed to be an efficient drying method. However, due to weather-dependent reasons and the fact that other countries do not have enough sunlight during a 24-h cycle, the use of solar dying is a disadvantage. Moreover, when the sample ceases to dehydrate, the nutrient value in the species dissipates [40]. Hence, other forms of drying emerged, such as microwave drying of algal species [41].

2.3.4. Oil Extraction

Oil extraction is the process where the algal product is separated from the dried biomass. Aside from dewatering and drying, this is process also entails considerable cost and effort. Different methods, such as chemical and physical methods, are available. However, the cost of extracting the algal product is much higher than that of extracting oil from palms [1].

3. Applications of Sustainable Technologies

The application of the technologies discussed in the previous section are critical for meeting our Sustainable Development Goals. In particular, they are seen as a means to achieve the obsolescence of the small and portable power-generating machines being used in rural communities, which consume fossil fuels and are environmentally inefficient. As most of these rural communities have access to biomass feedstock, which can be used in biofuel production, this could be a highly suitable alternative for their energy requirements. However, this is still dependent on the availability of the choice of technology and system design.

The number and variety of technologies available for producing algae-based biofuels call for a holistic and impartial means of comparing different methods of production. As previously mentioned, the Life Cycle Assessment (LCA) framework is one such method. A standard LCA has the following components: (1) goal and scope definition, (2) life cycle inventory, (3) impact assessment, and (4) interpretation. In goal and scope definition, a "functional unit" will be decided for the product in question. All material components and emissions generated by a product will be given in quantities of the functional unit (e.g., grams of apple fruit per box of apple juice). System boundaries will also be set, which will be the scope of the study. Although an LCA would typically entail that the scope be set from "cradle-to-grave", or raw material sourcing until the final disposal, there are some cases where it may not be practical to adopt a scope of this magnitude. In life cycle inventory, all material components and emissions are identified and quantified according to the functional unit. In this sense, the material components and emissions of a product should add up to the actual mass of the product itself. These will be converted into environmental impact in the impact assessment phase. The outcome is the impact on the quantified impact on the environment. This will be the basis for further analysis and solution development in the interpretation phase.

For biofuels, a cradle-to-biofuel scope can be adopted, starting from energy crop cultivation until biofuel production. This allows for the evaluation of the technologies discussed in the previous section. Whereas most LCA studies are typically from "cradle-to-grave", with grave indicating the disposal or any other form of expulsion from the system, this is impractical in the case of energy production, given that the energy can be used for a multitude of purposes. From an LCA, considering environmental impact alone, the most favorable alternative would be the use of natural solar power in cultivation and solar-assisted drying techniques following cultivation. Energy consumption is high under closed-culture cultivation, and the use of solar power can reduce energy consumption. However, it must be noted that the unreliability of solar power, particularly during inclement weather, may also prove to be a disadvantage.

In many cases, algal biorefineries producing biofuels alone are found to be economically infeasible, regardless of how efficient the current processing technologies are. However, the opportunity to produce co-products is a definite way of making these production systems feasible. Hydrogens, polymers, and biochemicals are just some examples of the kinds of co-products that can be produced. Some of the biofuel production processes identified in Table 4 yield co-products. Table 7 shows a summary of these co-products and their purpose in the biorefinery.

Table 7. Summary of co-products and purpose.

Processes	Co-Product	Purpose
Biochemical Conversion (Fermentation)	Carbon dioxide	Can be recycled for use in cultivation.
Chemical Reaction (Transesterification)	Glycerol and methanol	Co-products can be commercialized externally.

More comprehensive are the available biorefinery models published in literature and their corresponding co-products. These have been proven to be feasible from industrial, techno-economic, and socio-economic perspectives. A summary of the models developed is given in Table 8.

Table 8. Summary of biorefinery models with co-products.

Feedstock	Main Product	Co-Product(s)	Performance Measure	Source
Lignocellulosic, Macroalgae, and Microalgae	Biodiesel and bioethanol	1. Hydrogen 2. Polymers 3. Biochemicals	Research and development	[42]
Biomass	Biofuels	1. Power 2. Char 3. Bio-oil 4. Gaseous fuels	Technology	[43]
Biomass	Biofuels	1. Solvent surfactants 2. Petrol derivatives 3. Composites 4. Lubricants 5. Pastes	Industrial Metabolism	[44]
Macroalgae	Bioethanol	Energy deficit	Production	[45]
Microalgae: C. reinhardtii, C. kessleri, E. gracilis, A. plantesis, S. obliquus, D. salina	Biogas	Biohydrogen	Production	[46]
Microalgae	Electricity	1. Biofuels 2. Omega-3 3. Algal meal 4. Biochemicals	Sustainability and energy policy	[47]
Microalgae: Botryococcus braunii, Chlorella	Petrol fraction and Biodiesel	Jet fuel	Technology evaluation	[48]
Microalgae	Biodiesel	1. Algal meal 2. Omega-3 fatty acids 3. Glycerin	Economics and water footprint	[49]

4. Case Study: A Life-Cycle-Based Production of Biodiesel from Microalgae

4.1. The Life Cycle Assessment (LCA) Process

LCA is a comprehensive environmental evaluation tool for holistically examining the product life cycle. It consists of four major steps: (1) the goal and scope, (2) the life cycle inventory, (3) the life cycle impact assessment, and (4) the interpretation of results. The first step is to define the goal of the LCA study through the establishment of the functional unit. After defining the goal, the next step is to decide on the scope by illustrating the system boundary of the study. An example of the system boundary is the cradle-to-grave, which indicates that the analysis includes the initial process from the raw material extraction to the preparation, the processing, the utilization, and the disposal. The system boundary also provides the degree of depth of the impacts of the indirect materials used in processing the product under study. The second step is to gather all relevant information of the input and output of the processes and the analogous environmental impact to establish the life cycle inventories. The third step is to calculate the environmental impacts and quantify the environmental burden in terms of the various environmental impact categories. The last step is the interpretation of the results, which leads to the identification of the environmental bottleneck and the recommendations to address the identified concern.

4.1.1. The Goal and Scope

The functional unit for the LCA case study is 1000 kg of biodiesel. This case study considers algal biomass feedstock. The processes consist of the cultivation, harvesting, drying, oil extraction, and transesterification. The generic system boundary for the microalgae-based biodiesel production is shown in Figure 3.

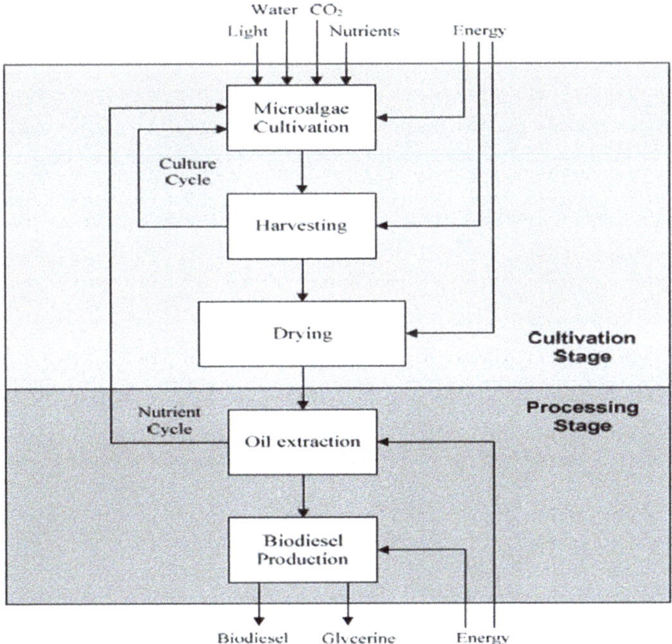

Figure 3. Flowchart of microalgae cultivation, harvesting, and drying to biodiesel production [3].

4.1.2. Life Cycle Inventory

The life cycle inventory of the microalgae biodiesel production is shown in Table 3. Table 3 presents the detailed input and output streams of the unit processes.

In Table 8, the unit processes of the microalgae biodiesel production were identified as cultivation, harvesting, drying, oil extraction, and transesterification. In order to grow the microalgae, the cultivation adopted in the study was an open pond from [50,51]. The electricity consumed for the cultivation stage is attributable to the lighting and aeration requirements for growing the biomass. After growing the microalgae, they are dewatered and harvested using a coagulation process adopted from [52]. The harvested microalgae are then dried to attain a moisture content less than 10%, using a microwave drying taken from [41]. The dried microalgae are then further processed to extract the oil, employing an oil yield of 30% based on [53]. Lastly, the extracted microalgae oil is converted to biodiesel through the conventional transesterification process adopted from [54].

In order to translate the impacts of the input and output of the unit processes of the microalgae biodiesel, the Ecoinvent LCA database from SimaPro was utilized as the source for the life cycle inventory for the study.

4.1.3. Life Cycle Impact Assessment

The life cycle impact assessment was performed using the Environmental Development of Industrial Products (EDIP), which employed the middle point method that accounted for the impacts as they was generated from the plant. The EDIP was utilized for the case study, as it provided a general categorization of the impact categories, enabling an easier interpretation of the results on the microalgae biodiesel production.

4.2. Sustainability Analysis

Microalgal biodiesel had a strong impact in 16 out of the 19 impact categories covered by EDIP. This is due to the fact that the production of biodiesel from microalgae required additional energy-intensive processes such as harvesting and drying, as shown in Figure 1. Different technologies have been explored for harvesting. A summary of these technologies is given in Table 9.

Table 9. Advantages and disadvantages of algae harvesting [55,56].

Technique	Advantages	Disadvantages
Coagulation/flocculation	• Fast and easy technique • Large scale usage • Less cell damage • Applied to vast range of species • Less energy requirements • Auto and bioflocculation may be inexpensive methods	• Chemicals may be expensive • Highly pH dependent • Difficult to separate the coagulant from harvested biomass • Efficiency depends upon the coagulant used • Culture medium recycling is limited • Possibility of mineral or microbial contamination
Flotation	• Suitable for large scale • Low cost and low space requirement • Short operation time	• Needs surfactants • Ozoflotation is expensive
Electrical-based processes	• Applicable to all microalgal species • No chemical required	• Metal electrodes required • High energy and equipment costs • Metal contamination
Filtration	• High recovery efficiency • Cost-effective • No chemical required • Low energy consumption • Low shear stress • Water recycles	• Slow, requires pressure or vacuum • Not suitable for small algae • Membrane fouling/clogging and high operational and maintenance cost • High energy consumption
Centrifugation	• Fast and effective technique • High recovery efficiency • Preferred for small scale and laboratory • Application to all microalgae	• Expensive technique with high energy requirement • High operational and maintenance cost • Appropriate for recovery of high-valued products • Time consuming and too expensive for large scale • Risk of cell destruction

Since the microalgae slurry biomass is considered as a wet biomass, it requires a special method for harvesting, and in this case the coagulation method was used [52]. To successfully generate the appropriate amounts of biodiesel, the moisture content of biomass prior oil extraction should be less than, or equal to, 10% [41]. Hence, the moisture content from the harvested microalgae was still high and required further moisture removal from drying. It was found that the energy requirement per mass of dried microalgae biomass was much lower for the microwave drying compared with conventional drying methods. Due to this reason, the microwave drying was utilized as the drying method for this case study.

It was recognized that in 16 out of the 19 impact categories, the microalgae cultivation process contributed the highest impact when compared with the other processes. The cultivation stage which required lighting and aeration dominated the other processes in terms of the impact on most of the different impact categories. The environmental impact of the cultivation process can be lessened by implementing cultivation technologies which do not require lighting [57]. The drying process was observed to follow the cultivation process as the next highest process contributor to the 16 impact categories. According to [58], the impact of microalgae drying can be significantly reduced by utilizing solar-assisted and solar dryers. To further avoid the energy-intensive drying process, hydrothermal liquefaction can be employed to directly extract the lipids from microalgae [59]. Recently, Felix et al. [60] proposed a direct in situ transesterification in producing biodiesel form microalgae. Their results have shown that the global warming potential (GWP) impact of the direct in situ transesterification is approximately nine times lower when compared to a conventional microalgae biodiesel pathway. The transesterification tallied the highest impact contribution for the hazardous waste (HW and ozone depletion (ODP) due to the consumption of methanol as a catalyst to produce biodiesel).

For microalgae biodiesel production to be environmentally viable in remote areas, process-centric improvements need to be developed and implemented, such as cultivation using natural solar lighting and the employment of solar-assisted and solar dryers. Access to sophisticated equipment such as hydrothermal liquefaction or direct in situ transesterification may post challenges for remote community areas in producing biodiesel from microalgae. In addition, techno-economic analysis is one of the major considerations for the commercialization of algal biofuels. Recent works have focused on the techno-economic analysis of algal biofuels. Ahmad Ansari et al. [61] performed a techno-economic analysis on the commercialization of Scenedesmus obliquus growth through an integrated fish and biofuel production. Beckstrom et al. [62] utilized techno-economic analysis to evaluate the viability of the production of bioplastics and biofuels from microalgae. Rajesh Banu et al. [63] employed a techno-economic analysis to evaluate the operation of various configurations of a biorefinery to generate algal biofuels. One of the challenges of algal biofuels is competing commercially with currently available biofuels. In order for algal biofuels to be competitive, the processing requires co-production with a high-valued product such as the one mentioned in these studies.

5. Sustainability Model for Microalgae Biofuel Production

The previous sections demonstrated the feasibility and effectiveness of algal biofuels as an energy source. The current section will consider how algal biofuels can be successfully introduced in society as a replacement for conventional energy sources. The sustainability of algal biofuels depends on their continued adoption and use by the industry sector. This makes them a case study for technology diffusion, capacity expansion, and environmental policy. The system dynamics framework can incorporate these aspects in a single model. It also allows the sustainability of algal biofuels to be modeled and assessed, even without a large amount of data on existing successful implementations. The current section adopts this framework in identifying sustainable pathways for biofuel production.

Demand and supply are the primary actors in the diffusion of algal biofuels. The initial supply and the growth rate of supply are critical in facilitating the adoption of this energy source. Likewise, for any product, if the provider (of algal biofuels) fails to secure all demand, competitors (other energy sources, including the current energy source being patronized by the industry) may service the demand. Figure 4 demonstrates this as a causal loop diagram, according to the system dynamics methodology. The consequence of this for biofuels is that the supply gap may discourage potential and current adopters. This would put to waste the existing biorefineries and resources invested in biofuel production.

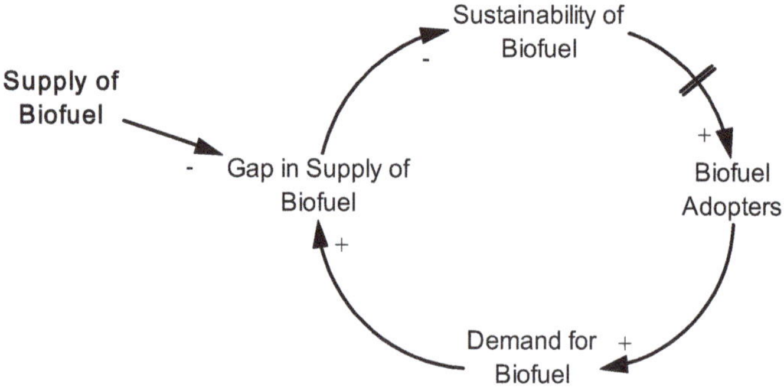

Figure 4. Demand and Supply Loop for Biofuels.

Increased demand can be met with capacity expansion. Capacity expansion may be driven by government initiatives. These initiatives may contribute a direct increase in capacity, through facilities funded by the government; or these may be in the form of incentives and penalties that encourage capacity expansion. These can give algal biofuels a strong entry into the industry sector. However, these initiatives also function as a short-term resource for capital expansion. The environmental issues, or interest in the technology, will diminish over time as the initiative begins to generate positive results (see Figure 5). This may be one way for demand to eventually exceed supply, in the long run.

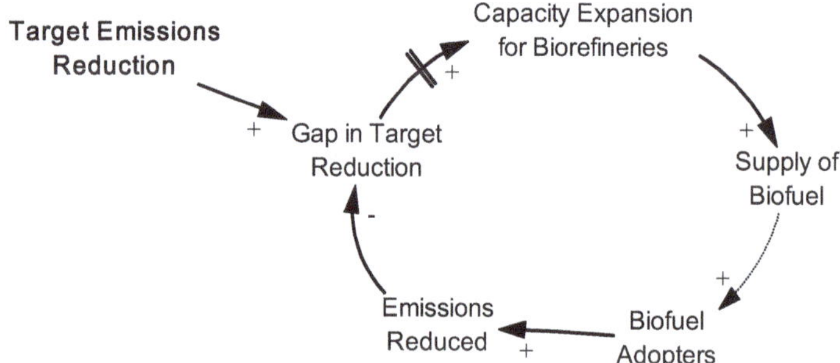

Figure 5. Environmental Incentives Loop for Biofuel Production.

Capacity expansion can also be driven by demand, through investors from the industry itself. As demand grows, algal biofuels proves to be a valuable industry in itself, and a fruitful investment. By basing capacity expansion on the funding from external investors, capacity grows continually, although whether or not demand can be met depends on the parameters of demand and supply growth (see Figure 6). In particular, energy yield and the rate of capacity expansion relative to demand, being the parameters which can be controlled, will determine the possibility of shortage, and consequently the sustainability of algal biofuels.

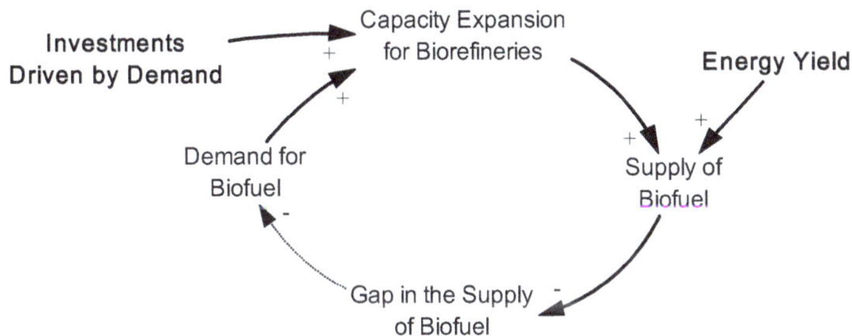

Figure 6. Investment Loop for Biofuel Production.

Aside from funding, other resources will be consumed in capacity expansion. This may create a competition for resources, as aggressive expansion will disrupt the balance of supply chains for these resources (see Figure 7). The efficient use of resources, with optimized energy yield, would be critical to ensuring that the rate of supply is able to grow in proportion to demand.

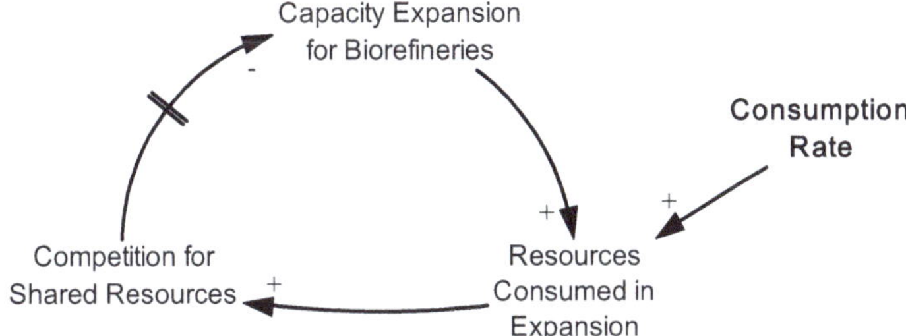

Figure 7. Resource Competition Loop for Biofuel Production.

It is important to note that the problems discussed in this section all grow in magnitude alongside demand. These can be addressed by streamlining capacity expansion. With this strategy, the following pathways toward sustainability are proposed:

1. Channel funding toward research and development.

Research and development (R&D), targeted at increasing the energy yield of plants and all subsequent expansions, will be more efficient in ensuring that supply will be able to meet demand. Without R&D, the growth rate of supply will be constant, whereas with R&D, it increases over time. The higher growth rate is important in overcoming the barriers against expansion associated with demand (i.e., resource competition, supply gaps). A simulation, based on the system dynamics framework, demonstrates higher consistency and less chance for collapse in biofuel adoption (see Figure 8).

2. Partner with the industry for expansion

Another means of increasing the growth rate of supply is through investments from the industry, as demonstrated in Figure 6. Investments are important as they provide a direct link between demand and capacity expansion, and hence a smoother growth in biofuel adoption in general (see Figure 9). The implications of the smoother curve are less barriers against entry for potential adopters, and also less openings for other energy sources which may fill the gap in supply. For greater efficiency, this solution may be coupled with further investments in R&D.

Biofuel Adoption

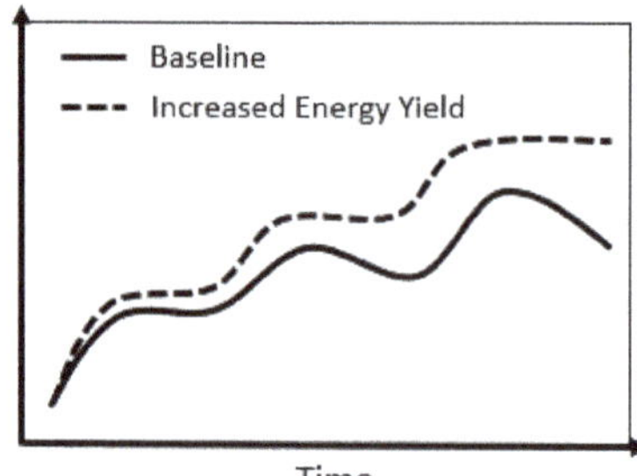

Figure 8. Simulated Impact of R&D on Biofuel Adoption.

Biofuel Adoption

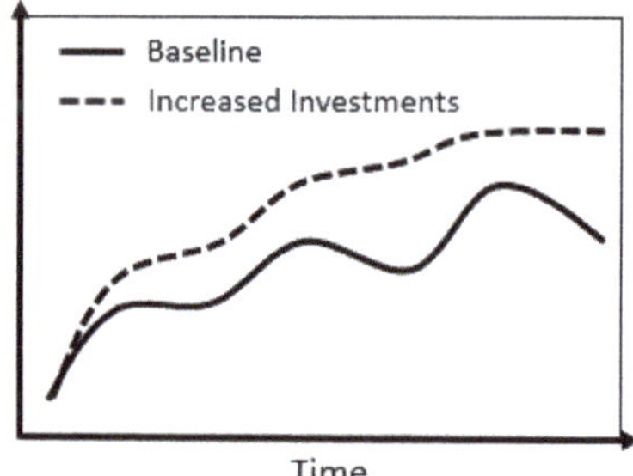

Figure 9. Simulated Impact of Industry Partners on Biofuel Adoption.

3. Scalable design for algal biorefineries

The outcomes achieved from investments can be replicated by changing the design of algal biorefineries. Generally, the bottleneck against expansion is the large amount of capital expenditure required to construct and operate algal biorefineries. By reinterpreting the increase in investment associated with demand to more capacity growth in proportion to the investments being made. This would also address resource competition to some extent. Land, one of the likeliest areas for resource competition, can be used more efficiently through biorefineries that are smaller in scale and strategically dispersed within an area.

Author Contributions: Conceptualization, A.B.C. and A.T.U.; methodology, A.T.U., W.-H.C. and J.-S.C.; software, P.M.L.C.; validation, W.-H.C. and J.-S.C.; formal analysis, A.T.U. and P.M.L.C.; investigation; A.B.C., A.T.U. and P.M.L.C.; resources, A.B.C.; writing—original draft preparation, P.M.L.C.; writing—review and editing, P.M.L.C.; project administration, A.B.C. All authors have read and agreed to the published version of the manuscript.

Funding: This research was supported by the Office of the Vice Chancellor of Research and Innovation of De La Salle University (DLSU), and the DLSU Science Foundation.

Acknowledgments: The authors acknowledge Engr. Roberto Louis Moran for the figures of the microalgal cultivation systems.

Conflicts of Interest: The authors declare no conflict of interest.

References

1. Chisti, Y. Biodiesel from microalgae. *Biotechnol. Adv.* **2007**, *25*, 294–306. [CrossRef]
2. De Bhowmick, G.; Koduru, L.; Sen, R. Metabolic pathway engineering to-wards enhancing microalgal lipid biosynthesis for biofuel application—A review. *Renew. Sustain. Energy Rev.* **2015**, *50*, 1239–1253. [CrossRef]

3. Mata, T.M.; Martins, A.A.; Caetano, N.S. Microalgae for biodiesel production and other applications: A review. *Renew. Sustain. Energy Rev.* **2010**, *14*, 217–232. [CrossRef]

4. Tan, R.R.; Culaba, A.B.; Purvis, M.R.I.; Tanchuco, J.Q. Life Cycle Design, Planning, and Assessment'. In *Web-Based Green Products Life Cycle Management Systems: Reverse Supply Chain Utilization*; Information Science Reference: New York, NY, USA, 2009; pp. 1–15.

5. Ueda, R.; Hirayama, S.; Sugata, K.; Nakayama, H. Process for the Production of Ethanol from Microalgae. Mitsubishi Heavy Industries Ltd. U.S. Patent 5,578,472, 26 November 1996.

6. Harman-Ware, A.E.; Morgan, T.; Wilson, M.; Crocker, M.; Zhang, J.; Liu, K.; Debolt, S. Microalgae as a renewable fuel source: Fast pyrolysis of Scenedesmus sp. *Renew. Energy* **2013**, *60*, 625–632. [CrossRef]

7. Babich, I.V.; Van der Hulst, M.; Lefferts, L.; Moulijn, J.A.; O'Connor, P.; Seshan, K. Catalytic pyrolysis of microalgae to high-quality liquid bio-fuels. *Biomass Bioenergy* **2011**, *35*, 3199–3207. [CrossRef]

8. Pan, P.; Hu, C.; Yang, W.; Li, Y.; Dong, L.; Zhu, L.; Fan, Y. The direct pyrolysis and catalytic pyrolysis of Nannochloropsis sp. residue for renewable bio-oils. *Bioresour. Technol.* **2010**, *101*, 4593–4599. [CrossRef] [PubMed]

9. Nguyen, M.T.; Choi, S.P.; Lee, J.; Lee, J.H.; Sim, S.J. Hydrothermal Acid Pretreatment of *Chlamydomonas reinhardtii* Biomass for Ethanol Production. *J. Microbiol. Biotechnol.* **2009**, *19*, 161–166.

10. Choi, O.; Das, A.; Yu, C.P.; Hu, Z. Nitrifying bacterial growth inhibition in the presence of algae and cyanobacteria. *Biotechnol. Bioeng.* **2010**, *107*, 1004–1011. [CrossRef]

11. Chew, K.W.; Yap, J.Y.; Show, P.L.; Suan, N.H.; Juan, J.C.; Ling, T.C.; Chang, J.S. Microalgae biorefinery: High value products perspectives. *Bioresour. Technol.* **2017**, *229*, 53–62. [CrossRef]

12. Dutta, K.; Daverey, A.; Lin, J.G. Evolution retrospective for alternative fuels: First to fourth generation. *Renew. Energy* **2014**, *69*, 114–122. [CrossRef]

13. Yu, M.; Zhang, Y.; Tang, I.C.; Yang, S.T. Metabolic engineering of Clostridium tyrobutyricum for n-butanol production. *Metab. Eng.* **2011**, *13*, 373–382. [CrossRef] [PubMed]

14. Shen, C.R.; Liao, J.C. Metabolic engineering of Escherichia coli for 1-butanol and 1-propanol production via the keto-acid pathways. *Metab. Eng.* **2008**, *10*, 312–320. [CrossRef] [PubMed]

15. Hashemi, A.; Moslemi, M.; Pajoum Shariati, F.; Delavari Amrei, H. Beta-carotene production within Dunaliella salina cells under salt stress condition in an indoor hybrid helical-tubular photobioreactor. *Can. J. Chem. Eng.* **2020**, *98*, 69–74. [CrossRef]

16. Helisch, H.; Keppler, J.; Detrell, G.; Belz, S.; Ewald, R.; Fasoulas, S.; Heyer, A.G. High density long-term cultivation of Chlorella vulgaris SAG 211–12 in a novel microgravity-capable membrane raceway photobioreactor for future bioregenerative life support in SPACE. *Life Sci. Space Res.* **2020**, *24*, 91–107. [CrossRef]

17. Khalekuzzaman, M.; Alamgir, M.; Islam, M.B.; Hasan, M. A simplistic approach of algal biofuels production from wastewater using a Hybrid Anaerobic Baffled Reactor and Photobioreactor (HABR-PBR) System. *PLoS ONE* **2019**, *14*, e0225458. [CrossRef]

18. Brennan, L.; Owende, P. Biofuels from microalgae—A review of technologies for production, processing, and extractions of biofuels and co-products. *Renew. Sustain. Energy Rev.* **2010**, *14*, 557–577. [CrossRef]

19. Huang, G.; Chen, G.; Wei, D.; Zhang, X.W. Biodiesel production by micro-algal biotechnology. *Appl. Energy* **2010**, *87*, 38–46. [CrossRef]

20. Ranjbar, S.; Inoue, R.; Katsuda, R.; Yamaji, T.; Katoh, H. High efficiency production of astaxanthin in an airlift photobioreactor. *J. Biosci. Bio Eng.* **2008**, *106*, 204–207. [CrossRef]

21. Pulz, W.; Gross, O. Valuable products. *Biotechnol. Appl. Microb.* **2004**, *65*, 635–648. [CrossRef]

22. Olaizola, M. Commercial development of microalgal biotechnology: From test tube to the market place. *Biomol. Eng.* **2003**, *20*, 459–466. [CrossRef]

23. Molina, G. Mass culture methods. In *Encyclopedia of Bioprocess Technology: Fermentation, Biocatalysis, and Bioseparation*; Wiley: Hoboken, NJ, USA, 1999.

24. Borowitzka, M.A. Commercial production of microalgae: Ponds, tanks, tubes and fermenters. *J. Biotechnol.* **1999**, *70*, 313–321. [CrossRef]

25. Ugwu, C.U.; Aoyagi, H.; Uchiyama, H. Photobioreactors for mass cultivation of algae. *Bioresour. Technol.* **2008**, *99*, 4021–4028. [CrossRef]

26. Bitog, J.P.; Lee, I.B.; Lee, C.G.; Kim, K.S.; Hwang, H.S.; Hong, S.W.; Mostafa, E. Application of computational fluid dynamics for modeling and designing photobioreactors for microalgae production: A review. *Comput. Electron. Agric.* **2011**, *76*, 131–147. [CrossRef]

27. Becker, E. *Microalgae: Biotechnology and Microbiology*; Cambridge University Press: Cambridge, UK, 1994.

28. Slade, R.; Bauen, A. Micro-algae cultivation for biofuels: Cost, energy balance, environmental impacts and future prospects. *Biomass Bioenergy* **2013**, *53*, 29–38. [CrossRef]

29. Tredici, G.; Zittelli, M. Efficiency of sunlight utilization: Tubular versus flat photobioreactors. *Biotechnol. Bioeng.* **1998**, *57*, 187–197. [CrossRef]

30. Devi, T.E.; Parthiban, R. Hydrothermal liquefaction of Nostoc ellipsosporum biomass grown in municipal wastewater under optimized conditions for bio-oil production. *Can. J. Chem. Eng.* **2020**, *98*, 69–74. [CrossRef]

31. Dandamudi, K.P.R.; Muhammed Luboowa, K.; Laideson, M.; Murdock, T.; Seger, M.; McGowen, J.; Lammers, P.J.; Deng, S. Hydrothermal liquefaction of Cyanidioschyzon merolae and Salicornia bigelovii Torr.: The interaction effect on product distribution and chemistry. *Fuel* **2020**, *277*, 118146. [CrossRef]

32. Arun, J.; Gopinath, K.P.; SundarRajan, P.; Malolan, R.; Adithya, S.; Sai Jayaraman, R.; Srinivaasan Ajay, P. Hydrothermal liquefaction of Scenedesmus obliquus using a novel catalyst derived from clam shells: Solid residue as catalyst for hydrogen production. *Bioresour. Technol.* **2020**, *310*, 123443. [CrossRef]

33. Uduman, A.; Qi, N.; Danquah, Y.; Forde, M.K.; Hoadley, G.M. Dewatering of microalgal cultures: A major bottleneck to algae-based fuels. *J. Renew. Sustain. Energy* **2010**, *2*, 012701. [CrossRef]

34. Amaro, F.X.; Guedes, H.M.; Malcata, A.C. Advances and perspectives in using microalgae to produce biodiesel. *Appl. Energy* **2011**, *88*, 3402–3410. [CrossRef]

35. Grima, A.R.; Fernandez, E.M.; Medina, F.A. Downstream Processing of Cell-mass and Products. In *Handbook of Microalgal Culture: Biotechnology and Applied Phycology*; Wiley: Hoboken, NJ, USA, 2004.

36. Singh, S.K.; Patidar, G. Microalgae harvesting techniques: A review. *J. Environ. Manag.* **2018**, *217*, 499–508. [CrossRef]

37. Lardon, O.; Helias, L.; Sialve, A.; Stayer, B.; Barnard, J.P. Life-cycle as-sessment of bio-diesel production from microalgae. *Environ. Sci. Technol.* **2009**, *43*, 6475–6481. [CrossRef]

38. Viswanathan, M.; Mani, T.; Das, S.; Chinnasamy, K.C.; Bhatnagar, S.; Singh, A.; Singh, R.K. Effect of cell rupturing methods on the drying characteristics and lipid compositions of microalgae. *Bioresour. Technol.* **2012**, *126*, 131–136. [CrossRef]

39. Prakash, R.; Pushparaj, J.; Carlozzi, B.; Torzillo, P.; Montaimi, G.; Materassi, E. Microalgal biomass drying by a simple solar device. *Int. J. Sol. Energy* **1997**, *18*, 303–311. [CrossRef]

40. Show, J.S.; Lee, K.Y.; Tay, D.J.; Lee, J.H.; Chang, T.M. Microalgal drying and cell distruption—Recent advances. *Bioresour. Technol.* **2015**, *184*, 258–266. [CrossRef]

41. Villagracia, A.R.C.; Mayol, A.P.; Ubando, A.T.; Biona, J.B.M.M.; Arboleda, N.B.; David, M.Y.; Tumlos, R.B.; Lee, H.; Lin, O.H.; Espiritu, R.A.; et al. Microwave drying characteristics of microalgae (Chlorella vulgaris) for biofuel production. *Clean. Technol. Environ. Policy* **2016**, *18*, 2441–2451. [CrossRef]

42. Mathews, J.A. Biofuels, climate change and industrial development: Can the tropical South build 2000 biorefineries in the next decade? *Biofuelsbioprod. Biorefin.* **2008**, *2*, 103–125. [CrossRef]

43. Demirbas, A. Biorefineries: Current activities and future developments. *Energy Convers. Manag.* **2009**, *50*, 2782–2801. [CrossRef]

44. Octave, S.; Thomas, D. Biorefinery: Toward an industrial metabolism. *Biochimie* **2009**, *91*, 659–664. [CrossRef]

45. Goh, C.S.; Lee, K.T. A visionary and conceptual macroalgae-based third generation bioethanol (TGB) in Sabah, Malaysia as an underlay for renewable and sustainable development. *Renew. Sustain. Energy Rev.* **2010**, *14*, 842–848. [CrossRef]

46. Mussgnug, J.H.; Klassen, V.; Schuter, A.; Kruse, O. Microalgae as substrate for fermentive biogas production in a combined biorefinery concept. *J. Biotechnol.* **2010**, *150*, 51–56. [CrossRef] [PubMed]

47. Subhadra, B.G. Sustainability of algal biofuel production using integrated renewable energy park (IREP) and algal biorefinery approach. *Energy Policy* **2010**, *38*, 5892–5901. [CrossRef]

48. Tran, N.H.; Bartlett, J.R.; Kannangara, G.S.K.; Milev, A.S.; Volk, H.; Wilson, M.A. Catalytic upgrading of biorefinery oil from micro-algae. *Fuel* **2010**, *89*, 265–274. [CrossRef]

49. Subhadra, B.; Edwards, M. An integrated renewable energy park approach for algal biofuel production in United States. *Energy Policy* **2010**, *38*, 4897–4902. [CrossRef]

50. Rawat, I.; Bhola, V.; Kumar, R.R.; Bux, F. Improving the feasibility of producing biofuels from microalgae using wastewater. *Environ. Technol.* **2013**, *34*, 1765–1775. [CrossRef]

51. Ubando, A.T.; Culaba, A.B.; Aviso, K.B.; Tan, R.R.; Cuello, J.L.; Ng, D.K.; El-Halwagi, M.M. Fuzzy mixed integer non-linear programming model for the design of an algae-based eco-industrial park with prospective selection of support tenants under product price variability. *J. Clean. Prod.* **2016**, *136*, 183–196. [CrossRef]

52. Khoo, H.H.; Sharratt, P.N.; Das, P.; Balasubramanian, R.K.; Naraharisetti, P.K.; Shaik, S. Life cycle energy and CO_2 analysis of microalgae-to-biodiesel: Preliminary results and comparisons. *Bioresour. Technol.* **2011**, *102*, 5800–5807. [CrossRef]

53. Rodolfi, L.; Chini Zittelli, G.; Bassi, N.; Padovani, G.; Biondi, N.; Bonini, G.; Tredici, M.R. Microalgae for oil: Strain selection, induction of lipid synthesis and outdoor mass cultivation in a low-cost photobioreactor. *Biotechnol. Bioeng.* **2009**, *102*, 100–112. [CrossRef]

54. Hou, Y.; Liang, W.; Zhang, L.; Cheng, S.; He, F.; Wu, Z. Freshwater algae chemotaxonomy by high-performance liquid chromatographic (HPLC) analysis. *Front. Environ. Sci. Eng. China* **2011**, *5*, 84–91. [CrossRef]

55. Abdelaziz, P.C.; Leite, A.E.; Hallenback, G.B. Addressing the challenges for sustainable production of algal biofuels: II. Harvesting and conversion to biofuels. *Environ. Technol.* **2013**, *13–14*, 1807–1836. [CrossRef]

56. Barros, J.C.; Goncalves, A.I.; Simones, A.L.; Pires, M. Harvesting techniques applied to microalgae: A review. *Renew. Sustain. Energy Rev.* **2015**, *41*, 1489–1500. [CrossRef]

57. Patel, A.K.; Joun, J.M.; Hong, M.E.; Sim, S.J. Effect of light conditions on mixotrophic cultivation of green microalgae. *Bioresour. Technol.* **2019**, *282*, 245–253. [CrossRef]

58. Azadi, Y.; Yazdanpanah, M.; Mahmoudi, H. Understanding smallholder farmers' adaptation behaviors through climate change beliefs, rsk perception, trust, and psychological distance: Evidence from wheat growers in Iran. *J. Environ. Manag.* **2019**, *250*, 109456. [CrossRef] [PubMed]

59. Ubando, A.T.; Chen, W.H.; Ashokkumar, V.; Chang, J.S. Iron oxide reduction by torrefied microalgae for CO_2 capture and abatement in chemical-looping combustion. *Energy* **2019**, *186*, 115903. [CrossRef]

60. Felix, C.; Ubando, A.; Madrazo, C.; Gue, I.H.; Sutanto, S.; Tran-Nguyen, P.L.; Chen, W.H. Non-catalytic in-situ (trans) esterification of lipids in wet microalgae Chlorella vulgaris under subcritical conditions for the synthesis of fatty acid methyl esters. *Appl. Energy* **2019**, *248*, 526–537. [CrossRef]

61. Ahmad Ansari, F.; Nasr, M.; Guldhe, A.; Kumar Gupta, S.; Rawat, I.; Bux, F. Techno-economic feasibility of algal aquaculture via fish and biodiesel production pathways: A commercial-scale application. *Sci. Total Environ.* **2020**, *704*, 135259. [CrossRef] [PubMed]

62. Beckstrom, B.D.; Wilson, M.H.; Crocker, M.; Quinn, J.C. Bioplastic feedstock production from microalgae with fuel co-products: A techno-economic and life cycle impact assessment. *Algal Res.* **2020**, *46*, 101769. [CrossRef]

63. Rajesh Banu, J.; Preethia; Kavitha, S.; Gunasekaran, M.; Kumar, G. Microalgae based biorefinery promoting circular bioeconomy-techno economic and life-cycle analysis. *Bioresour. Technol.* **2020**, *302*, 122822. [CrossRef]

Article

Core Competitiveness Evaluation of Clean Energy Incubators Based on Matter-Element Extension Combined with TOPSIS and KPCA-NSGA-II-LSSVM

Guangqi Liang [1], Dongxiao Niu [1] and Yi Liang [2],*

[1]   School of Economics and Management, North China Electric Power University, Beijing 102206, China; 92494@163.com (G.L.); ndx@ncepu.edu.cn (D.N.)
[2]   School of Management, Hebei Geo University, Shijiazhuang 050031, China
*   Correspondence: louisliang@hgu.edu.cn

Received: 17 October 2020; Accepted: 15 November 2020; Published: 17 November 2020

Abstract: Scientific and accurate core competitiveness evaluation of clean energy incubators is of great significance for improving their burgeoning development. Hence, this paper proposes a hybrid model on the basis of matter-element extension integrated with TOPSIS and KPCA-NSGA-II-LSSVM. The core competitiveness evaluation index system of clean energy incubators is established from five aspects, namely strategic positioning ability, seed selection ability, intelligent transplantation ability, growth catalytic ability and service value-added ability. Then matter-element extension and TOPSIS based on entropy weight is applied to index weighting and comprehensive evaluation. For the purpose of feature dimension reduction, kernel principal component analysis (KPCA) is used to extract momentous information among variables as the input. The evaluation results can be obtained by least squares support vector machine (LSSVM) optimized by NSGA-II. The experiment study validates the precision and applicability of this novel approach, which is conducive to comprehensive evaluation of the core competitiveness for clean energy incubators and decision-making for more reasonable operation.

Keywords: clean energy incubator; core competitiveness evaluation; matter-element extension; TOPSIS; KPCA; NSGA-II; LSSVM

---

1. Introduction

Clean energy plays an important role in reducing carbon emissions, so more and more countries vigorously develop clean energy [1]. As a promotion agency of clean energy industrialization, clean energy incubator has become a crucial part of innovation system. With the burgeoning development of clean energy incubators in recent years, governments all over the world are introducing policies to actively encourage their establishment [2]. Whereas, these incubators are faced with the problems of valuing quantity and neglecting quality. Some incubators still stay in the stage of property leasing and simple incubation service, but still enjoy the preferential policies of the government and waste public resources [3]. In addition, the planning of many clean energy business incubators is mainly oriented towards the layout of hardware and space, and ignores the planning of soft capabilities such as incubation ability [4]. In order to further play the role of incubators and realize their sustainable development, it is necessary to make the factors clear which affect the core competitiveness and propose a feasible evaluation model. The research on the core competitiveness evaluation of clean energy incubator can help managers find out the deficiencies, and then help managers to improve these deficiencies. Thus this study can provide suggestions for enhancing competitiveness of clean energy incubators.

In 1990, C.K. Prahalad and Gary Hamel, professors of Michigan University and London Business School, respectively, formally raised the concept of "core competitiveness" for the first time. They pointed out that core competitiveness was the ability to coordinate different production skills as well as organically combine various technological schools [5]. Literature [6] discussed how to integrate technological innovation into the core competitiveness of enterprises. The research held that it was of great theoretical and practical significance to put the core competitiveness evaluation into force for sustainable development. Reference [7] considered that environmental adaptability was an important component of the core competitiveness in photovoltaic module supply enterprises. Thereby, it provided suggestion in business decision-making. Thirteen influential factors of core competitiveness were selected in modern circulation industry from market expansion, innovation, circulation efficiency, sustainable development and circulation effect in reference [8]. The evaluation was achieved through study on both independent and synergistic effect of the factors. Hence, it becomes a good entry point to analyze the key resources and capabilities demanded by the sustainable development of enterprises.

At present, scholars have published their significant work on core competitiveness evaluation of incubators, but few focus on the clean energy enterprise. There are some similarities among these research subjects, such as the content and components of kernel competence. Thus the existing study also can provide reference for the research. Reference [9] developed Principal component analysis (PCA) and SPSS based model to evaluate scientific enterprise incubators of nine cities in Pearl River Delta. The results showed the ranking and classification of incubators and put forward development plan accordingly. Literature [10] conducted a questionnaire survey on six chain incubators. The relationship between quantitative performance and importance was revealed through IPA in competitiveness evaluation. Reference [11] outlined the assessment approach of enterprise incubator based on PCA and DEA from the perspective of service and operation efficiency. A comprehensive technique that integrates questionnaire survey with grey correlation assesses the national incubators regarding knowledge service capability in [12]. Reference [13] paid close attention to social benefits in core competitiveness assessment of incubators. The evaluation is executed from five aspects, that is, public service, innovation and development, social education, labor employment and enterprise cultivation.

Evaluation methods can be separated into two categories: traditional assessment technique and modern intelligent algorithm. The former needs complex calculation despite its mature development and accurate results, such as fuzzy comprehensive evaluation method [14] matter-element extension assessment [15] and TOPSIS [16]. Reference [14] developed an approach to reveal the relationship between exposure, sensitivity and adaptive capacity for better flood vulnerability assessment, based on the fuzzy comprehensive evaluation method. Reference [15] utilized the improved fuzzy matter-element extension assessment model to evaluate urban water ecosystem health effectively. Reference [16] used the improved TOPSIS model to evaluate the sustainability of power grid construction projects. Modern intelligent approaches chiefly consist of back propagation neural network [17], extreme learning machine [18], support vector machine [19], least squares support vector machine (LSSVM) [20]. Thereby, this paper intends to combine the traditional evaluation model and intelligent algorithm to assess the core competitiveness of clean energy incubators. Concretely, matter-element extension assessment is first applied to obtain the results. Then, LSSVM is employed for intelligent evaluation. However, owing to blind selection of penalty and kernel coefficients, it's essential to use an appropriate method for optimization [21]. NSGA-II is a multi-objective model based on the first generation of non dominated sorting genetic algorithm with the advantage of excellent performance and high efficiency [22]. For example, reference [23] used the NSGA-II algorithm to obtain the multi-objective optimisation of an interactive buildings-vehicles energy sharing network with high energy flexibility, which achieved good optimization results. To this end, this paper exploits NSGA-II to automatically decide the appropriate values in LSSVM.

It is worth noting that there exist many influencing factors on the core competitiveness of clean energy incubators. Dimension reduction is critical for accuracy on account of strong coupling, nonlinearity and redundancy in the indicators [24]. PCA, as a multivariate statistical approach, has the advantage over dealing with indexes with strong linear relationship [25]. Nevertheless, the relationship among the core competitiveness indicators of clean energy incubators is mostly nonlinear. PCA is not able to acquire high-order features which ignores the nonlinear information while reducing the dimension [26]. As a result, kernel principal component analysis (KPCA) is used in this paper to map the initial input variables to high-dimensional feature space through nonlinear transformation so that the input dimension can be reduced on the premise of retaining nonlinear information among variables [27].

Therefore, this paper establishes the core competitiveness index system of clean energy incubator and proposes a hybrid model integrated matter-element extension assessment and KPCA-NSGA-II-LSSVM for evaluation. The rest of paper is organized as follows: Section 2 designs the indicator system including strategic positioning ability, seed selection ability, intelligent transplantation ability, growth catalytic ability as well as service value-added ability, and introduces the preprocessing method. Section 3 presents a brief description of the proposed evaluation technique. In Section 4, a case study is carried out to validate the established model. Section 5 concludes the paper.

2. Establishment of Evaluation Index System for Core Competitiveness of Clean Energy Incubator

2.1. Selection of Evaluation Index

The kernel function of clean energy incubator is to cultivate corresponding enterprises. Whether incubator owns core competitiveness resolves its survival rate and growth. The main workflow includes seed business selection, related incubation service and tracking. Thus, specific conditions can be employed for judgement whether incubated enterprises are able to graduate or be obsoleted. These links are closely relevant and interact with each other. The core competitiveness is dependent on the crucial resource and capabilities required by these parts.

It is obvious that the seed quality is significant. Clean energy incubators must make a choice subjected to the resource. Furthermore, incubated service is an interactive process limited both by the ability of incubator and enterprise itself. Enterprises with no development potential can hardly survive or grow slowly even if they devote quantities of resources. Consequently, the quality of incubated seed is quite important which is principally determined by strategic positioning ability and seed selection ability [4].

The service provided by clean energy incubator aims at making the enterprises realize burgeoning development. Entrepreneurs are mostly confronted with lack resource and systematic knowledge, weak management capability as well as experience. Service with core competitiveness can be measured by the following conditions: An entrepreneurial team can become an independent and growing enterprise after incubation. It means incubator should possess the ability to transfer or graft the pivotal resource needed by entrepreneurs. In particular, it mainly refers to intelligent transplantation ability, growth catalytic ability, and service value-added ability.

In brief, the core competitiveness of clean energy incubator is chiefly decided by: strategic positioning ability, seed selection ability, intelligent transplantation ability, growth catalytic ability, service value-added ability. Figure 1 shows the established index system.

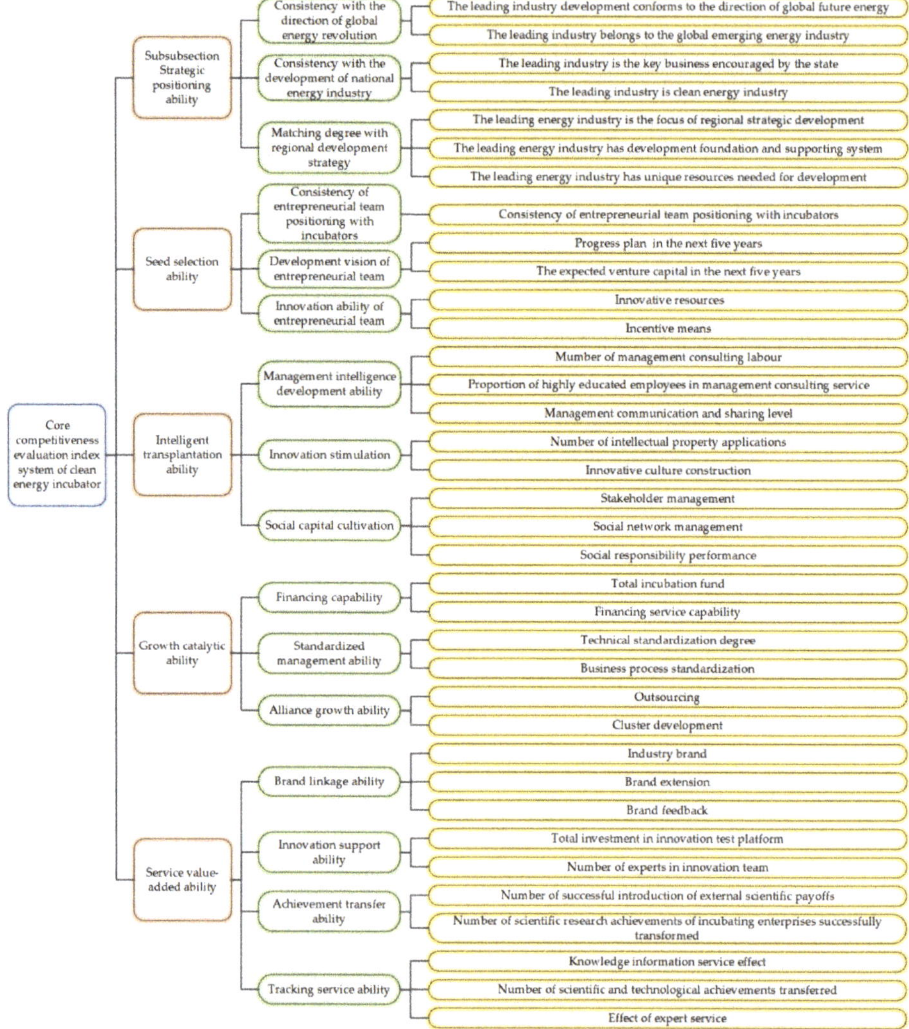

Figure 1. Core competitiveness evaluation index system of clean energy incubator.

2.1.1. Subsubsection Strategic Positioning Ability

Strategic positioning ability refers to the location capability of clean energy incubators in industries, regions and entrepreneurial groups. Incubators should first make strategic positioning for the target enterprises or customers. The more technology opportunities and better development prospects the incubators face, the more they match the regional progress. Additionally, strategic positioning ability displays the positive change with the potential the entrepreneurial group.

(1) Consistency with the direction of global energy revolution: judge whether the leading business development conforms to the direction of global future energy and whether it belongs to the global emerging energy industry via the literature on key progress of clean energy assembled from the United States, the United Kingdom, the European Union, Japan and other developed countries.

(2) Consistency with the development of national energy industry: considering national economic planning and policy, identify the key development of clean energy industry in China. Determine whether the leading industry of the incubator is the key business encouraged by the state.

(3) Matching degree with regional development strategy: provincial and regional energy planning as well as relevant materials can be used as the foundation to distinguish strategic development. Thus, matching degree is able to be estimated from the perspective of supporting system and unique resources through enterprise research and expert interview.

2.1.2. Seed Selection Ability

Seed selection ability represents the capability of incubator to identify the prospects of clean energy enterprises. If the seed is identical with incubator positioning, development as well as innovation ability, it declares the seed selection capability is strong.

(1) Consistency of entrepreneurial team positioning with incubators.

(2) Development vision of entrepreneurial team: progress plan and the expected venture capital in the next five years.

(3) Innovation ability of entrepreneurial team: innovative resources and incentive means.

2.1.3. Intelligent Transplantation Ability

Intelligent transplantation ability stands for the capability to improve intellectual capital of incubated enterprises including management intelligence development, innovation stimulation and social capital cultivation.

(1) Management intelligence development ability: number of management consulting labour, proportion of highly educated employees in management consulting service, management communication and sharing level.

(2) Innovation stimulation: number of intellectual property applications, innovative culture construction.

(3) Social capital cultivation: stakeholder management, social network management, social responsibility performance.

2.1.4. Growth Catalytic Ability

Growth catalytic ability can judge whether the incubated enterprises grow rapidly and expand well which includes financing capability, standardized management ability as well as alliance development ability.

(1) Financing capability: total incubation fund, financing service capability.

(2) Standardized management ability: technical standardization degree, business process standardization.

(3) Alliance growth ability: outsourcing, cluster development.

2.1.5. Service Value-Added Ability

Service value-added ability reflects the increase of value that incubator brings itself and enterprises, including brand linkage ability, innovation support ability, achievement transfer ability and tracking service ability.

(1) Brand linkage ability: industry brand, brand extension, brand feedback.

(2) Innovation support ability: total investment in innovation test platform, number of experts in innovation team.

(3) Achievement transfer ability: number of successful introduction of external scientific payoffs, number of scientific research achievements of incubating enterprises successfully transformed.

(4) Tracking service ability: knowledge information service effect, number of scientific and technological achievements transferred, effect of expert service.

2.2. Preprocessing of Evaluation Index

There mainly exist three kinds in the indicator system, namely extremely large, extremely small and interval indexes. For example, extremely large indicators change in accordance with core competitiveness. In order to make comparison with schemes and obtain the valuation results, normalization processing should be implemented. In this paper, all the indicators are assigned into extremely large indexes.

Due to distinguished attribute and quantity level of original data, it is essential to make the indicators dimensionless as Equation (1):

$$x_{ij}^* = \frac{x_{ij}}{\sqrt{\sum_{i=1}^{m} x_{ij}^2}}, \ i = 1, 2, \cdots, m; j = 1, 2, \cdots, n \tag{1}$$

where m and n represent the number of incubators and indexes, respectively.

3. Methodology

3.1. Entropy Weight Method

As a common approach in objective weight calculation, the principle of entropy weight method derives from thermodynamics. Entropy is a measure of the system uncertainty. The greater entropy means more chaotic system and less information [28]. Suppose the system locates in n states, the probability of each state is set as $P_i(i = 1 \rightarrow n)$. The entropy of the system is described as Equation (2)

$$E = -\sum_{i=1}^{n} P_i \ln p_i \tag{2}$$

where $P_i(i = 1 \rightarrow n)$ satisfies: $0 \le P_i \le 1$, $\sum_{i=1}^{n} P_i = 1$.

The following procedures are listed:

(1) Establishment of standardized judgment matrix

In accordance with standardized data obtained from Equation (1), the standardized judgment matrix is acquired.

(2) Calculate the information entropy of each index

$$H_j = -k \sum_{i=1}^{m} f_{ij} \ln f_{ij} \tag{3}$$

$$f_{ij} = \frac{x_{ij}^*}{\sum_{i=1}^{m} x_{ij}^*} \tag{4}$$

$$k = \frac{1}{\ln m} \tag{5}$$

(3) Index weighting

$$w_j = \frac{1 - H_j}{\sum_{j=1}^{n}\left(1 - H_j\right)} \tag{6}$$

where $0 \le w_j \le 1$, $\sum_{j=1}^{n} w_j = 1$.

3.2. Evaluation Model Integrated Matter-Element Extension with TOPSIS

TOPSIS, also known as approximate ideal solution, firstly draws up two schemes. One of the scheme presents the best attribute values, called positive ideal solution, while the attributes of the other plan are all the worst, named negative ideal solution. Then, compare each scheme with the positive as well as negative ideal solutions, respectively. The optimal programme is selected which is close to the positive ideal solution and far away from the negative one simultaneously [29]. Hence, this best scheme shows the highest core competitiveness. Whereas, TOPSIS can not judge the competitiveness level of each plan, that is, it can't be decided whether the level belongs to very high, high, general or low [30].

Matter-element extension evaluation method combined matter-element theory, extension set with correlation degree for quantitative evaluation. This model divides the data interval of the evaluated target into several orders and determine their levels. Correlation degree is calculated between each plan and the grade. The larger the correlation degree is, the higher the membership extent present. The level of the evaluated target depends on the grade of data interval with the highest membership degree [31].

Thus, this paper integrates matter-element extension model with TOPSIS for core competitiveness assessment. TOPSIS is firstly used to determine the positive and negative ideal solutions and divide the interval into several grades by equal distance. Each grade corresponds to the given level of core competitiveness. Then calculate the correlation degree between each scheme and the grades. Hence, the core competitiveness level of the plan can be accordingly judged.

The main measures can be described as follows:

(1) Establishment of standardized index matrix

Suppose the scheme set including n evaluation indicators is $M = (M_1, M_2, \cdots, M_m)$. The assessment indexes are presented as $D = (D_1, D_2, \cdots, D_n)$. The standardized index matrix C is derived from consistent and dimensionless processing as Equation (7)

$$C = \left(c_{ij} \right)_{m \times n} = \begin{bmatrix} c_{11} & c_{12} & \cdots & c_{1n} \\ c_{21} & c_{22} & \cdots & c_{2n} \\ \cdots & \cdots & \cdots & \cdots \\ c_{m1} & c_{m2} & \cdots & c_{mn} \end{bmatrix} \tag{7}$$

where $c_{ij} = (i = 1, 2, \cdots, m; j = 1, 2, \cdots, n)$ is the standardized indicator of D_i in scheme M_i.

(2) Establishment of weighted standardized matrix

The entropy weight method is applied to calculate the weight of indexes w_j. The weighted standardized matrix is shown as Equation (8)

$$X = \left(x_{ij} \right)_{m \times n} = \begin{bmatrix} w_1 c_{11} & w_2 c_{12} & \cdots & w_n c_{1n} \\ w_1 c_{21} & w_2 c_{22} & \cdots & w_n c_{2n} \\ \cdots & \cdots & \cdots & \cdots \\ w_1 c_{m1} & w_2 c_{m2} & \cdots & w_n c_{mn} \end{bmatrix} \tag{8}$$

(3) Determine positive and negative ideal solutions of each scheme

The positive ideal solution is:

$$X^+ = \left\{ \max_{1 \leq i \leq m} x_i(j) \Big| j \in J^+, \min_{1 \leq i \leq m} x_i(j) \Big| j \in J^- \right\} \tag{9}$$

The negative ideal solution is:

$$X^- = \left\{ \min_{1 \leq i \leq m} x_i(j) \Big| j \in J^+, \max_{1 \leq i \leq m} x_i(j) \Big| j \in J^- \right\} \tag{10}$$

(4) Division of extremum interval and calculation of closeness degree

The extremum interval composed of negative positive ideal solutions is divided into N layers. namely, $H_{it} = (h_{it}^1, h_{it}^2), i = 1, 2, \cdots, n; t = 1, 2, \cdots, N, x_i^- \le h_{it}^1 \le x_i^+, x_i^- \le h_{it}^2 \le x_i^+, [x_i^-, x_i^+]$ consists of h_{it}^1 and h_{it}^2.

The closeness degree between each index and evaluation interval in standardized decision matrix can be obtained according to Equation (11):

$$E(N_i) = \left| x_{ij} - \frac{h_{jt}^1 + h_{jt}^2}{2} \right| \tag{11}$$

Thus, the weighted closeness degree of each evaluation plan can be further derived:

$$Q_j(N_i) = 1 - \sum_{j=1}^{n} w_j E(N_i) \tag{12}$$

(5) Determine the grade of each scheme

The level that the maximum of $Q_j(N_i)$ belongs to can be regarded as the evaluation grade of the scheme.

3.3. KPCA

PCA is able to merge the original features and reduce the dimension to simplify computation, especially aiming at strong linear indicators. However PCA is difficult to grab high-order features, so it ignores nonlinear information during dimension reduction. With consideration of nonlinearity in core competitiveness of clean energy incubators, kernel principal component analysis (KPCA) is taken advantage of to extract key factors in this study [32].

This approach can validly reduce the dimension of the input in condition of retaining main nonlinear information. It is achieved by using nonlinear transformation to map initial input into high-dimensional feature space. Apparently, KPCA is able to compress the information contained in a large number of indexes into some comprehensive indicators. The basic steps are presented as follows:

Set a random vector $X = \{x_1, x_2, \ldots, x_n\}^T$, $x_k \in R^N (k = 1, 2, \ldots, m)$, m is the number of input, namely the original input can be expressed as $M = [a_1, a_2, \ldots, a_n]^T = [b_1, b_2, \ldots, b_m]$. The dataset is projected into the space F via nonlinear mapping Φ, $M = [\phi(b_1), \phi(b_2), \ldots, \phi(b_m)]$, $\sum_{k=1}^{m} \widetilde{\phi}(b_i) = 0$.

The covariance matrix is shown as follows:

$$C^F = \begin{bmatrix} C_{11}, C_{12}, \ldots, C_{1n} \\ \cdots\cdots \\ C_{n1}, C_{n2}, \ldots, C_{nn} \end{bmatrix} \tag{13}$$

$$C = \frac{1}{m} \sum_{i}^{m} \phi(b_i)\phi(b_i)^T \tag{14}$$

The eigenvalue and eigenvector can be obtained as Equation (15)

$$C^F W^F = \lambda^F W^F \tag{15}$$

where λ^F is the eigenvalue, $W^F = \sum_{k=1}^{M} \alpha_k \widetilde{\phi}(x_k)$ represents the corresponding eigenvector. The symmetric matrix $\mathbf{K}$ is shown in Equation (16):

$$\mathbf{K}(x_k, x_j) = (\phi(x_k) \cdot \phi(x_j)) \tag{16}$$

$\widetilde{\mathbf{K}}$ can be acquired via matrix centralization:

$$\widetilde{\mathbf{K}} = \mathbf{K} - \mathbf{I}_n\mathbf{K} - \mathbf{K}\mathbf{I}_n + \mathbf{I}_n\mathbf{K}\mathbf{I}_n \tag{17}$$

where $\mathbf{I}_n$ is $n \times n$ matrix, $\mathbf{I}_{i,j} = 1/n$, Equation (15) can be simplified as:

$$m\lambda^F M = \widetilde{\mathbf{K}}M \tag{18}$$

Thus the kernel principal component can be calculated with reference to extraction technique of traditional PCA.

3.4. NGSA-II

NSGA-II optimization algorithm is a novel genetic algorithm on the foundation of NSGA. Despite Pareto optimum can be derived from individual classification in accordance with non dominated sorting via NSGA, it needs complicated calculation and given shared radius value. Therefore, fast non-dominated sorting technique, crowding degree and elite strategy are introduced into NSGA, namely NSGA-II, to promote the robustness and operation speed [33].

The specific procedures of NSGA-II are described as follows:

(1) Generate initial population P_t and achieve non-dominated sorting as well as given Pareto optimum. The genetic operation P_t is carried out to form the offspring Q_t.
(2) Aamalgamate P_t with Q_t to generate the new species R_t. Implement non-dominated ranking on R_t and calculate the crowding degree d_i of r_i.
(3) A new generation P_{t+1} comes into being in line with non-dominated ranking and crowding degree d_i.
(4) If t does not reach the maximum number of iteration $t_{\max}$, $t = t + 1$. Otherwise, the calculation will be terminated and the output will be obtained.

The crowding degree ($d_{i.}$) refers to the density of other individuals around each personality in the population. The individual crowding degree at the boundary is defined as infinite, the crowding degree at other positions can be calculated as follows:

$$d_i = \sum_{j=1}^{m} \frac{\left|f_{j,i+1} - f_{j,i-1}\right|}{f_{j\max} - f_{j\min}} \tag{19}$$

where $d_{i.}$ represents the crowding degree of i, m equals the number of objective functions, $f_{j\max}, f_{j\min}$ are the maximum and minimum of j-th objective function.

3.5. LSSVM

LSSVM is an extension of SVM. This method makes use of equality constraints to replace the inequality constraints and employs kernel functions to transform prediction into equation problems, which contributes to the improvement of evaluation accuracy and speed [34].

In LSSVM, the sample is set as $T = \{(x_i, y_i)\}_{i=1}^{N}$, N is the number of samples. The regression model can be established as Equation (20)

$$y(x) = \boldsymbol{w}^T \cdot \varphi(x) + b \tag{20}$$

where $\varphi(*)$ maps the training samples into a highly dimensional space. $\boldsymbol{w}$ and b are the weight vector and bias, partly.

The converted optimization problem is shown as follows:

$$\min \frac{1}{2}\boldsymbol{w}^T\boldsymbol{w} + \frac{1}{2}\gamma\sum_{i=1}^{N}\xi_i^2 \tag{21}$$

$$y_i = w^T \phi(x_i) + b + \xi_i, i = 1, 2, 3, \ldots N \tag{22}$$

where γ equal the regularization parameter applicable to balance the complexity and accuracy in LSSVM. ξ_i is the error.

The Lagrange function is defined to address the problem:

$$L(w, b, \xi_i, \alpha_i) = \frac{1}{2} w^T w + \frac{1}{2}\gamma \sum_{i=1}^{N} \xi_i^2 - \sum_{i=1}^{N} \alpha_i \left[w^T \varphi(x_i) + b + \xi_i - y_i \right] \tag{23}$$

where α_i is Lagrange multipliers. In the light of the Karush-Kuhn-Tucker (KKT) conditions, Equation (24) is derived:

$$\begin{cases} \frac{\partial L}{\partial w} = 0 \rightarrow w = \sum_{i=1}^{N} \alpha_i \varphi(x_i) \\ \frac{\partial L}{\partial b} = 0 \rightarrow \sum_{i=1}^{N} \alpha_i = 0 \\ \frac{\partial L}{\partial \xi} = 0 \rightarrow \alpha_i = \gamma \xi_i \\ \frac{\partial L}{\partial \alpha} = 0 \rightarrow w^T + b + \xi_i - y_i = 0 \end{cases} \tag{24}$$

The optimization problem is converted into the following matter through eliminating the variables of w and ξ_i:

$$\begin{bmatrix} 0 & e_n^T \\ e_n & \Omega + \gamma^{-1} \cdot I \end{bmatrix} \cdot \begin{bmatrix} b \\ a \end{bmatrix} = \begin{bmatrix} 0 \\ y \end{bmatrix} \tag{25}$$

where

$$\Omega = \varphi^T(x_i)\varphi(x_i) \tag{26}$$

$$e_n = [1, 1, \ldots, 1]^T \tag{27}$$

$$\alpha = [\alpha_1, \alpha_2, \ldots, \alpha_n] \tag{28}$$

$$y = [y_1, y_2, \ldots, y_n]^T \tag{29}$$

The final form of LSSVM is displayed in Equation (30):

$$y(x) = \sum_{i=1}^{N} \alpha_i K(x_i, x) + b \tag{30}$$

where $K(x_i, x)$ equals the kernel function that satisfies Mercer's condition. The paper selects radial basis function (RBF) as the kernel function of LSSVM on account of its wide convergence range and application, as described in Equation (31):

$$K(x_i, x) = \exp\left\{ -\|x - x_i\|^2 / 2\sigma^2 \right\} \tag{31}$$

where σ^2 is the width of the kernel width.

Evidently, the performance of LSSVM is chiefly decided by two parameters: kernel parameter γ and regularization parameter σ^2.

3.6. Approaches of Matter-Element Extension Combined with TOPSIS and KPCA-NSGA-II-LSSVM

Based on the approach that combines matter-element extension, the core competitiveness evaluation model incorporating KPCA, NSGA-II, and LSSVM is constructed for clean energy incubators where KPCA is utilized for the determination of input and NSGA-II is exploited for parameter optimization in LSSVM. The flowchart of the novel evaluation technique is illustrated in Figure 2.

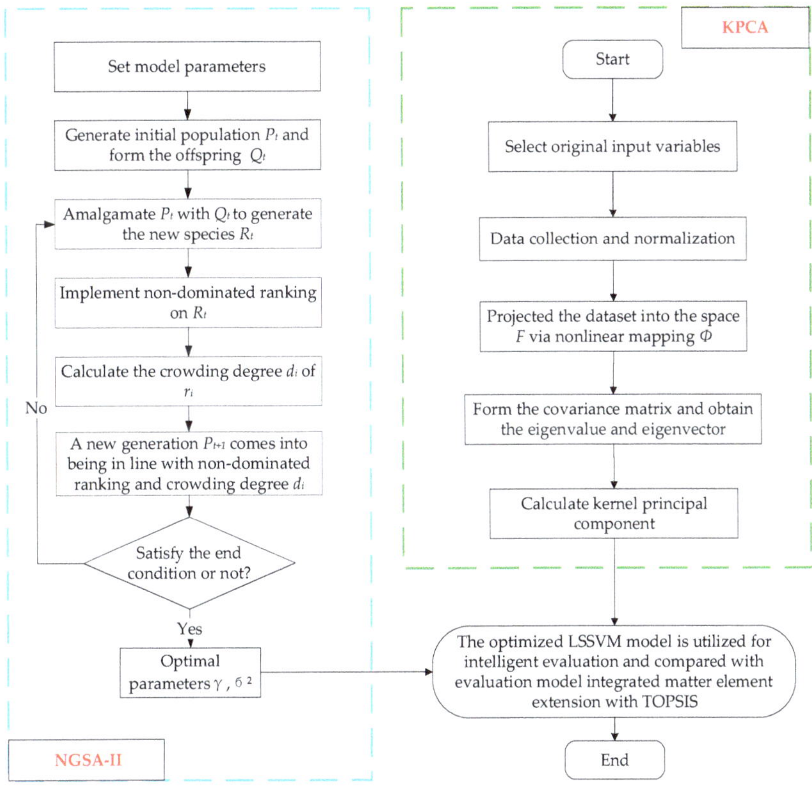

Figure 2. The flowchart of the proposed model.

Step 1: Parameters initialization and data preprocessing. Suppose the input $X = \{x_i,\ i = 1, 2, \cdots, n\}$ consists of the aforementioned core competitiveness evaluation indexes of clean energy incubators, as Table 1 shows. Consistency and standardized processing are implemented on the original input x_i.

Step 2: Weight Determination. The entropy weight method is used here. Besides, evaluation results can be obtained based on matter-element extension integrated with TOPSIS.

Step 3: Input selection. Crucial factors in original input X are extracted via KPCA. Gaussian kernel function is selected for nonlinear mapping as shown in Equation (32). The kernel principal components whose cumulative variance contribution is higher than 95% form the new input matrix.

$$k(x, y) = \exp\left\{ -\frac{\|x - y\|^2}{2\sigma^2} \right\} \tag{32}$$

Step 4: Parameter optimization. Initialize the parameters in LSSVM and NSGA-II. Due to the influence of parameters on training and learning ability in LSSVM, this paper applies NSGA-II to optimize the two parameters, that is, γ and σ^2. Circulation ends at the maximum number of iteration $t_{\max}$, thus the optimal parameters can be substituted into LSSVM. Through continuous retraining and testing, the two parameters are adjusted again and the optimal core competitiveness evaluation model for clean energy incubator is established.

Step 5: Output the evaluation results. According to the characteristics of the core competitiveness level of clean energy incubators, the category labels are set as 1, 2, 3 and 4, that corresponds to the grades of core competitiveness, namely very high, high, general and low, respectively.

Table 1. The input of evaluation model.

Input	Index	Input	Index
x_1	Leading industry is in line with the global future energy development direction	x_{19}	Social network management
x_2	Leading industry belongs to the global emerging energy industry	x_{20}	Social responsibility performance
x_3	Leading industry is attributed to key industries encouraged by the state	x_{21}	Total incubation fund
x_4	Leading industry is ascribed to clean energy industry	x_{22}	Financing service capability
x_5	Leading energy industry is the focus of regional strategic progress	x_{23}	Technical standardization degree
x_6	Leading energy industry possesses growth foundation and supporting system	x_{24}	Business process standardization
x_7	Leading energy industry owns unique resources	x_{25}	Outsourcing
x_8	Positioning consistency between entrepreneurial team and incubator	x_{26}	Cluster development
x_9	Future five-year plan	x_{27}	Industry brand
x_{10}	Future five-year expected venture capital investment	x_{28}	Brand extension
x_{11}	Innovation resources	x_{29}	Brand feedback
x_{12}	Innovative incentives	x_{30}	Total investment in innovation test platform
x_{13}	Number of management consulting labour	x_{31}	Number of experts in innovation team
x_{14}	Proportion of highly educated employees in management consulting service	x_{32}	Number of successful introduction of external scientific payoffs
x_{15}	Management communication and sharing level	x_{33}	Number of scientific research achievements of incubating enterprises successfully transformed
x_{16}	Number of intellectual property applications	x_{34}	Knowledge information service effect
x_{17}	Innovative culture construction	x_{35}	Number of scientific and technological achievements transferred
x_{18}	Stakeholder management	x_{36}	Effect of expert service

4. Experiment Study

4.1. Input Selection and Preprocessing

In order to verify the performance of the established approach, this paper conducts a study on core competitiveness evaluation base on 20 clean energy incubators. The type and representative quantitative index data of 20 incubators are shown in Table 2.

According to the evaluation index system for the core competitiveness of clean energy incubator, Table 3 shows the indicator properties and attributes. Through field investigation and data collection, the relevant data are assembled and sorted out. Simultaneously, 15 experts are invited to score the qualitative indexes in [0, 100]. The corresponding mean value can be considered as the stationary criteria.

In the light of above-mentioned data preprocessing method, the standardized results of the 20 samples are derived as listed in Tables 4 and 5.

Table 2. The type and representative quantitative index data of 20 incubators.

Clean Energy Incubator	Type	Future Five-Year Expected Venture Capital Investment (Million Chinese Yuan)	Proportion of Highly Educated Employees in Management Consulting Service (%)	Total Incubation Fund (Million Chinese Yuan)
A	Solar, hydro, biomass	59	91	111
B	Solar	79	61	98

Table 2. *Cont.*

Clean Energy Incubator	Type	Future Five-Year Expected Venture Capital Investment (Million Chinese Yuan)	Proportion of Highly Educated Employees in Management Consulting Service (%)	Total Incubation Fund (Million Chinese Yuan)
C	Wind, solar, biomass	73	49	68
D	Wind, solar, hydro, biomass	88	95	92
E	Solar	87	64	83
F	Biomass	71	69	105
G	Hydro	46	47	70
H	Wind, Solar, biomass	58	84	86
I	Solar, biomass	81	89	77
J	Wind, solar, hydro	49	72	87
K	Wind, biomass	71	63	74
L	Wind, solar, hydro, biomass	80	72	103
M	Wind, solar, biomass	120	42	98
N	Wind, solar, hydro, biomass	56	73	84
O	Wind, solar	59	79	96
P	Wind, solar	84	54	101
Q	Wind, biomass	79	80	69
R	Wind, solar, hydro, biomass	58	96	94
S	Wind, hydro	165	79	94
T	Wind, solar, hydro, biomass	69	57	156

Table 3. Evaluation index properties for core competitiveness of clean energy incubator.

Index	Index Properties
Leading industry is in line with the global future energy development direction	qualitative indicator
Leading industry belongs to the global emerging energy industry	qualitative indicator
Leading industry is attributed to key industries encouraged by the state	qualitative indicator
Leading industry is ascribed to clean energy industry	qualitative indicator
Leading energy industry is the focus of regional strategic progress	qualitative indicator
Leading energy industry possesses growth foundation and supporting system	qualitative indicator
Leading energy industry owns unique resources	qualitative indicator
Positioning consistency between entrepreneurial team and incubator	qualitative indicator
Future five-year plan	qualitative indicator
Future five-year expected venture capital investment	quantitative indicator
Innovation resources	qualitative indicator
Innovative incentives	qualitative indicator
Number of management consulting labour	quantitative indicator
Proportion of highly educated employees in management consulting service	quantitative indicator
Management communication and sharing level	qualitative indicator
Number of intellectual property applications	quantitative indicator
Innovative culture construction	qualitative indicator
Stakeholder management	qualitative indicator
Social network management	qualitative indicator
Social responsibility performance	qualitative indicator
Total incubation fund	quantitative indicator
Financing service capability	qualitative indicator
Technical standardization degree	qualitative indicator
Business process standardization	qualitative indicator
Outsourcing	qualitative indicator
Cluster development	qualitative indicator
Industry brand	qualitative indicator
Brand extension	qualitative indicator
Brand feedback	qualitative indicator

Table 3. *Cont.*

Index	Index Properties
Total investment in innovation test platform	quantitative indicator
Number of experts in innovation team	quantitative indicator
Number of successful introduction of external scientific payoffs	quantitative indicator
Number of scientific research achievements of incubating enterprises successfully transformed	quantitative indicator
Knowledge information service effect	qualitative indicator
Number of scientific and technological achievements transferred	quantitative indicator
Effect of expert service	qualitative indicator

Table 4. Data processing results of core competitiveness evaluation indexes for clean energy incubator (A–J).

Index	Clean Energy Incubator									
	A	B	C	D	E	F	G	H	I	J
x_1	0.1759	0.1699	0.2176	0.2653	0.1580	0.1699	0.1550	0.2176	0.1669	0.2116
x_2	0.2106	0.2106	0.2244	0.2494	0.2411	0.2217	0.1690	0.2549	0.2522	0.1607
x_3	0.1946	0.1828	0.2270	0.2830	0.2329	0.2299	0.2329	0.2299	0.1533	0.2005
x_4	0.1611	0.1845	0.2577	0.2109	0.2314	0.1669	0.1992	0.2577	0.1845	0.1962
x_5	0.2542	0.1636	0.1928	0.2571	0.1461	0.1870	0.2308	0.2717	0.2747	0.1636
x_6	0.2367	0.2033	0.2200	0.2144	0.1392	0.1922	0.1810	0.1782	0.2367	0.2646
x_7	0.1713	0.2156	0.1624	0.1565	0.1477	0.2805	0.2126	0.1595	0.2599	0.1624
x_8	0.1923	0.2358	0.2203	0.2637	0.1830	0.1954	0.2575	0.2854	0.2078	0.2296
x_9	0.2744	0.1547	0.2481	0.2568	0.1693	0.2043	0.1868	0.2481	0.1780	0.1985
x_{10}	0.1580	0.2534	0.2117	0.2653	0.2743	0.2295	0.1520	0.1818	0.2236	0.1580
x_{11}	0.1849	0.2333	0.1992	0.2561	0.1451	0.2162	0.2134	0.2703	0.2532	0.2504
x_{12}	0.2156	0.1983	0.2529	0.2616	0.1696	0.2414	0.2299	0.1955	0.2443	0.2817
x_{13}	0.1532	0.1763	0.2746	0.2659	0.1705	0.2544	0.2804	0.2341	0.2544	0.1445
x_{14}	0.2644	0.1968	0.1783	0.2829	0.2060	0.1906	0.1568	0.2460	0.2644	0.2306
x_{15}	0.1948	0.1791	0.2765	0.1948	0.1885	0.2137	0.1760	0.2168	0.2639	0.1571
x_{16}	0.2401	0.2075	0.2579	0.2519	0.1838	0.1838	0.1512	0.2312	0.2579	0.2579
x_{17}	0.1985	0.1496	0.1438	0.2761	0.1668	0.2617	0.2013	0.2128	0.1841	0.2646
x_{18}	0.2753	0.1639	0.2227	0.2691	0.1639	0.1547	0.1639	0.1949	0.2011	0.2815
x_{19}	0.2199	0.1789	0.2785	0.2140	0.1554	0.2375	0.2610	0.2140	0.2698	0.2170
x_{20}	0.1850	0.1577	0.2395	0.2426	0.1910	0.2335	0.1941	0.1546	0.2426	0.2880
x_{21}	0.2783	0.2416	0.1529	0.2294	0.1927	0.2661	0.1591	0.2111	0.1743	0.2111
x_{22}	0.1675	0.1764	0.1645	0.2871	0.1884	0.1615	0.2183	0.2422	0.2392	0.1884
x_{23}	0.1934	0.1876	0.1817	0.2726	0.1671	0.1700	0.2579	0.1817	0.2608	0.2491
x_{24}	0.1527	0.2363	0.2392	0.2709	0.2248	0.2162	0.2796	0.1787	0.2248	0.1643
x_{25}	0.1630	0.2673	0.2804	0.2413	0.1924	0.1728	0.2869	0.1663	0.1695	0.2739
x_{26}	0.2383	0.2085	0.1519	0.2234	0.1758	0.1847	0.2800	0.2949	0.1639	0.1609
x_{27}	0.2686	0.1618	0.2773	0.2195	0.2571	0.2282	0.2369	0.2109	0.1878	0.2253
x_{28}	0.1572	0.1659	0.1688	0.2678	0.1717	0.2853	0.1979	0.2445	0.2125	0.2649
x_{29}	0.2057	0.2299	0.2541	0.2814	0.2420	0.2511	0.1694	0.2602	0.2965	0.1634
x_{30}	0.2547	0.2179	0.2009	0.2122	0.1641	0.1472	0.1953	0.2405	0.2377	0.2377
x_{31}	0.1662	0.1543	0.1484	0.2463	0.1721	0.2018	0.2967	0.2077	0.2166	0.2196
x_{32}	0.2060	0.1752	0.2060	0.2367	0.1537	0.1722	0.2336	0.1906	0.2582	0.2859
x_{33}	0.2935	0.1585	0.1585	0.2641	0.1761	0.1819	0.1790	0.1643	0.2406	0.2524
x_{34}	0.2185	0.1742	0.2805	0.2923	0.2067	0.1801	0.1506	0.1978	0.1713	0.2244
x_{35}	0.2219	0.1849	0.1941	0.2435	0.2712	0.1602	0.2496	0.2219	0.2157	0.1849
x_{36}	0.2709	0.1777	0.1602	0.2855	0.2564	0.2156	0.2069	0.2797	0.2535	0.2709

Table 5. Data processing results of core competitiveness evaluation indexes for clean energy incubator (K–T).

Index	Clean energy Incubator									
	K	**L**	**M**	**N**	**O**	**P**	**Q**	**R**	**S**	**T**
x_1	0.1908	0.2832	0.2534	0.2623	0.2414	0.2414	0.2265	0.2891	0.1997	0.2891
x_2	0.1746	0.2328	0.2716	0.2383	0.2078	0.1773	0.2106	0.2688	0.2189	0.2328
x_3	0.2801	0.2948	0.1739	0.1533	0.2299	0.1651	0.1946	0.2859	0.2211	0.2270
x_4	0.1464	0.2929	0.1581	0.1523	0.2460	0.2753	0.2665	0.2812	0.2197	0.2812
x_5	0.2600	0.2308	0.2425	0.2484	0.1753	0.1812	0.1899	0.2659	0.2338	0.2308
x_6	0.2116	0.2673	0.2590	0.1671	0.1894	0.2701	0.2172	0.2089	0.2701	0.2785
x_7	0.2510	0.2599	0.2451	0.2185	0.2835	0.2392	0.2303	0.2776	0.2717	0.1683
x_8	0.2327	0.1923	0.2047	0.1861	0.1861	0.2078	0.2482	0.1706	0.2761	0.2482
x_9	0.2860	0.2919	0.2131	0.2247	0.1722	0.2919	0.1868	0.1605	0.2423	0.1956
x_{10}	0.2146	0.2534	0.2862	0.1848	0.1878	0.2534	0.2206	0.1789	0.2921	0.2117
x_{11}	0.1679	0.2418	0.2731	0.2361	0.1536	0.1849	0.1792	0.2731	0.1878	0.2760
x_{12}	0.2328	0.2472	0.2070	0.2213	0.1667	0.1437	0.1725	0.2817	0.2414	0.2041
x_{13}	0.2544	0.2197	0.2833	0.2601	0.1532	0.1503	0.2457	0.2486	0.1792	0.1590
x_{14}	0.1753	0.2306	0.1599	0.2029	0.2460	0.1753	0.2737	0.2952	0.2491	0.1630
x_{15}	0.1791	0.2828	0.2074	0.1791	0.2168	0.2671	0.2734	0.2514	0.2357	0.2514
x_{16}	0.2193	0.2549	0.1838	0.1719	0.1778	0.1927	0.2342	0.1541	0.2964	0.2845
x_{17}	0.1956	0.2531	0.1639	0.2560	0.2100	0.2704	0.2330	0.1783	0.2819	0.2790
x_{18}	0.2103	0.3062	0.2475	0.2073	0.1980	0.1918	0.1732	0.3093	0.2506	0.1825
x_{19}	0.2082	0.2199	0.1965	0.2698	0.2404	0.1613	0.2639	0.1701	0.1730	0.2580
x_{20}	0.2304	0.2971	0.1607	0.1789	0.1789	0.2850	0.2213	0.2850	0.2426	0.1698
x_{21}	0.1743	0.2600	0.2478	0.1958	0.2386	0.2569	0.1560	0.2355	0.2325	0.2845
x_{22}	0.2452	0.2093	0.1525	0.2452	0.2452	0.2692	0.2063	0.2632	0.2303	0.2931
x_{23}	0.2638	0.2520	0.1465	0.2052	0.2579	0.2403	0.2550	0.2110	0.2257	0.2286
x_{24}	0.1556	0.1989	0.2248	0.2709	0.2767	0.1441	0.1787	0.2680	0.1960	0.2796
x_{25}	0.2641	0.2999	0.1858	0.2282	0.2250	0.1826	0.2152	0.2021	0.1663	0.1989
x_{26}	0.1490	0.2234	0.2443	0.2830	0.2115	0.2443	0.1996	0.2800	0.1966	0.2651
x_{27}	0.2455	0.2744	0.1618	0.1820	0.1762	0.2658	0.2484	0.2195	0.1733	0.1878
x_{28}	0.2009	0.1892	0.2387	0.2212	0.2678	0.1921	0.2503	0.2212	0.2824	0.2009
x_{29}	0.2481	0.2299	0.1967	0.1785	0.1573	0.1815	0.2693	0.2299	0.1936	0.1513
x_{30}	0.2660	0.2660	0.2462	0.1783	0.2717	0.2462	0.2009	0.2151	0.2405	0.1783
x_{31}	0.1721	0.2285	0.2819	0.2819	0.2404	0.1899	0.2522	0.2255	0.2344	0.2552
x_{32}	0.2859	0.2675	0.2183	0.1906	0.1937	0.2798	0.1875	0.2336	0.1845	0.2429
x_{33}	0.1614	0.2348	0.2377	0.1878	0.2817	0.2582	0.2758	0.2612	0.2465	0.1585
x_{34}	0.2510	0.2658	0.1742	0.1919	0.1536	0.2599	0.2923	0.2628	0.2303	0.2038
x_{35}	0.1664	0.2928	0.2589	0.2003	0.2619	0.2280	0.2157	0.2743	0.1695	0.1911
x_{36}	0.1632	0.2185	0.1894	0.2156	0.2098	0.1777	0.2331	0.2506	0.1748	0.1923

4.2. Evaluation and Analysis of Core Competitiveness of Clean Energy Incubator Based on Matter-Element Extension Model Integrated with TOPSIS

Table 6 presents the indicator weights derived from Equations (2)–(5).

The weighted standardized matrix X can be further obtained as Equation (33):

$$X = \begin{bmatrix} 0.0058 & 0.0087 & 0.0078 & \cdots & 0.0088 \\ 0.0027 & 0.0036 & 0.0042 & \cdots & 0.0036 \\ 0.0078 & 0.0082 & 0.0048 & \cdots & 0.0063 \\ \vdots & \vdots & \vdots & \cdots & \vdots \\ 0.0040 & 0.0053 & 0.0046 & \cdots & 0.0047 \end{bmatrix} \tag{33}$$

Table 6. Indicator weights.

Index	Weight	Index	Weight
x_1	0.0306	x_{19}	0.0230
x_2	0.0156	x_{20}	0.0327
x_3	0.0278	x_{21}	0.0262
x_4	0.0364	x_{22}	0.0278
x_5	0.0256	x_{23}	0.0229
x_6	0.0236	x_{24}	0.0327
x_7	0.0353	x_{25}	0.0312
x_8	0.0165	x_{26}	0.0327
x_9	0.0306	x_{27}	0.0224
x_{10}	0.0285	x_{28}	0.0246
x_{11}	0.0273	x_{29}	0.0290
x_{12}	0.0223	x_{30}	0.0195
x_{13}	0.0401	x_{31}	0.0283
x_{14}	0.0295	x_{32}	0.0241
x_{15}	0.0232	x_{33}	0.0354
x_{16}	0.0280	x_{34}	0.0314
x_{17}	0.0326	x_{35}	0.0226
x_{18}	0.0358	x_{36}	0.0243

Table 7 manifests the positive ideal solution Z^+ as well as the negative one calculated from Equations (9) and (10). The extremum interval composed of the above results is divided into four levels, namely low, general, high and very high. Here, x_1 is taken as an example, the corresponding positive and negative ideal solutions are 0.0088 and 0.0047, respectively. The interval can be expressed as Equation (34). This technique is applicable for other indexes.

$$
\begin{aligned}
H_{11} &= [0.0047, 0.0058] \\
H_{12} &= [0.0058, 0.0068] \\
H_{13} &= [0.0068, 0.0078] \\
H_{14} &= [0.0078, 0.0088]
\end{aligned}
\tag{34}
$$

Table 7. Positive and negative ideal solutions.

Index	Positive Ideal Solution	Negative Ideal Solution	Index	Positive Ideal Solution	Negative Ideal Solution
x_1	0.0088	0.0047	x_{19}	0.0064	0.0036
x_2	0.0042	0.0025	x_{20}	0.0097	0.0051
x_3	0.0082	0.0043	x_{21}	0.0075	0.0040
x_4	0.0107	0.0053	x_{22}	0.0081	0.0042
x_5	0.0070	0.0037	x_{23}	0.0062	0.0034
x_6	0.0066	0.0033	x_{24}	0.0091	0.0047
x_7	0.0100	0.0052	x_{25}	0.0094	0.0051
x_8	0.0047	0.0028	x_{26}	0.0096	0.0049
x_9	0.0089	0.0047	x_{27}	0.0062	0.0036
x_{10}	0.0083	0.0043	x_{28}	0.0070	0.0039
x_{11}	0.0075	0.0040	x_{29}	0.0086	0.0044
x_{12}	0.0063	0.0032	x_{30}	0.0053	0.0029
x_{13}	0.0114	0.0058	x_{31}	0.0084	0.0042
x_{14}	0.0087	0.0046	x_{32}	0.0069	0.0037
x_{15}	0.0066	0.0036	x_{33}	0.0104	0.0056
x_{16}	0.0083	0.0042	x_{34}	0.0092	0.0047
x_{17}	0.0092	0.0047	x_{35}	0.0066	0.0036
x_{18}	0.0111	0.0055	x_{36}	0.0069	0.0039

Furthermore, the closeness degree $E(N_i)$ between each index in standardized decision matrix and four evaluation intervals is computed as Equations (35)–(38).

$$E(N_1) = \begin{bmatrix} 0.0001 & 0.0001 & 0.0014 & \cdots & 0.0036 \\ 0.0006 & 0.0006 & 0.0008 & \cdots & 0.0009 \\ 0.0007 & 0.0003 & 0.0016 & \cdots & 0.0016 \\ \vdots & \vdots & \vdots & \cdots & \vdots \\ 0.0023 & 0.0000 & 0.0004 & \cdots & 0.0004 \end{bmatrix} \tag{35}$$

$$E(N_2) = \begin{bmatrix} 0.0009 & 0.0011 & 0.0004 & \cdots & 0.0026 \\ 0.0001 & 0.0001 & 0.0003 & \cdots & 0.0005 \\ 0.0003 & 0.0007 & 0.0006 & \cdots & 0.0006 \\ \vdots & \vdots & \vdots & \cdots & \vdots \\ 0.0015 & 0.0007 & 0.0011 & \cdots & 0.0004 \end{bmatrix} \tag{36}$$

$$E(N_3) = \begin{bmatrix} 0.0019 & 0.0021 & 0.0006 & \cdots & 0.0015 \\ 0.0003 & 0.0003 & 0.0001 & \cdots & 0.0000 \\ 0.0013 & 0.0016 & 0.0004 & \cdots & 0.0004 \\ \vdots & \vdots & \vdots & \cdots & \vdots \\ 0.0008 & 0.0015 & 0.0019 & \cdots & 0.0011 \end{bmatrix} \tag{37}$$

$$E(N_4) = \begin{bmatrix} 0.0030 & 0.0031 & 0.0017 & \cdots & 0.0005 \\ 0.0007 & 0.0007 & 0.0005 & \cdots & 0.0004 \\ 0.0023 & 0.0026 & 0.0014 & \cdots & 0.0014 \\ \vdots & \vdots & \vdots & \cdots & \vdots \\ 0.0000 & 0.0022 & 0.0027 & \cdots & 0.0019 \end{bmatrix} \tag{38}$$

Figure 3 and Table 8 illustrate the weighted closeness degree of each evaluation scheme. The level of the maximum *of $Q_j(N_i)$* can be determined as the level of the evaluation object.

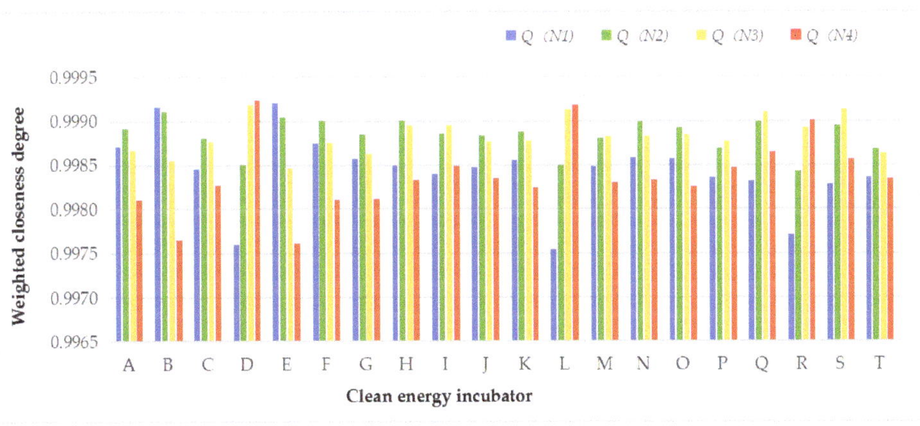

Figure 3. The weighted closeness degree of each evaluation scheme.

Table 8. The weighted closeness degree and valuation level of each evaluation scheme.

Clean Energy Incubator	$Q(N_1)$	$Q(N_2)$	$Q(N_3)$	$Q(N_4)$	Max Value	Valuation Level
A	0.9987	0.9989	0.9987	0.9981	0.9989	general
B	0.9992	0.9991	0.9986	0.9977	0.9992	low
C	0.9985	0.9988	0.9988	0.9983	0.9988	general
D	0.9976	0.9985	0.9992	0.9992	0.9992	very high
E	0.9992	0.9990	0.9985	0.9976	0.9992	low
F	0.9988	0.9990	0.9988	0.9981	0.9990	general
G	0.9986	0.9988	0.9986	0.9981	0.9988	general
H	0.9985	0.9990	0.9990	0.9983	0.9990	general
I	0.9984	0.9989	0.9990	0.9985	0.9990	high
J	0.9985	0.9988	0.9988	0.9984	0.9988	general
K	0.9986	0.9989	0.9988	0.9983	0.9989	general
L	0.9975	0.9985	0.9991	0.9992	0.9992	very high
M	0.9985	0.9988	0.9988	0.9983	0.9988	high
N	0.9986	0.9990	0.9988	0.9983	0.9990	general
O	0.9986	0.9989	0.9988	0.9983	0.9989	general
P	0.9984	0.9987	0.9988	0.9985	0.9988	high
Q	0.9983	0.9990	0.9991	0.9986	0.9991	high
R	0.9977	0.9984	0.9989	0.9990	0.9990	very high
S	0.9983	0.9990	0.9991	0.9986	0.9991	high
T	0.9984	0.9987	0.9986	0.9983	0.9987	general

4.3. Evaluation and Analysis of Core Competitiveness of Clean Energy Incubator Based on KPCA-NSGA-II-LSSVM

Exact though it is, the aforementioned techniques are not suitable to address massive data for their weakness of complex calculation and low efficiency. Thus, the intelligent evaluation model is proposed in this paper.

4.3.1. Growth Catalytic Ability

In this part, the above 20 clean energy incubators are taken as training sample, and the other 20 incubators are used as testing samples. PCA and KPCA are applied to 36 indicators to attain the variance contribution rates and cumulative variance contribution rate, as shown in Tables 9 and 10. Figures 4 and 5 exhibit the results more intuitively.

Table 9. Variance contribution rates and cumulative variance contribution rate of PCA.

Component	Variance Contribution Rate/%	Cumulative Variance Contribution Rate/%
1	25.68	25.68
2	16.67	42.35
3	8.66	51.01
4	8.56	59.57
5	6.56	66.13
6	6.02	72.15
7	5.22	77.37
8	3.65	81.02
9	3.02	84.04
10	2.66	86.7
11	2.56	89.26
12	2.36	91.62
13	1.88	93.5
14	1.83	95.33
15	1.16	96.49
16	0.96	97.45
17	0.88	98.33

Table 9. *Cont.*

Component	Variance Contribution Rate/%	Cumulative Variance Contribution Rate/%
18	0.52	98.85
19	0.38	99.23
20	0.22	99.45
21	0.18	99.63
22	0.16	99.79
23	0.09	99.88
24	0.06	99.94
25	0.03	99.97
26	0.01	99.98
27	0.01	99.99
28	0.01	100
29	0	100
30	0	100
31	0	100
32	0	100
33	0	100
34	0	100
35	0	100
36	0	100

Table 10. Variance contribution rates and cumulative variance contribution rate of KPCA.

Component	Variance Contribution Rate/%	Cumulative Variance Contribution Rate/%
1	69.56	69.56
2	18.58	88.14
3	7.56	95.7
4	2.82	98.52
5	0.99	99.51
6	0.26	99.77
7	0.18	99.95
8	0.03	99.98
9	0.02	100
10	0	100
11	0	100
12	0	100
13	0	100
14	0	100
15	0	100
16	0	100
17	0	100
18	0	100
19	0	100
20	0	100
21	0	100
22	0	100
23	0	100
24	0	100
25	0	100
26	0	100
27	0	100
28	0	100
29	0	100
30	0	100

Table 10. *Cont.*

Component	Variance Contribution Rate/%	Cumulative Variance Contribution Rate/%
31	0	100
32	0	100
33	0	100
34	0	100
35	0	100
36	0	100

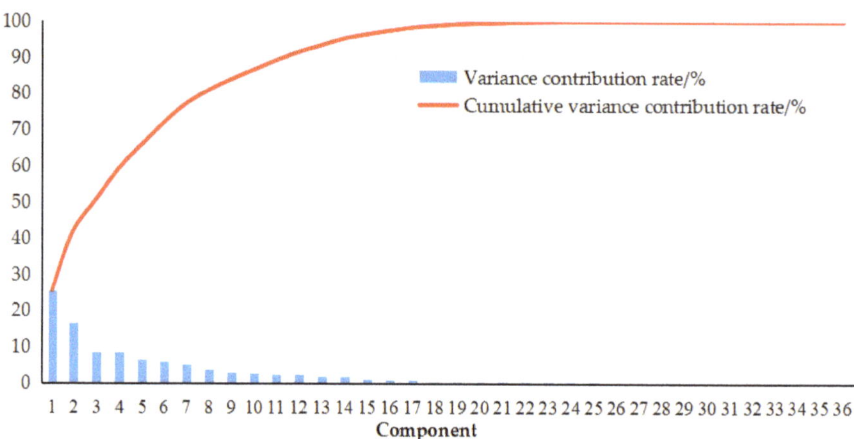

Figure 4. Total variance of PCA.

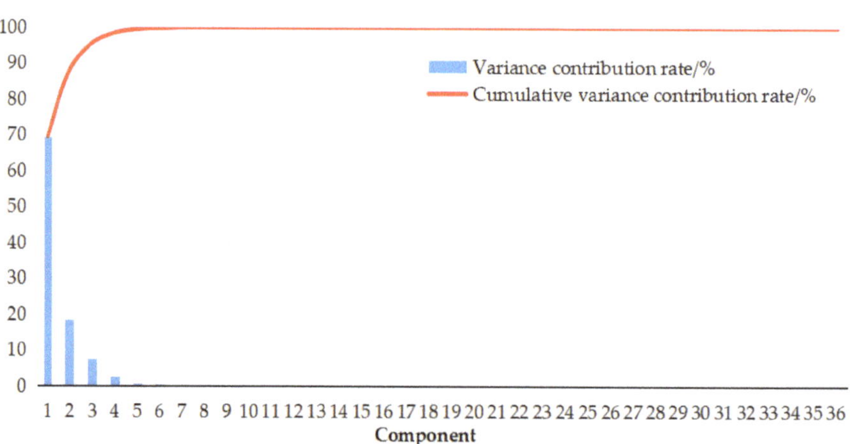

Figure 5. Total variance of KPCA.

From Tables 9 and 10, Figures 4 and 5, it can be seen that the first principal component of KPCA explains 69.56% of the factors, while the contribution rate of the first principal component in PCA is only 25.68%. In addition, it needs to extract 14 principal components in PCA or 3 principal components in KPCA to ensure cumulative contribution rate more than 95%, which can retain enough original data information. Therefore, the dimension reduction and simplification of KPCA is superior to PCA.

The selected principal components coefficients are acquired through dividing the principal component load vectors by the arithmetic square root of the eigenvalues. The output component matrix is presented in Table 11.

Table 11. Component matrix of KPCA.

Factor	Component		
	1	2	3
x_1	−0.903	−0.933	0.598
x_2	0.267	−0.452	0.948
x_3	0.471	0.700	0.352
x_4	0.862	−0.548	0.821
x_5	−0.060	0.708	0.861
x_6	0.673	−0.721	0.440
x_7	0.443	−0.199	−0.368
x_8	0.227	−0.669	−0.607
x_9	0.174	−0.645	0.547
x_{10}	−0.399	0.181	0.487
x_{11}	−0.508	−0.852	0.475
x_{12}	−0.167	0.401	0.438
x_{13}	−0.690	0.548	0.379
x_{14}	−0.385	0.568	−0.053
x_{15}	0.750	−0.054	−0.055
x_{16}	−0.125	−0.007	−0.809
x_{17}	−0.475	0.350	0.336
x_{18}	0.958	0.696	0.610
x_{19}	0.901	−0.248	0.948
x_{20}	0.066	−0.182	0.663
x_{21}	0.033	0.991	0.125
x_{22}	0.144	0.006	−0.919
x_{23}	0.378	0.491	0.527
x_{24}	0.518	0.784	0.144
x_{25}	−0.561	0.964	0.333
x_{26}	−0.960	−0.256	−0.828
x_{27}	0.659	0.961	−0.572
x_{28}	−0.814	0.307	0.110
x_{29}	0.717	0.876	0.236
x_{30}	−0.295	−0.077	−0.607
x_{31}	0.600	0.987	−0.400
x_{32}	0.029	0.147	−0.289
x_{33}	0.227	−0.445	0.886
x_{34}	0.867	0.105	0.006
x_{35}	−0.025	−0.826	−0.299
x_{36}	0.546	−0.771	−0.841

4.3.2. Result Analysis

This paper performs on Matlab R2014a with Intel Core i5-6300U, 4G memory and 500G hard disk. It should be noted that important parameters of the proposed model are obtained by NSGA algorithm to guarantee the precision. In NSGA-II, the population size is 300, the maximum iteration number equals 200 and crossover and mutation probability is set as 2. In this way, the optimized parameters of LSSVM are $\gamma = 36.6786$ and $\sigma^2 = 16.1816$.

The experiment selects several techniques to make comparison so as to examine the performance of the proposed method. The involved approaches include NSGA-II-LSSVM, GA-LSSVM, LSSVM and SVM. In LSSVM, γ and σ^2 equal 9.8568 and 16.2657 by cross validation, partly. In SVM, penalty parameter c is 8.659, kernel function parameter g equals 0.299 and loss function parameter p is 2.869.

The three principal components extracted by KPCA are exploited as input samples. The results of the test samples are displayed in Table 12 and Figure 6. In order to show the results clearly, the samples are divided into four groups, which are shown in (a–d) of Figure 6. Here, the relative errors in Table 12 are the results of the test samples and the number 1, 2, 3, 4 in Figure 6 represent the level of very high, high, general and low, respectively. In addition, the accuracy of 5 models is shown in Figure 7.

Table 12. Results of 5 models in core competitiveness valuation of clean energy incubators.

Clean Energy Incubator	Valuation Level	KPCA-NSGA-II-LSSVM	NSGA-II-LSSVM	GA-LSSVM	LSSVM	SVM
A	general	general	very high	general	low	general
B	low	low	low	low	low	general
C	general	general	general	low	general	general
D	very high	very high	very high	very high	general	very high
E	low	low	low	general	low	low
F	general	high	general	general	general	general
G	general	general	low	general	high	general
H	general	general	general	general	general	general
I	high	high	high	high	high	low
J	general	general	general	general	general	general
K	general	general	general	general	general	low
L	very high	very high	very high	very high	general	very high
M	high	high	high	high	high	general
N	general	general	general	general	general	high
O	general	general	general	general	general	general
P	high	high	high	high	high	high
Q	high	high	high	very high	high	high
R	very high	very high	very high	very high	very high	very high
S	high	high	high	high	very high	high
T	general	general	general	general	general	high
Relative error	-	5%	10%	15%	25%	30%

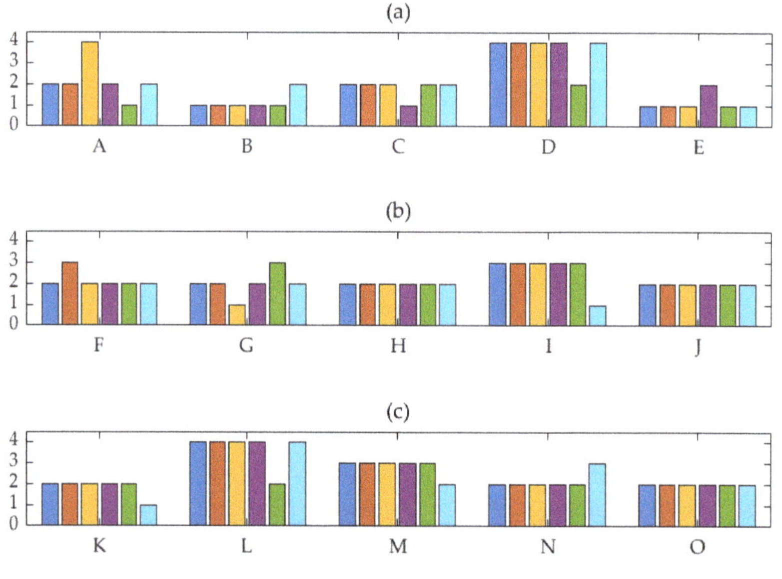

Figure 6. *Cont.*

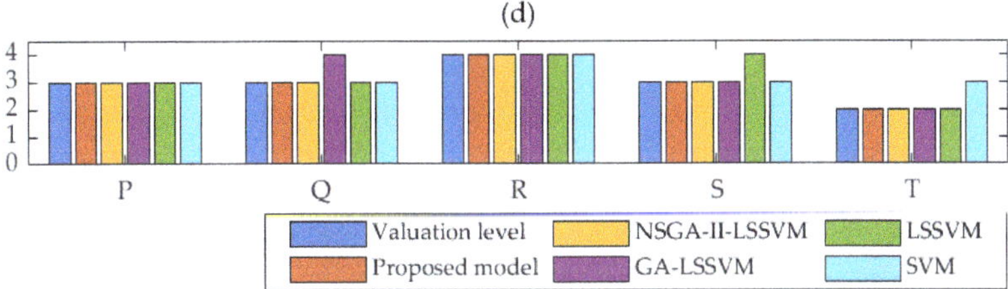

Figure 6. Results of 5 models in core competitiveness valuation of clean energy incubators. Note: (**a**) shows the results from sample A to E; (**b**) shows the results from sample F to J; (**c**) shows the results from sample K to O; (**d**) shows the results from sample P to T.

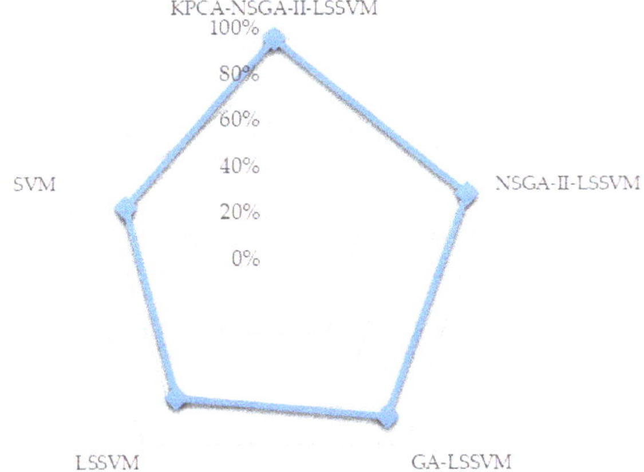

Figure 7. The accuracy of the proposed model and the comparison models.

In terms of method performance, the relative error is minimum, only 5% between KPCA-NSGA-II-LSSVM and matter-element extension model combined with TOPSIS. That is, only one of the 20 clean energy incubators presents diverse level. Whereas, the relative errors of NSGA-II-LSSVM, GA-LSSVM, LSSVM and SVM. In LSSVM are 10%, 20%, 25% and 30%, respectively. As a consequence, the established model in this paper outperforms other approaches with reference to relative error as well as classification accuracy. Compared with NSGA-II-LSSVM, KPCA overcomes the adverse effects of redundant factors on LSSVM training. The GA optimization part is inferior to NSGA-II of the parameters' setting in LSSVM in aspect of improving its generalization ability and classification precision. In comparison with SVM, the categorization accuracy is enhanced due to the transformation of forecasting problems into equations through kernel function. Overall, the evaluation performance can be ranked from superior to inferior as follows: KPCA-NSGA-II-LSSVM, NSGA-II-LSSVM, GA-LSSVM, LSSVM, SVM. What's more, we can also observe the advantages of the proposed model through Figure 7, where the accuracy of the proposed model is the highest.

Hence, KPCA-NSGA-II-LSSVM is able to be utilized in the field of core competitiveness of clean energy incubators scientifically and effectively. On the foundation of extension matter-element model integrated with TOPSIS, the introduction of auxiliary intelligent algorithm can achieve fast calculation and support the decision-making of relevant investors.

5. Conclusions

This paper designs an evaluation index system for core competitiveness of clean energy incubators and puts forward an assessment approach based on matter-element extension model integrated with TOPSIS as well as KPCA-NSGA-II-LSSVM. The indicator system is built from five aspects, that is strategic positioning ability, seed selection ability, intelligent transplantation ability, growth catalytic ability and service value-added ability to respect its core competitiveness. The paper takes advantage of extension matter-element model and TOPSIS to weight the selected evaluation indexes objectively and acquires the result from the perspective of traditional approaches. KPCA is manipulated in response to feature dimension so as to extract important information. Then LSSVM optimized by NSGA-II outputs the evaluation level concerning modern intelligence. The case study demonstrates the scientificity and accuracy of the proposed model in this paper, thereinto the traditional technique is able to obtain exact results while the intelligent evaluation methods can achieve fast calculation and decision-making support. Research on the evaluation of the core competitiveness of clean energy incubator will help managers to understand the level and deficiency of their core competitiveness, and then help managers to carry out more targeted work. To sum up, this research is expected to provide decision-making basis for more reasonable operation mode of clean energy incubators. However, more sample data is needed for verification. At the same time, using more intelligent models to evaluate the core competitiveness of clean energy incubators is also our next work.

Author Contributions: Conceptualization, Y.L.; Data curation, G.L. and Y.L.; Funding acquisition, Y.L.; Investigation, G.L. and Y.L.; Methodology, Y.L.; Project administration, D.N. and Y.L.; Resources, Y.L.; Software, G.L.; Supervision, Y.L. and D.N.; Validation, G.L.; Writing—original draft, G.L. and Y.L.; Writing—review & editing, Y.L. All authors have read and agreed to the published version of the manuscript.

Funding: This work is supported by the 2018 Key Projects of Philosophy and Social Sciences Research, Ministry of Education, China (Project No. 18JZD032), Natural Science Foundation of Hebei Province, China (Project No. G2020403008) and Humanities and Social Science Research Project of Hebei Education Department, China (Project No. SQ201097).

Conflicts of Interest: The authors declare no conflict of interest.

References

1. Mazzoni, S.; Ooi, S.; Nastasi, B.; Romagnoli, A. Energy storage technologies as techno-economic parameters for master-planning and optimal dispatch in smart multi energy systems. *Appl. Energy* **2019**, *254*, 113682. [CrossRef]
2. Ullal, H.; Mitchell, R.; Keyes, B.; VanSant, K.; Von Roedern, B.; Symko-Davies, M.; Kane, V. Progress of the PV Technology Incubator project towards an enhanced U.S. manufacturing base. In Proceedings of the 2011 37th IEEE Photovoltaic Specialists Conference, Seattle, WA, USA, 19–24 June 2011; IEEE: Seattle, WA, USA, 2011; pp. 003287–003291.
3. Brun, E.C. Understanding a Business Incubator as a Start-Up Factory: A Value Chain Model Perspective. *Int. J. Innov. Technol. Manag. (IJITM)* **2019**, *16*, 1950025. [CrossRef]
4. Wang, Z.; He, Q.; Xia, S.; Sarpong, D.; Xiong, A.; Maas, G. Capacities of business incubator and regional innovation performance. *Technol. Forecast. Soc. Chang.* **2020**, *158*, 120125. [CrossRef]
5. Prahalad, C.K.; Hamel, G. The Core Competence of the Corporation. *Harv. Bus. Rev.* **2010**, *68*, 275–292.
6. Zhao, X. Evaluation and analysis on core competitiveness of China cotton textile enterprises with intuitionistic fuzzy information. *Int. J. Knowl. Based Intell. Eng. Syst.* **2019**, *23*, 9–13. [CrossRef]
7. Wu, Y.; Ke, Y.; Xu, C.; Li, L. An integrated decision-making model for sustainable photovoltaic module supplier selection based on combined weight and cumulative prospect theory. *Energy* **2019**, *181*, 1235–1251. [CrossRef]
8. Zhang, T.; Pan, M.Z. Evaluation of core competitiveness of modern circulation industry considering coupling effect. *Res. Commer. Econ.* **2020**, *16*, 9–11.
9. Changhong, L. Research on comprehensive competitiveness evaluation of science and technology business incubator in Pearl River Delta. *J. Intell.* **2010**, *29*, 147–149.
10. Qiang, M.; Zecong, Y.; Jing, S. Competitive Evaluation of Chain Incubator Based on IPA Analysis. *Sci. Technol. Manag. Res.* **2019**, *11*, 78–84.
11. Zhimei, T. Sustainable Development Ability Evaluation Research about Business Incubation Based on Principal Component Analysis and DEA Method. *Sci. Technol. Manag. Res.* **2016**, *36*, 88–94.
12. Ning, A.; Chen, G. Knowledge service capacity evaluation of incubators based on satisfaction of incubating startups. *Sci. Technol. Manag.* **2016**, *18*, 82–87.

13. Shunyi, H. Construction of social benefit evaluation index system and weight analysis of business incubator. *Pioneer. Sci. Technol. Mon.* **2016**, *29*, 75–79.

14. Yang, W.; Xu, K.; Lian, J.; Bin, L.; Ma, C. Multiple flood vulnerability assessment approach based on fuzzy comprehensive evaluation method and coordinated development degree model. *J. Environ. Manag.* **2018**, *213*, 440. [CrossRef]

15. Han, H.; Li, H.; Zhang, K. Urban Water Ecosystem Health Evaluation Based on the Improved Fuzzy Matter-Element Extension Assessment Model: Case Study from Zhengzhou City, China. *Math. Probl. Eng.* **2019**, *2019*, 7502342. [CrossRef]

16. Niu, D.; Li, Y.; Dai, S.; Kang, H.; Xue, Z.; Jin, X.; Song, Y. Sustainability Evaluation of Power Grid Construction Projects Using Improved TOPSIS and Least Square Support Vector Machine with Modified Fly Optimization Algorithm. *Sustainability* **2018**, *10*, 231. [CrossRef]

17. Sun, W.; Xu, Y. Financial security evaluation of the electric power industry in China based on a back propagation neural network optimized by genetic algorithm. *Energy* **2016**, *101*, 366–379. [CrossRef]

18. Liu, D.; Liu, C.; Fu, Q.; Li, T.; Imran, K.M.; Cui, S.; Abrar, F.M. ELM evaluation model of regional groundwater quality based on the crow search algorithm. *Ecol. Indic.* **2017**, *81*, 302–314. [CrossRef]

19. Ren, D.; Du, J. Marine Foreign Trade Economic Zone Industry Investment Risk Evaluation Model under the Background of the Belt and Road. *J. Coast. Res.* **2018**, *83*, 212–216. [CrossRef]

20. Niu, D.; Li, S.; Dai, S. Comprehensive Evaluation for Operating Efficiency of Electricity Retail Companies Based on the Improved TOPSIS Method and LSSVM Optimized by Modified Ant Colony Algorithm from the View of Sustainable Development. *Sustainability* **2018**, *10*, 860. [CrossRef]

21. Wu, Q.; Lin, H. Short-Term Wind Speed Forecasting Based on Hybrid Variational Mode Decomposition and Least Squares Support Vector Machine Optimized by Bat Algorithm Model. *Sustainability* **2019**, *11*, 652. [CrossRef]

22. Zhang, Y.; Liu, M. Adaptive directed evolved NSGA2 based node placement optimization for wireless sensor networks. *Wirel. Netw.* **2020**, *26*, 3539–3552. [CrossRef]

23. Zhou, Y.; Cao, S.; Kosonen, R.; Hamdy, M. Multi-objective optimisation of an interactive buildings-vehicles energy sharing network with high energy flexibility using the Pareto archive NSGA-II algorithm. *Energy Convers. Manag.* **2020**, *218*, 113017. [CrossRef]

24. Tsamardinos, I.; Borboudakis, G.; Katsogridakis, P.; Pratikakis, P.; Christophides, V. A greedy feature selection algorithm for Big Data of high dimensionality. *Mach. Learn.* **2019**, *108*, 149–202. [CrossRef]

25. Polutchko, S.K.; Stewart, J.J.; Demmig-Adams, B.; Adams, W.W. Evaluating the link between photosynthetic capacity and leaf vascular organization with principal component analysis. *Photosynthetica* **2018**, *56*, 1–12. [CrossRef]

26. Papaioannou, A.; Anastasiadis, A.; Kouloumvakos, A.; Paassilta, M.; Vainio, R.; Valtonen, E.; Belov, A.; Eroshenko, E.; Abunina, M.; Abunin, A. Nowcasting Solar Energetic Particle Events Using Principal Component Analysis. *Sol. Phys.* **2018**, *293*, 1–23. [CrossRef]

27. Zhou, Z.; Du, N.; Xu, J.; Li, Z.; Wang, P.; Zhang, J. Randomized Kernel Principal Component Analysis for Modeling and Monitoring of Nonlinear Industrial Processes with Massive Data. *Ind. Eng. Chem. Res.* **2019**, *58*, 10410–10417. [CrossRef]

28. Bayram, B.Ç. Evaluation of Forest Products Trade Economic Contribution by Entropy-TOPSIS: Case Study of Turkey. *Bioresources* **2020**, *15*, 1419–1429. [CrossRef]

29. Tang, H.; Shi, Y.; Dong, P. Public blockchain evaluation using entropy and TOPSIS. *Expert Syst. Appl.* **2018**, *117*, 204–210. [CrossRef]

30. Huang, W.; Shuai, B.; Sun, Y.; Wang, Y.; Antwi, E. Using entropy-TOPSIS method to evaluate urban rail transit system operation performance: The China case. *Transp. Res. Part A Policy Pract.* **2018**, *111*, 292–303. [CrossRef]

31. Chen, Y. Study on the Risk Evaluation of Government Purchasing Public Service Based on Matter Element Extension Model. *Open J. Soc. Sci.* **2018**, *6*, 127–143. [CrossRef]

32. Soh, W.; Kim, H.; Yum, B.-J. Application of kernel principal component analysis to multi-characteristic parameter design problems. *Ann. Oper. Res.* **2018**, *263*, 69–91. [CrossRef]

33. Ardakan, M.A.; Rezvan, M.T. Multi-objective optimization of reliability-redundancy allocation problem with cold-standby strategy using NSGA-II. *Reliab. Eng. Syst. Saf.* **2018**, *172*, 225–238. [CrossRef]

34. Song, X.; Zhao, J.; Song, J.; Dong, F.; Xu, L.; Zhao, J. Local Demagnetization Fault Recognition of Permanent Magnet Synchronous Linear Motor Based on S-Transform and PSO–LSSVM. *IEEE Trans. Power Electron.* **2020**, *35*, 7816–7825. [CrossRef]

Publisher's Note: MDPI stays neutral with regard to jurisdictional claims in published maps and institutional affiliations.

 *sustainability*

Article

Effects of a Battery Energy Storage System on the Operating Schedule of a Renewable Energy-Based Time-of-Use Rate Industrial User under the Demand Bidding Mechanism of Taipower

Cheng-Ta Tsai [1], Yu-Shan Cheng [2], Kuen-Huei Lin [1] and Chun-Lung Chen [1,*]

[1] Department of Marine Engineering, National Taiwan Ocean University (NTOU), No. 2, Beining Rd., Zhongzheng Dist., Keelung City 202, Taiwan; 20766004@mail.ntou.edu.tw (C.-T.T.); godss4508@gmail.com (K.-H.L.)

[2] Department of Electrical Engineering, National Taiwan Ocean University (NTOU), No. 2, Beining Rd., Zhongzheng Dist., Keelung City 202, Taiwan; yscheng@mail.ntou.edu.tw

* Correspondence: cclung@mail.ntou.edu.tw; Tel.: +886-2-2462-2192 (ext. 7107)

Abstract: Due to the increased development of the smart grid, it is becoming crucial to have an efficient energy management system for a time-of-use (TOU) rate industrial user in Taiwan. In this paper, an extension of the direct search method (DSM) is developed to deal with the operating schedule of a TOU rate industrial user under the demand bidding mechanism of Taipower. To maximize the total incentive obtained from the Taiwan Power Company (TPC, namely Taipower), several operational strategies using a battery energy storage system (BESS) are evaluated in the study to perform peak shaving and realize energy conservation. The effectiveness of the proposed DSM algorithm is validated with the TOU rate industrial user of the TPC. Numerical experiments are carried out to provide a favorable indication of whether to invest in a BESS for the renewable energy-based TOU rate industrial user in order to execute the demand bidding program (DBP).

Keywords: smart grid; time-of-use; demand bidding program; battery energy storage system; direct search method

Citation: Tsai, C.-T.; Cheng, Y.-S.; Lin, K.-H.; Chen, C.-L. Effects of a Battery Energy Storage System on the Operating Schedule of a Renewable Energy-Based Time-of-Use Rate Industrial User under the Demand Bidding Mechanism of Taipower. *Sustainability* **2021**, *13*, 3576. https://doi.org/10.3390/su13063576

Academic Editors: Tomonobu Senjyu, Wei-Hsin Chen, Alvin B. Culaba, Aristotle T. Ubando and Steven Lim

Received: 9 February 2021
Accepted: 20 March 2021
Published: 23 March 2021

Publisher's Note: MDPI stays neutral with regard to jurisdictional claims in published maps and institutional affiliations.

1. Introduction

Due to the soaring prices of fossil fuels and the rising awareness of environmental protection, renewable energy resources have attracted more and more attention from the utility industry. In Taiwan, the development of the hybrid generation system composed of different renewable energy sources (RESs) has been rapidly growing, and currently, it is widely applied for the time-of-use (TOU) rate industrial users [1]. Despite the benefits provided by the RES, the intermittency and unpredictability of renewable power generation may cause operational issues and waste usable capacity when the installation of the RES increases [2]. The power dispatch gap caused by the intermittency of renewable power generation can be compensated for by the battery energy storage systems [3]. The uncertainties posed by renewable power generation also require the scheduling of additional generation reserves to compensate for power fluctuations in the RES system [4,5]. However, the percentage of the reserve margin of generators in Taiwan has been decreasing year by year, which may result in a high-risk situation for the system. To ensure the security and reliability of a power system, a better understanding of the required operating reserves with larger renewable penetrations is needed [6]. Moreover, in the smart grid system, a variety of issues may arise for renewable energy-based TOU rate customers, particularly in system operating and planning. Hence, the investigation into energy management has become very important in recent decades [7,8].

In terms of energy management, it is commonly divided into three aspects: demand side management, peak load regulation, and carbon emission reduction. In the last decade, studies have focused on load clipping with time-of-use rates [9], real-time pricing [10,11], and demand bidding [12]. These topics were proposed either to increase society's benefit from the use of electrical power or reduce the cost of electricity for TOU rate users. The TOU rate is usually regarded as a load management strategy to further smooth the demand curve for the utility grid and enable a reduced cost for industrial customers. Recently, a program has been implemented to encourage more TOU rate users to get involved with load management. This program, named "demand bidding", has been developed by the Taiwan Power Company (TPC) [13]. For the TPC, the demand bidding program (DBP) was designed to make a collaboration between customers and suppliers on demand response (DR) to prevent the risks of energy shortage and reduce the operating cost of expensive generators [14]. For certain TOU rate users with significant demand for electricity, electricity bills account for a substantial proportion of their overall expenses. In fact, the electricity bills include the total energy cost, the contract cost for the demand capacity, and the penalty bills. In order to minimize the total electricity cost for a TOU rate customer, a variety of energy storage technologies, such as a battery energy storage system (BESS), refrigeration storage (RS), compressed air energy storage (CAES), etc., have been investigated. For TOU rate users, the BESS is one of the most promising technologies for reducing electricity expenses [15]. Promising results have been reached in most studies in terms of electricity savings [16,17]. Therefore, when and how much power to charge and discharge turns out to be a critical problem for maximizing the benefits provided by a BESS. Many studies have focused on developing advanced algorithms for the DBP to increase the electricity incentives received from the power utility.

The energy management system plays a crucial role in implementing the DBP for the TOU customers in the smart grid. The aim of this study is to evaluate the operating strategy of a BESS in a hybrid generation system for a TOU rate industrial user under the demand bidding mechanism of Taipower. Many mathematical programming analysis methods and random search optimization techniques were used to solve the extended generation scheduling problem, such as multi-pass dynamic programming (MPDP) [15], the direct search method (DSM) [18], genetic algorithm (GA) [19], and particle swarm optimization [20]. Among them, the DSM method is especially of interest due to its simple architecture, high-quality solution, and fast convergence. In this paper, an extension of the DSM is developed to solve the optimal generation schedule problem in a TOU system for executing the DBP. To deal effectively with the coupling constraints of a system operation problem, the three-state dynamic programming (DP) is also incorporated into the DSM to augment the direct stochastic search procedure. The developed DSM computation tool can be used for addressing the key BESS integration issues. The developed DSM software can also be used to maximize the contribution of a RES and a BESS for reducing the electricity cost for a TOU rate industrial user. Test results are provided to assess the impact and economic benefits of the installation of a BESS for executing the DBP.

In general, the technical novelty and contribution of this paper can be presented as follows:

1. A demand bidding mechanism is designed to make a collaboration between customers and suppliers on demand response to perform peak shaving and realize energy conservation.
2. An improved DSM incorporated with a three-state DP is proposed to solve the operating schedule of a TOU rate industrial user under the demand bidding mechanism of Taipower.
3. Several operational strategies of a BESS are evaluated for a TOU rate industrial user to maximize the total incentive obtained from the TPC.
4. Numerical results are provided to assess the impact and economic benefits of the installation of a BESS for executing the DBP. The proposed DSM is also efficient and suitable for practical applications.

The remaining parts of this paper are presented in the following sequences. The mathematical formulation of the demand bidding mechanism of Taipower is expressed in

Section 2. Section 3 describes the extension of the DSM to coordinate the PV/wind, utility grid, and battery generation scheduling. Detailed descriptions of the operational strategy of the BESS are also provided for executing the DBP. Section 4 presents the simulation results and the conclusions are drawn in Section 5.

2. Problem Formulation and System Modeling

2.1. Notation

TOC	: Total electricity cost of the TOU rate industrial user (NT\$).
$CBL(d^*)$	: Customer baseline load for the day d^* (kW).
$PU_{max}(d^*)$	: Maximum demand during DR-execution time for the day d^* (kW).
$P_D^{bt}(d^*)$	: Load demand in period bt for the day d^* (kW).
$BDT(d^*)$	: DR-execution time (2 h or 4 h) for the day d^* (hours).
$ABP(d^*)$	: Actual load-reduction amount for the day d^* (kW).
$F_{BD}(d)$	: Total incentive for the day d (NT\$/h).
$F_{PE}(t,d)$	: Cost of the purchased power at interval t for the day d (NT\$/h).
$C_{BD}(d)$	: Bidding price during the DR-execution time for the day d (NT\$/ kWh).
$C_{PE}(t,d)$	: Tariff of the purchased power at interval t for the day d (NT\$/kWh).
d	: Index for day intervals (one day).
D	: Total observation days (days).
t	: Index for time intervals (15 min interval).
T	: Number of time intervals (one day).
j	: Index for non-dispatchable units.
ND	: Number of non-dispatchable units in system.
$P_D(t,d)$	: System load demand at interval t for the day d (kW).
$P_{NDj}(t,d)$	: Power of non-dispatchable unit j at interval t for the day d (kW).
$P_{grid}(t,d)$	: Power of purchased from utility grid at interval t for the day d (kW).
P_{grid}^{max}	: Maximum output power from utility grid (namely, the contract capacity) (kW).
$P_{bat}(t,d)$	: Charging/discharging power of battery storage system at interval t for the day d (positive for discharging and negative for charging) (kW).
P_{bat}^{max}	: Maximum power from the battery storage system (kW).
$SOC(t,d)$	: State of charge of the battery at interval t for the day d (kWh).
SOC_{min}	: Minimum battery state of charge (kWh).
SOC_{max}	: Maximum battery state of charge (kWh).
η_B	: Battery round-trip efficiency.
Δt	: Sampling time interval.
$P_{Wj}^*(t,d)$	: Available power of wind power generation unit j at interval t for the day d (kW).
P_{Wj}^{max}	: Maximum power of wind power generation unit j (kW).
$\phi_j(\bullet)$	: Wind power curve of wind power generation unit j (kW).
$v(t,d)$	: Wind speed at interval t for the day d.
v_{Ij}	: Cut in wind speed for wind power generation unit j.
v_{Rj}	: Rated wind speed for wind power generation unit j.
v_{Oj}	: Cut out wind speed for wind power generation unit j.
$P_{PVj}^*(t,d)$	: Available power of solar power generation unit j at interval t for the day d (kW).
$\delta_j(\bullet)$	: Radiation/ambient temperature power curve of solar power generation unit j (kW).
P_{PVj}^{max}	: Maximum power of solar power generation unit j (kW).
$S_r(t,d)$	: Intensity of solar radiation at interval t for the day d.
$T_r(t,d)$	: Ambient temperature at interval t for the day d.
SD	: Minimum intensity of solar radiation.
SU	: Maximum intensity of solar radiation.
$P_{vir}(bt+l)$	: Virtual price at period $bt+l$ (NT\$/kWh).
α	: Coefficient of virtual price.
P_D^{max} / P_D^{min}	: Maximum/minimum load demand during the DR-executing time (kW).
PLC	: Price of peak load periods (NT\$/kWh).
LLC	: Price of off-peak load periods (NT\$/kWh).

2.2. Demand Bidding Mechanism of Taipower

To increase the reliability of the power grid, the DBP is designed to encourage heavy electricity consumers to alter their usage pattern and remove the peak demand by giving rewards or bonuses. Recently, in many countries, power grid operators have applied the DBP in practice. For instance, Southern California Edison (SCE) and Pacific Gas and Electric Company (PG&E) have carried out DBPs [12,21]. SCE opens the DBP for customers with at least one service account with a demand of 200 kW or greater. Customers are able to participate in the DBP event from 12 p.m. to 8 p.m. and bid for at least 2 consecutive hours to earn bonus or rewards. The minimum bid is required at 30 kWh/hour. The payment is 50 cents per 1 kWh reduction minus the hourly price of energy. Apart from SCE, Mashhad Electric Energy Distribution Company (MEEDC) in Iran also employed the DBP [22]. Heavy electricity consumers with a demand of 100 kW can participate in the DBP. In the period from 11:00 a.m. to 3:00 p.m., the DBP will last for a minimum of 2 h and a maximum of 4 h. In the period of collaboration, the DBP will be less than 200 h. It is necessary for customers to eliminate the power consumption for more than 15% of their hourly baseline value. As for the power baseline for customers, it is computed by averaging the maximum load for 2 months before the start of the participation period. The reward provided by MEEDC can be 2500 rial (Iran's currency) per kWh in constant.

In Taiwan, with the demand bidding program, TOU rate users are able to determine the amount of load for peak shaving and the bidding price for their available time. The winning customers are then determined by the TPC according to the system marginal cost. The demand bidding mechanisms of Taipower are categorized by economical type, reliable type, and aggregated type [13]. In this study, the economical type is of particular interest due to the incentive for customers. The general rule of economical type is described as follows. Firstly, the TOU rate user can determine which month to conduct DR in and how much the monthly minimum capacity for load reduction is in their contact with the TPC. Next, the TOU user can decide the duration for DR implementation. Either 2 h or 4 h in a single day is available as an option. In addition, the entire implementation time within a month is not allowed to be more than 36 h. Thirdly, the customer baseline load (CBL) is obtained by averaging the power in the DR implementation period in the previous five days, except for weekends, off-peak days, holidays, and load-reduction days. Eventually, the load reduction can be computed by the difference between the maximum demand and CBL within the same period of the DR. If the amount of load reduction is less than the minimum contract value (50 kW), it is treated as 0 without a penalty bill. Figure 1 gives an explanation and it can be formulated as follows for the load-reduction day ($d*$):

$$CBL(d*) = \frac{\sum_{x=d*-5}^{d*-1} PU_{max}(x)}{5} \tag{1}$$

$$PU_{max}(d*) = Max\left\{P_D^{bt}(d*), P_D^{bt+1}(d*), \ldots \ldots, P_D^{bt+h}(d*)\right\} \quad h = \left\{ \begin{array}{lll} 8 & if & BDT(d*) = 2 \\ 16 & if & BDT(d*) = 4 \end{array} \right. \tag{2}$$

$$ABP(d*) = CBL(d*) - PU_{max}(d*) \tag{3}$$

$$ABP(d*) = \left\{ \begin{array}{ll} ABP(d*), & if\ ABP(d*) \geq 50\ \text{kW} \\ 0, & else \end{array} \right. \tag{4}$$

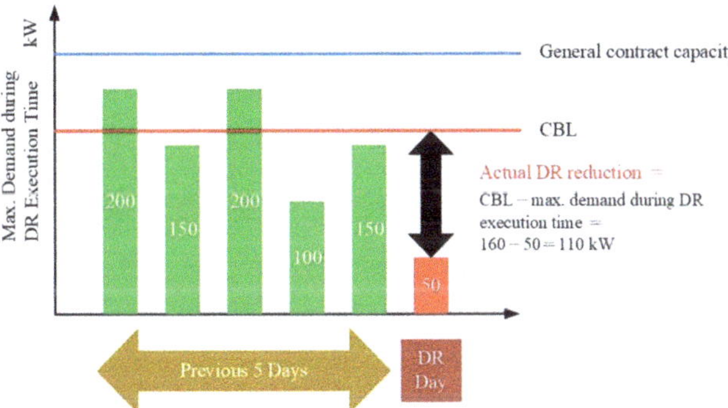

Figure 1. An exemplar figure to show the computation of actual load reduction.

2.3. Objective Function

The objective function is formulated as in (5) to minimize the total expenses for a TOU rate industrial user. Meanwhile, it is important to satisfy the operational constraints of the RES and BESS. Thus, the given scheduling problem can be presented in a mathematical model as follows:

$$TOC = Minimize \sum_{d=1}^{D} \sum_{t=1}^{T} F_{PE}(t,d) - \sum_{d=1}^{D} F_{BD}(d) \tag{5}$$

$$F_{PE}(t,d) = \begin{cases} C_{PE}(t,d) \times \left(P_D(t,d) - P_{bat}(t,d) - \sum_{j=1}^{ND} P_{NDj}(t,d)\right), & if\ P_{grid}(t,d) \geq 0 \\ 0, & else \end{cases} \tag{6}$$

$$F_{BD}(d) = \begin{cases} C_{BD}(d) \times ABP(d) \times BDT(d), & if\ d \in load-reduction\ day \\ 0, & else \end{cases} \tag{7}$$

2.4. Operational Constraints

To model the investigated system, mathematical modeling is utilized to mimic the operations of generation sources and the BESS. Similar approaches can also be found in [23,24] where multiple energy storage units are taken into account and logical variables are used to distinguish the charging/discharging operations of each energy storage unit to ensure security. In this study, it is noted that the BESS is taken as one energy storage unit. The power command for the BESS will be determined from the proposed DP-based power dispatch method. Limited by the rated power of a converter, the power command of the BESS $P_{bat} > 0$ represents discharging, while $P_{bat} < 0$ implies charging. On the other hand, references [25,26] show that loads could be classified into non-controllable, controllable comfort-based loads. However, this paper focuses on the power dispatch problem, the investigated load demand profile is statistical data from TPC customers and is regarded as an uncontrollable load. The operational constraints of the hybrid system with a RES and BESS are introduced as below.

2.4.1. System Constraints

- Power balance constraint

$$\sum_{j=1}^{ND} P_{NDj}(t,d) + P_{grid}(t,d) + P_{bat}(t,d) = P_D(t,d) \tag{8}$$

2.4.2. Non-Dispatchable Unit Constraints

- Wind power curve constraints

$$P^*_{Wj}(t,d) = \begin{cases} 0 & v(t,d) \leq v_{Ij} \text{ or } v(t,d) > v_{Oj} \\ \phi_j(v(t,d)) & v_{Ij} \leq v(t,d) \leq v_{Rj} \\ P^{\max}_{Wj} & v_{Rj} \leq v(t,d) \leq v_{Oj} \end{cases} \tag{9}$$

- Solar radiation/ambient temperature power curve constraints

$$P^*_{PVj}(t,d) = \begin{cases} 0 & S_r(t,d) \leq SD \\ \delta_j(S_r(t,d),\ T_r(t,d)) & SD \leq S_r(t,d) \leq SU \\ P^{\max}_{PVj} & S_r(t,d) \geq SU \end{cases} \tag{10}$$

2.4.3. Battery Constraints

- Limits of charge/discharge power

$$-P^{\max}_{bat} \leq P_{bat}(t,d) \leq P^{\max}_{bat} \tag{11}$$

- Upper and lower limits for state of charge

$$SOC_{\min} \leq SOC(t,d) \leq SOC_{\max} \tag{12}$$

- State of charge for battery storage system

$$SOC(t,d) = \begin{cases} SOC(t-1,d) - P_{bat}(t,d) \times \eta_B \times \Delta t & if \ P_{bat}(t,d) < 0 \\ SOC(t-1,d) - P_{bat}(t,d) \times \frac{\Delta t}{\eta_B} & if \ P_{bat}(t,d) \geq 0 \end{cases} \tag{13}$$

2.4.4. Constraints of the Utility Grid

- Limit of the purchased power

$$0 \leq P_{grid}(t,d) \leq P^{\max}_{grid} \tag{14}$$

3. Evaluation of Operating Policy for the TOU Rate Industrial User

3.1. Development of the DSM Software

To assess the potential of making a profit from the TPC, the DSM [27] is updated to deal with the scheduling problem of a TOU rate industrial user under the demand bidding mechanism of Taipower. Similar to other stochastic techniques, the main drawback of the conventional DSM is its tendency to be easily trapped in a local optimal solution, particularly when handling generating scheduling problems with a high number of local optima and heavy constraints. The solutions obtained from the DSM would rely heavily on the parameter selection, such as initial random starting points, values of the initial step size S_1, and reduced factor K. The previous work on the DSM approaches used a larger initial step size S_1 for effective search, and the step size was then successively refined until the calculation step was less than the predetermined resolution. Clearly, the DSM with a coarse convergence step can enhance the global exploration ability; however, it causes an incapability to find nearby extreme points (exploitation problem). By contrast, the DSM with a refined convergence step can enhance the local exploitation ability; however, it is easily trapped in local minima (exploration problem). Consequently, the standard DSM cannot guarantee that the solutions are optimal, or even close to the optimal, due to the deficiency in the balance between global exploration and local exploitation. Providing a well-balanced mechanism between these abilities is critical to avoid early convergence.

To elevate the global searching capability, a novel heuristic strategy is proposed to employ the stochastic searching mechanism and make good use of the exploration and exploitation capabilities. Generally, the initial candidate solutions are usually far from the global optimum; hence, a larger calculation step SP may prove beneficial. However, making all candidate solutions take the same calculation step SP in a convergence level is unreasonable. In the study, to balance the global and local exploration abilities, the selection of step size SP for candidates are different. A large calculation step SP enables the DSM to explore globally, and a small calculation step SP enables the DSM to explore locally. Clearly, the reduced factor K prevents the premature convergence. Generally, as the number of convergence level increases, the balance of exploration and exploitation abilities is enhanced and the solution quality is improved. The proposed DSM algorithm is outlined in Figure 2. Typically, the DSM with a high S_1 and a low K is recommended. From our numerical experience, S_1 = 10% of upper limit for BESS and K = 5 are applied in this study.

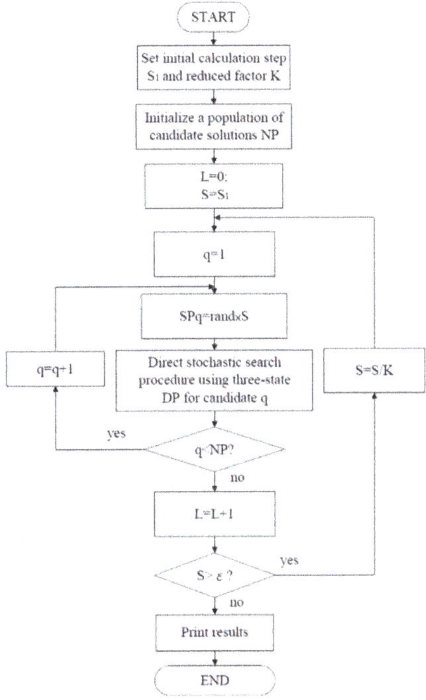

Figure 2. Flow chart of the proposed direct search method (DSM) approach.

In this study, the energy stored in the BESS is taken as the state variables and they are initialized stochastically. The constraints represented by (8)–(14) will be treated in different ways. The operating constraints (8) and (11)–(13) are handled during the direct search procedure. The available renewable power generation can be obtained from the wind speed, solar radiation intensity, and ambient temperature by applying Equations (9) and (10). To deal with the power limits of the BESS given in (11) and the limit of purchased power from utility grid in (14), the non-negative penalty terms are integrated with the electricity cost to penalize the violation of any constraint. In addition, considering the coupling constraints of the power dispatch problem, three-state dynamic programming is combined with the DSM to enhance the performances of the direct stochastic search. As shown in Figure 3, three states of BESS are defined as follows: 1 implies charge, 0 implies idle, and –1 implies discharge. An exemplar trajectory in Figure 3 illustrates the transition from one state for a certain time interval to another state for the next time interval. Thus, this transition may require charging and discharging of the BESS. In this way, the forward dynamic programming can be employed to deal with the power dispatch problem. The accumulated electricity cost is evaluated along with each trajectory recursively. Then, the path with the least cost will be backtracked at the final stage to obtain the optimum solution.

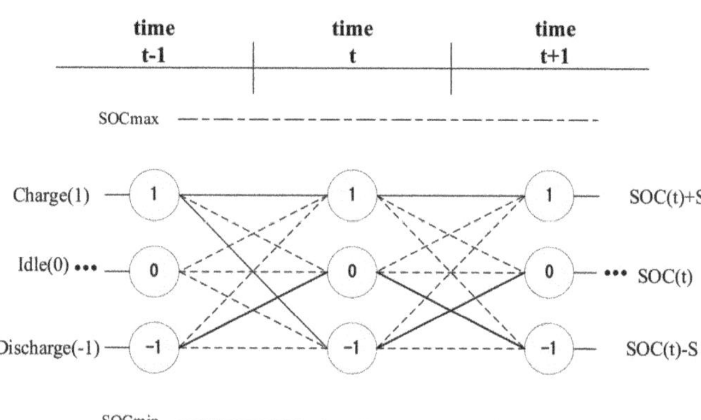

Figure 3. State transition diagram of the battery energy storage system (BESS) for dynamic programming (DP).

3.2. Assessment of Operational Strategy for Executing the DBP

To realize the minimum total electricity costs of a TOU rate industrial user, several operating strategies of the BESS are evaluated with the DBP to further reduce the peak load demand and achieve energy conservation. Given that there is no DBP, it is widely recognized that the BESS can store electrical power during the off-peak load periods because of the low purchasing price of the energy provided by the utility grid. The BESS system then discharges randomly in the periods of peak load demand as the high electricity price. In this way, electricity costs can be saved and a penalty bill caused by exceeding contracts can also be prevented. However, a more advanced operating strategy is required to perform DBP to eliminate the power shortage of the BESS in a DR load-reduction duration. To maximize the total incentive obtained from the TPC on a load-reduction day, the best operating strategy for the BESS is to fully discharge at maximum power output during the DR-executing time. This mechanism can significantly reduce the electricity costs, increasing the economic benefits of energy generated by the BESS. In the study, a virtual electricity price is designed during the DR-executing time and the recommended value is chosen as follows:

$$Pvir(bt+l) = PLC + \alpha \times \frac{P_D(bt+l) - P_D^{\min}(bt+l)}{P_D^{\max}(bt+l) - P_D^{\min}(bt+l)} \quad l=1,2,\ldots,h; \quad h = \begin{cases} 8 & if \quad BDT(d*) = 2 \\ 16 & if \quad BDT(d*) = 4 \end{cases} \quad (15)$$

But on a non-load-reduction day, another operating strategy is necessary for increasing the actual load-reduction amount. According to the computation of load reduction, it is found that electricity cost savings depend on the degree of CBL. With the larger CBL, the TOU customers are prone to earn incentives received from the TPC for DR execution. To raise the baseline load by using the BESS, the best operating strategy for the BESS is to fully discharge during the peak-load periods except for DR-executing time. The BESS system would then stop discharging at the predictive maximum demand period (or some higher load periods) of DR implementation during the five days prior to the DR event. Although this process may not be an economic policy for current non-load-reduction days, it has more potential to decrease electricity costs in the future load-reduction days. The aim of the study is to evaluate the dispatch strategy of the BESS for the TOU rate industrial user to minimize the system electricity costs. In the study, a virtual electricity price is also designed during the DR-executing time and the recommend value is chosen as follows:

$$Pvir(bt+l) = PLC - (PLC - LLC) \times \frac{P_D(bt+l) - P_D^{\min}(bt+l)}{P_D^{\max}(bt+l) - P_D^{\min}(bt+l)} \quad l = 1,2,\ldots,h \,; \quad h = \begin{cases} 8 & if \quad BDT(d*) = 2 \\ 16 & if \quad BDT(d*) = 4 \end{cases} \tag{16}$$

4. Numerical Examples

A chemical industrial customer of the TPC is used as an example to show the effectiveness and feasibility of the proposed DSM algorithm [27]. The pricing structure of three-section TOU rates is considered for the high-voltage customer. The energy costs of peak load (10:00–12:00 and 13:00–17:00), medium load (07:30–10:00, 12:00–13:00, and 17:00–22:30), and light load (00:00–07:30 and 22:30–24:00) periods are 4.67, 2.90, and 1.32 NT\$/kWh, respectively. The contract capacity, namely, the maximum power purchased from the utility grid, is assumed to be 350 kW. In the studied case, the DR-execution duration is chosen to be 4 h (13:00–17:00) by the users. The bidding price is assumed to be 10 NT\$/kWh during the DR-executing time. The load forecasting of a typical day in the summer season is given in Figure 4. The minimum and maximum loads for the study period of 24 h are 125 kW and 250 kW, respectively. Figure 5 shows the investigated system consisting of wind farm, solar PV array, BESS, and utility grid. The wind farm includes two wind turbine generators (WTGs) and the total capacity of wind power installed is 40 kW. The capacity of solar PV models is 37.8 kW. As illustrated in Figure 5, the solar PV modules and BESS are connected to a step-up transformer via an inverter. The efficiency of the inverter is 0.95. Based on Equations (9) and (10), the available power of the RES for a typical day in summer can be obtained as given in Figure 6. To compensate the depleted and surplus power in the system, a BESS with the capacity and power rated at 180 kWh/30 kW is simulated. The battery round-trip efficiency is 0.9. In addition, the initial and end of SOC are set at 66.67%, and the lower operating limit is set to SOC = 20%. The parameters of DSM are selected as: the number of initial solutions NP = 1, the initial calculation step S_1 = 18 kW, the reduced factor K = 5, and the predetermined resolution ε = 0.01 kW. All the computation is performed on a PC Intel(R) Core(TM) i5-4570 CPU, up to 3.2 GHz. Several scenarios are taken into account and discussed as follows:

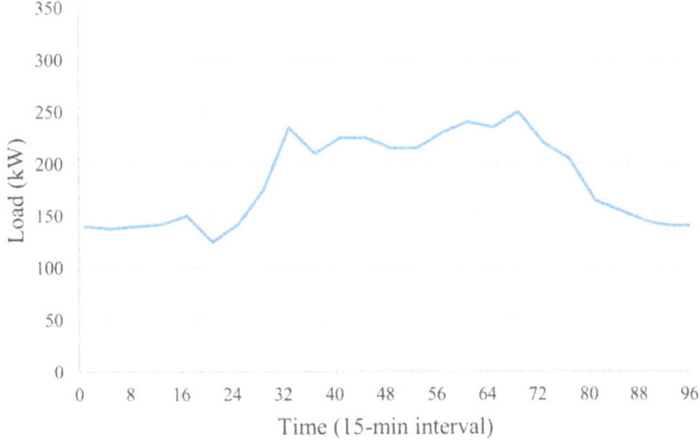

Figure 4. Load curve of a typical day in the summer season.

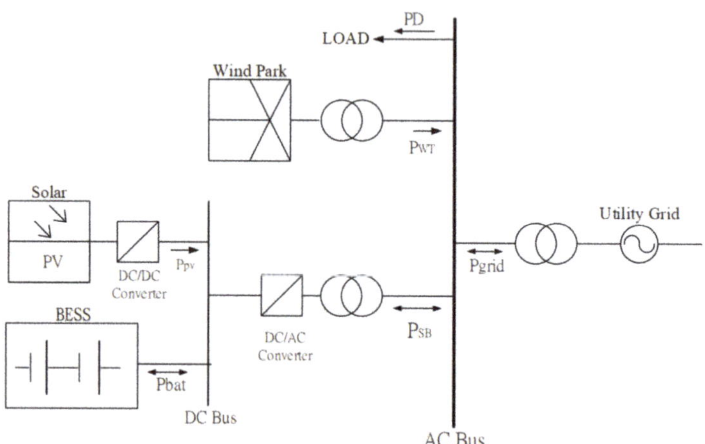

Figure 5. A system diagram of a time-of-use (TOU) rate customer.

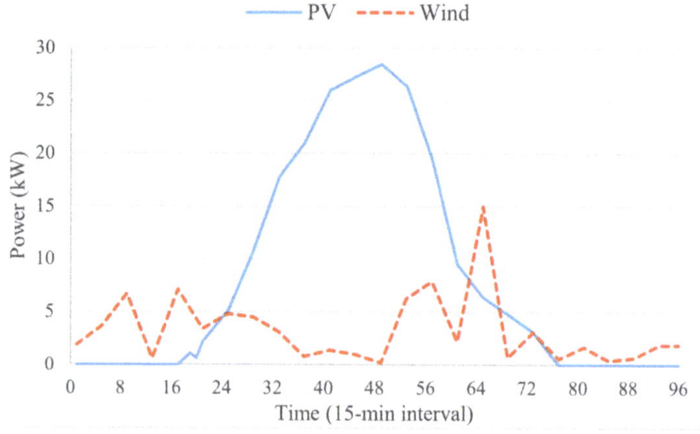

Figure 6. Power profiles of PV and wind power generation for a typical day in the summer season.

4.1. Performance of the Proposed DSM Algorithm

A good convergence of the proposed DSM algorithm is presented in Table 1. The iteration numbers and electricity costs of different cases of convergence level are compared. The results show that the proposed DSM program is able to achieve an advantageous operation schedule for the TOU industrial customer while satisfying all the constraints simultaneously. It is also observed that the total electricity cost is not sensitive to the calculation step S. Ignoring the BESS integration, the total electricity cost is about NT$12,975.646 in this test case. When the BESS is integrated into this customer system, the total electricity charge is reduced to NT$12,544.588. It will save 3.32% of the electricity cost within a day. To analyze the influences of the initial values on the final results, several random numbers are taken as the initial values in the DSM approach. The corresponding results of the 10 trial tests are given in Table 2. The satisfactory solutions can be obtained in approximately 0.02 s with the proposed DSM. In fact, several different cases were studied and the results demonstrated the merit of the proposed algorithm.

Table 1. Comparison of iterations and total electricity cost (TOC) under various S in the TOU system.

Convergence	Iterations	TOC (NT$)
Initialization	-	12,975.646
S_1 = 18 kW	0	12,975.646
S_2 = 3.6 kW	23	12,611.493
S_3 = 0.72 kW	24	12,556.483
S_4 = 0.144 kW	24	12,545.985
S_5 = 0.0288 kW	12	12,544.726
S_6 = 0.00576 kW	7	12,544.588

Table 2. Results of DSM after ten runs.

Run	Initialization (NT$)	TOC (NT$)
1	31,211,025	12,544.537
2	27,560,846	12,544.554
3	22,997,452	12,544.547
4	27,190,036	12,544.555
5	25,530,881	12,544.548
6	25,004,720	12,544.545
7	24,011,364	12,544.560
8	24,318,074	12,544.539
9	20,842,510	12,544.548
10	27,207,412	12,544.546

4.2. Prediction of Electricity Cost Savings for Executing the DBP

To evaluate the economic benefits of the installation of the BESS, the developed DSM software is applied and validated as a useful tool for the TOU rate industrial users to predict the cost savings. Table 3 gives a good indication to help understand the effects of the BESS on the total cost savings for executing the DBP. In the previous TOU system, when the BESS was excluded in the system, the total electricity cost was NT$12,975.646 in Case 1. As given in Case 2, a 3.32% reduction in electricity cost is achieved when the TOU system includes the BESS. Obviously, the installation of the BESS enables a reduction in the electricity cost of 19.49% for executing the DBP in Case 3 when the bidding price is chosen to be 10 NT$/kWh. Numerical results certainly provide valuable information and verify that the installation of the BESS enables a reduction in the electricity cost in the TOU system. Thus, different amounts of the BESS can be added to the original system to evaluate the significant benefits of annual electricity cost savings. In this way, the economic penetration limit of the optimal capacity of the BESS into a given TOU system can be determined.

Table 3. Comparison of the electricity cost saved by different simulation scenarios.

Case	BESS	DBP (Load-Reduction Day)	DBP (Non-Load-Reduction Day)	TOC (NT$)	Saving (%)
1	Without	Without	Without	12,975.646	-
2	With	Without	Without	12,544.546	3.32%
3	With	With	Without	10,445.745	19.49%
4	With	Without	With	12,544.546	3.32%

4.3. Effects of BESS on the Operating Schedule for Load-Reduction Day

To demonstrate the performances of integrating the BESS into the TOU system for a load-reduction day. Figure 7 shows the energy profiles of the BESS during a typical daily load. The optimal power dispatch of the BESS can also be observed from Figure 8. Without the DBP (Case 2), the BESS was charged in low load demand periods when the electricity price is low (1.32 NT$/kWh). During heavy load demand periods, namely 10:00–12:00

and 13:00–17:00, the BESS was discharged randomly when the electricity price is high (4.67 NT\$/kWh). However, a more advanced operational strategy of the BESS is necessary to curtail the peak demand for the load-reduction day when the DBP is considered (Case 3). As shown in Figure 8, it is more cost-effective not to discharge at high system load times (11:30–12:00) and keep the maximum power outputs (30 kW) of the BESS during the DR-executing time (13:00–17:00). As shown in Table 3, it is found that there is a reduction in electricity cost of 19.49% for executing the DBP (Case 3). The DSM can be used to test the user system in many load conditions under different seasons, summarizing the test results to develop expert knowledge for the BESS controller design. The developed DSM software is therefore a useful tool for the TOU rate industrial user to maximize the benefits of the BESS for reducing the electricity cost of grid dispatch on the load-reduction day.

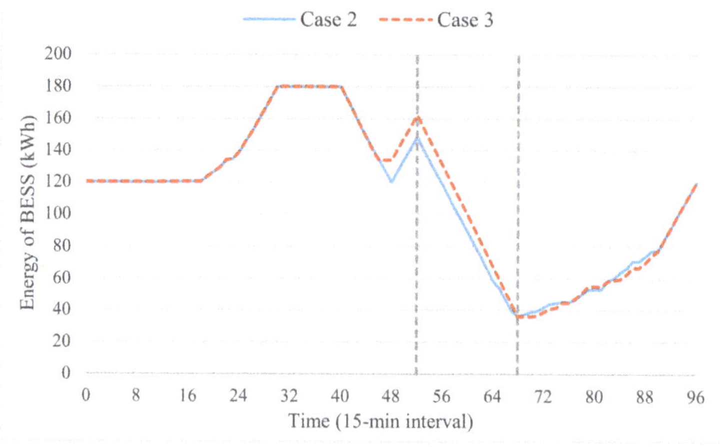

Figure 7. Electrical energy changes in the BESS (Case 2 and Case 3).

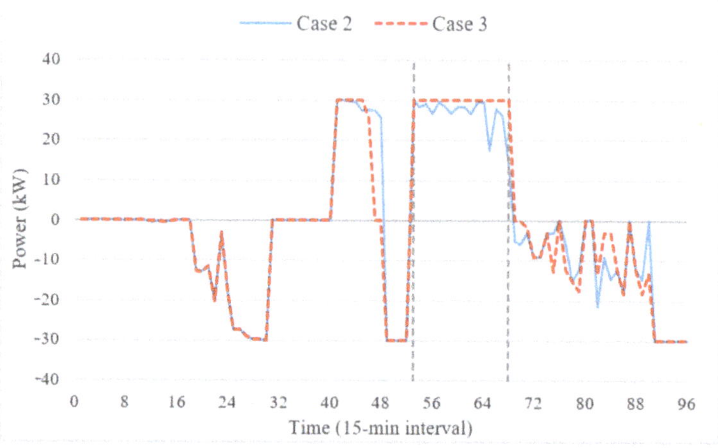

Figure 8. Power profiles of the BESS during a typical daily load (Case 2 and Case 3).

4.4. Effects of BESS on the Operating Schedule for Non-Load-Reduction Day

To earn more incentive for DR execution, the developed DSM software is also a useful tool for the non-load-reduction day to increase the actual load-reduction amount. To show the effects of utilizing the BESS in the TOU system, Figure 9 shows the electrical energy changes in the BESS on the non-load-reduction day, and the power outputs are shown in Figure 10. Without the DBP (Case 2), it can be seen that the operating strategy of the BESS is to discharge randomly during peak load hours (10:00–12:00 and 13:00–17:00) for cost savings. When the DBP is considered (Case 4), it is necessary to update the energy flow control strategies from the BESS to fully explore the TOU rate customer system benefits. The results show that the maximum power outputs (30 kW) of the BESS is kept for peak load duration (10:00–12:00), and the BESS system stops discharging at some higher load periods (15:00–15:15 and 16:45–17:00) of DR execution to raise the baseline load. In Case 4, the baseline load (CBL) can be raised from 228.0 kW to 257.7 kW by using the BESS. This mechanism can significantly reduce the electricity charges in the future load-reduction day, increasing the economic benefits of energy generated by the BESS. As shown in Table 3, it is found that the total electricity cost is NT$12,544.546 in Case 4, that is, identical to those obtained in Case 2. The feasibility of the algorithm is confirmed and it is an effective power dispatch solution for the BESS.

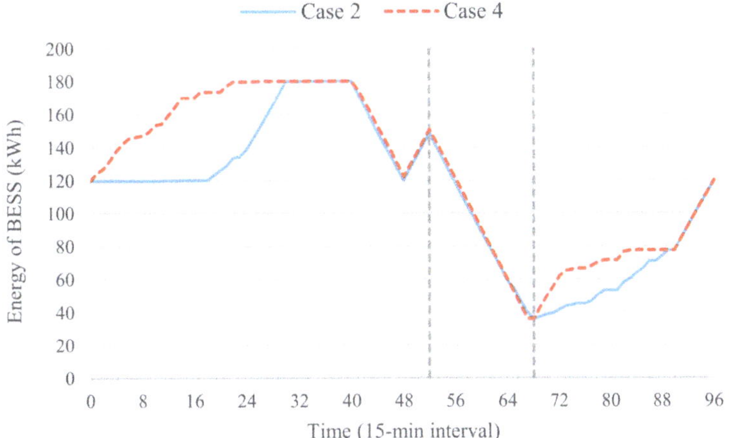

Figure 9. Electrical energy changes in the BESS (Case 2 and Case 4).

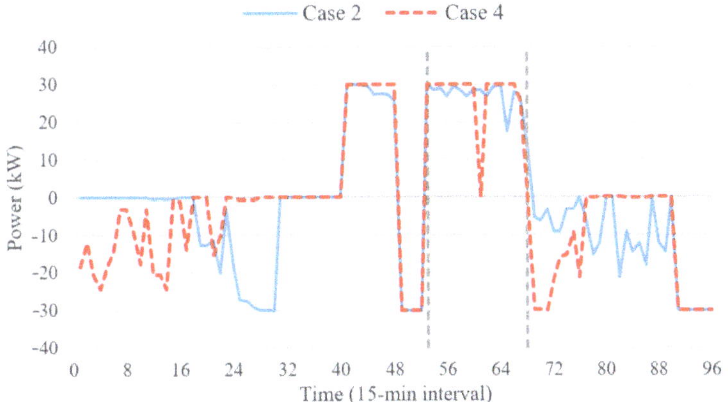

Figure 10. Power profiles of BESS during a typical daily load (Case 2 and Case 4).

5. Conclusions

In a smart grid, it is crucial to have efficient energy management that provides a reliable and beneficial scheduling solution for the TOU rate industrial customers. To maximize the total incentive obtained from the TPC, an extended DSM was developed to solve the scheduling problem of a TOU system under the demand bidding mechanism of Taipower. The operations of the BESS was investigated and discussed with the proposed DSM software. Several operational strategies of the BESS were also evaluated to curtail the peak load demand and achieve energy conservation. The results demonstrated that the BESS enables a reduction of the electricity cost of a TOU rate custom system for executing the DBP. The proposed strategy is validated as a useful tool to determine the capacity of the BESS in the TOU system. Numerical experiments were conducted to provide valuable information for both operational and planning problems for the TOU rate industrial customers. In real-time application, the proposed DSM can be used to determine the optimal operating policy of the next time stage. This function can save on energy costs and reduce the risk of the BESS running out of energy in a peak-demand reduction application. In off-line application, the proposed DSM can also be used to evaluate the economic benefits of the BESS. The computer program developed is currently being experimentally added to a TOU management system as auxiliary software to support TOU rate users. Although this study was based on the TPC rate structure, it can easily be modified to satisfy other TOU rate structures.

Author Contributions: Conceptualization, C.-L.C.; Data curation, C.-T.T. and K.-H.L.; Formal analysis, C.-T.T. and Y.-S.C.; Funding acquisition, C.-L.C.; Investigation, C.-T.T. and C.-L.C.; Methodology, C.-T.T.; Project administration, K.-H.L.; Resources, Y.-S.C.; Software, C.-T.T.; Supervision, C.-L.C.; Validation, C.-T.T. and Y.-S.C.; Visualization, C.-T.T.; Writing—original draft, C.-T.T. and K.-H.L.; Writing—review & editing, Y.-S.C. and C.-L.C. All authors have read and agreed to the published version of the manuscript.

Funding: This research was funded by the Ministry of Science and Technology, Taiwan, under grants MOST 108-2221-E-019-031.

Conflicts of Interest: The authors declare no conflict of interest.

References

1. Chen, F.; Lu, S.M.; Wang, E.; Tseng, K.T. Renewable energy in Taiwan. *Renew. Sustain. Energy Rev.* **2010**, *14*, 2029–2038. [CrossRef]
2. Bouzguenda, M.; Rahman, S. Value analysis of intermittent generation sources from the system operations perspective. *IEEE Trans. Energy Convers.* **1993**, *8*, 484–490. [CrossRef]
3. Alqunun, K.; Guesmi, T.; Albaker, A.F.; Alturki, M.T. Stochastic Unit Commitment Problem, Incorporating Wind Power and an Energy Storage System. *Sustainability* **2020**, *12*, 10100. [CrossRef]
4. Su, W.; Wang, J.; Roh, J. Stochastic energy scheduling in microgrids with intermittent renewable energy resources. *IEEE Trans. Smart Grid* **2013**, *5*, 1876–1883. [CrossRef]
5. Behabtu, H.A.; Messagie, M.; Coosemans, T.; Berecibar, M.; Anlay Fante, K.; Kebede, A.A.; Mierlo, J.V. A Review of Energy Storage Technologies' Application Potentials in Renewable Energy Sources Grid Integration. *Sustainability* **2020**, *12*, 10511. [CrossRef]
6. Chen, C.L. Optimal wind–thermal generating unit commitment. *IEEE Trans. Energy Convers.* **2008**, *23*, 273–280. [CrossRef]
7. Ramli, M.A.M.; Bouchekara, H.R.E.H.; Alghamdi, A.S. Efficient Energy Management in a Microgrid with Intermittent Renewable Energy and Storage Sources. *Sustainability* **2019**, *11*, 3839. [CrossRef]
8. Ejaz, W.; Naeem, M.; Shahid, A.; Anpalagan, A.; Jo, M. Efficient energy management for the internet of things in smart cities. *IEEE Commun. Mag.* **2017**, *55*, 84–91. [CrossRef]
9. Sundt, S.; Rehdanz, K.; Meyerhoff, J. Consumers' Willingness to Accept Time-of-Use Tariffs for Shifting Electricity Demand. *Energies* **2020**, *13*, 1895. [CrossRef]
10. Chen, Z.; Wu, L.; Fu, Y. Real-time price-based demand response management for residential appliances via stochastic optimization and robust optimization. *IEEE Trans. Smart Grid* **2012**, *3*, 1822–1831. [CrossRef]
11. Yu, Z.; Jia, L.; Murphy-Hoye, M.C.; Pratt, A.; Tong, L. Modeling and stochastic control for home energy management. *IEEE Trans. Smart Grid* **2013**, *4*, 2244–2255. [CrossRef]
12. Tarasak, P.; Chai, C.C.; Kwok, Y.S.; Oh, S.W. Demand bidding program and its application in hotel energy management. *IEEE Trans. Smart Grid* **2014**, *5*, 821–828. [CrossRef]

13. Taiwan Power Company, Demand Bidding Measures. Available online: https://dbp.taipower.com.tw/TaiPowerDBP/Portal/proj_data/%E9%9C%80%E9%87%8F%E7%AB%B6%E5%83%B9%E6%8E%AA%E6%96%BDDM.pdf (accessed on 20 February 2020).
14. Yao, L.; Lim, W.H. Optimal purchase strategy for demand bidding. *IEEE Trans. Power Syst.* **2017**, *33*, 2754–2762. [CrossRef]
15. Lee, T.Y. Operating schedule of battery energy storage system in a time-of-use rate industrial user with wind turbine generators: A multipass iteration particle swarm optimization approach. *IEEE Trans. Energy Convers.* **2007**, *22*, 774–782. [CrossRef]
16. Cheng, Y.S.; Liu, Y.H.; Hesse, H.C.; Naumann, M.; Truong, C.N.; Jossen, A. A pso-optimized fuzzy logic control-based charging method for individual household battery storage systems within a community. *Energies* **2018**, *11*, 469. [CrossRef]
17. Samuel, O.; Javaid, S.; Javaid, N.; Ahmed, S.H.; Afzal, M.K.; Ishmanov, F. An Efficient Power Scheduling in Smart Homes Using Jaya Based Optimization with Time-of-Use and Critical Peak Pricing Schemes. *Energies* **2018**, *11*, 3155. [CrossRef]
18. Chen, C.L. Non-convex economic dispatch: A direct search approach. *Energy Convers. Manag.* **2007**, *48*, 219–225. [CrossRef]
19. Kazarlis, S.A.; Bakirtzis, A.G.; Petridis, V. A genetic algorithm solution to the unit commitment problem. *IEEE Trans. Power Syst.* **1996**, *11*, 83–92. [CrossRef]
20. Selvakumar, A.I.; Thanushkodi, K. A new particle swarm optimization solution to nonconvex economic dispatch problems. *IEEE Trans. Power Syst.* **2007**, *22*, 42–51. [CrossRef]
21. Demand Bidding Program, Southern California Edison. Available online: https://www.sce.com/sites/default/files/inline-files/0804_Com.pdf (accessed on 9 March 2021).
22. Delavaripour, H.; Khazaee, A.; Ghasempoor, M.; Hooshmandi, H. Reduced peak-time energy use by the demand bidding program in Iran. *Cired-Open Access Proc. J.* **2017**, *2017*, 1959–1962. [CrossRef]
23. Hosseini, S.M.; Carli, R.; Dotoli, M. Robust optimal energy management of a residential microgrid under uncertainties on demand and renewable power generation. *IEEE Trans. Autom. Sci. Eng.* **2020**, 1–20. [CrossRef]
24. Sperstad, I.B.; Korpås, M. Energy storage scheduling in distribution systems considering wind and photovoltaic generation uncertainties. *Energies* **2019**, *12*, 1231. [CrossRef]
25. Carli, R.; Dotoli, M. Decentralized control for residential energy management of a smart users' microgrid with renewable energy exchange. *IEEE/CAA J. Autom. Sin.* **2019**, *6*, 641–656. [CrossRef]
26. Scarabaggio, P.; Grammatico, S.; Carli, R.; Dotoli, M. Distributed Demand Side Management With Stochastic Wind Power Forecasting. *IEEE Trans. Control Syst. Technol.* **2021**, 1–16. [CrossRef]
27. Lu, T.C.; Huang, C.Y.; Chen, Y.Y.; Chen, C.L.; Lee, T.Y. Efficient Energy Management of a Time-of-Use Rate Industrial User for Smart Cities. In Proceedings of the 2018 International Conference on System Science and Engineering (ICSSE), New Taipei, Taiwan, 28–30 June 2018; pp. 1–6.

 *sustainability*

Article

Torrefaction Thermogravimetric Analysis and Kinetics of Sorghum Distilled Residue for Sustainable Fuel Production

Shih-Wei Yen [1], Wei-Hsin Chen [2,3,4,*], Jo-Shu Chang [1,3,5,*], Chun-Fong Eng [2], Salman Raza Naqvi [6] and Pau Loke Show [7]

[1]   Department of Chemical Engineering, National Cheng Kung University, Tainan 701, Taiwan; n38071029@ncku.edu.tw
[2]   Department of Aeronautics and Astronautics, National Cheng Kung University, Tainan 701, Taiwan; ecfong94@gmail.com
[3]   Research Center for Smart Sustainable Circular Economy, Tunghai University, Taichung 407, Taiwan
[4]   Department of Mechanical Engineering, National Chin-Yi University of Technology, Taichung 411, Taiwan
[5]   Department of Chemical and Materials Engineering, College of Engineering, Tunghai University, Taichung 407, Taiwan
[6]   School of Chemical and Materials Engineering (SCME), National University of Sciences and Technology (NUST), H-12, Islamabad 44000, Pakistan; salman.raza@scme.nust.edu.pk
[7]   Department of Chemical and Environmental Engineering, Faculty of Science and Engineering, University of Nottingham Malaysia, Semenyih 43500, Selangor Darul Ehsan, Malaysia; PauLoke.Show@nottingham.edu.my
*   Correspondence: chenwh@mail.ncku.edu.tw (W.-H.C); changjs@mail.ncku.edu.tw (J.-S.C.)

Citation: Yen, S.-W.; Chen, W.-H.; Chang, J.-S.; Eng, C.-F.; Raza Naqvi, S.; Show, P.L. Torrefaction Thermogravimetric Analysis and Kinetics of Sorghum Distilled Residue for Sustainable Fuel Production. *Sustainability* **2021**, *13*, 4246. https://doi.org/10.3390/su13084246

Academic Editor: Algirdas Jasinskas

Received: 15 March 2021
Accepted: 7 April 2021
Published: 11 April 2021

Publisher's Note: MDPI stays neutral with regard to jurisdictional claims in published maps and institutional affiliations.

Abstract: This study investigated the kinetics of isothermal torrefaction of sorghum distilled residue (SDR), the main byproduct of the sorghum liquor-making process. The samples chosen were torrefied isothermally at five different temperatures under a nitrogen atmosphere in a thermogravimetric analyzer. Afterward, two different kinetic methods, the traditional model-free approach, and a two-step parallel reaction (TPR) kinetic model, were used to obtain the torrefaction kinetics of SDR. With the acquired 92–97% fit quality, which is the degree of similarity between calculated and real torrefaction curves, the traditional method approached using the Arrhenius equation showed a poor ability on kinetics prediction, whereas the TPR kinetic model optimized by the particle swarm optimization (PSO) algorithm showed that all the fit qualities are as high as 99%. The results suggest that PSO can simulate the actual torrefaction kinetics more accurately than the traditional kinetics approach. Moreover, the PSO method can be further employed for simulating the weight changes of reaction intermediates throughout the process. This computational method could be used as a powerful tool for industrial design and optimization in the biochar manufacturing process.

Keywords: sorghum distilled residue; thermogravimetric analysis; torrefaction kinetics; biomass and bioenergy; particle swarm optimization (PSO); biochar

1. Introduction

Since the mid-20th century, the prosperous growth of human economic activities has caused a significant increase in greenhouse gas (GHG) emissions. The greenhouse effect continues to drive up the global average temperature rapidly, causing global warming and climate change simultaneously. These strong counterattacks from Mother Nature have badly been affecting human civilization and development. Therefore, reducing carbon emissions has become an urgent issue and critical action. Renewable energy, including biochar, is regarded as one of the sustainable and clean energies that can gradually replace traditional fossil fuels.

With the advantages of high heating value and great combustion quality as solid fuel, biochar has been widely used as bioenergy and biofuel [1]. It is an important part of renewable energy and is regarded as one of the alternatives to fossil fuels [2]. Compared with traditional coal fuels, biochar is a new generation of environmentally sustainable energy.

Reviewing the research literature on biochar, there are many types of sources that have been used as feedstocks, including algae, crops, forest waste, etc. [1,3–11]. However, there is hardly any research on SDR upgrades by torrefaction [11]. Sorghum liquor, also known as Baijiu, is the most consumed spirit in the world. While producing more than 10 billion liters of sorghum liquor annually, it also produces more than 66.1 billion kg of distilled residue, SDR, every year. This by-product is usually processed into animal feed. However, owing to oversupply, it is not easy to dispose of all of it, so the price is inevitably low. SDR also causes environmental hygiene problems. In the application of SDR for bioenergy, there are many studies related to microbial fermentation in the literature. For example, distilled grain waste (DGW, i.e., SDR) can be converted into biogas through anaerobic fermentation by methanogen or anaerobic microorganisms [12,13]. In addition, Saccharomyces cerevisiae can be used to ferment sorghum liquor waste (or SDR) to produce bioethanol [14]. In terms of thermochemical research, Ye et al. [15] evaluated the potential of pyrolysis on converting Baijiu Diuzao (Chinese liquor industry waste, i.e., SDR) into bioenergy for the first time and used the artificial neural networks (ANN) model to validate the pyrolytic behavior. The research results showed that the temperature between 130 °C and 373 °C would be the best condition for converting SDR into chemicals and energy. This study aims at evaluating two different kinetic approaches to the torrefaction process of SDR and trying to determine its process parameters, which could provide useful references for the SDR-biochar industry in the future.

Pyrolysis is a kind of thermal decomposition reaction in an oxygen-deficient or inert gas environment. The reaction temperature varies with the material being used, roughly between 200–1200 °C. It is widely used in the industry to produce carbon black, syngas, pyroligneous acid, or biofuels such as biodiesel. Torrefaction, which has been called mild pyrolysis [16], is a thermal treatment technology in a relatively low-temperature range of 200–300 °C [1,2,10,17], being carried out in an inert gas environment or oxygen-deficient atmosphere, which aims to upgrade biomass to a homogeneous and hydrophobic biofuel with increased energy density [6], heating value, better grindability, and superior combustion characteristics [2,16,18–21]. In this study, SDR was chosen as a cheap feedstock to carry out the torrefaction process at five different temperatures isothermally in a thermogravimetric analyzer under an inert gas environment. The torrefaction severity index (TSI) chart [3], a unique three-dimensional profile that depicts the relationship between the degree of thermal decomposition, torrefaction time, and temperature, could be a practical tool for providing the process parameters of torrefaction.

Simply speaking, there are two approaches to study the weight loss kinetics, according to the presence or absence of its reaction kinetic model, namely the model-free method [3] and the model-based method [6,18]. The first method does not need any model to find the thermal decomposition reaction rate equation of SDR and various parameters applicable throughout the range from 200 to 300 °C. In the second approach, usually, a two-step reaction kinetic model is used with artificial intelligence method, e.g., particle swarm optimization (PSO) to simulate the weight loss curves of real torrefaction and the kinetic parameters, as well as each fit quality [22], which is a degree of similarity between simulation and real torrefaction curve. Reviewing the literature, PSO has been used for the prediction of thermal decomposition kinetics of a variety of biomass materials, including wood, microalgae, crops, and forest waste. A two-step parallel reaction (TPR) model was coupled with PSO for simulating the behavior of pyrolysis or torrefaction [6,18,22]. As mentioned above, only Ye et al. [15] used ANN to study the pyrolysis behavior of SDR for the first time. However, the kinetics of torrefaction reaction using SDR as biomass has not been discussed in depth. Therefore, in addition, to using PSO to optimize the torrefaction kinetics of SDR for the first time, this research also compared and evaluated the traditional model-free method with the model-based PSO method, which can provide us with more inspiration when studying torrefaction kinetics of SDR.

The evolutionary algorithm (EA) is a subset of evolutionary computation (EC) in artificial intelligence (AI). It has successively developed various technologies, to name a few, including genetic algorithm (GA), differential evolution (DE), ant colony optimization (ACO), and particle swarm optimization (PSO). PSO is a well-regarded, simple but powerful technique to optimize the pyrolysis kinetics of microalgae [22]. In this study, to achieve global optimization, PSO was used to simulate the isothermal torrefaction kinetics of SDR. For further evaluating the traditional model-free method and the evolutionary algorithm PSO, the fit qualities of the two methods were calculated to compare the degree of similarity between predicted and actual torrefaction kinetics.

2. Materials and Methods

2.1. Materials

Sorghum distilled residue (SDR) was the raw material selected for this study. It is a by-product of the liquor production process in a distillery located in southern Taiwan. SDR was ground into powder after drying at 105 °C for 6 h and stored in a dry-keeper until performing the thermogravimetric analysis (TGA). The analyses of various basic properties are detailed in Table 1. The proximate analysis was based on the standard procedure of the American Society for Testing and Materials. The content of crude carbohydrate, crude protein, and crude lipid in the sample was obtained by the phenol-sulfuric acid method, Kjeldahl method, and Soxhlet-extract method, respectively. The calorific value was measured by a bomb calorimeter (IKA C5000).

Table 1. Basic properties of sorghum distilled residue (SDR).

Biomass	**SDR**
Photograph	
Composition analysis (%)	
Crude carbohydrate	68.5
Crude protein	12.7
Crude lipid	4.5
Others	14.3
Proximate analysis (wt%)	
Volatile matter (VM)	68.97
Fixed carbon (FC)	16.28
Moisture	5.68
Ash	9.08
HHV (MJ·kg-1, dry basis)	17.386

2.2. Thermogravimetric Analysis

A thermogravimetric analyzer was used to analyze the SDR pyrolysis and torrefaction characteristics and the torrefaction kinetics was conducted accordingly. Each time about 5 mg of the biomass powder was weighed and loaded inside an aluminum crucible. Sample weight was continuously measured under a nitrogen flow rate of 100 mL/min. The sample was first heated from room temperature to 105 °C at a heating rate of 20 °C/min and then held at 105 °C for 10 min to remove moisture. After that, the temperature was further increased to 850 °C at a heating rate of 20 °C/min. In this way, the distribution curves of

TGA and DTG were obtained to understand the characteristics of the biomass throughout the whole pyrolysis process.

In the isothermal torrefaction experiment, the samples were heated from room temperature to 105 °C at a heating rate of 20 °C/min and then held up at 105 °C for 10 min for moisture removal. Next, they were further heated to 5 different temperatures at the same heating rate of 20 °C/min; they were 200 °C, 225 °C, 250 °C, 275 °C, and 300 °C and then subjected to isothermal torrefaction at these temperatures for 60 min, followed by heating to 850 °C at a heating rate of 20 °C/min to end the experiments. To ensure a high degree of both accuracy and precision of the experimental data, a thermogravimetric analyzer is regularly calibrated. Each of the same experimental conditions was carried out more than twice to confirm the excellent reproducibility. The relative error was controlled within 3% for the TGA measurement.

2.3. Isothermal Torrefaction Kinetics

Equation (1) is the kinetic modeling equation [5,23]. This formula can be used to macroscopically express the rate of thermal degradation to study the isothermal torrefaction kinetics.

$$\frac{dC}{dt} = k(1-C)^n \tag{1}$$

where C is the conversion of the sample, k is the reaction rate constant and n is the reaction order. The conversion C can be defined as

$$C = \frac{W_i - W}{W_i - W_f} \tag{2}$$

where W, W_i, W_f are the instant sample weight, initial sample weight, and final sample weight, respectively. In this study, W_i represents the weight of SDR after removing moisture at 105 °C for 10 min and W_f represents the weight of SDR at the final temperature of 800 °C. In the process of isothermal torrefaction, the relationship between conversion and heating time is expressed as [24]

$$\begin{cases} \ln\left(\frac{1-C_0}{1-C}\right) = k(t-t_0) & \text{if } n = 1 \\ (1-C)^{1-n} - (1-C_0)^{1-n} = k(n-1)(t-t_0) & \text{if } n \neq 1 \end{cases} \tag{3}$$

where C_0 is the conversion at the onset (t = t_0) of isothermal torrefaction. In Equation (3), when n = 1, which means that the thermal degradation is a first-order reaction, by plotting ln(1—C)-1 versus heating time (t–t_0), a straight line can be obtained and the slope of this straight line will be the reaction rate constant k, while n ≠ 1, the plot of (1–C)1-n versus torrefaction time t gives a straight line with a slope of (n−1) k. Therefore, regardless of which situation it is in, by processing the data from torrefaction experiments via Equations (2) and (3), the reaction rate constant k can be obtained.

To further discuss the reaction rate constant k, assuming that the torrefaction reaction obeys the Arrhenius equation (Equation (4)), which shows that the chemical rate constant varies as a negative exponential of the reciprocal absolute temperature.

$$k = A\exp\left(-\frac{Ea}{RT}\right) \tag{4}$$

where A is the pre-exponential factor, Ea is the activation energy, R is the universal gas constant and T stands for the absolute temperature (in kelvins). Taking the logarithm of both sides in Equation (4) to get the following Equation (5).

$$\ln k = \ln A - \frac{Ea}{RT} \tag{5}$$

Through the aforementioned approach, five reaction rate constants at different temperatures can be obtained. Furthermore, by plotting ln(k) versus 1/T, a straight line can be obtained with a slope of -Ea/R and an intercept of lnA. Therefore, two important constants, Ea and A, applicable to 200–300 °C can be obtained. Finally, the equation of the thermal degradation conversion in the temperature range of 200–300 °C can be obtained and it can be used to describe the actual weight loss curves of SDR in the process of torrefaction.

2.4. Two-Step Reaction Mechanism

Prins et al. [6] used a two-step reaction mechanism to simulate the kinetics of torrefaction (as given in Equation (6)). First, reactant A is converted into intermediate I and the volatiles V_1, and intermediate I is further converted into the final product B and the volatiles V_2.

$$\begin{array}{ccc} & V_1 & V_2 \\ k_{V1} \nearrow & k_{V2} \nearrow & \\ A \rightarrow & I \rightarrow B & \\ k_1 & k_2 & \end{array} \tag{6}$$

where k_1, k_{V1}, k_2, k_{V2}, respectively, represent four reaction rate constants.

Assuming that all reactions obey first-order kinetics, the differential rate equations for each of the components are given by Equations (7)–(11)

$$\frac{dW_A}{dt} = -(k_1 + k_{V1})W_A \tag{7}$$

$$\frac{dW_I}{dt} = k_1 W_A - (k_2 + k_{V2})W_I \tag{8}$$

$$\frac{dW_B}{dt} = k_2 W_I \tag{9}$$

$$\frac{dW_{V1}}{dt} = k_{V1} W_A \tag{10}$$

$$\frac{dW_{V2}}{dt} = k_{V2} W_I \tag{11}$$

where W_X is the weight of component X (X = A, I, B, V1, V2).

2.5. Particle Swarm Optimization (PSO)

In Section 2.3, the Arrhenius equation was adopted to study the kinetics in the torrefaction process. It assumes that the two important factors in the Arrhenius equation, Ea and A, are temperature-independent constants. In reality, though, Ea and A are dependent on temperature. Therefore, if the "temperature range is large", the Arrhenius equation is not suitable for describing the kinetics of chemical reaction without being modified. In addition to the Arrhenius equation, many other phenomenological formulas about activation energy have been proposed. For example, the deformed Arrhenius equation (DAE), generalized Mott law (GML), modified Arrhenius equation (MAE) and curved Arrhenius plot (CAP) all appropriately describe how Ea varies with temperature [25]. Nevertheless, the Chi-square Test and coefficient of determination found that these corrections still cannot accurately describe the real torrefaction behavior [25]. In summary, in Section 2.3, the set of Ea and A obtained in the temperature range (200–300 °C) may not accurately describe the kinetics at each isothermal torrefaction temperature. Therefore, a more precise approach is needed and the evolutionary algorithm can be used to simulate the kinetics of these five different isothermal torrefaction temperatures. The torrefaction temperatures in the simulation are not a range but five fixed values and the simulated weight loss curve of each group may be more in line with the actual weight loss curve.

In this section, momentum-type PSO [26] is applied to achieve global optimization and the equation is given by Equations (12) and (13).

$$\vec{v}_i^{\,k+1} = \beta_c \times \Delta \vec{v}_i^{\,k+1} + \varphi_1 \text{rand}() \left(\text{pbest}_i - \vec{x}_i^{\,k} \right) + \varphi_2 \text{rand}() \left(\text{gbest} - \vec{x}_i^{\,k} \right) \tag{12}$$

$$\vec{x}_i^{\,k+1} = \vec{x}_i^{\,k} + \alpha_c \times \vec{x}_i^{\,k+1}, \, i = 1, 2, \ldots, N_{particle} \tag{13}$$

where v_i is the velocity of the i-th particle, x_i is the position of the i-th particle; α_C, β_C are momentum constants; φ_1 and φ_2 are cognitive learning rate and social learning rate, respectively; $\text{rand}()$ is a random number in the range of [0,1]; pbest_i is the best position of the i-th particle; gbest is the global best position; $N_{particle}$ is the number of particles searching for the optimal solution.

To predict the weight loss curve, the least-squares method is applied and the target function (TAR) to be calculated is given by Equation (14)

$$\text{TAR}^T = \sum_j \left(W_{j,\,exp}^T - W_{j,\,cal}^T \right)^2 \tag{14}$$

where T is the temperature at which the isothermal torrefaction take place; $W_{j,\,exp}^T$ is the experimental sample weight of the j-th datum at a certain reaction temperature; $W_{j,\,cal}^T$ is the calculated sample weight of the j-th datum at a certain reaction temperature.

The flowchart of the PSO algorithm is shown in Figure 1. In addition, the fit quality (F) shown below is used to indicate the similarity between the calculated weight loss curve and the actual weight loss curve (Equation (15)). The PSO algorithm in this study was compiled with the Fortran programming language. First, a set of rough values, Ea and A, must be chosen simply based on relevant literature. If Ea and A were given appropriately at the beginning, it may reduce the number of iterations and have a better chance to converge. After initializing them, the program continues to process them with formulas to obtain the TAR and F values and then update a set of speed and position values of particles based on this result. Afterward, the PSO algorithm repeats the iterative cycle until the optimized criterion is met. In this study, the criterion is that TAR is smaller than 10^{-5}. When the criterion is met the iteration stops and Ea and A converge as the globally best values, so the fit quality can also be found. In the process of optimizing Ea and A values, i.e., particles, the velocity and position of particle i determine the position of the next particle i + 1, which is the value of the next set of Ea and A.

$$F(\%) = \left[1 - \frac{\sqrt{\frac{TAR}{N}}}{\left(W_{exp} \right)_{max}} \right] \times 100\%. \tag{15}$$

where $\left(W_{exp} \right)_{max}$ is the maximum sample weight in the experiment.

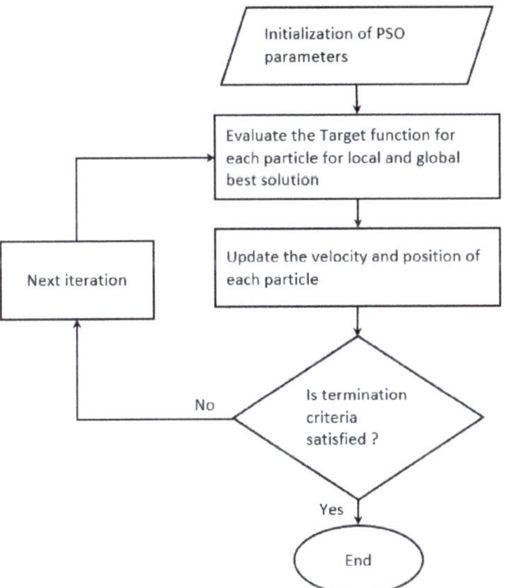

Figure 1. Flowchart of particle swarm optimization (PSO) algorithm.

3. Results and Discussion

3.1. Basic Properties and Thermogravimetric Analysis

Excluding other components from the results of the composition analysis in Table 1, the composition could be arranged in descending order according to the amount of content and gives that carbohydrate > protein > lipid. It is speculated that the biggest chunk of the composition is derived from the starch in the endosperm of the sorghum grain. Surprisingly, after the fermentation process of the liquor production, a large amount of starch remains in the SDR without being utilized by the microorganisms. In the approximate analysis, the volatile matter is as high as 68.97%, indicating that the reactivity of the SDR is high [3]. The content of ash in the residue is 9.08% by weight. From the perspective of gasification, the slagging phenomenon in the gasifier can be reduced by blending the residues with coal [3].

The thermogravimetric analysis (TGA) and derivative thermogravimetric (DTG) analysis of SDR are plotted in Figure 2. The whole pyrolysis process can be roughly divided into 4 stages. Dehydration is the first stage (25–200 °C), the second stage is the decomposition of proteins and carbohydrates (200–350 °C) [3,27], the third stage is the decomposition of lipids (350–550 °C) [3,8] and, finally, the fourth stage is the decomposition of other components (550–800 °C). Hardly any weight change occurs after 800 °C. Obviously, the sample weight drops drastically in the second stage and the DTG curve shows the fastest thermal decomposition rate at 317 °C. It is known from Table 1 that the total content of carbohydrates and proteins exceeds 80%; thus, the depolymerization, decarboxylation, and cracking reactions [28,29] of carbohydrates and proteins [27] dominate the entire pyrolysis process of SDR.

3.2. Isothermal Torrefaction

The curves of TGA and heating temperature throughout the entire SDR pyrolysis process are shown in Figure 3. In the temperature holding interval, the two curves of 200 and 225 °C almost overlap, as a consequence of light torrefaction [3] and the heating time almost does not affect the thermal decomposition of SDR. On the contrary, when torrefying at 300 °C, most of the carbohydrates and proteins in SDR decompose severely and cause the sample weight to fall to a large extent at the beginning and then after about 10 min, the sample weight decreases much slightly. In the middle torrefaction temperatures of 250 and 275 °C, the weight loss gradually increases with the torrefaction time.

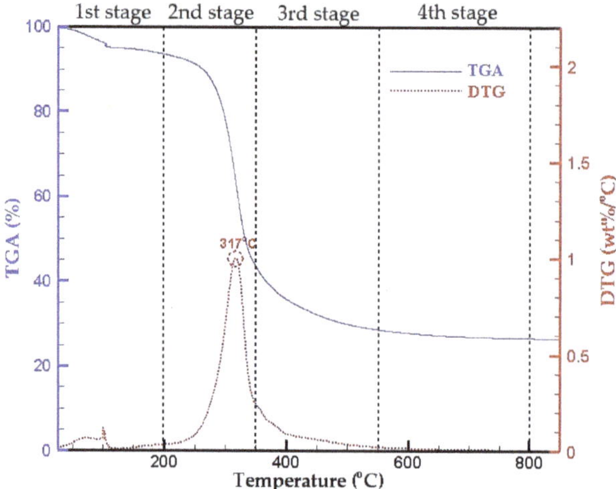

Figure 2. Distributions of thermogravimetric analysis (TGA) and derivative thermogravimetric (DTG) curves of sorghum distilled residue (SDR) in 4 different stages of pyrolysis at a heating rate of 20 °C/min in nitrogen. (1st stage: dehydration; 2nd stage: decomposition of proteins and carbohydrates; 3rd stage: decomposition of lipids; and 4th stage: decomposition of other components).

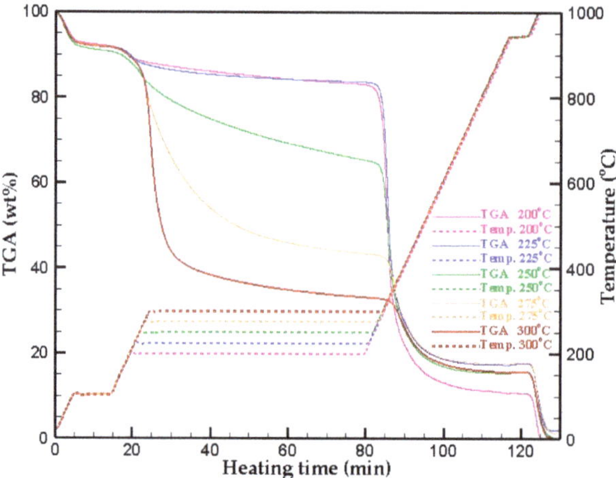

Figure 3. Distributions of thermogravimetric analysis (TGA) and heating temperature curves of 5 different torrefaction temperatures (200, 225, 250, 275, and 300 °C of SDR).

In Figure 3, the isothermal torrefaction started at the point when the temperature reached 200, 225, 250, 275, or 300 °C. Then, these five temperatures were held for 60 min to isothermally torrefy the SDR. Figure 4 graphs only the data for the 60 min isothermal period and plots the weight loss increment against the torrefaction time. For the torrefaction temperature below 225 °C, as the temperature time, the weight loss is not significant. For the medium-temperature torrefaction, i.e., 250 and 275 °C, obviously, the longer the torrefaction time is, the greater the weight loss. For the case of 300 °C, the weight loss increases sharply in the beginning, which indicates that most of the thermal decomposition occurs in the beginning and within a short time.

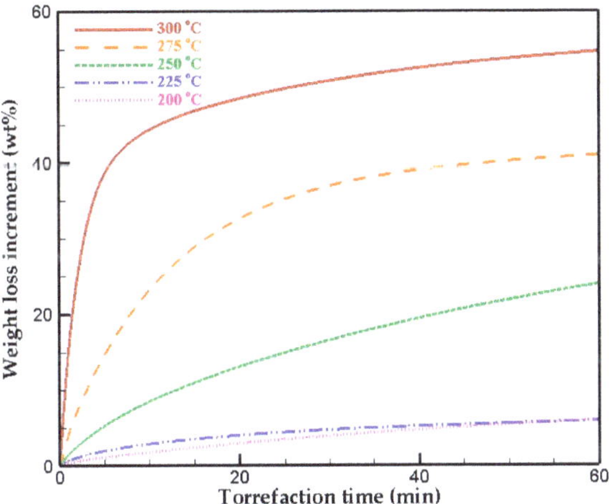

Figure 4. Distributions of weight loss increment of SDR.

Torrefaction severity index (TSI) is a dimensionless parameter [3], which was developed to account for the degree of torrefaction. The definition of TSI is given by Equation (16).

$$TSI = \frac{\Delta WI}{\Delta WI_{max}} \tag{16}$$

where ΔWI stands for the weight loss increment at a certain temperature and duration; ΔWI_{max} is the weight loss increment at 300 °C for 1 h.

By plotting TSI, torrefaction time, and temperature, a three-dimensional diagram can be obtained as shown in Figure 5. From this three-dimensional profile, it can be pointed out that when torrefaction at the temperature below 225 °C, SDR has only a small degree of thermal decomposition. At medium temperature, TSI is relatively sensitive to the duration, so it is not quite suitable for quality control of biochar. For torrefaction at 300 °C, however, the TSI value could exceed 0.8 within 10 min. The 3D graph of TSI-temperature-duration can be used as a useful tool for SDR upgrading to biochar through torrefaction. For example, if the criterion of TSI for biochar is 0.8, the recommended operating conditions would be 300 °C and 10–15 min. Therefore, the unstable biochar quality caused by the sensitive curve in the first 10 min can be avoided and the energy cost can be reduced by ending up the process in about 15 min.

3.3. Torrefaction Kinetics from the Traditional Model-Free Approach

The experimental data obtained from five different isothermal torrefaction conditions can be brought into Equation (3) to obtain the rate constant k. Afterward, ln(k) is plotted against $1/T$ and the coefficient of determination (R^2) of each n can be obtained, as shown in Table 2. When n is equal to 3, R^2 reaches a maximum, and thereby the best regression line is obtained. In the case of n = 3, Ea and A can be obtained from the slope and intercept of the straight line in Figure 6. A list of the important kinetics parameters of the torrefaction reaction of SDR in the range of 200–300 °C is detailed in Table 3.

To evaluate the traditional model-free approach, these optimal kinetics parameters were brought into Equation (3) (n = 3) to obtain the optimal weight loss curves, to compare with the actual curves (Figure 7). It shows that the curve predicted by model-free kinetics is not very consistent with the actual TGA curve. In addition, the fit qualities of five different temperatures are in the range between 92 and 97%.

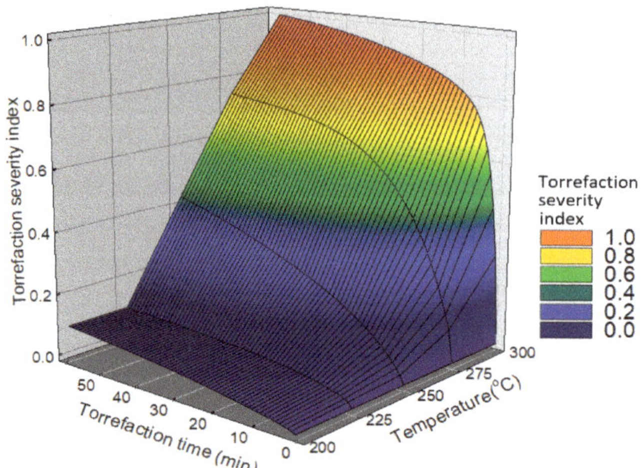

Figure 5. Three-dimensional diagram of torrefaction severity index (TSI) of SDR.

Table 2. Coefficient of determination (R^2) of the isothermal kinetics of SDR.

n	R^2
1	0.8469
2	0.8931
3	0.9115 (Max.)
4	0.9083
5	0.9062
6	0.9012
7	0.8975
8	0.8938
9	0.8909

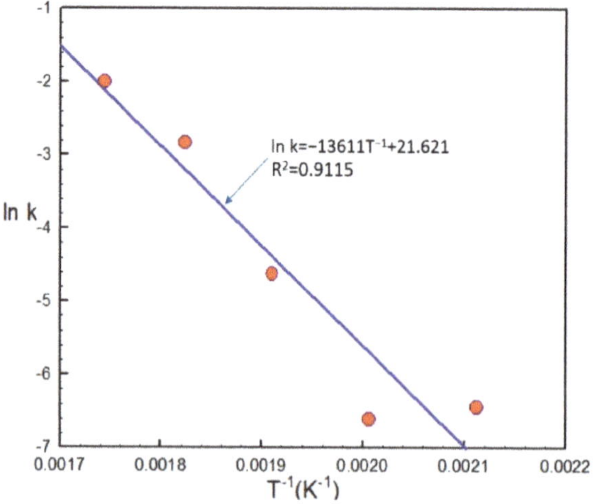

Figure 6. Linear regression of the isothermal kinetics ($n = 3$) of SDR.

Table 3. The regression line and kinetics parameters of the thermal decomposition of SDR.

Parameters/Equation	Amount/Expression
R^2	0.9115
Regression line	ln k = −13,611/T+21.621
n	3
Ea (kJ·mol^{-1})	113.1619
A(min^{-1})	2,454,036,361

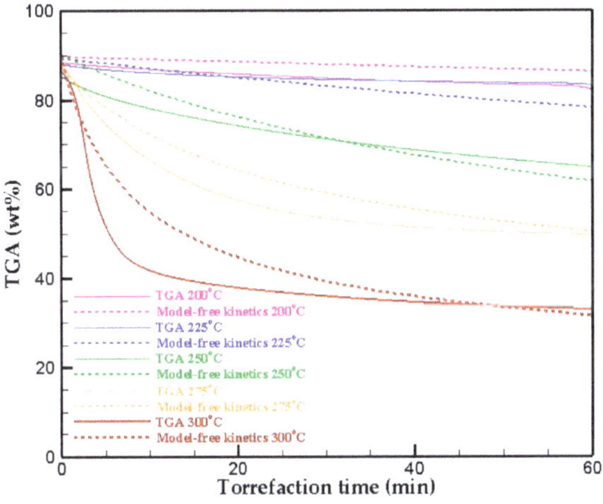

Figure 7. Experimental TGA curve (solid line) and model-free kinetics curve (dotted line) of SDR.

3.4. Torrefaction Kinetics from PSO Approach

In this section, a two-step reaction mechanism, shown in Equation (6), was adopted and it was assumed that each step was a first-order reaction. The weight loss curves during the isothermal torrefaction process were simulated with the PSO approach and the results are plotted in Figure 8. From the figure, it is clear that the simulated weight-loss curves and the experimental weight-loss curves are very close to each other. That is, they almost overlap and have very high fit qualities (>99.25). The fit qualities of each temperature are listed in Table 4, while Table 5 details the kinetic parameters at each temperature.

PSO approach was further adopted to simulate the weight change of each component in the two-step reaction mechanism and the results are shown in Figure 9, where reactant A represents SDR; reaction intermediate I represents the partially reacted SDR; product B represents the biochar after SDR upgrading; V1 and V2 represent two volatiles in step 1 and step 2 of the mechanism, respectively. The sum of curves A, I and B is equal to the simulation curve.

Comparing the Figure 9a,b, their total solid weight distributions (the sum of A, I, and B) are very similar, where the 225 °C case has a higher final biochar B, yet a smaller amount of the partial-reacted SDR (i.e., I). This could be attributed to a relatively higher extent of reaction due to the higher temperature. Figure 9c indicates that the amount of intermediate I, representing the partial-reacted SDR, is almost unchanged at a high level, which is similar to the curves I at 200 °C and 225 °C. However, the amounts of V1 and V2 are higher than those two low-temperature curves, so they decrease the final amount of biochar B. Comparing Figure 9d,e, the higher temperature of 300 °C make the reaction of step 1 completed, producing a large amount of gas V1 and intermediate solid I. Meanwhile, the reaction in step 2 is gentler and produces less biochar B. In the case of 275 °C, the temperature is milder than 300 °C. Step 2 fully reacts, so that curve I drops at the initial stage of the reaction and a large amount of intermediate I is continuously converted into biochar B. Therefore, as far as the quality of SDR is concerned, the operating temperature of 275 °C is the best, because biochar B is very high (about 60%) and intermediate I that has not been upgraded is very little (about 2%).

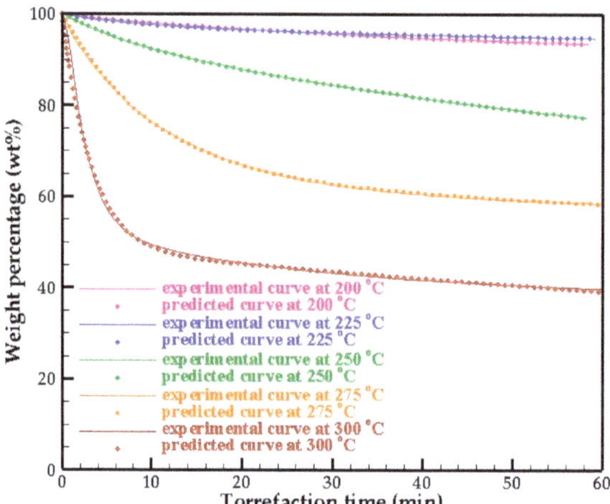

Figure 8. Experimental TGA curve (solid line) and PSO-simulated curve (dotted line) of SDR.

Table 4. Fit quality for each temperature of SDR.

Temperature (°C)	Fit Quality, F (%)
200	99.98
225	99.97
250	99.91
275	99.77
300	99.28

Table 5. Kinetic parameters at each temperature.

Parameters	k_1	k_{V1}	k_2	k_{V2}
200 °C				
Ea	4.21×10^4	3.88×10^4	1.67×10^5	9.27×10^4
A	2.42×10^2	1.05	8.13×10^{14}	4.45×10^5
k	5.41×10^{-3}	5.49×10^{-5}	2.65×10^{-4}	2.57×10^{-5}
225 °C				
Ea	4.52×10^4	4.04×10^4	1.75×10^5	9.83×10^4
A	1.45×10^2	1.09	8.50×10^{14}	3.23×10^5
k	2.64×10^{-3}	6.41×10^{-5}	3.72×10^{-4}	1.59×10^{-5}
250 °C				
Ea	4.53×10^4	3.69×10^4	1.86×10^5	9.56×10^4
A	8.43×10	8.78×10^{-1}	6.49×10^{14}	2.78×10^5
k	2.54×10^{-3}	1.83×10^{-4}	1.80×10^{-4}	7.93×10^{-5}
275 °C				
Ea	5.21×10^4	3.24×10^4	1.89×10^5	9.43×10^4
A	2.63×10^2	8.10×10^{-1}	9.07×10^{14}	3.07×10^5
k	2.85×10^{-3}	6.65×10^{-4}	7.84×10^{-4}	3.13×10^{-4}
300 °C				
Ea	5.17×10^4	3.00×10^4	2.02×10^5	1.03×10^5
A	1.38×10^2	1.50	6.15×10^{14}	2.25×10^5
k	2.68×10^{-3}	2.77×10^{-3}	2.22×10^{-4}	9.31×10^{-5}

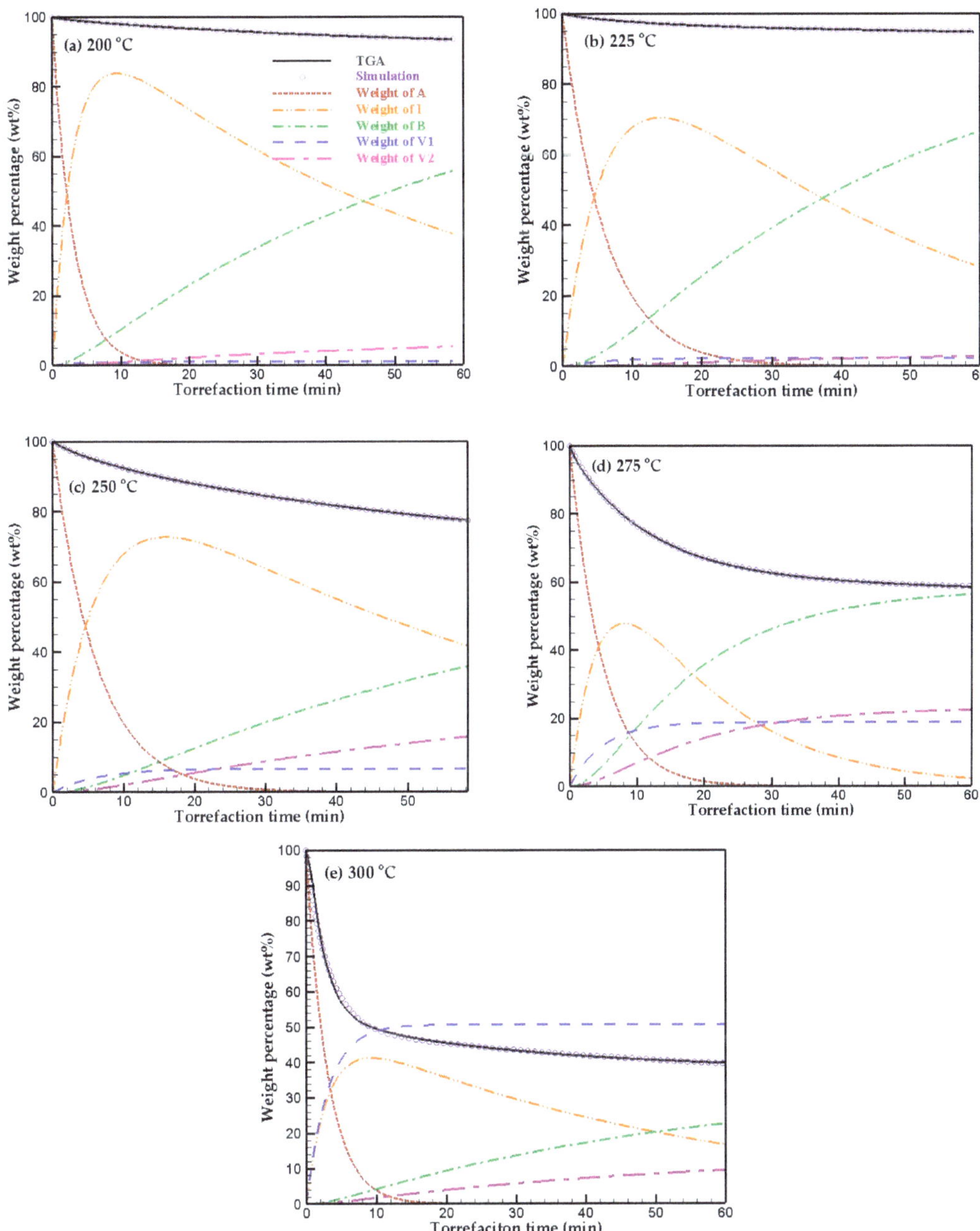

Figure 9. Weight distribution of each component over torrefaction time at (**a**) 200, (**b**) 225, (**c**) 250, (**d**) 275 and (**e**) 300 °C.

4. Conclusions

The results of composition analysis show that SDR remains a lot of carbohydrates and proteins that are not utilized by microorganisms. Therefore, in the TGA diagram, a violent thermal decomposition occurs around 317 °C. This could be attributed to the reactions of depolymerization, decarboxylation, and cracking of carbohydrates and proteins. The

3D TSI is a useful reference tool for upgrading SDR to biochar. The analysis reveals that the torrefaction should be held at 300 °C for 10–15 min to avoid the difficulty of quality control of biochar caused by the sensitive curve in the first 10 min and to reduce the energy cost by ending up the process in 15 min. In the Arrhenius equation, Ea and A can be regarded as constants only on an important assumption that the reaction takes place in a situation where the "reaction temperature is a constant". The optimal parameters obtained by the conventional kinetic approach can not accurately predict the torrefaction behavior and the fit qualities are at a relatively low level of 92–97%. However, the PSO algorithm can accurately simulate the real torrefaction behavior, all the fit qualities are higher than 99%. PSO algorithm is not only superior to the traditional Arrhenius equation in terms of kinetics behavior approach, it can also be further adopted to simulate the weight change distribution curves of each component in the two-step reaction mechanism. For example, the results from Table 5 show that k_{V1} increases with increasing temperature. Therefore, Figure 9a–e shows that V1 increases significantly with increasing temperature. When the torrefaction temperature reaches 300 °C, half of the SDR converted into gas phase V1, so that product B is at a low level of 22%. Another example is that the results in Table 5 show that the maximum value of k_2 appears at 275 °C, so a large amount of impure intermediate I is converted into product B. Figure 9d shows that I is only 2% at 275 °C. If higher biochar quality is required, the operating temperature of 275 °C is better than 300 °C, because the reaction in step 2 is relatively complete, the final SDR-biochar has a high product B yield and a low partial-reacted intermediate I content. This study shows that the PSO algorithm is a promising tool that can be used to accurately predict the torrefaction kinetics and even predict reaction intermediates that affect product quality.

Author Contributions: S.-W.Y. performed the experiments, analyzed data and wrote the paper. W.-H.C. created the research concept, organized the work, designed the experiments, analyzed data and provided facilities and instruments for the research. J.-S.C. provided advice and research strategy. C.-F.E. assisted in operating the PSO simulation program. S.R.N. analyzed data. P.L.S. analyzed data. All authors have read and agreed to the published version of the manuscript.

Funding: This research was funded by the Ministry of Science and Technology, Taiwan, R.O.C., under the contracts MOST 109-2221-E-006-040-MY3 and MOST 109-3116-F-006-016-CC1.

Institutional Review Board Statement: Not applicable.

Informed Consent Statement: Not applicable.

Data Availability Statement: Not applicable.

Acknowledgments: The authors acknowledge the financial support of the Ministry of Science and Technology, Taiwan, R.O.C., under the contracts MOST 109-2221-E-006-040-MY3 and MOST 109-3116-F-006-016-CC1 for this research.

Conflicts of Interest: The authors declare no conflict of interest.

References

1. Adeleke, A.A.; Odusote, J.K.; Ikubanni, P.P.; Lasode, O.A.; Malathi, M.; Paswan, D. Essential basics on biomass torrefaction, densification and utilization. *Int. J. Energy Res.* **2020**, *45*, 1375–1395. [CrossRef]
2. Chen, W.-H.; Lin, B.-J.; Lin, Y.-Y.; Chu, Y.-S.; Ubando, A.T.; Show, P.L.; Ong, H.C.; Chang, J.-S.; Ho, S.-H.; Culaba, A.B.; et al. Progress in biomass torrefaction: Principles, applications and challenges. *Prog. Energy Combust. Sci.* **2021**, *82*, 100887. [CrossRef]
3. Chen, W.H.; Huang, M.Y.; Chang, J.S.; Chen, C.Y. Thermal decomposition dynamics and severity of microalgae residues in torrefaction. *Bioresour. Technol.* **2014**, *169*, 258–264. [CrossRef] [PubMed]
4. Chen, W.H.; Lu, K.M.; Tsai, C.M. An experimental analysis on property and structure variations of agricultural wastes undergoing torrefaction. *Appl. Energy* **2012**, *100*, 318–325. [CrossRef]
5. Lu, K.-M.; Lee, W.-J.; Chen, W.-H.; Lin, T.-C. Thermogravimetric analysis and kinetics of co-pyrolysis of raw/torrefied wood and coal blends. *Appl. Energy* **2013**, *105*, 57–65. [CrossRef]
6. Prins, M.J.; Ptasinski, K.J.; Janssen, F.J.J.G. Torrefaction of wood. *J. Anal. Appl. Pyrolysis* **2006**, *77*, 28–34. [CrossRef]
7. Rousset, P.; Aguiar, C.; Labbe, N.; Commandre, J.M. Enhancing the combustible properties of bamboo by torrefaction. *Bioresour. Technol.* **2011**, *102*, 8225–8231. [CrossRef]

8.   Chen, Y.-D.; Liu, F.; Ren, N.-Q.; Ho, S.-H. Revolutions in algal biochar for different applications: State-of-the-art techniques and future scenarios. *Chin. Chem. Lett.* **2020**, *31*, 2591–2602. [CrossRef]

9.   Li, Y.H.; Kuo, W.C. The study of optimal parameters of oxygen-enriched combustion in fluidized bed with optimal torrefied woody waste. *Int. J. Energy Res.* **2020**, *44*, 7416–7434. [CrossRef]

10.   Chen, C.Y.; Chen, W.H. Co-torrefaction followed by co-combustion of intermediate waste epoxy resins and woody biomass in the form of mini-pellet. *Int. J. Energy Res.* **2020**, *44*, 9317–9332. [CrossRef]

11.   Ubando, A.T.; Felix, C.B.; Chen, W.H. Biorefineries in circular bioeconomy: A comprehensive review. *Bioresour. Technol.* **2020**, *299*, 122585. [CrossRef]

12.   Wang, T.T.; Sun, Z.Y.; Huang, Y.L.; Tan, L.; Tang, Y.Q.; Kida, K. Biogas Production from Distilled Grain Waste by Thermophilic Dry Anaerobic Digestion: Pretreatment of Feedstock and Dynamics of Microbial Community. *Appl. Biochem. Biotechnol.* **2018**, *184*, 685–702. [CrossRef]

13.   Luo, T.; Huang, H.; Mei, Z.; Shen, F.; Ge, Y.; Hu, G.; Meng, X. Hydrothermal pretreatment of rice straw at relatively lower temperature to improve biogas production via anaerobic digestion. *Chin. Chem. Lett.* **2019**, *30*, 1219–1223. [CrossRef]

14.   Su, M.Y.; Tzeng, W.S.; Shyu, Y.T. An analysis of feasibility of bioethanol production from Taiwan sorghum liquor waste. *Bioresour. Technol.* **2010**, *101*, 6669–6675. [CrossRef] [PubMed]

15.   Ye, G.; Luo, H.; Ren, Z.; Ahmad, M.S.; Liu, C.-G.; Tawab, A.; Al-Ghafari, A.B.; Omar, U.; Gull, M.; Mehmood, M.A. Evaluating the bioenergy potential of Chinese Liquor-industry waste through pyrolysis, thermogravimetric, kinetics and evolved gas analyses. *Energy Convers. Manag.* **2018**, *163*, 13–21. [CrossRef]

16.   Chen, W.-H.; Peng, J.; Bi, X.T. A state-of-the-art review of biomass torrefaction, densification and applications. *Renew. Sustain. Energy Rev.* **2015**, *44*, 847–866. [CrossRef]

17.   Devos, P.; Commandre, J.M.; Brancheriau, L.; Candelier, K.; Rousset, P. Modeling mass loss of biomass by NIR-spectrometry during the torrefaction process. *Int. J. Energy Res.* **2020**, *44*, 9787–9797. [CrossRef]

18.   Bach, Q.V.; Chen, W.H.; Chu, Y.S.; Skreiberg, O. Predictions of biochar yield and elemental composition during torrefaction of forest residues. *Bioresour. Technol.* **2016**, *215*, 239–246. [CrossRef]

19.   Chew, J.J.; Doshi, V. Recent advances in biomass pretreatment—Torrefaction fundamentals and technology. *Renew. Sustain. Energy Rev.* **2011**, *15*, 4212–4222. [CrossRef]

20.   Ciolkosz, D.; Wallace, R. A review of torrefaction for bioenergy feedstock production. *Biofuels Bioprod. Biorefining* **2011**, *5*, 317–329. [CrossRef]

21.   Van der Stelt, M.J.C.; Gerhauser, H.; Kiel, J.H.A.; Ptasinski, K.J. Biomass upgrading by torrefaction for the production of biofuels: A review. *Biomass Bioenergy* **2011**, *35*, 3748–3762. [CrossRef]

22.   Chen, W.-H.; Chu, Y.-S.; Liu, J.-L.; Chang, J.-S. Thermal degradation of carbohydrates, proteins and lipids in microalgae analyzed by evolutionary computation. *Energy Convers. Manag.* **2018**, *160*, 209–219. [CrossRef]

23.   Chen, W.-H.; Eng, C.F.; Lin, Y.-Y.; Bach, Q.-V. Independent parallel pyrolysis kinetics of cellulose, hemicelluloses and lignin at various heating rates analyzed by evolutionary computation. *Energy Convers. Manag.* **2020**, *221*, 113165. [CrossRef]

24.   Chen, W.-H.; Kuo, P.-C. Isothermal torrefaction kinetics of hemicellulose, cellulose, lignin and xylan using thermogravimetric analysis. *Energy* **2011**, *36*, 6451–6460. [CrossRef]

25.   Aquilanti, V.; Mundim, K.C.; Cavalli, S.; De Fazio, D.; Aguilar, A.; Lucas, J.M. Exact activation energies and phenomenological description of quantum tunneling for model potential energy surfaces. The F+H2 reaction at low temperature. *Chem. Phys.* **2012**, *398*, 186–191. [CrossRef]

26.   Liu, J.-L.; Lin, J.-H. Evolutionary computation of unconstrained and constrained problems using a novel momentum-type particle swarm optimization. *Eng. Optim.* **2007**, *39*, 287–305. [CrossRef]

27.   Rizzo, A.M.; Prussi, M.; Bettucci, L.; Libelli, I.M.; Chiaramonti, D. Characterization of microalga Chlorella as a fuel and its thermogravimetric behavior. *Appl. Energy* **2013**, *102*, 24–31. [CrossRef]

28.   Peng, W.M.; Wu, Q.Y.; Tu, P.G.; Zhao, N.M. Pyrolytic characteristics of microalgae as renewable energy source determined by thermogravimetric analysis. *Bioresour. Technol.* **2001**, *80*, 1–7. [CrossRef]

29.   Peng, W.M.; Wu, Q.Y.; Tu, P.G. Pyrolytic characteristics of heterotrophic Chlorella protothecoides for renewable bio-fuel production. *J. Appl. Phycol.* **2001**, *13*, 5–12. [CrossRef]

 *sustainability*

Article

Spatiotemporal Comparison of Drivers to CO_2 Emissions in ASEAN: A Decomposition Study

Edwin Bernard F. Lisaba, Jr. and Neil Stephen A. Lopez *

Mechanical Engineering Department, De La Salle University, Manila 0922, Philippines; edwin_lisaba@dlsu.edu.ph
* Correspondence: neil.lopez@dlsu.edu.ph

Abstract: The Southeast Asian region is one of the most vulnerable to climate change given its geographical location and economic situation. This study aims to conduct a combination of spatial and temporal analyses in order to understand differences between member nations in terms of driving factors to changing emissions. The logarithmic mean Divisia index (LMDI) method was used in order to estimate carbon dioxide emissions due to population, economic activity, economic structure, and energy intensity effects from the year 1990 to 2018. In conducting the study, spatial analysis showed that Singapore was the only country to effectively lessen carbon emissions, due to population and energy intensity, in comparison to the others. Additionally, temporal analysis showed that the ASEAN initially developed at the same rate, before countries such as Singapore, Malaysia, and Thailand, started becoming more economically active, as shown by their economic activity. Finally, results have shown that some countries, especially the Philippines and Indonesia, have undergone significant changes in economic structure, which significantly affected carbon emissions. The results also highlight the increasing per capita emissions as income levels rise. The paper concludes by presenting a summary of the findings and some policy recommendations.

Keywords: LMDI decomposition; spatiotemporal analysis; ASEAN; climate change; CO_2 emissions

Citation: Lisaba, E.B.F., Jr.; Lopez, N.S.A. Spatiotemporal Comparison of Drivers to CO_2 Emissions in ASEAN: A Decomposition Study. *Sustainability* **2021**, *13*, 6183. https://doi.org/10.3390/su13116183

Academic Editors: Wei-Hsin Chen, Alvin B. Culaba, Aristotle T. Ubando and Steven Lim

Received: 30 April 2021
Accepted: 28 May 2021
Published: 31 May 2021

Publisher's Note: MDPI stays neutral with regard to jurisdictional claims in published maps and institutional affiliations.

1. Introduction

Recently, the world has been experiencing a rapid increase in temperature per year. More specifically, the National Oceanic and Atmospheric Administration (NOAA) released a Global Climate Report in the year 2020, stating that the combined temperature of both land and sea increased at an average of about 0.08 degrees Celsius (0.13 °F) per decade until 1880 [1]. However, starting 1881, the combined average temperature has been increasing by about 0.18 degrees Celsius (0.32 °F). This increase in temperature corresponds to climate change, which is a growing concern in more recent years. In the Fourth Assessment Report, published in 2007, the International Panel on Climate Change (IPCC) has related the increase in temperature to the rapid melting of polar ice caps and increase in the frequency of natural disasters [2]. This, in turn, leads to the global warming effect, which is accelerated by the discharge of greenhouse gases (GHG) such as methane (CH_4), nitrous oxide (N_2O), and carbon dioxide (CO_2). In 2014, the IPCC published the Fifth Assessment Report, which determined that carbon dioxide was the most prominent driver for global warming [3]. Furthermore, the IPCC has warned that, if GHG emissions were not abated, sea levels would rise by about 18 to 59 cm, and the average temperature rise around the world would increase from 1.8 °C to 4 °C. Additionally, natural disasters would occur more frequently in later years.

Given the correlation between carbon dioxide emissions and global warming, it is therefore recommended to find solutions that involve mitigating the discharge of greenhouse gases into the atmosphere. The World Health Organization (WHO) states that global warming will have a significant impact on human and environmental health [4]. It is estimated that about 250,000 people will die between the years 2030 and 2050 due to climate change. Most alarming, however, is that the WHO predicts that areas with a weak infrastructure on health, which are especially found in developing countries, would be the least able to cope to climate change. This, therefore, highlights the necessity of finding alternative solutions in order to abate carbon dioxide emissions.

This therefore leads to the Southeast Asian countries, which are Cambodia, Myanmar, Lao PDR, Vietnam, Brunei Darussalam, Indonesia, Malaysia, Singapore, Thailand, and the Philippines. This collection of 10 countries accounts for the Association of Southeast Asian Nations (ASEAN), established in August 1967. The ASEAN is meant for member countries to collaboratively promote cultural development and accelerate economic growth within the region. Additionally, each member country is encouraged to actively support, and cooperate, in projects concerning culture, education, science, and economics [5].

According to Hill, the Southeast Asian region is exceptionally complex due to their diversity, both culturally and economically [6]. Countries like Singapore and Indonesia have exceptional economies compared to the poverty in Myanmar and Cambodia [7,8]. Despite massive economic and technological diversity, the aforementioned countries are still encouraged to support one another, as outlined in the Treaty of Amity and Cooperation in Southeast Asia (TAC). The TAC states that, although unity and support between member countries are encouraged, there still must be mutual respect for their individual territories, independence, and national identities [9].

As mentioned earlier, global warming correlates to the increase in frequency and intensity of natural disasters, such as typhoons, droughts, and floods. It must therefore be noted that the Southeast Asian region is located to the west of the Pacific Ocean, where about one-third of the world's annual tropical cyclones form [10]. Given the fact that a majority of the member countries are archipelagic, the effects of climate change will be more evident and devastating, should precautionary measures not be implemented sooner. Furthermore, the Southeast Asian region has areas of extreme poverty, and a majority of the countries heavily rely on agriculture. Additionally, the region is dependent on forestry and natural resources.

The effects of climate change are already apparent and have already devastated some parts of the region. In terms of temperature, the Asian Development Bank stated that in the ASEAN, the average temperature between the years 1951 and 2000 increased by about 0.1 and 0.3 °C [11]. Moreover, it was stated that rainfall lessened from 1960 to 2000, and sea levels have risen by about one to three millimeters. Given this, Nunti et al. further stated that climate change has adverse effects on areas that rely on agriculture by using the copula-based stochastic frontier approach [12]. These unfavorable conditions therefore increase the risk imposed on numerous rural and impoverished areas in terms of survival and livelihood. For example, Vietnam was hit by typhoons Goni and Vamco on November 2020, thus causing more than 200 deaths and around 1.5 billion US dollars in damages [13]. Another example would be the extreme flooding in the Philippines caused by typhoon Vamco, which resulted in more than 200 deaths and around 400 million US dollars in damages.

Given that the Southeast Asian region is extremely vulnerable to various natural disasters, it must be known that the ASEAN nations have already started taking precautionary measures. They have not only cooperated on a number of programs, but also enacted policies in order to abate the effects of climate change. Such policies involved the mitigation of carbon dioxide emissions in order to lessen risks to life and property. According to the ASEAN, a number of member states have voluntarily pledged CO_2 mitigation targets [14]. For example, Indonesia pledged to reduce carbon dioxide emissions by 26% from business-as-usual (BAU) by 2020. With the addition of international assistance, this could even be improved to 41%. Furthermore, 2020 was also set as the deadline for Malaysia, Philippines, and Singapore, who have all pledged greenhouse gas reduction by 40%, 20%, and 16%, respectively.

This present study determines drivers for greenhouse gas emissions in the ASEAN, specifically CO_2. In doing so, historical data on past policies will also be discussed in order to assess their efficacy and feasibility for the upcoming years. The study will first highlight the most effective and impressive policies enacted by the most successful countries, and then derive or infer other possible policies based on each country's performance.

A decomposition analysis approach will be used to estimate the drivers to CO_2 emissions per sector per country. According to de Boer and Rodrigues, the decomposition analysis method makes it possible to determine main drivers that caused changes in energy, and the environment, as time passes by [15]. The same source states that decomposition analysis can be further subdivided into two: namely, index decomposition analysis and structural decomposition analysis. First, index decomposition analysis (IDA) explores the relationship between a particular impact and production due to demand [16]. In contrast, structural decomposition analysis (SDA) examines the correlation between impact and the consumption of a product [17]. The impact can vary, such as energy and environmental impacts, due to the production or consumption of a particular mix or item. Both decomposition analyses have been increasingly used in the more recent years, particularly in energy- and environment-related studies. An example would be a study, conducted by Sumabat et al., wherein the logarithmic mean Divisia index method was used to determine CO_2 emissions stemming from the consumption of fuel for both power and electricity generation in the Philippines [18]. Another example would be a study by Nie and Kemp, where the researchers used a combination of index and structural decomposition analyses in order to examine energy fluctuations within a certain time period [19]. There was also a study conducted by Jurkėnaitė and Baležentis, where the index decomposition method was used to determine drivers for the average growth of farms in the European Union [20]. Additionally, other studies involving decomposition analysis exist for various ASEAN countries. Such countries include Malaysia [21], Thailand [22], and Singapore [23].

Other studies based on the Asian region have also used varying methodologies in combination with decomposition methods. For example, Li et al. have conducted decoupling elasticity and decoupling index methods on acquired LMDI results in order to examine the correlation between economic growth and CO_2 emissions in Central Asia [24]. Another study based on Central Asia rather used the environmental Kuznets curve hypothesis in order to examine the same pattern [25]. Finally, a study was made on the context of China and the ASEAN using decomposition and decoupling analyses [26]. It can therefore be seen that a common methodology between these studies is that a temporal approach has been taken in the LMDI decomposition processes. Additionally, these studies have shown that carbon emissions are intrinsically linked with economic growth. However, it would be insightful to understand the general growth of each ASEAN member nation in relation to each other, as well as determine the actions these individual countries have taken in order to create such change in carbon emissions coming from a variety of stages in economic growth.

In addition to the traditional temporal analysis, the present study will integrate a spatial comparison of driving factors to CO_2 emissions within the region. When investigating the aggregated effects of various countries, an integrated spatiotemporal approach can conveniently provide novel insights to policymakers. Therefore, the research gap, filled-in by this study, is the need for a comprehensive spatiotemporal comparison of indices and drivers to CO_2 emissions in the ASEAN. The ASEAN countries will not only be analyzed by how their indices changed over the years, in response to climate change, but will also be compared with each other to reveal their development trajectory and investigate their policy efficacy, in response to the mitigation of CO_2 emissions. Using this approach, the researchers will be able to benchmark effectively across the ASEAN nations. This study aims to provide a new perspective with regards to the cross-cutting evolution of emission drivers through space and time. Finally, in lieu of the ASEAN Economic Integration, the study aims to provide research and insight by benchmarking the environmental and sustainability practices for each region [27].

For this particular study, the spatial variant of the logarithmic mean Divisia index method (LMDI) will be implemented on seven (7) different time periods from 1990 to 2018. The paper proceeds as follows: Section 2 will provide the methodology, including equations and data; Section 3 will present the results and collated data; Section 4 will discuss the proposed policies and overall conclusions.

2. Materials and Methods

2.1. Fuel Consumption Data

The fuel consumption data per country was acquired from the International Energy Agency [28]. As stated earlier, the data gathered ranged between the years of 1990 and 2018 by intervals of five. The year 2018 was added to simply showcase the most recent performance and standing of each country. Furthermore, the total fuel consumption was further subdivided into eight fuel types; namely crude oil, oil products, coal, natural gas, biofuels and waste, nuclear, hydro, wind, and solar. Figure 1 below shows the total fuel consumption per fuel type in the ASEAN. Since it is a percentage share, the figure also depicts the average fuel consumption. The average is important since it gives an outline of the benchmark country which will be used for comparison and analysis later on. It must be noted that the fuel consumption is in terms of kilotons of oil equivalent (ktoe).

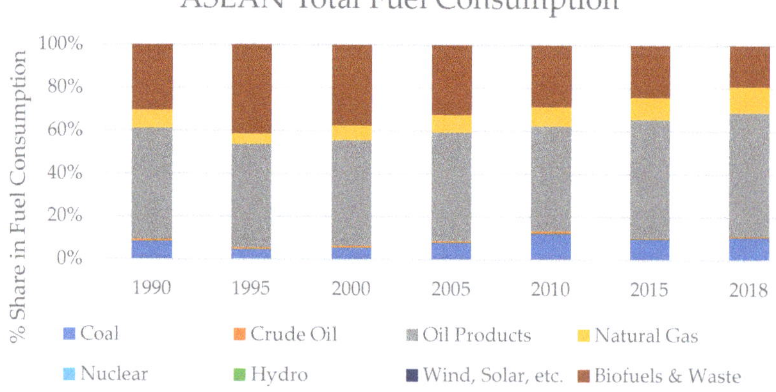

Figure 1. ASEAN total fuel consumption.

2.2. Carbon Dioxide Emissions Data

After obtaining the total fuel consumption per country, the carbon dioxide emissions from using each fuel type need to be calculated. This can be solved for by using Equation (1).

$$C_i^j = T_i^j F_i, \tag{1}$$

where C stands for the amount of carbon dioxide emitted from using a particular fuel type i in the year j. The variable C is in terms of kilotons of carbon dioxide, or ktCO2. Variable T, in terms of kilotons of oil equivalent (ktoe), is the amount of fuel type i consumed in a particular year j. Finally, F is the emission factor of a certain fuel type i. The values for F for each fuel type were taken from various literature, namely the Intergovernmental Panel for Climate Change [29], Environmental Protection Agency [30], and the U.S. Energy Information Association [31]. Please note that in using Equation (1), the emission factor needs to be in terms of ktCO2 per ktoe. Therefore, conversion of factors was necessary in order to have consistency across all fuel types.

2.3. Logarithmic Mean Divisia Index

For this study, the logarithmic mean Divisia index method was used in order to estimate drivers to carbon dioxide emissions in the ASEAN region. In order to have a better understanding of the process, a study by Ang provides a well-defined guide that helped streamline the process [32]. It shows the steps for formulating the identity function, as well as the general equations to be used for the method. Stated earlier, the study employed an index decomposition analysis, where the correlation between effects and the production of a certain aggregate was studied. In this case, the aggregate is carbon dioxide emissions.

After determining the aggregate, the identity function must then be formulated. In this study, the effects due to population, economic activity, economic structure, and energy intensity were considered. It was therefore necessary to obtain data regarding the population and gross domestic product (GDP) for each country. Additionally, the GDP per country must be further subdivided into three sectors; namely Agriculture, Industry, and Services for the economic structure effect. The data for the population, GNI per capita, overall GDP, and GDP per sector were taken from the World Bank [33]. Finally, in order to take inflation into account, the GDP was kept at a constant 2010 USD value for all countries. The GDP was also expressed in terms of million USD. After taking these factors into consideration, the identity function was then determined, which is shown in Equation (2).

$$C = \sum_{ijk} C_{ijk} = P_j \times \frac{G_j}{P_j} \times \sum_{ijk} \frac{G_{jk}}{G_j} \times \frac{E_{ijk}}{G_{jk}} \times \frac{C_{ijk}}{E_{ijk}} = P_j \times A_j \times \sum_{ijk} S_{jk} \times I_{ijk} \times F_i . \tag{2}$$

which also further expands the identity function into separate variables. It must be noted that the subscript i refers to the fuel type, j refers to the year, and k refers to the sector. The variables in the expanded notation represent the various effects to be analyzed. Therefore, P_j refers to the population of a country in a particular year j, and G_j represents the gross domestic product (GDP) in a year j. This fraction refers to the economic activity, and it can be simplified by the variable A_j. Similarly, the variable G_{jk} denotes the gross domestic product of a country in a particular year j and in a specific sector k. The ratio G_{jk}/G_j therefore refers to the economic structure effect S_{jk}. Furthermore, E_{ijk} refers to the total energy consumed when using a particular fuel type i during a year j within a specific sector k. Together with G_{jk}, this forms the energy intensity I_{ijk}. Finally, the last fraction is simply an altered form of Equation (1).

After outlining the effects to be studied as well as the identity function, it is then necessary to determine the actual change in carbon dioxide emissions due to the aforementioned effects. These can be solved for using Equations (3)–(6) as shown.

$$\Delta C_{pop}^R = \sum_j \frac{C_{ijk}^R - C_{ijk}^\mu}{ln C_{ijk}^R - ln C_{ijk}^\mu} \left(\frac{P_j^R}{P_j^\mu} \right) \tag{3}$$

$$\Delta C_{act}^R = \sum_j \frac{C_{ijk}^R - C_{ijk}^\mu}{ln C_{ijk}^R - ln C_{ijk}^\mu} \left(\frac{A_j^R}{A_j^\mu} \right) \tag{4}$$

$$\Delta C_{struc}^R = \sum_{jk} \frac{C_{ijk}^R - C_{ijk}^\mu}{ln C_{ijk}^R - ln C_{ijk}^\mu} \left(\frac{S_{jk}^R}{S_{jk}^\mu} \right) \tag{5}$$

$$\Delta C_{int}^R = \sum_{jk} \frac{C_{ijk}^R - C_{ijk}^\mu}{ln C_{ijk}^R - ln C_{ijk}^\mu} \left(\frac{I_{ijk}^R}{I_{ijk}^\mu} \right) \tag{6}$$

where the variables ΔC_{pop}, ΔC_{act}, ΔC_{struc}, and ΔC_{int}, refer to the change in CO_2 emissions due to the population, economic activity, economic structure, and energy intensity effects, respectively.

The effects in the study were analyzed based on how they affect the amount of CO_2 produced by a fuel type in a particular year and sector. First, the variable ΔC_{pop} connected the increase or decrease of the population to the amount of CO_2 released to the atmosphere. It also relates carbon emissions to human activity. Next, ΔC_{act} referred to the economic growth or decline of a country given the increase in population. Furthermore, the economic structure effect referred to the overall contribution of a particular sector to the country's overall GDP. Denoted by ΔC_{struc}, this variable measured the increase in carbon emissions due to the growth of a particular sector. It could also mean that a particular sector has had a change in mix, where the country has conducted efforts to switch from one sector to another. Switching the base sector of a country results in lesser carbon emissions, and can even bring about negative values for the economic structure effect. Finally, ΔC_{int} referred to the overall increase or decrease in energy used in order to produce the same amount of product. It must be noted that, should the energy intensity decrease, then it signifies that either the method in which these items are produced has changed, or the production process has gotten more efficient. As mentioned earlier, a benchmark country was necessary in order to properly compare the performance of a particular country with the rest. A hypothetical benchmark country was created for this study based on the average of the entire ASEAN in terms of population, overall GDP per year, GDP per year per sector, and fuel consumption per year per sector. With respect to Equations (3)–(6) above, the benchmark country is denoted by a μ in the superscript. Similarly, the particular country is denoted by the superscript R.

Additionally, given the gross national income (GNI) per capita, each ASEAN country was classified into different income levels; namely low, lower-middle, upper-middle, and high-income countries. These were then plotted with respect to each country's overall carbon dioxide emissions. Given this, it must be noted that the GNI per capita was based off the most recent year with complete data, which is 2018. The researchers then analyzed the resulting scatter plot for possible trends that could help further support the collated data.

3. Results

Given that this study has taken a spatiotemporal approach to the LMDI method, it was one of the goals of the researchers to both analyze the performance and emissions evolution of a country and benchmark these nations with one another. In order to do so, the different effects were plotted against each other. However, for brevity of the paper, only the ones which show notable findings are presented in this paper. Figures 2a, 3a, 4a and 5a show the progression of effects for each ASEAN member nation from 1990 to 2018. The benchmark country was used as a basis to see how each country faired in relation to each other. Finally, the pink dots show the plot point for each country in the year 1990, while the bright red dots show each country's position at 2018. In order to make the charts legible, the country's abbreviations, based off the ISO 3166 standard, were added only for these two dates so as to show the initial and most recent positions of each country. Therefore, these charts will be used to conduct the spatial analysis part of the study.

It must be noted that, in the year 1990, the International Energy Agency had no available data regarding the energy consumption of Cambodia and Lao PDR. Additionally, the World Bank did not have data for the total GDP and value-added percentage share of each sector to the GDP for Myanmar and Cambodia. Therefore, the researchers have simply plotted these three countries on the (0,0) coordinate on all four charts for the year 1990. The researchers have found that this does not actually affect the data much, since their respective data points in the year 1995 were not that far from the origin.

Finally, in order to more properly see how the ASEAN nation has evolved over time, trendlines were created on the spatial data for each year. This more properly illustrates the evolution of the region. These will be shown in Figures 2b, 3b, 4b and 5b. In order to make the graphs more readable, the researchers have opted to remove the data points from the spatial analysis and, instead, only show the trendline for each year. These specific charts will be used to do the temporal analysis on the collated data.

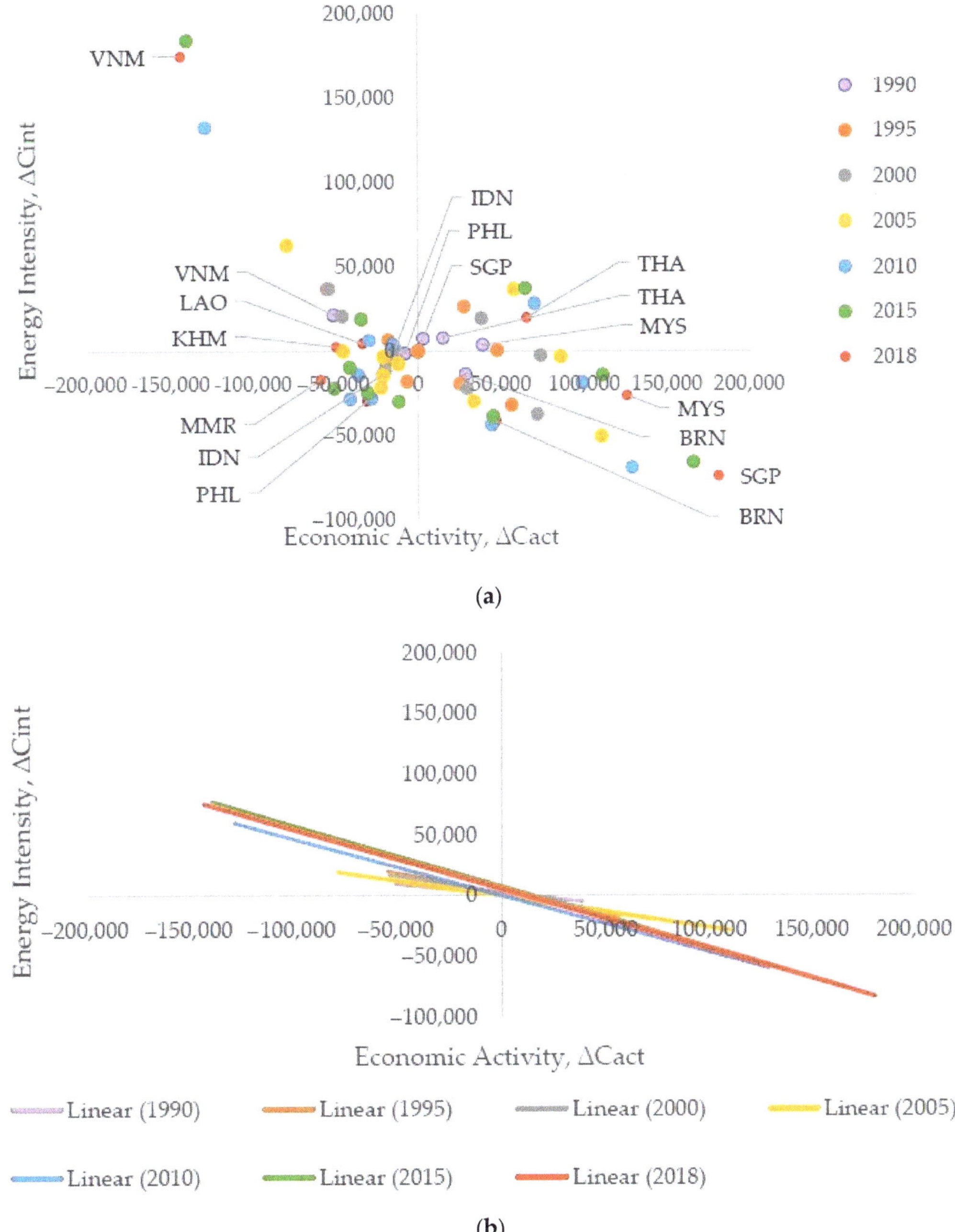

(**a**)

(**b**)

Figure 2. (**a**) ΔC_{act} vs. ΔC_{int} (Spatial Analysis). Legend: Brunei Darussalam (BRN), Cambodia (KHM), Indonesia (IDN), Lao PDR (LAO), Malaysia (MYS), Myanmar (MMR), Philippines (PHL), Singapore (SGP), Thailand (THA), Vietnam (VNM). (**b**) ΔC_{act} vs. ΔC_{int} (Temporal Analysis).

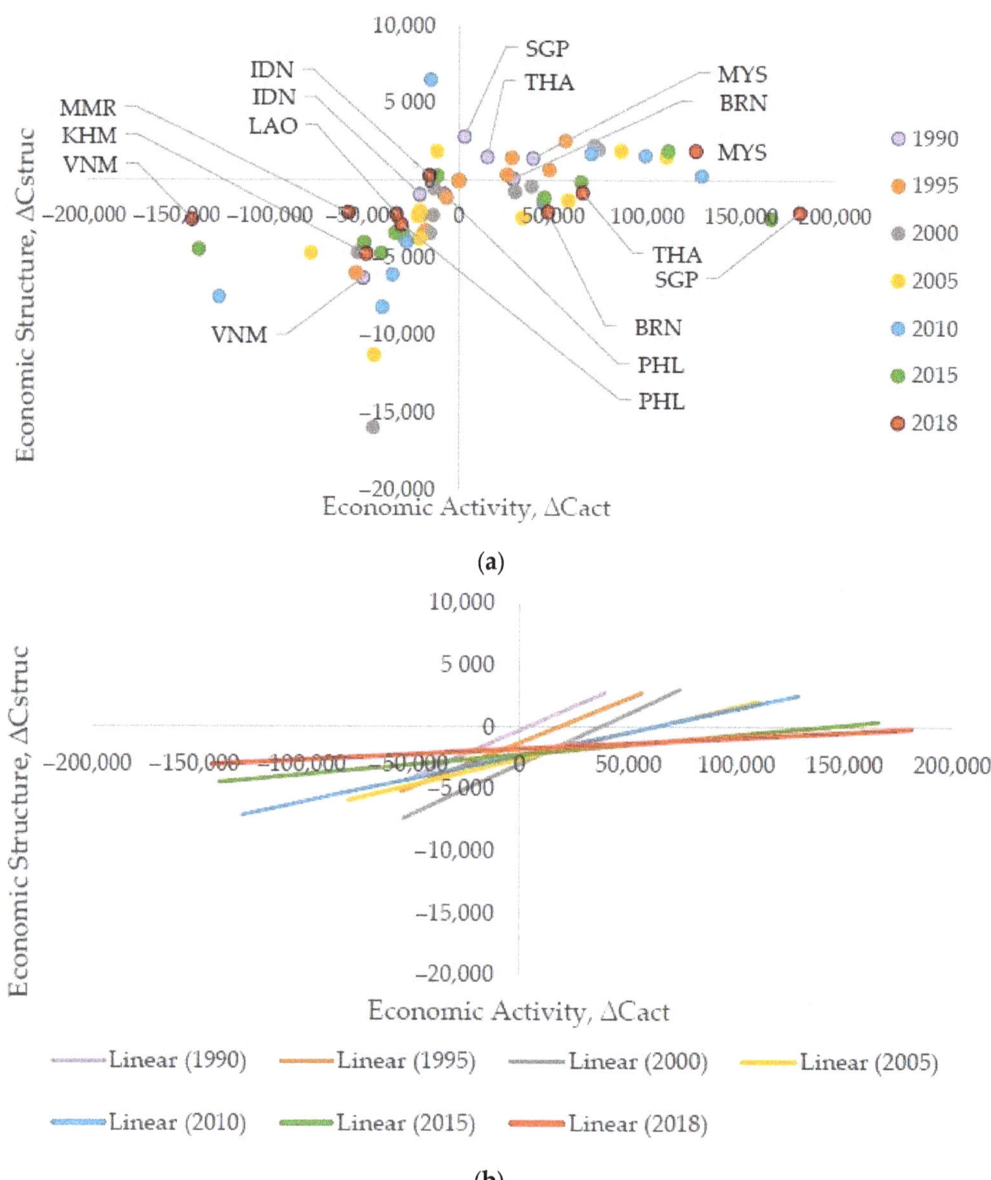

(**a**)

(**b**)

Figure 3. (**a**) ΔC_{act} vs. ΔC_{struc} (Spatial Analysis). (**b**) ΔC_{act} vs. ΔC_{struc} (Temporal Analysis).

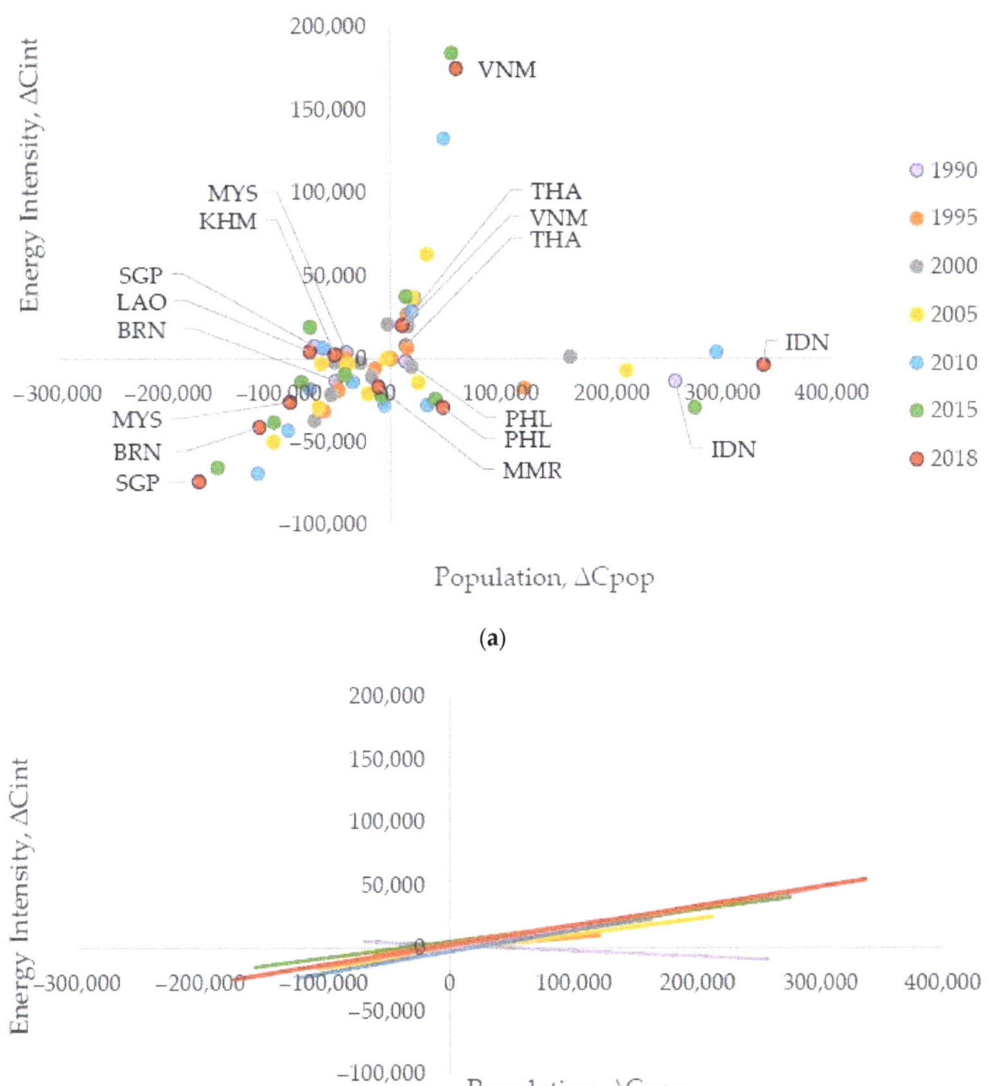

(a)

(b)

Figure 4. (a) ΔC_{pop} vs. ΔC_{int} (Spatial Analysis). (b) ΔC_{pop} vs. ΔC_{int} (Temporal Analysis).

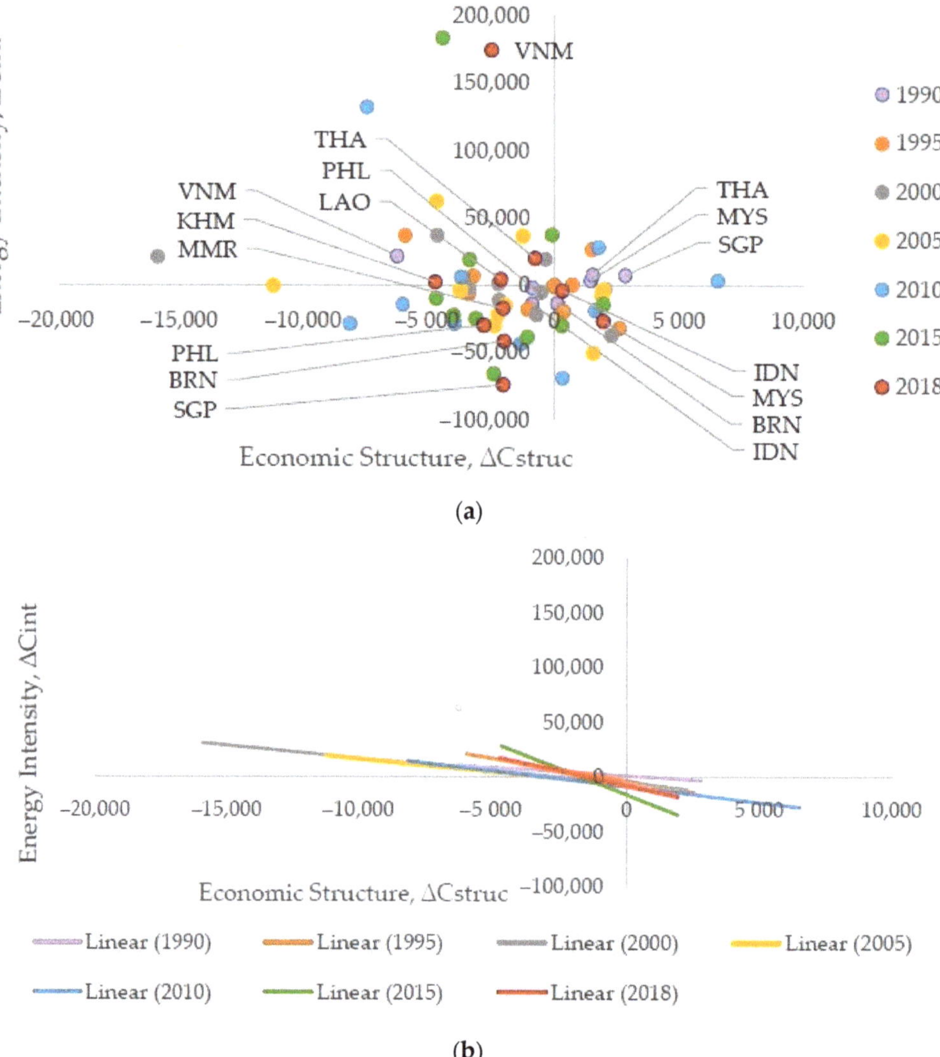

(a)

(b)

Figure 5. (a) ΔC_{struc} vs. ΔC_{int} (Spatial Analysis). (b) ΔC_{struc} vs. ΔC_{int}. (Temporal Analysis).

3.1. Economic Activity vs. Energy Intensity (ΔC_{act} vs. ΔC_{int})

3.1.1. Spatial Analysis

Figure 2a plots the economic activity effects against the energy intensity. It can be seen that, at the year 1990, a majority of the countries were located close to the origin, with Thailand, Malaysia, Vietnam, and Brunei Darussalam being the farthest. Besides Brunei Darussalam and Vietnam, the rest of the countries straddled and spread about the *x*-axis, meaning that these countries were not as energy intensive. It can also be said that countries to the right of the *y*-axis emitted more carbon dioxide, due to their economic activity, while those to the left emitted less. With this, Thailand and Malaysia emitted more carbon in relation to the rest. In contrast, the Philippines and Indonesia emitted slightly less. The most notable countries in the year 1990 were Vietnam and Brunei Darussalam. It can be seen that Vietnam was more energy intensive than the rest of the countries, however, it also emitted the least amount of carbon dioxide due to its economic activity. The exact opposite can be seen for Brunei Darussalam however. It is the least energy intensive, however, it is almost the same as Malaysia in carbon emissions due to economic activity.

The different colored dots show the progression of each country as they transition to 2018. Though each country was not specifically marked during the years in-between, they provide an accurate graph as to how a majority of these countries shifted over time. That being said, Singapore, Malaysia, and Vietnam show the greatest change towards the year 2018. To begin with, Singapore had actually lowered its carbon emissions already due to improved energy intensity by a significant amount. However, this has come with a notable increase in economic activity, making it emit the most carbon due to that effect. Similarly, Malaysia and Brunei Darussalam seemed to have also followed the same pattern but to a smaller degree. Thailand gradually became more energy intensive and economically active, while the Philippines did the exact opposite. Myanmar, Lao PDR, and Cambodia were still not as energy intensive, however, and they managed to lessen carbon emissions from their economic activity. The most notable find was Vietnam, where a majority of its carbon emissions came from its energy intensity, while managing to significantly lessen emissions from their economic activity. The country with the least change was found to be Indonesia, with a slight decrease in carbon emissions from both energy intensity and economic activity.

3.1.2. Temporal Analysis

The scatter plot used for the temporal analysis is shown in Figure 2b. For this section, only the general pattern of growth is studied, which will represent the general direction the ASEAN is heading to. Please note that the overall performance of the ASEAN region does not entirely reflect that of its individual countries.

It can be seen that the ASEAN nations usually follow a negative slope. This means that some parts of the region have more carbon emissions from their energy intensity compared to their economic activity, while other countries have a more impactful economic activity effect compared to their respective energy intensity. It is also interesting to note that the trendline slopes seemed to oscillate at a certain range before settling at a more constant value. The least steep slopes were found in the years of 1990, 2000, and 2005. The data makes sense given the fact that less energy was needed in the earlier years compared to what the region consumes now. Additionally, the economic activity during those years contributed more to the increase or decrease of carbon emissions, which can be seen in the graph. The carbon emissions came more from economic activity and less from energy intensity, thus resulting to a more horizontal slope. However, as the years progressed, it can be seen that the region has begun to show more pronounced changes in energy intensity, and some countries have either increased or decreased their carbon emissions for that particular effect. For the economic activity, a majority of the countries seem to just spread out more along the x-axis, given that their respective populations and gross domestic products have increased. From the years 2010 to 2018, the slopes tended to gradually reach a constant value.

3.2. Economic Activity vs. Economic Structure (ΔC_{act} vs. ΔC_{struc})

3.2.1. Spatial Analysis

Figure 3a plots the economic activity against the economic structure effect. At first glance, the dots seemed to create a natural curve from left to right. In the year 1990, it can be seen that Singapore, Thailand, Malaysia, and Brunei Darussalam were in the first quadrant, meaning that their carbon emissions due to both the economic structure and economic activity effects were significant. In contrast, the Philippines, Indonesia, and Vietnam, which were found in the third quadrant, displayed a significant decrease in carbon emissions due to both effects. Malaysia was the most economically active, while Singapore had the highest economic structure effect at the time. This goes in accordance with the fact that Singapore's GDP per capita actually surpassed that of South Korea, Portugal, and Israel [34]. Meanwhile,

Malaysia experienced labor shortages during the early 1990s, thus causing an influx of foreign workers in all sectors, as stated in a study by Lee [35]. The country managed to keep up with the labor demand, which enabled its economy to remain significantly active. In contrast, Vietnam managed to decrease carbon emissions from both effects by a significant amount. As written by Cima, Vietnam, in 1990, was considered to be one of the poorest countries in the world, with high rates of unemployment, wavering food supplies, and increasing inflation [36]. Relating to the study, it makes sense that the country did not emit as much carbon dioxide compared to the rest of the countries with respect to its economy at the time.

In the year 2018, however, it can be seen that there has been a huge change for all countries. First, the country with the greatest difference in economic activity was Singapore. The economic activity effect indicate that the country's CO_2 emissions have notably increased. However, it has managed its carbon emissions due to its economic structure by a significant amount. Similar to Singapore, Thailand, and Brunei Darussalam have followed the same pattern. Malaysia has displayed an almost constant economic structure effect, with a notable increase in carbon emissions due to its economic activity. For the other half of the x-axis, it can be seen that Indonesia has had a slight increase in carbon emissions due to its economic structure. On the other hand, the Philippines, Vietnam, Lao PDR, Cambodia, and Myanmar have shown an improvement in decreasing carbon emissions from both effects. Vietnam, most notably, has increased CO_2 emissions due to its economic structure, but has significantly managed its carbon emissions due to its economic activity. According to the World Bank, Vietnam has actually shifted labor away from the agricultural sector, and has allocated it to their industry and service sectors [37]. This change in economic structure has mitigated carbon emissions, due to their economic activities, even further. Combined with its booming economy in 2018, it also addresses the slight increase in carbon emissions due to the economic structure effect.

3.2.2. Temporal Analysis

Figure 3b shows the temporal analysis of the same scatter plot. It is notable to see that as the years progressed, a pattern can be more clearly seen. From 1990 to 2000, the y-intercept decreased at an almost constant interval. The slope gradually decreased as well. It is only by the year 2005 that the slope changed drastically. Finally, starting from 2010 to 2018, the slope gradually became less steep, until it was almost horizontal. From this chart, it can be seen how the ASEAN countries have collectively lessened total carbon emissions with regards to each country's economic structure. The increase or decrease of carbon emissions actually came from the economic activity effect. According to the Asian Development Bank (ADB), the ASEAN region has some of the world's most rapidly improving economies [38]. However, it was stated that the region has various policies that actually permit and encourage the emissions of greenhouse gases, which will be discussed later on in this paper. This region's collective GDP, and GDP per sector, have managed to mitigate carbon emissions. However, it is the technical inefficiencies and permitted emissions, present in the economic activity, that caused a massive increase in the overall carbon emissions as stated by the ADB.

From the temporal analysis, it can be observed that, in the years of 1990 to 2000, the ASEAN countries were developing together at the same pace. It should be noted that during this time, significant shifts in the economic structure were not as evident. However, in the years of 2010 to 2018, the trendline was observed to pivot. This therefore suggests that the countries were no longer developing at the same pace. It can be observed that some countries were developing faster than the others. This can be seen with countries like Singapore, Malaysia, and Thailand, who's rapidly increasing carbon emissions, due to economic activity, suggested rapid economic growth. The plot also further illustrates that, even though some countries like Vietnam and Cambodia were lagging in terms of economic development, it can be said that they are putting in aggressive efforts to restructure their economy before moving forward.

3.3. Population vs. Energy Intensity (ΔC_{pop} vs. ΔC_{int})

3.3.1. Spatial Analysis

Figure 4a shows the population effect plotted against the energy intensity effect. In the year 1990, it can be seen that the countries were not as energy intensive as they are now, since they straddled the x-axis. First, it can be observed that Indonesia had the greatest carbon emissions due to the population. The Philippines, Thailand, and Vietnam came second, with a slight increase in CO_2 emissions due to population growth. Vietnam, however, emitted more due to increasing energy intensity effects. In contrast, the countries of Malaysia, Singapore, and Brunei Darussalam had a negative change in carbon emissions during that time. From the three, Brunei Darussalam was the least energy intensive. According to Sjahrir, Indonesia, during the 1990s, had a series of constant changes in economic structure for production and export [39]. This series of changes in production managed increases in carbon emissions due to energy intensity, as stated earlier. Therefore, a majority of its own emissions had come from the country's population growth. Indonesia's main driver for carbon emissions seemed to be human activity.

In the year 2018, however, it can be seen that Vietnam has skyrocketed in carbon emissions due to the energy intensity effect, which can also be seen in the previous figures. It can also be observed that there was a change for Indonesia's carbon emissions, with respect to its population, despite being slightly more energy intensive. Furthermore, the Philippines had mitigated carbon emissions due to its energy intensity, with a slight increase in emissions due to population growth. Thailand has an almost constant population effect value, and has become slightly more energy intensive. Additionally, Singapore and Malaysia have significantly decreased carbon emissions for both effects, despite having started from a positive energy intensity value. Finally, Brunei Darussalam, Cambodia, Myanmar, and Lao PDR seemed to have simply decreased carbon emissions due to both energy intensity and population effects.

3.3.2. Temporal Analysis

It can be seen from Figure 4b that the ASEAN countries followed a more positive slope in the more recent years. It must be noted, however, that only the initial year of 1990 showed a negative slope. This is probably due to the lack of data from Myanmar, Lao PDR, and Cambodia during this year. Despite this, it can be seen that a majority of the ASEAN countries exhibited fluctuations in negative values for their respective energy intensity and population effects. This can be especially seen in the third quadrant of the figure. This shows that the majority have low carbon emissions due to population and energy intensity. From Figure 4a, the only countries which did not follow this trend were Vietnam, Thailand, the Philippines, and Indonesia. This actually coincides with a report by the Asian Development Bank, where these four countries, along with Malaysia, accounted for about 90% of greenhouse gas emissions in the ASEAN region [38]. Relating to this study, the aforementioned countries seem to have their emissions stem from both energy intensity and population effects.

3.4. Economic Structure vs. Energy Intensity (ΔC_{struc} vs. ΔC_{int})

3.4.1. Spatial Analysis

The last figure for spatial analysis is Figure 5a. In this scatter plot, it can be seen that most of the countries evenly surrounded the origin, and this pattern can be more easily seen in the earlier years. However, as time passed, it can be noted that the countries radiated outward in all directions. In the year 1990, the countries of Thailand, Singapore, and Malaysia had a slight increase in carbon emissions due to their energy intensity. Of the three, Malaysia had the highest increase in emissions due to changes in economic structure. Brunei Darussalam appeared to have almost no change in its carbon emissions for ΔC_{struc}, even though it had a negative change in its ΔC_{int} value. Finally, Vietnam, Indonesia, and the Philippines appeared to have a negative change in economic structure. Of the three, Indonesia had a decrease in carbon emissions, due to its energy intensity, while the exact opposite can be said for Vietnam. As can be seen from previous figures, the countries were not as energy intensive back then, and thus they were relatively closer to the *x*-axis.

In the year 2018, however, the countries have stayed relatively closer to the origin compared to previous graphs involving the energy intensity. As usual, Vietnam had a dramatic increase in carbon emissions due to its energy intensity effect. In contrast, Malaysia managed to maintain a positive change in carbon emissions due to economic structure, while it decreased emissions from its energy intensity. Thailand slightly increased emissions because of its energy intensity, but had mitigated emissions through its economic structure. Cambodia, Lao PDR, Myanmar, and the Philippines all had decreased their emissions from both effects, with Cambodia having the greatest decrease in carbon emissions because of its economic structure and Myanmar having the greatest decrease in emissions because of its energy intensity. However, Indonesia had increased carbon emissions from both effects. Finally, the most notable and beneficial change can be seen for Singapore. It managed to reach a negative change in emissions for both effects by a significant amount, despite having almost the same ΔC_{struc} value as Malaysia in 1990. According to the pledge and document, released by the National Climate Change Secretariat in Singapore, the country had already pledged to reduce emission intensity by 36% starting from 2015 [40]. This pledge was supported by its results in the year 2012. During this time, its economy grew by 5.7% per annum, but the greenhouse gas emissions grew at a much lower rate of 2.1% compared to previous years.

3.4.2. Temporal Analysis

Figure 5b shows the trendlines exhibited by the ASEAN region from 1990 to 2018. From the graph, it can be seen that the trendline slope in the year 1990 was almost parallel to the *x*-axis, with a slight drop to the right of the plane. However, in the year 1995, the negative slope became more prominent. The slope almost stayed constant from 1995 to 2010, before notably dipping in 2015. Finally, in 2018, the region had returned back to the same slope it had in 2010. It can therefore be observed that the ASEAN region follows a negative slope, similar to Figure 2b. Combined with the distribution of countries and their position on the scatter plot, this simply means that a majority of the ASEAN nations have a more influential energy intensity effect than the economic structure effect. Additionally, they collectively had a decrease in carbon emissions because of changes in economic structure, compared to their energy intensity. It can be seen that the ASEAN eventually stabilized to a particular slope, where an increase in energy intensity correlated to a decrease in economic structure and vice versa.

3.5. Income Level vs. CO_2 Emissions (GNI per Capita vs. tCO_2 per Capita)

Figure 6 shows each country's GNI per capita plotted against their total CO_2 emissions per capita for the year 2018. It can be seen that there are six ASEAN members that classify as lower-middle income countries; namely Indonesia, Vietnam, Philippines, Cambodia, and Lao PDR. Thailand and Malaysia have upper-middle income economies, while Singapore and Brunei Darussalam are considered high income countries. The scatter plot clearly shows a significant divide between countries of high and low income, as well as the carbon emissions per capita for each. The lower-middle income countries are shown to have a less per capita emissions, and this increases as the income level increases. This finding is very interesting considering the massive overpopulation in countries like the Philippines and Indonesia. This suggests that, in low-middle income countries, the driver for emissions is more of industrial and economic activity, while it shifts more to human activity in higher-income countries. This emphasizes the importance of behavioral solutions to mitigating emissions, in addition to measures focusing on industrial and economic activities. This also suggests that policies need to be revisited and evolved, as the country's economy grows.

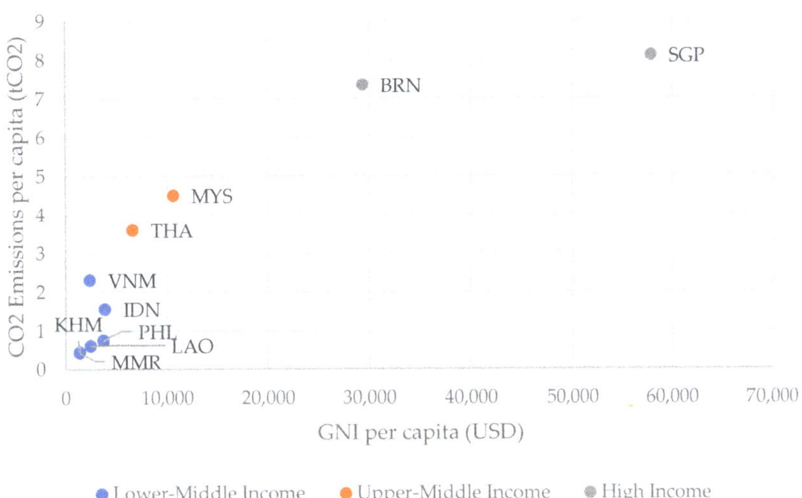

Figure 6. GNI per Capita vs. CO_2 emissions.

It must be noted that Indonesia has recently been reclassified to be an upper-middle income economy in the year 2019 [41]. However, given that there is no complete data regarding the energy consumption of all countries in the International Energy Agency as of writing this paper, analysis is only be limited to the year 2018.

Figure 6 above shows each country's GNI per capita plotted against their total CO_2 emissions per capita for the year 2018. It can be seen that there are six ASEAN members that classify as lower-middle income countries; namely Indonesia, Vietnam, Philippines, Cambodia, and Lao PDR. Thailand and Malaysia have upper-middle income economies, while Singapore and Brunei Darussalam are considered high income countries. The scatter plot clearly shows a significant divide between countries of high and low income, as well as the carbon emissions per capita for each. The lower-middle income countries are shown to have a less per capita emissions, and this increases as the income level increases. This finding is very interesting considering the massive overpopulation in countries like the Philippines and Indonesia. This suggests that, in low-middle income countries, the driver for emissions is more of industrial and economic activity, while it shifts more to human activity in higher-income countries. This emphasizes the importance of behavioral solutions to mitigating emissions, in addition to measures focusing on industrial and economic activities. This also suggests that policies need to be revisited and evolved, as the country's economy grows.

It must be noted that Indonesia has recently been reclassified to be an upper-middle income economy in the year 2019 [41]. However, given that there is no complete data regarding the energy consumption of all countries in the International Energy Agency as of writing this paper, analysis is only limited to the year 2018.

4. Discussion

The drivers to changing carbon emissions in each ASEAN country are summarized in Table 1. The values obtained from the calculations are listed down. For brevity, only the years of 2000 and 2018 are shown. This is because of the lack of data from Cambodia, Lao PDR, and Myanmar in 1990 and 1995. In the table, the highest and lowest values for each index are highlighted in red and green, respectively.

Table 1. Findings in the Study.

Country	2000				2018			
	ΔC_{pop}	ΔC_{act}	ΔC_{struc}	ΔC_{int}	ΔC_{pop}	ΔC_{act}	ΔC_{struc}	ΔC_{int}
Brunei Darussalam	−53,900	29,766	−725	−21,878	−119,689	47,230	−2023	−41,287
Cambodia	−17,088	−19,738	−2210	−10,981	−50,666	−49,485	−4752	2534
Indonesia	162,458	−13,949	−2230	1403	337,352	−15,896	323	−3990
Lao PDR	−28,034	−15,496	−3423	−2608	−73,450	−33,618	−2146	4455
Malaysia	−50,662	73,978	2036	−2551	−91,378	125,691	1950	−26,252
Myanmar	−2832	−45,832	−15,980	20,816	−11,282	−58,886	−2061	−16,935
Philippines	19,228	−13,358	−517	−4931	47,315	−30,947	−2820	−29,502
Singapore	−69,159	71,790	2281	−37,326	−174,585	181,263	−2046	−73,805
Thailand	15,561	38,437	−355	19,440	10,596	65,663	−779	20,045
Vietnam	21,556	−53,960	−4667	36,923	59,867	−143,680	−2514	174,717

Please note that these values were taken in comparison to the average of all countries, which was set as the hypothetical benchmark country. A high positive value simply means that a particular country is emitting more because of that index compared to the regional average. The same logic applies for a low, negative value in the change in carbon emissions. With this in mind, it can be seen that only Singapore in 2018 was able to reduce carbon emissions, due to population and energy intensity effects, compared to the rest of the ASEAN. The countries with the least carbon emissions were found to be Brunei Darussalam, Lao PDR, Myanmar, and Cambodia. These countries managed to have negative values for a majority of the effects as time progressed. Despite this, it was found that Indonesia, Malaysia, Vietnam, Thailand, and Philippines have the most notable contributions to greenhouse gas emissions as their indices range in the positive tens of thousands to even hundreds of thousands. Compared to the rest of the indices, the economic structure does not seem to contribute as much to the overall carbon emissions in the region. The most influential drivers in the ASEAN were found to be the energy intensity, economic activity, and population effects.

Based on the results found however, it can be seen that a majority of the DA5 countries such as Indonesia, Malaysia, Vietnam, the Philippines, and Thailand, have significantly decreased carbon emissions due to a few indices. Despite this progress, there was still a massive increase in carbon emissions as these countries have started shifting to one or two particular effects as can be seen in the temporal analysis. Given this, more focus should be done on Indonesia, Thailand, and Vietnam.

To begin with, Indonesia's population has been steadily increasing, up to more than 270 million people in the year 2019. This increase in population can lead to an increase in the need for more space and land for cultivation, as well as the building of facilities and buildings to accommodate for the increase in demand of supplies [42]. This issue, combined with an attractive demand of timber from China and Japan, has led to rampant illegal deforestation within the country [43]. Additionally, the burning of forests is a common practice, which destroys carbon sinks and thus emits greenhouse gases into the air [44]. This practice has made Indonesia the eighth largest emitter of carbon dioxide in the world in 2020, according to the World Resources Institute [45]. It is therefore recommended, including the rest of the ASEAN countries, to actively make illegal logging and deforestation much less attractive in terms of concessions, ease, and income. It is further recommended to monitor existing regulations regarding deforestation, as well as implementing stricter tax and penalties to those who repeatedly offend the law. Additionally, it is best to remind the countries, especially those in the DA5, of their obligations and targets in order to reduce emissions from deforestation and forest degradation, or REDD+, as stated in the UN [46].

Thailand's population, economic activity, and energy intensity have significant contributions to its carbon emissions. Though not as high as Indonesia's, these values are more evenly distributed. Despite the country having successfully decreased its emissions in relation to later years and to the ASEAN, improvements can still be made. According to the UNFCC, Thailand's efforts in mitigating carbon emissions involve increasing the areas of permanent trees by 72,000 hectares, and reducing land for open burning by 24,000 hectares [47]. The same source states that cooperation with the Kyoto Protocol's Clean Development Mechanism led to the approval of more than 30 projects, which have the potential of reducing carbon emissions by about 2 million tons. Additionally, it was found that subsidizing the acquisition and use of fossil fuels is actually inefficient, as the cost is actually greater to the individual consumer compared to society [38] (p. 104). It is, therefore, recommended to increase mitigation targets and ambition levels in order to effectively hasten the reduction in GHG emissions since it does not only benefit the world as a whole, but it also costs less than current energy policies towards fossil fuels.

Vietnam has repetitively shown a massive increase in energy intensity in more recent years. This is backed by a study by Le, where it was stated that pre-existing government policies indirectly helped the energy sector by funding fossil energy costs through the use of government-owned businesses and facilities [48]. This has resulted in the inefficient use of energy in the production processes, thus increasing the amount of carbon emissions. The ADB even further stated that the largest overall source of mitigating carbon emissions is through efficient energy use [38]. It is therefore imperative to conduct research and studies in order to analyze ongoing energy use patterns and economic behavior in order to, not only formulate appropriate management practices that could effectively allocate end-use energy for each sector, but to also implement suitable environmental and energy policies for all countries.

Finally, the performance of Singapore must be emulated and acknowledged. The country has actually become a center of research and development for Southeast Asia [49]. As can be seen from the study, Singapore was able to drastically lessen carbon emissions from both population and energy intensity. According to the National Climate Change Secretariat in Singapore, the country's manufacturing sector has cooperated with the National Environment Agency (NEA) and the Economic Development Board (EDB) in order to improve its energy efficiency by about 1% to 2% per year [40]. Additionally, efforts have been done with existing policies, such as the Energy Conservation Act (ECA). This policy keeps energy intensive facilities in check and reviewed regularly in order to monitor energy use patterns. This helps the government and various sectors not only predict future energy trends, but also actively adjust and make informed decisions about energy allocation and use. Additionally, research and development departments within each country should implement, and standardize, CO_2 emission limits that are feasible with respect to the aforementioned pledges and obligations. It is recommended to study each sector and implement emission limits based on the nature of their activities. An example would be the EU implementing emissions standards for the transportation sector such as private cars and light vehicles [50]. Overall, further studies would help the country make their energy use much more efficient and release less carbon dioxide than they need to. It is therefore recommended for the rest of the ASEAN countries to work closely with their existing institutions and abide by their implemented policies strictly. Should these not work, it is then advised to revise policies and allocate government funding to the formation of agencies that monitor the country's economic activity, structure, and energy consumption.

In terms of the effects studied, it can be seen that more successful countries, like Singapore, Brunei Darussalam, and Malaysia, all have low energy intensity effects with exception to Thailand. The high population effects of Indonesia and the Philippines, as well as the high energy intensity effects of Vietnam, are clearly reflected in the plot as well. The trend exhibits a behavior similar to the findings in a study by Tucker [51], where it was found that the rate of carbon emissions decreases as the income per capita increases. More interestingly, it was stated that higher income countries have a relatively easier process of implementing GHG mitigation projects. In contrast, other developing countries may have a more difficult implementation, given that they may have to concentrate on other projects that would, instead, accelerate economic growth. The results, therefore, further highlight the disparity between the ASEAN member nations in terms of development, as well as their individual trends and directions for economic growth.

In relation to previously conducted regional-level studies, the results corroborate the fact that changes in energy mix have a strong impact on carbon dioxide emissions, as can be seen in a study by Fernández-Gonzales et al. in the European Union [52]. More particularly, changes in energy mix led by switching to cleaner fuels was also proven to significantly lessen carbon emissions. A similar decomposition analysis on energy consumption and economy was also done on the Eastern, Western, Northern, and Southern parts of Europe [53]. Given that these studies were also conducted on a group of countries, albeit in different regions, they also came to the conclusion that it is necessary to formulate more specific policies for each country. This is because each country has their own strengths and weaknesses that dictate how well they could respond to climate changes and implement environmental policies. Given that some countries in the ASEAN region are still in the process of development, it is best to consider how well they would adapt to the required climate goals.

Additionally, the study conducted shows a more graphical representation of each country's evolution, in relation to each other and with respect to time. Previous studies on the Latin America and Caribbean (LAC) region [54], South Africa [55], and Organization for Economic Cooperation and Development (OECD) member countries [56], all came to the conclusion that population and intensity effects were main drivers to carbon emissions through temporal analyses combined with Tapio decoupling methods. This present study was able to easily determine the significance of both effects, in addition to important insights on economic activity effects, through simple decomposition methods. The researchers, therefore, believe that the proposed spatiotemporal methodology could be a simpler, and quicker, alternative to studying drivers to carbon emissions on a regional and international level.

5. Conclusions

In this section, the policy recommendations that were discussed are summarized in bullet form below, followed by the conclusion that talks about the key findings in the study.

With the looming dangers of climate change, it is imperative to research energy patterns and build upon existing policies. In conducting the study, a spatiotemporal analysis was used based on the logarithmic mean Divisia index method in order to determine drivers to carbon emissions in the ASEAN region. With this, the performance and emissions of each country were compared with each other. Additionally, the overall evolution of carbon emissions in the entire region was analyzed in order to, not only, examine its general direction but also to prepare and be aware of the possible courses of action to either abate emissions or maintain low emission values.

The most influential drivers to carbon emissions in more recent years are the population, economic activity, and energy intensity effects. The ASEAN nations managed to decrease carbon emissions through changes in their respective economic structures. Singapore had the greatest decrease from both population and intensity effects, while the majority of its emissions come from its economic activity. Indonesia was found to have an alarming contribution due to its population growth, suggesting significant human activity. Additionally, it was found that Vietnam was the most energy intensive country in ASEAN up until 2018. The spatial and temporal analyses also show that Brunei Darussalam, Myanmar, Cambodia, and Lao PDR emit the least carbon due to all indices in relation to the rest of the ASEAN. Finally, the Philippines, Malaysia, and Thailand, which are members of the DA5, seem to be on a transition period to less carbon intensive economic sectors.

Both spatial and temporal analyses have provided much needed insight on the overall performance of each country as well as the general evolution of the region's emissions.

From these insights, the researchers were not only able to benchmark the best and worst performing countries, but also derive possible courses of action for the region.

1. Reiterating commitments and obligations in reducing emissions from deforestation and forest degradation (REDD+) and their Intended Nationally Determined Contributions (INDC) as submitted to the UN;
2. Implementing strict penalties and tougher measures [57] for various sectors and industries that continually emit more than the recommended emission limits with respect to past obligations and pledges;
3. Actively disincentivizing illegal logging and deforestation efforts in terms of ease, concessions, and income, given that these actions greatly contribute to GHG emissions in the ASEAN;
4. Increasing levels of ambition in mitigating carbon emissions, in order to hasten its reduction and make countries more economically efficient;
5. Conducting in-depth analyses and studies on each country's energy use patterns and economic behavior in order to facilitate energy efficiency and formulate timely carbon reduction policies and procedures;
6. Encouraging research and development efforts which monitor consumption patterns and revise policies accordingly.

Alternatively speaking, it is recommended for countries to play to their strengths by studying their own energy consumption and economic performance. Through this, specific practices and policies can be made. The results obtained from the study support the previous sentiment from related literature that each country is unique, with their own strengths and weaknesses.

Additionally, aside from their differences, each country is in different stages of development. Singapore and Brunei are in the developed stage already, while countries like Vietnam and Thailand are still in the rapidly developing stage. Other countries were observed to follow different trajectories as well. In conclusion, the researchers therefore hope that the ASEAN would recognize that these different countries would need different kinds of support. Reiterating the Treaty of Amity and Cooperation in Southeast Asia, member nations are encouraged to actively cooperate with, and support, neighboring countries in order to guide the region in the transition to sustainable practices.

Author Contributions: Conceptualization, N.S.A.L. and E.B.F.L.J.; methodology, N.S.A.L. and E.B.F.L.J.; formal analysis, N.S.A.L. and E.B.F.L.J.; writing—original draft preparation, E.B.F.L.J.; writing—review and editing, N.S.A.L. All authors have read and agreed to the published version of the manuscript.

Funding: The APC was funded by the Office of the Vice Chancellor for Research and Innovation of De La Salle University, Manila, Philippines. At the time of writing, Edwin Bernard F. Lisaba was receiving scholarship funding from the Department of Science and Technology—Engineering Research and Development for Technology Program for his Master of Science in Mechanical Engineering degree.

Institutional Review Board Statement: Not applicable.

Informed Consent Statement: Not applicable.

Data Availability Statement: The data presented in this study are openly available from the World Bank and the International Energy Agency.

Conflicts of Interest: The authors declare no conflict of interest.

References

1. NOAA National Centers for Environmental Information. State of the Climate: Global Climate Report for Annual 2020. Available online: https://www.ncdc.noaa.gov/sotc/global/202013 (accessed on 30 April 2021).
2. Intergovernmental Panel on Climate Change. *Fourth Assessment Report*; Intergovernmental Panel on Climate Change: Geneva, Switzerland, 2007.
3. Intergovernmental Panel on Climate Change. *Fifth Assessment Report*; Intergovernmental Panel on Climate Change: Geneva, Switzerland, 2014.

4.   World Health Organization. Climate Change and Health. Available online: https://www.who.int/news-room/fact-sheets/detail/climate-change-and-health (accessed on 30 April 2021).

5.   Overview, Association of Southeast Asian Nations. Available online: https://asean.org/asean/about-asean/overview/ (accessed on 30 April 2021).

6.   Hill, H. Southeast Asian studies: Economics. *Int. Encycl. Soc. Behav. Sci.* **2001**, 14670–14676. [CrossRef]

7.   Bhaskaran, M. *Getting Singapore in Shape: Economic Challenges and How to Meet Them*; Lowy Institute for International Policy: Sydney, Australia, 2018.

8.   Findlay, R.; Park, C.Y.; Verbiest, J.P. Myanmar: Unlocking the potential—A strategy for high, sustained, and inclusive growth. *ADB Econ. Work. Paper Ser.* **2015**, *437*. [CrossRef]

9.   Treaty of Amity and Cooperation in Southeast Asia Indonesia, 24 February 1976, Association of Southeast Asian Nations. Available online: https://asean.org/treaty-amity-cooperation-southeast-asia-indonesia-24-february-1976/ (accessed on 30 April 2021).

10. NOAA National Centers for Environmental Information. Frequently Asked Questions. Available online: https://www.aoml.noaa.gov/hrd-faq/#1569507388495-a5aa91bb-254c (accessed on 30 April 2021).

11. Asian Development Bank. *The Economics of Climate Change in Southeast Asia: A Regional Review*; Asian Development Bank: Mandaluyong City, Philippines, 2008; p. 24.

12. Nunti, C.; Somboon, K.; Intapan, C. The impact of climate change on agriculture sector in ASEAN. In Proceedings of the Conference Series, Guilin, China, 25 November 2020.

13. National Disaster Risk Reduction and Management Council. NDRRMC Update: Sitrep No. 26 re Preparedness Measures and Effects for Typhoon "Ulysses" (I.N. Vamco). Available online: https://ndrrmc.gov.ph/attachments/article/4138/SitRep_no_26_re_TY_ULYSSES_as_of_06DEC2020.pdf (accessed on 30 April 2021).

14. ASEAN Cooperation on Climate Change, Association of Southeast Asian Nations. Available online: https://environment.asean.org/asean-working-group-on-climate-change/ (accessed on 30 April 2021).

15. De Boer, P.; Rodrigues. Decomposition analysis: When to use which method? *Econ. Syst. Res.* **2020**, *32*, 1–28. [CrossRef]

16. Ang, B.W.; Zhang, F.Q. A survey of index decomposition analysis in energy and environmental studies. *Energy* **2000**, *25*, 1149–1176. [CrossRef]

17. Hoekstra, R.; van den Bergh, J. Structural decomposition analysis of physical flows in the economy. *Environ. Resour. Econ.* **2002**, *23*, 357–378. [CrossRef]

18. Sumabat, A.K.; Lopez, N.S.; Yu, K.D.; Hao, H.; Li, R.; Geng, Y.; Chiu, A.S.F. Decomposition analysis of Philippine CO_2 emissions from fuel combustion and electricity generation. *Appl. Energy* **2016**, *164*, 795–804. [CrossRef]

19. Nie, H.; Kemp, R. Why did energy intensity fluctuate during 2000–2009? A combination of index decomposition analysis and structural decomposition analysis. *Energy Sustain. Dev.* **2013**, *17*, 482–488. [CrossRef]

20. Jurkėnaitė, N.; Baležentis, T. The 'pure' and structural contributions to the average farm size growth in the EU: The index decomposition approach. *Ecol. Indic.* **2020**, *117*, 117. [CrossRef]

21. Chong, C.H.; Tan, W.X.; Ting, Z.J.; Liu, P.; Ma, L.; Li, Z.; Ni, W. The driving factors of energy-related CO_2 emission growth in Malaysia: The LMDI decomposition method based on energy allocation analysis. *Renew. Sustain. Energy Rev.* **2019**, *115*, 115. [CrossRef]

22. Chontanawat, J.; Wiboonchutikula, P.; Buddhivanich, A. An LMDI decomposition analysis of carbon emissions in the Thai manufacturing sector. *Energy Rep.* **2020**, *6*, 705–710. [CrossRef]

23. Boey, A.; Su, B. Low-carbon Transport Sectoral Development and Policy in Hong Kong and Singapore. *Energy Proc.* **2014**, *61*, 313–317. [CrossRef]

24. Li, J.; Chen, Y.; Li, Z.; Huang, X. Low-carbon economic development in Central Asia based on LMDI decomposition and comparative decoupling analyses. *J. Arid Land* **2019**, *11*, 513–524. [CrossRef]

25. Wang, F.; Wang, C.; Chen, J.; Li, Z.; Li, L. Examining the determinants of energy-related carbon emissions in Central Asia: Country-level LMDI and EKC analysis during different phases. *Environ. Dev. Sustain.* **2020**, *22*, 7743–7769. [CrossRef]

26. Zhang, J.; Fan, Z.; Chen, Y.; Gao, J.; Liu, W. Decomposition and decoupling analysis of carbon dioxide emissions from economic growth in the context of China and the ASEAN countries. *Sci. Total Environ.* **2020**, *714*. [CrossRef]

27. Hill, H.; Menon, J. *ASEAN Economic Integration: Features, Fulfillments, Failures and the Future*; Asian Development Bank: Mandaluyong City, Philippines, 2010.

28. Data and Statistics, International Energy Agency. Available online: https://www.iea.org/data-and-statistics?country=WORLD&fuel=Energy%20supply&indicator=TPESbySource (accessed on 30 April 2021).

29. Intergovernmental Panel on Climate Change (IPCC). *1996 IPCC Guidelines for National Green House Gas Inventories*; Intergovernmental Panel on Climate Change: Geneva, Switzerland, 1996.

30. Environmental Protection Agency. *Emission Factors for Greenhouse Gas Inventories*; Environmental Protection Agency: Washington, DC, USA, 2014.

31. U.S. Energy Information Administration (EIA). How Much Carbon Dioxide is Produced When Different Fuels are Burned? 2020. Available online: https://www.eia.gov/tools/faqs/faq.php?id=73&t=11 (accessed on 30 April 2021).

32. Ang, B.W. The LMDI approach to decomposition analysis: A practical guide. *Energy Policy* **2005**, *33*, 867–871. [CrossRef]

33. World Bank Open Data, The World Bank. Available online: https://data.worldbank.org/ (accessed on 30 April 2021).

34. Vu, K. Sources of Singapore's economic growth 1990-2009. *Macroecon. Rev.* **2010**, *4*, 66–81. [CrossRef]

35. Lee, C. Globalization and economic development: Malaysia's experience. *ERIA Discuss. Paper Ser.* **2019**, *307*, 2–6.
36. Cima, R. Vietnam's economic reform: Approaching the 1990s. *Asian Surv.* **1989**, *29*, 786–799. [CrossRef]
37. The World Bank. Vietnam's Economic Prospects Improve Further, with GDP Projected to Expand by 6.8 Percent in 2018. Available online: https://www.worldbank.org/en/news/press-release/2018/06/14/vietnams-economic-prospect-improves-further-with-gdp-projected-to-expand-by-68-percent-in-2018#:~{}:text=Hanoi%2C%20June%2014%2C%202018%20%2D,accompanied%20by%20broad%20macroeconomic%20stability.&text=Real%20GDP%20expanded%20nearly%207.4,at%203.1%20percent%20in%202018 (accessed on 30 April 2021).
38. Raitzer, D.A.; Bosello, F.; Tavoni, M.; Orecchia, C.; Marangoni, G.; Samson, J.N.G. *Southeast Asia and the Economics of Global Climate Stabilization*; Asian Development Bank: Mandaluyong City, Philippines, 2015; pp. 21–24.
39. Sjahrir. The Indonesian economy facing the 1990s: Structural transformation and economic deregulation. *Southeast Asian Aff.* **1990**, 117–131. [CrossRef]
40. National Climate Change Secretariat. *Singapore's Climate Action Plan: Take Action Today, for a Carbon-Efficient Singapore*; Prime Minister's Office: Singapore, 2016; pp. 6–7.
41. Data Help Desk, The World Bank. Available online: https://datahelpdesk.worldbank.org/knowledgebase/articles/906519-world-bank-country-and-lending-groups (accessed on 23 May 2021).
42. Eldeeb, O.; Prochazka, P.; Maitah, M. Causes for deforestation in Indonesia: Corruption and palm tree plantation. *Asian Soc. Sci.* **2015**, *11*, 120. [CrossRef]
43. Makkarennu, M.; Nakayasu, A.; Osozawa, K.; Ichikawa, M. An analysis of the demand market of Indonesia plywood in Japan. *Int. J. Sustain. Future Hum. Secur.* **2015**, *2*, 2–7. [CrossRef] [PubMed]
44. Jaenicke, J.; Rieley, J.; Mott, C.; Kimman, P. Determination of the amount of carbon stored in Indonesian peatlands. *Geoderma* **2008**, *147*, 151–158. [CrossRef]
45. World Resources Institute. Available online: https://www.wri.org/insights/interactive-chart-shows-changes-worlds-top-10-emitters (accessed on 30 April 2021).
46. UN-REDD Programme. Available online: https://www.un-redd.org/ (accessed on 30 April 2021).
47. Office of Natural Resources and Environmental Policy and Planning. *Thailand's Second National Communication under the United Nations Framework Convention on Climate Change*; Ministry of Natural Resources and Environment: Bangkok, Thailand, 2011; pp. 15–16.
48. Le, P.V. Energy demand and factor substitution in Vietnam: Evidence from two recent enterprise surveys. *J. Econ. Struct.* **2019**, *8*, 35. [CrossRef]
49. Singapore: A Global Hub for Innovation, Forbes. Available online: https://www.forbes.com/custom/2018/08/13/singapore-a-global-hub-for-innovation/ (accessed on 30 April 2021).
50. International Council on Clean Transportation. EU CO_2 Emission Standards for Passenger Cars and Light-Commercial Vehicles. Available online: https://theicct.org/sites/default/files/publications/ICCTupdate_EU-95gram_jan2014.pdf (accessed on 30 April 2021).
51. Tucker, M. Carbon dioxide emissions and global GDP. *Ecol. Econ.* **1995**, *15*, 215–223. [CrossRef]
52. Fernández González, P.; Landajo, M.; Presno, M.J. Tracking European CO_2 emissions through LMDI (logarithmic-mean Divisia index) decomposition. The activity revaluation approach. *Energy* **2014**, *73*, 741–750. [CrossRef]
53. Moutinho, V.; Moreira, A.; Silva, P. The driving forces of change in energy-related CO_2 emissions in Eastern, Western, Northern and Southern Europe: The LMDI approach to decomposition analysis. *Renew. Sustain. Energy Rev.* **2015**, *50*, 1485–1499. [CrossRef]
54. De Oliveira-De Jesus, P. Effect of generation capacity factors on carbon emission intensity of electricity of Latin America & the Caribbean, a temporal IDA-LMDI analysis. *Renew. Sustain. Energy Rev.* **2019**, *101*, 516–526. [CrossRef]
55. Lin, S.; Beidari, M. Energy Consumption Trends and Decoupling Effects between Carbon Dioxide and Gross Domestic Product in South Africa. *Aerosol Air Qual. Res.* **2015**, *15*. [CrossRef]
56. Chen, J.; Wang, P.; Cui, L.; Huang, S.; Song, M. Decomposition and decoupling analysis of CO_2 emissions in OECD. *Appl. Energy* **2018**, *231*, 937–950. [CrossRef]
57. Vallés-Giménez, J.; Zárate-Marco, A. A Dynamic Spatial Panel of Subnational GHG Emissions: Environmental Effectiveness of Emissions Taxes in Spanish Regions. *Sustainability* **2020**, *12*, 2872. [CrossRef]

 *sustainability*

Article

Analysis of Power Generation for Solar Photovoltaic Module with Various Internal Cell Spacing

June Raymond L. Mariano [1,2], **Yun-Chuan Lin** [1], **Mingyu Liao** [3] **and Herchang Ay** [1,*]

[1] Mold and Die Engineering Department, National Kaohsiung University of Science and Technology, Kaohsiung City 807618, Taiwan; juneraymond_mariano@tup.edu.ph (J.R.L.M.); kylemyoho@gmail.com (Y.-C.L.)

[2] Mechanical and Allied Engineering Department, Technological University of the Philippines—Taguig, Taguig City 1630, Philippines

[3] Department of Business Administration, National Taipei University of Business, Taipei 100025, Taiwan; mygliao@ntub.edu.tw

* Correspondence: herchang@nkust.edu.tw

Abstract: Photovoltaic (PV) systems directly convert solar energy into electricity and researchers are taking into consideration the design of photovoltaic cell interconnections to form a photovoltaic module that maximizes solar irradiance. The purpose of this study is to evaluate the cell spacing effect of light diffusion on output power. In this work, the light absorption of solar PV cells in a module with three different cell spacings was studied. An optical engineering software program was used to analyze the reflecting light on the backsheet of the solar PV module towards the solar cell with varied internal cell spacing of 2 mm, 5 mm, and 8 mm. Then, assessments were performed under standard test conditions to investigate the power output of the PV modules. The results of the study show that the module with an internal cell spacing of 8 mm generated more power than 5 mm and 2 mm. Conversely, internal cell spacing from 2 mm to 5 mm revealed a greater increase of power output on the solar PV module compared to 5 mm to 8 mm. Furthermore, based on the simulation and experiment, internal cell spacing variation showed that the power output of a solar PV module can increase its potential to produce more power from the diffuse reflectance of light.

Keywords: light trapping; zero-depth concentrator; light reflection; internal-cell spacing

Citation: Mariano, J.R.L.; Lin, Y.-C.; Liao, M.; Ay, H. Analysis of Power Generation for Solar Photovoltaic Module with Various Internal Cell Spacing. *Sustainability* **2021**, *13*, 6364. https://doi.org/10.3390/su13116364

Academic Editor: Wei-Hsin Chen

Received: 8 May 2021
Accepted: 26 May 2021
Published: 3 June 2021

Publisher's Note: MDPI stays neutral with regard to jurisdictional claims in published maps and institutional affiliations.

1. Introduction

Photovoltaic (PV) systems directly convert solar energy into electricity. The main structure of a solar PV system is the PV cell, which is a semiconductor device that converts solar energy into direct current (DC) electricity. PV cells are interconnected to form a PV module, typically with a capacity of 290 watts (W) to almost 400 W. PV modules and other components such as inverters, energy storage, electrical and mechanical equipment are assembled to form a PV system [1]. PV systems are extremely modular such that a PV module can be linked together to provide a few watts to hundreds of megawatts (MW) of power [2]. Solar PV cells are grouped into modules with transparent glass in front, a weatherproof material for the back, and regularly surrounded by a frame [3]. When designing a solar cell module, it is necessary to consider the arrangement of the cell sheets. The solar cell module is composed of multiple interconnected solar cells connected in series and the module is designed and packaged. To increase the packing density of a solar module, the arrangement of solar cells to the module should be maximized so that the actual area occupied by the cells over the total area of the solar module must be high [4]. Thus, more occupied areas and less space mean more packing density. Cells in a module with a white rear surface can also provide marginal increases in output via the light reflection effect [5]. This phenomenon shows the power enhancement in solar PV module output due to the light scattered in the backsheet at angles greater than the

Sustainability **2021**, *13*, 6364. https://doi.org/10.3390/su13116364 https://www.mdpi.com/journal/sustainability

critical angle of the cover glass [6]. The solar PV cells that are interconnected together in the module are then combined to form modules, arrays, and systems.

The silicon solar cells have been identified as the most viable option suitable for large volume production [7]. However, it has been reported that the continual generation of electricity by PV modules, manufactured using this type of cell, in the field for a minimum life span of 20 years has been a concern [8–10]. One of the key challenges is the conversion efficiency improvement of the existing module structure while increasing the light absorption on reflection from the backsheet of the module. To use the benefits of theory, some module designs take advantage of the light passing between the cells. The scattered light reflection from the rear back of the module can still be absorbed by solar cells and contribute to both objectives at once. Several light-trapping methods have been implemented over the last few years to enhance the light absorption within the solar cell that is much smaller compared to the material's intrinsic absorption length [11–16]. The average efficiency of solar panels during this period has increased but mainly because of higher performance solar cells.

Previous research has analyzed the performance of photovoltaic modules with internal reflection, encapsulation, internal and external light-trapping structures, photonic nanostructure, front side metallization, and cell interconnection that increase both performances and efficiency of the solar cell [17–22]. Despite this, past research did not take into account the layout of the module rear side, such as the effect of different cell spacing on light absorption and increase in power output of the module. For this reason, the study proposed a different approach whereby light falling between the cells is deliberately steered using the common solar PV module layout. In this way, three solar PV modules with three different cell spacings will be compared for power output.

Therefore, the researchers used an optical simulation engineering program, solar PV module simulator, and outdoor testing facility to analyze light absorption and solar PV module power output. The researchers considered three different cell spacing such as 2 mm, 5 mm, and 8 mm that would fit the available module area. Then, an analysis of the scattered light absorption from the rear side of the module was reflected to the upper surface of the glass and air interface until it reached the solar cell top surface. At the same time, output power produced by three different modules was evaluated. In this manner, the power production capacity of solar PV modules was optimized.

This paper is arranged as follows: Section 2 briefly describes optical principles and solar cell measurement. Section 3 describes the general approach in methodology. Section 4 provides results and discussions of the simulation and experiments, followed by a conclusion in Section 5.

2. Principles of Optical Principles and Principles of Solar Cell Measurement

2.1. Basic Principle of Optics

The physical phenomena are related to the generation, transmission, processing, detection, and application of light, as well as technological developments that cover a wide range of fields. The optical processing model is composed of physical optics which simply refers to seeing light as optics in an electromagnetic wave. Its physical optics can establish a model of amplitude and phase through the optical system. By this principle, it can provide a qualitative interpretation of the linear and spherical propagation of light waves and derive the law of reflection and the law of refraction [23]. Then, geometrical optics describes the light propagation of rays. A ray in geometric optics is useful for approximating the path along which the light propagates under certain circumstances [24]. Lastly, quantum optics are energy packets called photons. Their energy (E) can be expressed by:

$$E = hf \tag{1}$$

where the Planck constant (h) is higher than the frequency of light (f). The concept of photons can be used to accurately explore the role of light and matter, optoelectronic semiconductor components and photodetectors, laser systems, and quantum theory [25].

In the application of optical principles, there are roughly applications such as reflection, refraction, total internal reflection (TIR), forward scatter, backward scatter, and absorption. Both belong to the field of optics of geometric optics. Reflection can be divided into three forms: specular reflection, spread reflection, and diffuse reflection [26,27]. Specular reflection refers to the angle of incidence of light equal to the angle of reflection, spread reflection occurs on an uneven surface and the reflected light exceeds an angle, and the reflected angle of all reflected light is the same as the incident angle. Diffuse-type reflection, sometimes referred to as "Lambertian scattering" or "Diffusion", occurs on rough or matte surfaces with many different angles of reflected light [28]. Specular reflection occurs when the light shines on a perfectly smooth surface obeying the law of reflection, the angle of incidence (the angle between the incident ray and the normal of the vertical surface, and incident angle θ_1 is equal to the angle of reflection), the angle between the reflected ray and the normal to the vertical surface, reflected angle θ_2. When the light reflects on a rough surface, the light will immediately reflect or penetrate in many different directions on the material of the backlight.

Refraction is when light travels from one material to another, such as from air to glass, the light is refracted, that is, the light bends and changes speed [29]. The refraction depends on two factors. One is the incident angle, represented by the symbol θ_1, the other is the refractive index of the material, denoted by the letter n, and expressed by:

$$n = \frac{c}{v},\tag{2}$$

denotes the refractive index equal to the speed of light in the vacuum (c), and the speed of light (v) in the material rather than the glazing.

The speed of light in air is almost the same as the speed of light in a vacuum, and the refractive index of almost all other substances is greater than 1 because the speed of light passing through these substances is reduced [30]. The relationship between the angle of incidence and the angle of reflection of light passing through the air and the refraction through the glass is called Snell's law that is expressed as:

$$n_1 sin\theta_1 = n_2 sin\theta_2,\tag{3}$$

where n_1 is the refractive index of medium 1, n_2 is the refractive index of medium 2, θ_1 is the incident angle of light, and θ_2 is the angle of refraction of light. Total reflection light travels from a material with a higher refractive index to a material with a lower refractive index, and as the angle of incidence increases, the refracted light will deviate from the normal. If the angle of incidence increases, all the light will be reflected into the interior of the material.

Absorption is a phenomenon when the incident light is absorbed, and many materials absorb light of a specific wavelength, called selective absorption [31]. The absorption of light by materials can be expressed by Lambert's law of absorption:

$$I = I_0 e^{-\alpha x},\tag{4}$$

where I_0 is the incident light intensity, I is the transmitted light intensity, α is the material absorption coefficient, and x is the material thickness. From Equation (4), homogenous materials of the same thickness have the same absorption rate of light.

2.2. Solar Cell Measurement

The voltammetric characteristic curve is when the operating current of a solar cell under short-circuit conditions, this is called short-circuit current density (I_{sc}). The short-circuit photocurrent is equal to the absolute number of a photon that is converted into electron-hole pairs [32]. The output voltage of the solar cell under open-circuit conditions is called open-circuit voltage. Open-circuit voltage (V_{oc}) is the voltage that the solar cell is exposed to when the load is infinite ($R{\sim}\infty$), the voltage that is measured when the external current is disconnected, which is the maximum voltage that the photocell can generate. Fill factor (FF) in the I-V characteristic curve under illumination, the maximum output power is equal to the product of the current density corresponding to the maximum voltage [33]. Photoelectric conversion efficiency is the total cell efficiency that represents the performance of a solar cell, defined as the ratio of the maximum output power of a solar cell to the power of incident light [34]. These solar cell measurements were used in the solar cell simulator.

Measurement Principle of Solar Cell

The thermal and electrical properties of solar cell modules are indicated in their module product specifications. This information is a value obtained from standard test conditions (STC): irradiance 1000 W-m^{-2}, module temperature 25 °C, and AM 1.5. However, the operation of the solar cell module power generation system erected on site is rarely able to measure the power production capacity under STC conditions [35]. The reasons are as follows:

1. Accurate solar cell operating temperature is unknown.
2. When the incident irradiance reaches 1000 W-m^{-2}, the solar cell temperature of the module is higher than 25 °C.
3. The incident sunlight is not perpendicular to the solar cell module.

Under actual operating conditions, the module output is affected by various environmental conditions such as irradiance, temperature, spectrum, and angle of incidence.

The main content includes all parameters related to the aging of the solar module board, and different qualification tests are carried out based on the weather resistance of the simulated materials. Among them, the measurement of the IEC 61215, section 10.4 measure of temperature coefficient and IEC 61215, section 10.5 Nominal Operating Cell Temperature (NOCT), determines the power generation characteristics of the solar cell module in outdoor operation and are also important parameters in expressing the relationship between electrical output and temperature change. The power generation system is designed in a real working environment, and the electrical output that can be obtained in the outdoor operation can be calculated according to the temperature coefficient and NOCT [36]. The electrical I-V characteristics of solar cells have a critical impact on their output power. Since the semiconductor is very sensitive to temperature, the overall efficiency is reduced for every 1 °C increase in temperature of the solar cell above 25 °C. Therefore, before establishing the I-V characteristics of the solar cell, it is necessary to understand the temperature coefficient of the solar cell module [37].

The solar cell module equivalent circuit, combined with the semiconductor P-N junction characteristics, can obtain the output voltage and current equation of the solar cell module. The photovoltaic module is a power generation principle that is based on the photovoltaic effect, which converts light energy into electrical energy, including a photo-current source, connected diodes, and series and parallel resistors, then connected to the load.

The maximum output power (P_{max}) of a solar cell is shown as:

$$P_{max} = V_{oc} I_{sc} FF, \tag{5}$$

when the voltage and current of the solar cell module change under different irradiation amounts, the open-circuit voltage of the same environment is almost unchanged, and the short-circuit current will change significantly, so the output power and the maximum power point will also change significantly. The equations show that the module voltage and power change under different irradiation amounts, the open-circuit voltage does not change much when the irradiation intensity changes. But the maximum current generated will vary considerably. The power and maximum power point will change accordingly. Under the fixed irradiation intensity, when the temperature increases, the open-circuit voltage of the solar cell will decrease, and the short circuit current will rise slightly. The output power will decrease slightly with increasing temperature, and the maximum power output will decrease with increasing temperature. The maximum power value will also change linearly corresponding to temperature change. The rise of module voltage and power in temperature will cause the solar cell output power to decrease, so the temperature of the working environment will directly affect the efficiency of the solar cell.

3. Method

The method is composed of three major stages that are shown in Figure 1. First is the optical simulation analysis of the structural part of the solar cell module. The second is the solar module design, fabrication, and testing in indoor environments. The third is the outdoor test platform construction, monitoring, and data analysis. The optical parameter is initially established by optical simulation software that is shown in Figure 2. The glass microstructure is observed using a scanning electron microscope (SEM). Then, simulation is used to explore the ray tracing inside the solar cell with the glass surface microstructure and with different cell spacing. The solar PV module is designed with a half-frame module that is composed of 30 solar cells and a series of tests at different temperatures is performed. After the experiment is completed, the module is packaged into a full-frame module for STC measurement. Finally, a test is conducted to analyze and monitor the effects of the outdoor photoelectric conversion of power at different angles and cell spacing.

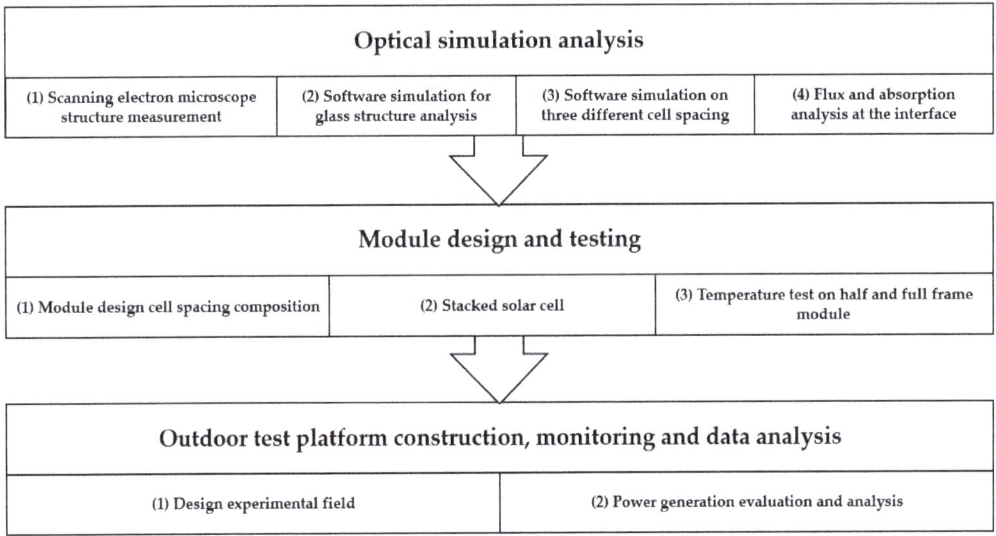

Figure 1. Research paradigm.

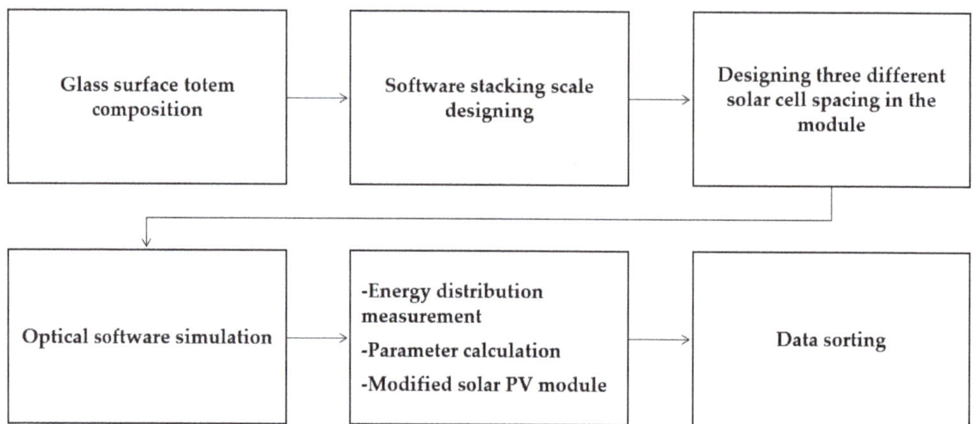

Figure 2. Optical simulation flow chart.

The multicrystalline cell with three busbars is used in the study. The dimension of the cell of 156 mm × 156 mm is shown in Figure 3. The solar cell will be connected in series to form a full-length module that is composed of 60 pieces of solar cell that are shown in Figure 4 and a half-frame module in Figure 5 that is composed of 30 pieces of a solar cell. This design is used to analyze under optical simulation, indoor testing, outdoor testing. The microstructure of the glass surface is measured by an electron microscope with a dimension of 0.8333 mm in length and 0.525 mm wide. It has an elliptical hemispherical protrusion surface facing the cell. The material parameters in Table 1 for the optical simulation software are used. Then, draw the solar cell array with the fixed backplate with different cell spacing to facilitate immediate simulation from replacing the glass, as shown in Figure 6. The actual design of the solar PV module is shown in Figure 7 that will be used in outdoor testing.

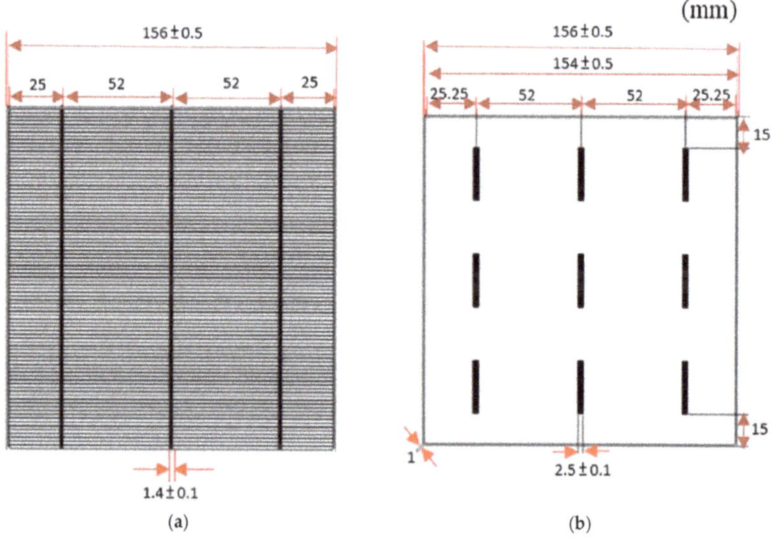

Figure 3. Multicrystalline solar cell size: (**a**) frontside; (**b**) backside.

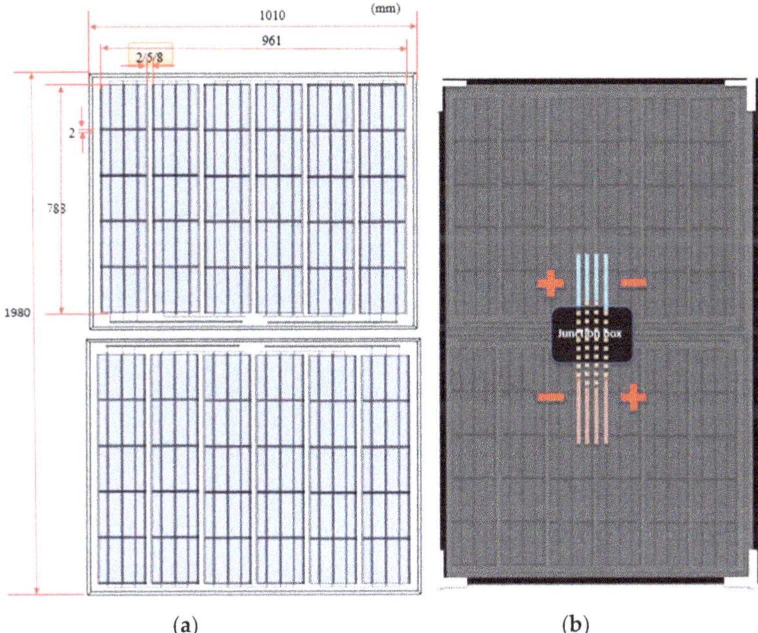

Figure 4. Full frame design solar photovoltaic (PV) module: (**a**) frontside; (**b**) backside.

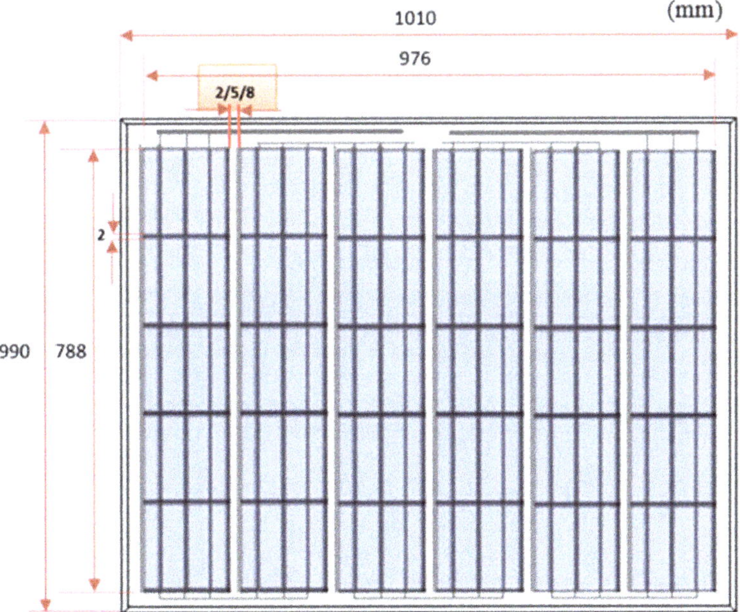

Figure 5. Half-frame design solar PV module.

Table 1. Solar cell structure materials and parameters.

Structure Material	Thickness (mm)	Refractive Index	Percent Transmittance (%)	Heat Transfer Coefficient K Value (W-m^{-1}-K^{-1})
Low iron glass	3.175	n/a	91.7	1.09
Ethylene Vinyl Acetate	0.46	1.49	91	0.35
Tedlar	0.335	n/a	n/a	0.35

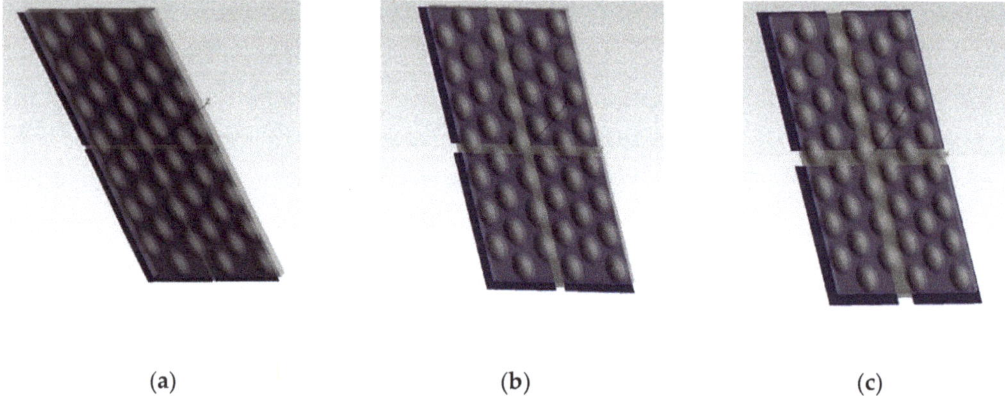

(a)   (b)   (c)

Figure 6. Solar PV cell design at different cell spacing: (**a**) 2 mm cell spacing, (**b**) 5 mm cell spacing, (**c**) 8 mm cell spacing.

Figure 7. Half and full-frame solar PV module.

4. Results

The result of optical simulation, indoor, and outdoor testing is presented and discussed in this section.

4.1. Optical Simulation Analysis of the Internal Structure of the Solar Cell Module

With the use of functional light tracing to observe the penetration and reflection of light in the glass, it was found that the high-energy red light in 2 mm cell spacing of the transparent glass could not be reflected, but directly penetrated the glass, which had the least reflected light. Figure 8 shows that the ray tracing of reflected light in 5 mm was more extreme than 2 mm cell spacing, while 8 mm is better than 5 mm cell spacing. Although the cell spacing and light source increase, the high-energy red light does not increase significantly, but slows down the secondary light and increases the energy. Blue light indicates low energy. When the cell spacing exceeds 5 mm, the reflected energy slows down. However, the reflective area is increased after the cell spacing is enlarged. The light at the surface is refracted back to the solar cell at a limited angle. The illumination of the light source is 1000 W-m^{-2}. Since the absorption range of the solar cell is 0.2–2 µm, the simulation was carried out at a wavelength of 0.55 µm, and the average acceptance ratio of the solar cell was calculated by irradiance analysis.

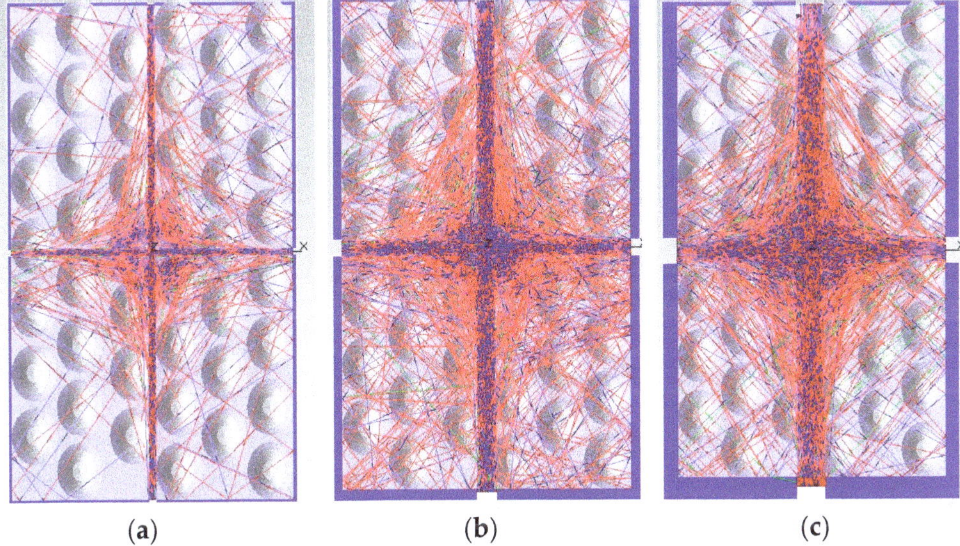

(a)   (b)   (c)

Figure 8. Ray tracing at different cell spacing: (**a**) 2 mm cell spacing; (**b**) 5 mm cell spacing; (**c**) 8 mm cell spacing.

The microstructure is shown in Figure 9 with a 2 mm cell spacing having a value of 0.00515% average absorption of irradiation, Figure 10 with a cell spacing of 5 mm having 0.01483% average absorption of irradiation, and Figure 11 with a cell spacing of 8 mm having 0.01667% average absorption of irradiation from the reflected light cause diffuse reflectance on the rear back of the module. These results show an increase of 0.689% average irradiation absorption on 2 mm to 5 mm while 5 mm to 8 mm cell spacing shows an increase in average irradiation absorption by 0.124% on the solar PV power output. Therefore, the increase in light reflection on 2 mm to 5 mm cell spacing was greater than the cell spacing of 5 mm to 8 mm. The results only show that the power generation increases, but the possibility of solar cells receiving light will gradually slow down after it exceeds 5 mm cell spacing.

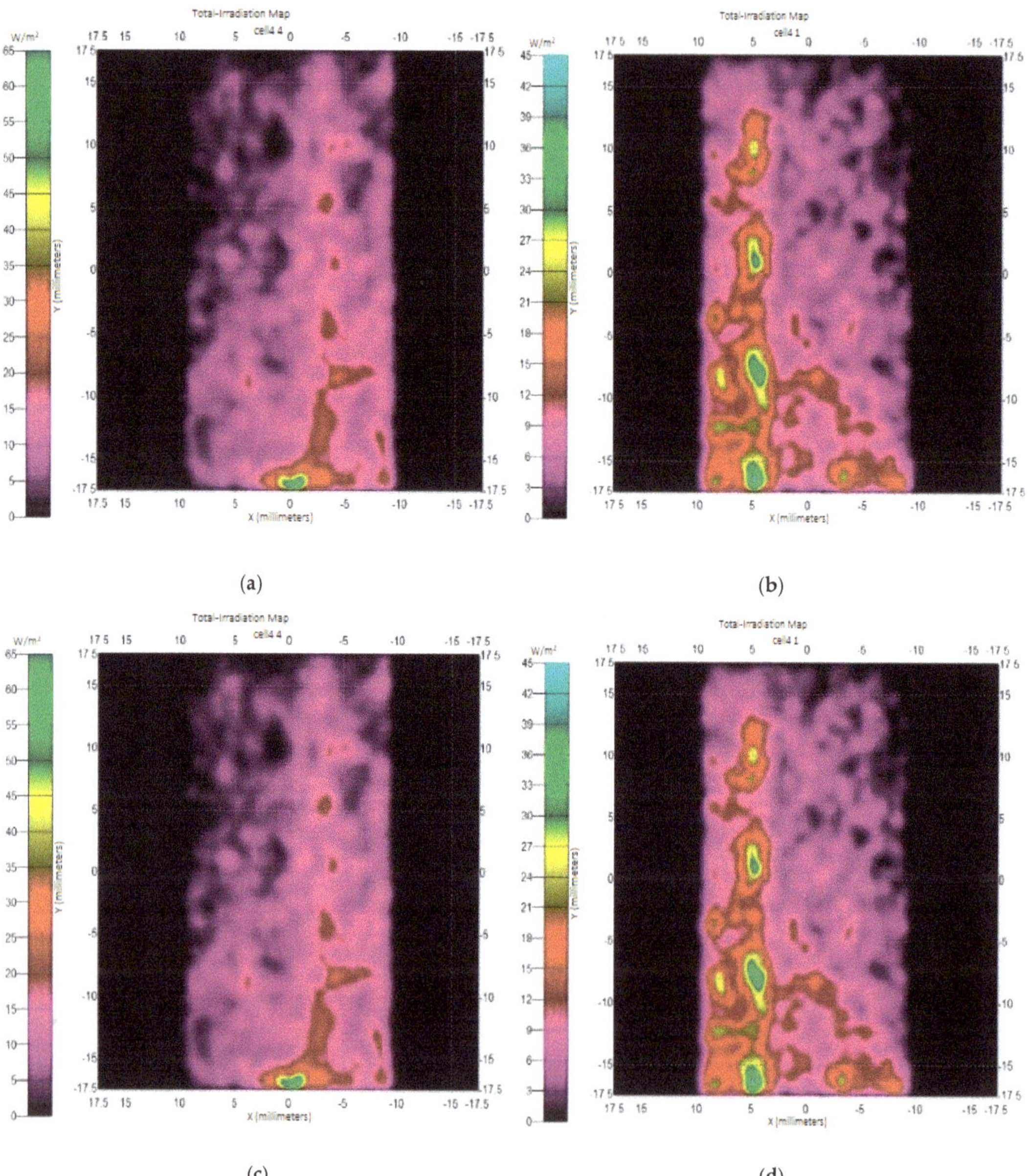

Figure 9. Total-irradiance map of solar cell microstructure with 2 mm cell spacing: (**a**) upper left, (**b**) upper right, (**c**) lower left, (**d**) lower right.

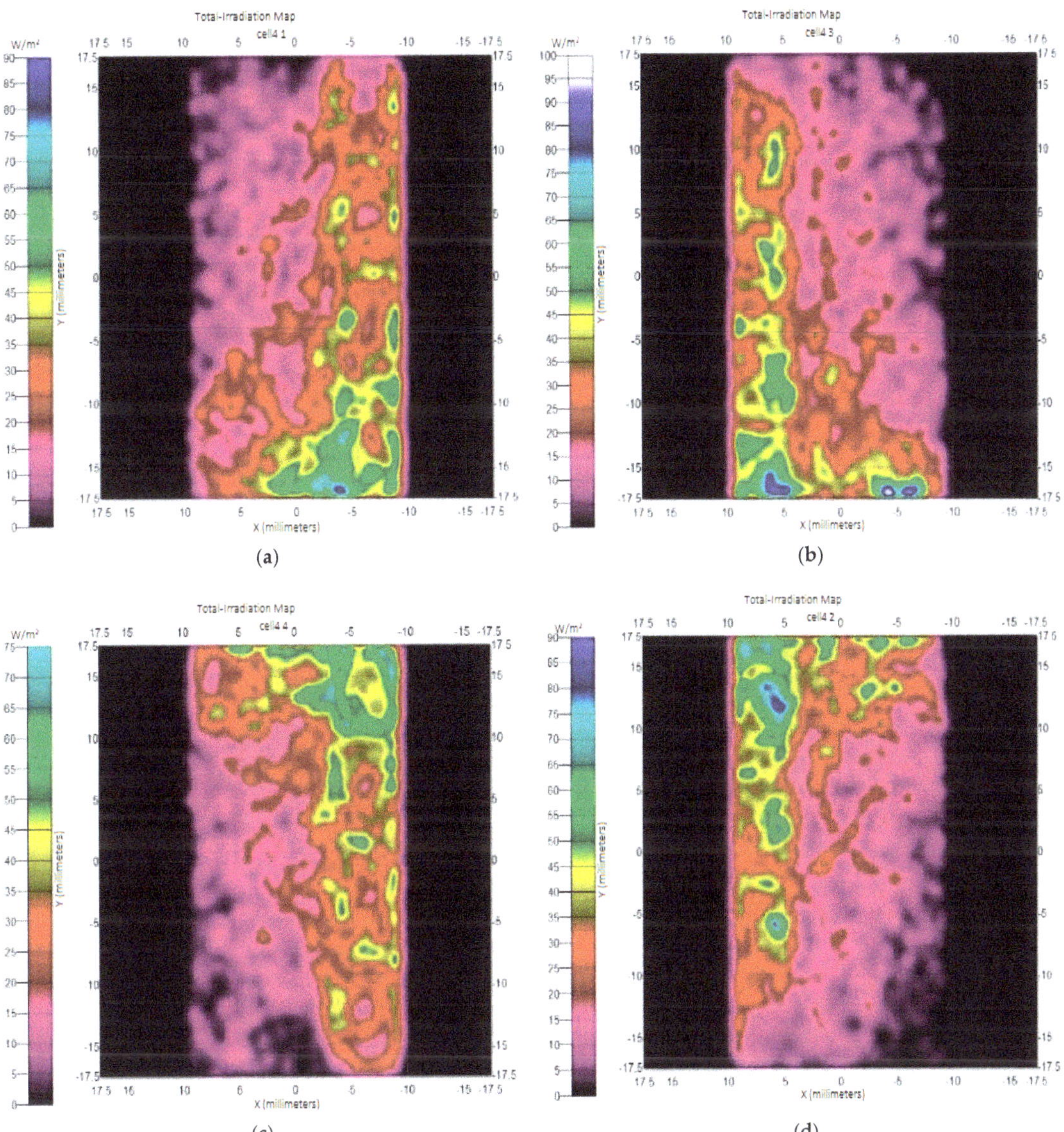

Figure 10. Total-irradiance map of solar cell microstructure with 5 mm cell spacing: (**a**) upper left, (**b**) upper right, (**c**) lower left, (**d**) lower right.

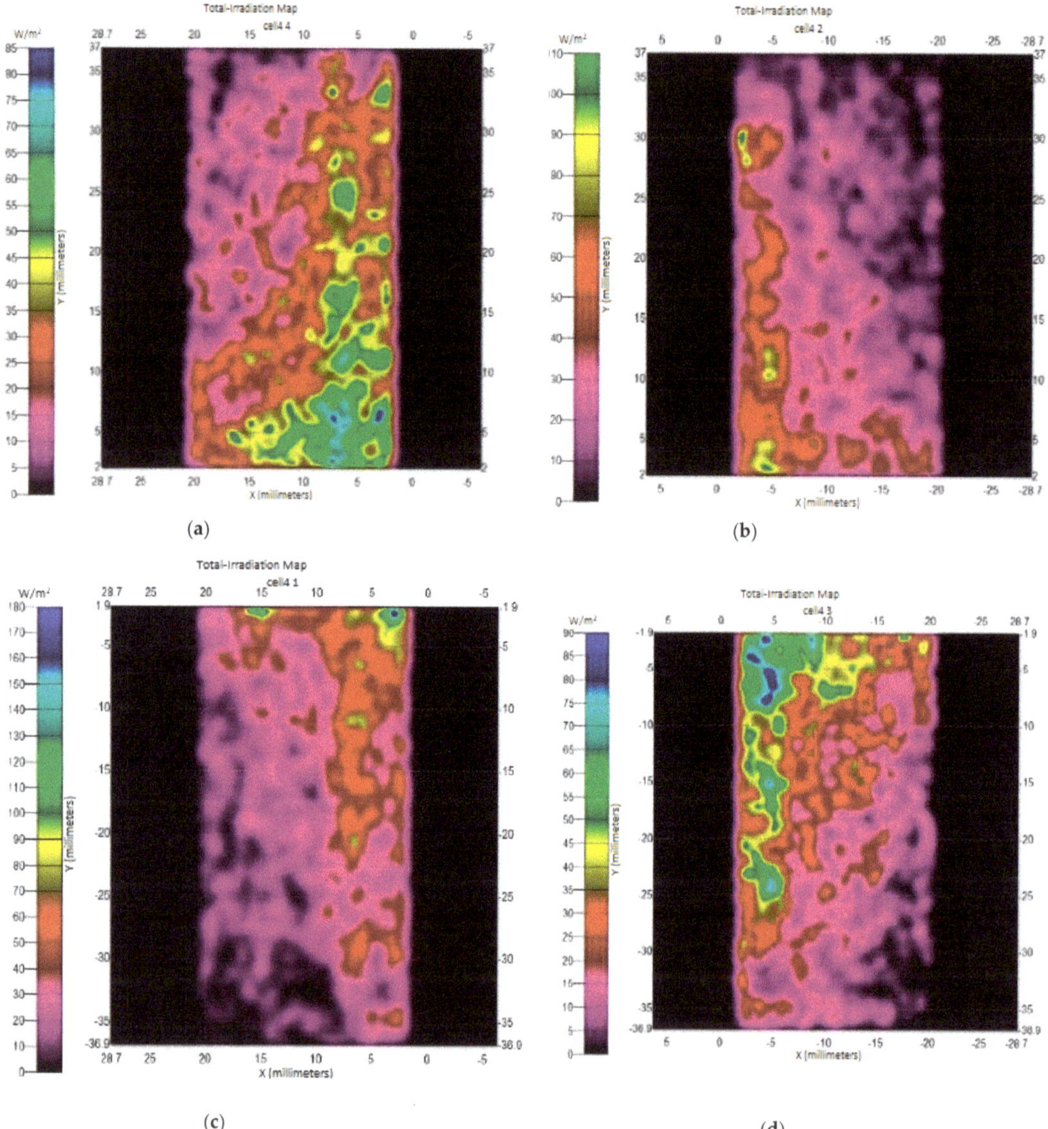

Figure 11. Total-irradiance map of solar cell microstructure with 8 mm cell spacing: (**a**) upper left, (**b**) upper right, (**c**) lower left, (**d**) lower right.

4.2. Influence of Indoor Temperature, Irradiation Amount, and Solar Photovoltaic (PV) Module Cell Spacing in Power Generation

In the field of solar photovoltaic applications, there are almost no single-use examples of solar cells, and most of them are used in series and pack solar cells with the same characteristics. Ideally, the solar cell in the module will exhibit the same characteristics of current and voltage that have the same curve as the current and voltage characteristics of the individual cell, but the scale of the corresponding coordinate axis of the curve will be different.

In actual situations, the cells in the module have slightly different characteristics from each other. Coupled with the irradiance unevenness and the contact resistance of the cell connection, the output characteristics of the module are not completely the coordinate change of the solar cell characteristic curve.

Since the photoelectric conversion of the solar cell module has a spectral response phenomenon, its I-V characteristic is related to the module temperature and irradiation intensity of sunshine, and the description of the IV characteristics of the solar cell module requires spectral irradiance distribution and module temperature. The solar simulator parameters were air mass (AM1.5), module temperature of 25 °C, and irradiance of 1000 W-m^{-2} to conduct a standard experiment.

The measurement results shown in Table 2 were the comparison of temperature and power generation on half-frame and full-frame modules. The cell spacing of 2 mm, 5 mm, and 8 mm in the upper and lower half-frame show a decreasing power while temperature increases. The cell spacing from 2 mm to 5 mm and 5 mm to 8 mm shows an increase of 0.95% and 0.35% of power generation in the upper and lower half-frame under 25 °C, respectively, while full-frame with 2 mm to 5 mm and 5 mm to 8 mm cell spacing has an increase of 0.98% and 0.14%, respectively.

Table 2. Comparison of temperature and power generation at 2 mm, 5 mm, and 8 mm.

Upper and Lower of the Half-Frame Design			
Temperature (°C)	2 mm (W)	5 mm (W)	8 mm (W)
25	122.79	123.97	124.41
35	115.6	116.53	117.02
45	110.29	111.33	111.63
55	105.24	106.24	106.54
65	100.45	101.45	101.87
75	95.481	97.435	97.916
Full-frame design			
25	250.837	253.32	253.68

4.3. Effect of Outdoor Irradiation and Solar PV Module Cell Spacing and Tilt Angle on Power Generation

The actual solar PV module shown in Figure 7 were the half and full-frame designs, and this is the basis of the experiment conducted in the study. There are two major factors in evaluating the use of solar cell modules and the performance of the system. They are the changes in the amount of solar radiation and temperature. In terms of temperature, the solar cell module package, the type of solar cell, and the solar cell module must be considered. The local climatic conditions, thermal properties of the materials after installation, and the irradiation amount were uncontrollable in the same environment. The efficiency of the solar cell depends on the operating temperature. The temperature of the solar cell during normal operation is 58 °C and its performance in power generation is reduced. The amount of irradiation is fixed for one day but will change with the influence of clouds or the atmosphere. The outdoor experiment time was from 15–25 June 2020, which is just the summer solstice period. Facing the south at 0° is ideal. At the beginning of the experiment, the temperature coefficient monitoring under IEC 61215, section 10.4, and section 10.5 of the NOCT was simulated. The test determined angle changes to 50°, 35°, 25°, 15°, and 0°.

The measurement tolerance was ±5° of the surfaces of the solar module. The sunshine intensity needs to be stable, and the wind speed was lower than 2 m·s^{-1}. Under stable sunlight intensity, the module rises in temperature as soon as it is illuminated by the sun and is electrically measured at the temperature point of interest. In the test environment, the solar cell module uses the board to block the sunlight, and the module and the environment reach a temperature balance state. The whole day of data monitoring is downloaded from the cloud to a personal computer for data sorting and experiment. After the operation and data acquisition, the analysis found that the trend in Figure 12 shows that as the angle increases from 0° to 50° as the accumulating power decreases at three different cell spacings. The average power generation in three different cell spacings at 0° reach 1.513 kW-hr^{-1}, 15° produces 1.485 kW-hr^{-1}, 25° produces 1.367 kW-hr^{-1}, 35° produces 1.232 kW-hr^{-1}, 45° produces 1.14 kW-hr^{-1}, and 50° produces 1.18 kW-hr^{-1} with accumulative irradiation of 5.56 kWh-m^{-2} to 5.63 kWh-m^{-2}. This only shows that the larger the tilt angle, the smaller the power generation.

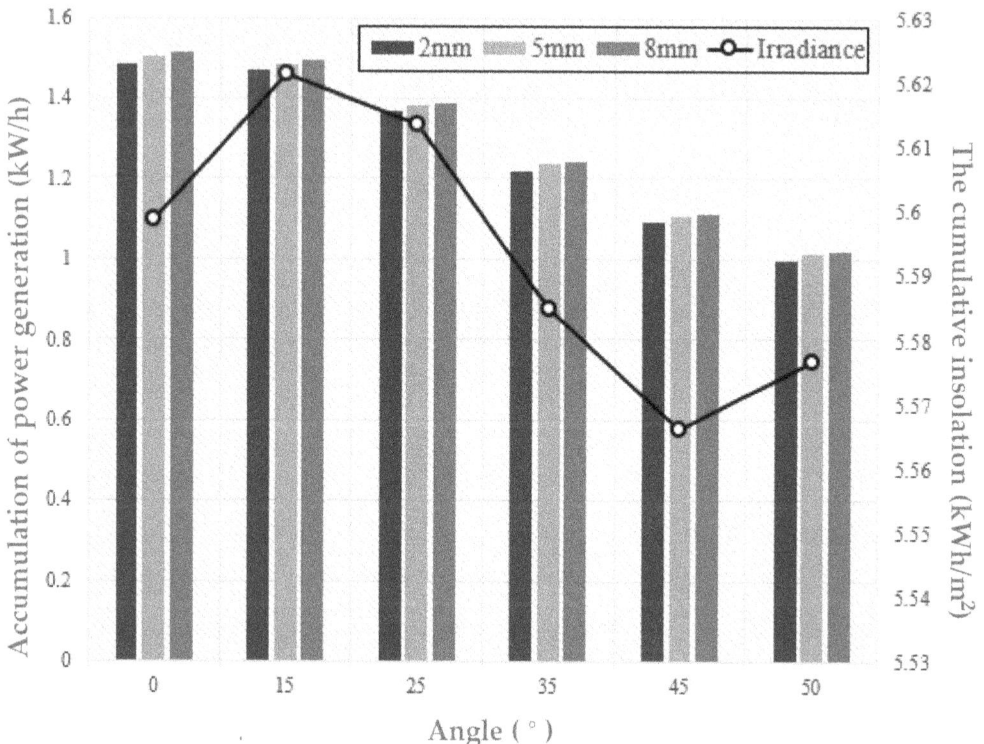

Figure 12. Comparison of accumulative power generation at various angles.

5. Conclusions

This study presents the analysis and effect of three different solar cell spacing designs in irradiation absorption from the diffuse reflectance on the backsheet as well as the power gain of the solar PV module. Optical simulation and indoor and outdoor tests showed that the diffuse reflectance of light from 2 mm to 5 mm cell spacing showed a greater increase in output power compared to 5 mm to 8 mm solar cell spacing in the solar PV module. It also showed a gradual increase in power generation when the solar cell spacing was greater than 5 mm. However, the output power of 8 mm cell spacing was greater than 2 mm and 5 mm cell spacing. The light absorption of the solar cell from the diffuse reflectance showed enhanced power output. Even though the study was successful, various limitations remain. As the solar PV module is dependent on solar irradiation, weather conditions were not included in the study which may affect its performance. For future evaluation, the experiment can be conducted for a longer period to consider the effects of weather conditions on power generation. Since renewable energy is the future of clean energy that will lessen the negative effect of climate change [38], the development of solar PV module design can contribute to a sustainable energy solution. Future research can employ methodical engineering and construction methods to optimize the solar PV module design and obtain the full potential of this study.

Author Contributions: Conceptualization, H.A., J.R.L.M., M.L. and Y.-C.L.; Methodology, H.A. J.R.L.M. and Y.-C.L.; Software, data curation, formal analysis, and investigation, J.R.L.M. and Y.-C.L.; Project administration, J.R.L.M. and Y.-C.L.; writing-original draft, J.R.L.M.; writing-review and editing, J.R.L.M., H.A. and M.L. All authors have read and agreed to the published version of the manuscript.

Funding: National Taipei University of Business, Project No.: 109051313453 "The research on improving the production efficiency of photovoltaic modules".

Institutional Review Board Statement: Not applicable.

Informed Consent Statement: Not applicable.

Data Availability Statement: Not applicable.

Acknowledgments: The authors appreciate the support from the National Kaohsiung University of Science and Technology, Ministry of Education of Taiwan, and National Taipei University of Business.

Conflicts of Interest: The authors declare no conflict of interest.

References

1. Chen, Y.; Altermatt, P.P.; Chen, D.; Zhang, X.; Xu, G.; Yang, Y.; Wang, Y.; Feng, Z.; Shen, H.; Verlinden, P.J. From Laboratory to Production: Learning Models of Efficiency and Manufacturing Cost of Industrial Crystalline Silicon and Thin-Film Photovoltaic Technologies. *IEEE J. Photovolt.* **2018**, *8*, 1531–1538. [CrossRef]
2. Wilson, G.M.; Al-Jassim, M.; Metzger, W.K.; Glunz, S.W.; Verlinden, P.; Xiong, G.; Mansfield, L.M.; Stanbery, B.J.; Zhu, K.; Yan, Y.; et al. The 2020 photovoltaic technologies roadmap. *J. Phys. D Appl. Phys.* **2020**, *53*, 493001. [CrossRef]
3. International Energy Agency. Photovoltaic Power Systems Program: Annual Report 2019. Available online: https://iea-pvps.org/wp-content/uploads/2020/05/IEA-PVPS-AR-2019-1.pdf (accessed on 23 May 2020).
4. Packing Density. Available online: http://pveducation.org/pvcdrom/modules/packing-density (accessed on 20 February 2020).
5. Chen, Z.; Jia, H.; Zhang, Y.; Fan, L.; Zhu, H.; Ge, H.; Cao, B.; Wang, S. Research on Performance Improvement of Photovoltaic Cells and Modules Based on Black Silicon. *Crystals* **2020**, *10*, 753. [CrossRef]
6. Mark, J.G.; Volk, C.H. *The Zero Depth Concentrator Phenomenon*; Jet Propulsion Laboratory: Pasadena, CA, USA, 1979; pp. 1–35.
7. Saga, T. Advances in crystalline silicon solar cell technology for industrial mass production. *NPG Asia Mater.* **2010**, *2*, 96–102. [CrossRef]
8. Jeong, J.; Park, N.; Hong, W.; Han, C. Analysis for the degradation mechanism of photovoltaic ribbon wire under thermal cycling. In Proceedings of the 2011 37th IEEE Photovoltaic Specialists Conference, Seattle, WA, USA, 19–24 June 2011; pp. 003159–003161. [CrossRef]
9. Gress, P.J.; Widenborg, P.I.; Varlamov, S.; Aberle, A.G. Wire bonding as a cell interconnection technique for polycrystalline silicon thin-film solar cells on glass: Wire bonding as a cell interconnection technique. *Prog. Photovolt. Res. Appl.* **2010**, *18*, 221–228. [CrossRef]
10. Klengel, R.; Petzold, M.; Schade, D.; Sykes, B. Improved Testing of Soldered Busbar Interconnects on Silicon Solar Cells. In Proceedings of the 18th European Microelectronics & Packaging Conference, Brighton, UK, 12–15 September 2011; p. 5.
11. Battaglia, C.; Hsu, C.-M.; Söderström, K.; Escarré, J.; Haug, F.-J.; Charrière, M.; Boccard, M.; Despeisse, M.; Alexander, D.T.L.; Cantoni, M.; et al. Light Trapping in Solar Cells: Can Periodic Beat Random? *ACS Nano* **2012**, *6*, 2790–2797. [CrossRef] [PubMed]
12. Zhu, J.; Hsu, C.-M.; Yu, Z.; Fan, S.; Cui, Y. Nanodome Solar Cells with Efficient Light Management and Self-Cleaning. *Nano Lett.* **2010**, *10*, 1979–1984. [CrossRef] [PubMed]
13. Kim, J.; Hong, A.J.; Nah, J.-W.; Shin, B.; Ross, F.M.; Sadana, D.K. Three-Dimensional a-Si:H Solar Cells on Glass Nanocone Arrays Patterned by Self-Assembled Sn Nanospheres. *ACS Nano* **2011**, *6*, 265–271. [CrossRef] [PubMed]
14. Kuang, Y.; Van Der Werf, K.H.M.; Houweling, Z.S.; Schropp, R.E.I. Nanorod solar cell with an ultrathin a-Si:H absorber layer. *Appl. Phys. Lett.* **2011**, *98*, 113111. [CrossRef]
15. Garnett, E.; Yang, P. Light Trapping in Silicon Nanowire Solar Cells. *Nano Lett.* **2010**, *10*, 1082–1087. [CrossRef] [PubMed]

16. Chang, H.-C.; Dai, Y.-A.; Wang, H.-H.; Lai, K.-Y.; Lin, C.-A.; He, J.-H. Nanowire arrays with controlled structure profiles for maximizing optical collection efficiency. *Energy Environ. Sci.* **2011**, *4*, 2863–2869. [CrossRef]
17. McIntosh, K.R.; Swanson, R.M.; Cotter, J.E. A simple ray tracer to compute the optical concentration of photovoltaic modules. *Prog. Photovolt. Res. Appl.* **2006**, *14*, 167–177. [CrossRef]
18. Grunow, P.; Krauter, S. Modelling of the Encapsulation Factors for Photovoltaic Modules. In Proceedings of the 2006 IEEE 4th World Conference on Photovoltaic Energy Conference, Waikoloa, HI, USA, 7–12 May 2006; pp. 2152–2155. [CrossRef]
19. Wenham, S.; Campbell, P.; Bowden, S.; Green, M. Improved optical design for photovoltaic cells and modules. In Proceedings of the Conference Record of the Twenty-Second IEEE Photovoltaic Specialists Conference—1991, Las Vegas, NV, USA, 7–11 October 1991; pp. 105–110. [CrossRef]
20. Amalathas, A.P.; Alkaisi, M.M. Nanostructures for Light Trapping in Thin Film Solar Cells. *Micromachines* **2019**, *10*, 619. [CrossRef] [PubMed]
21. Feng, N.-N.; Michel, J.; Zeng, L.; Liu, J.; Hong, C.-Y.; Kimerling, L.C.; Duan, X. Design of Highly Efficient Light-Trapping Structures for Thin-Film Crystalline Silicon Solar Cells. *IEEE Trans. Electron Devices* **2007**, *54*, 1926–1933. [CrossRef]
22. Witteck, R.; Schulte-Huxel, H.; Holst, H.; Hinken, D.; Vogt, M.; Blankemeyer, S.; Köntges, M.; Bothe, K.; Brendel, R. Optimizing the Solar Cell Front Side Metallization and the Cell Interconnection for High Module Power Output. *Energy Procedia* **2016**, *92*, 531–539. [CrossRef]
23. Fischer, R.E.; Tadic-Galeb, B.; Yoder, P.R. *Optical System Design*, 2nd ed.; Thoroughly Updated; McGraw-Hill: New York, NY, USA, 2008.
24. Lahiri, A. Foundations of Ray Optics. In *Basic Optics*; Elsevier BV: Amsterdam, The Netherlands, 2016; pp. 141–202.
25. Sweeney, S.J.; Mukherjee, J. Optoelectronic Devices and Materials. In *Springer Handbook of Electronic and Photonic Materials*; Kasap, S., Capper, P., Eds.; Springer International Publishing: Cham, Switzerland, 2017; pp. 1–1.
26. Choudhury, A.K.R. Characteristics of light sources. In *Principles of Colour and Appearance Measurement*; Elsevier: Amsterdam, The Netherlands, 2014; pp. 1–52.
27. Ryer, A. *Light Measurement Handbook*; International Light: Newburyport, MA, USA, 1997; pp. 13–16.
28. Vain, A.; Kaasalaine, S. Correcting Airborne Laser Scanning Intensity Data. In *Laser Scanning, Theory and Applications*; Wang, C.-C., Ed.; IntechOpen: London, UK, 2011.
29. Schejbal, V.; Fiser, O.; Zavodny, V. Study of Refraction Effects for Propagation over Terrain. In *Waveguide Technologies in Photonics and Microwave Engineering*; IntechOpen: London, UK, 2019.
30. Shrivastava, A. Plastic Properties and Testing. In *Introduction to Plastics Engineering*; Elsevier BV: Amsterdam, The Netherlands, 2018; pp. 49–110.
31. Gong, J.; Krishnan, S. Mathematical Modeling of Dye-Sensitized Solar Cells. In *Dye-Sensitized Solar Cells*; Elsevier: Amsterdam, The Netherlands, 2019; pp. 51–81.
32. Duran, E.; Andújar, J.M.; Enrique, J.M.; Pérez-Oria, J.M. Determination of PV GeneratorI-V/P-VCharacteristic Curves Using a DC-DC Converter Controlled by a Virtual Instrument. *Int. J. Photoenergy* **2012**, *2012*, 1–13. [CrossRef]
33. Lindholm, F.; Fossum, J.; Burgess, E. Application of the superposition principle to solar-cell analysis. *IEEE Trans. Electron Devices* **1979**, *26*, 165–171. [CrossRef]
34. Yu, J.; Fan, J.; Lv, K. Anatase TiO2 nanosheets with exposed (001) facets: Improved photoelectric conversion efficiency in dye-sensitized solar cells. *Nanoscale* **2010**, *2*, 2144–2149. [CrossRef] [PubMed]
35. Cubas, J.; Pindado, S.; De Manuel, C. Explicit Expressions for Solar Panel Equivalent Circuit Parameters Based on Analytical Formulation and the Lambert W-Function. *Energies* **2014**, *7*, 4098–4115. [CrossRef]
36. Tsai, H.-L.; Tu, C.-S.; Su, Y.-J. Development of Generalized Photovoltaic Model Using MATLAB/SIMULINK. In Proceedings of the world congress on Engineering and computer science, San Francisco, CA, USA, 22–24 October 2008; p. 7.
37. Salmi, T.; Bouzguenda, M.; Gastli, A.; Masmoudi, A. *MATLAB/Simulink Based Modelling of Solar Photovoltaic Cell*; International Journal of Renewable Energy Research: Ankara, Turkey, 2012; p. 7.
38. Wang, C.-N.; Dang, T.-T.; Tibo, H.; Duong, D.-H. Assessing Renewable Energy Production Capabilities Using DEA Window and Fuzzy TOPSIS Model. *Symmetry* **2021**, *13*, 334. [CrossRef]

 *sustainability*

Article

Environmental and Health Co-Benefits of Coal Regulation under the Carbon Neutral Target: A Case Study in Anhui Province, China

Wu Xie [1], Wenzhe Guo [1], Wenbin Shao [1], Fangyi Li [1,*] and Zhipeng Tang [2,*]

[1] School of Management, Hefei University of Technology, Hefei 230009, China;
2018110768@mail.hfut.edu.cn (W.X.); 2018110773@mail.hfut.edu.cn (W.G.);
2015010099@mail.hfut.edu.cn (W.S.)

[2] Institute of Geographic Sciences and Natural Resources Research, Chinese Academy of Sciences, Beijing 100101, China

* Correspondence: fyli@hfut.edu.cn (F.L.); tangzp@igsnrr.ac.cn (Z.T.)

Abstract: Coal regulation has been implemented throughout China. However, the potential benefits of pollution abatement and the co-benefits of residents' health were rarely assessed. In this study, based on the analysis of historical coal consumption and multiple coal regulation measures in Anhui Province, China, four scenarios (Business as Usual (BU), Structure Optimization (SO), Gross Consumption Control (GC), and Comprehensive Measures (CM)) were constructed to indicate four different paths from 2020 to 2060, which is a vital period for realizing carbon neutrality. The results show that reductions of SO_2, PM_{10}, and $PM_{2.5}$ emissions in the SO scenario are higher than those in the GC scenario, while the reduction of NO_x emission is higher in the GC scenario. Compared with the BU scenario, residents' health benefits from 2020 to 2060 are 8.3, 4.8, and 4.5 billion USD in the CM, GC, and SO scenarios, respectively, indicating that the achievements of coal regulation are significant for health promotion. Therefore, the optimization and implementation of coal regulation in the future is not only essential for the carbon neutrality target, but also a significant method to yield environmental and health co-benefits.

Keywords: energy system; coal regulation; pollution abatement; environmental benefits; health benefits

Citation: Xie, W.; Guo, W.; Shao, W.; Li, F.; Tang, Z. Environmental and Health Co-Benefits of Coal Regulation under the Carbon Neutral Target: A Case Study in Anhui Province, China. *Sustainability* **2021**, *13*, 6498. https://doi.org/10.3390/su13116498

Academic Editors: Wei-Hsin Chen, Alvin B. Culaba, Aristotle T. Ubando and Steven Lim

Received: 3 May 2021
Accepted: 4 June 2021
Published: 7 June 2021

Publisher's Note: MDPI stays neutral with regard to jurisdictional claims in published maps and institutional affiliations.

1. Introduction

Coal production and consumption provide strong support for a rapid and stable economic development and resident living in China. In 2019, China consumed 2809.99 Mt of standard coal, accounting for 51.7% of global coal consumption [1]. Coal consumption accounted for approximately 57.7% of the total primary energy consumption in China, a much higher percentage than in developed countries [2]. Compared with renewable energies and natural gas, the rough way uses of coal caused serious environmental pollution [3]. In addition, a large number of epidemiological studies have confirmed that the increase in the concentration of SO_2, NO_x, and particulate matter (PM) in the air will greatly increase the probability of people suffering from various respiratory diseases, and even cause death [4–7]. Faced with this situation, the China State Council issued the "Air Pollution Prevention and Control Action Plan (APPCAP)" in 2013, which focuses on the prevention and control of air pollutants with a focus on fine PM. In 2018, after the completion of the APPCAP, the State Council released the "Three-Year Action Plan for Winning the Battle of Blue Sky", a project that is in line with the APPCAP to further improve the atmospheric environment in China. In addition, the Chinese government has also issued a series of pollution prevention policies, such as the "Guiding Opinions on Promoting Electricity Substitution". In terms of coal regulation, it can be generally summarized as the following two points: first, promote the optimization of the coal consumption structure, accelerate the construction of "coal to electricity" and other related projects, reduce the use of scattered coal, and increase the proportion of thermal coal in coal consumption; second, control the gross of coal consumption, actively develop wind power, hydropower, photovoltaic power generation, and other new types of primary energy, and increase the proportion of non-fossil energy consumption. In 2020, China pledged to reach a carbon peak by 2030 and become carbon neutral by 2060, indicating that the country has to increase its solar and wind capacity massively to substitute coal consumption [8].

In future, the structure optimization and total quantity control of coal consumption will be promoted with more efforts to realize the carbon peak and carbon neutrality. It is significant to assess the potential and effects of coal-related measures in the future, especially the benefits on residents' health in different scenarios, so as to identify the optimized path to achieve an ecological civilization.

Previous studies on coal consumption focus mainly on the following three aspects: (1) greenhouse gases (GHGs) emissions caused by coal consumption [9–11]; (2) the impacts of coal consumption on the atmospheric environment [12–14]; (3) the effects of the coal consumption control policies [15,16]. In terms of GHGs emissions from coal consumption, certain studies only focus on the current estimate of GHG emissions and analyze the influence factors of carbon emissions [17]. Most studies focused on predicting GHG emissions have used the bottom-up method [18], the co-integration method [19], or the system dynamics approach [20]. Other studies attempted to explore ways to limit greenhouse gas emissions in the future through the trade of carbon emission permits and environmental taxes [21–23]. In terms of pollutant emission from coal consumption, most studies focused on analyzing the impacts of coal consumption on the environment from a single sector, due to the difference of emissions from coal utilization in various industries, especially energy-intensive industries, such as thermal power generation [24], building materials [25], and the steel and chemical industry [26,27]. Some studies assessed the potential of industry to reduce pollutants by technological improvements [28]. With the introduction of coal regulation in China, the research on coal regulation has gradually attracted the attention of scholars. Certain studies have explored how the current coal regulation policy has improved the air environment since the release of the APPCAP in China [29,30]. Other studies use game models, a system simulation approach, and other methods to estimate the emission reduction effect of future coal regulation on GHGs and air pollutants based on the simulation of future coal control policies [31–33]. In addition, a few studies have explored the effects of coal regulation policies from the perspective of the health risks to residents [34].

In summary, most of the researches on coal regulation focused on the abatement of pollutant emission, and only a few evaluated the effects of coal regulation on residents' health. In the future, residents' demand for a better living environment is also an important factor to be considered in coal regulation, which deserves more attention.

Anhui Province of China, a typical region to implement coal regulation policies currently and in the future, is selected as the research area. As an inland province in the Yangtze River Delta Region, it has great potential for future development. In addition, Anhui is one of the 14 key coal bases in China due to its huge coal production. The economic development of Anhui is highly dependent on coal. Therefore, assessing the effects of coal regulation is not only essential for Anhui, but also a significant reference for energy transition in other provinces in China. The rest of this paper is organized as follows. Section 2 introduces the research methods and materials, Section 3 gives the research results, Section 4 is the discussion, and Section 5 presents the conclusions of this paper.

2. Methods and Materials

Based on the coal consumption and the structure, economic development, and related coal regulation policies in Anhui in the past 20 years, four coal consumption scenarios were established against the background of carbon neutrality in the future, and the air pollutant emissions caused by coal consumption during 2020–2060 were assessed. Then, we further analyzed the health benefits of coal regulation to residents.

2.1. Methods

2.1.1. Air Pollutant Emissions

There are three methods to calculate the emission of air pollutants: the actual measurement method, the material conservation method, and the empirical calculation method (emission factor method) [35,36]. Considering that the actual measurement method is not easy to undertake and the material conservation method cannot reflect the differences of pollutant emissions from coal consumption in various industries, this study uses the emission factor method to estimate the emissions of various types of air pollutants caused by coal consumption. It mainly considers the emission of SO_2, NO_x, PM_{10}, and $PM_{2.5}$ air pollutants. The air pollutant emissions can be calculated as follows [36,37]:

$$APE_j = \sum_i CC_i \times EF_{ij} \tag{1}$$

where APE_j is the emission of air pollutant j (t); CC_i is the coal consumption in sector/industry i (t); EF_{ij} is the emission factor of air pollutant j in sector/industry i (t/t).

2.1.2. Health Benefits Estimation

The air pollutant concentration model is used to assess health risks caused by air quality. However, this type of model usually needs to consider a series of factors, such as time, climate, temperature, humidity, population density, etc., which cannot be determined based on long-term forecasts. On this basis, some scholars adopted a simplified modeling method by assuming the uniform diffusion of air pollutants in the study area and combining the population intake fraction (IF) to assess the health risks of residents [38,39]. Bennett et al., defined the IF as the proportion of the intake of a certain type of pollutant by the population [40], which is expressed by following equation:

$$Dose_j = APE_j \times IF_j \tag{2}$$

where $Dose_j$ represents the amount of inhaled pollutant j by an individual (t); IF_j represents the population intake fraction of air pollutant j.

This study considered four health outcomes due to SO_2, NO_x, PM_{10}, and $PM_{2.5}$: death, respiratory disease, cardiovascular disease, and asthma, respectively. The health outcomes are assessed by the human inhalation dose and concentration-response relations, which can be determined from Equation (3). The health benefits are estimated by the unit economic valuation of health outcomes h and health benefits as shown in Equation (4).

$$HO_{hj} = Dose_n \times \frac{CR_{hj} \times f_{hj} \times 10^{12}}{365 \times BR} \tag{3}$$

$$HB_{hj} = HO_{hj} \times UV_h \tag{4}$$

where HO_{hj} represents the number of cases with health outcome h caused by air pollutants j (case); CR_{hj} represents the concentration-response coefficient of the health outcome h caused by air pollutants j (case $\times$ m^3/ug); f_{hj} represents the baseline of mortality or morbidity incidence rate for the health outcome h caused by air pollutants j; BR is the respiratory rate, and the standard value is 20 m^3/d. HB_{hj} represents the total health benefit lost by the health outcome h caused by the pollutant j (USD), and UV_h is the unit value for the health outcome h (USD/case).

2.2. Materials

2.2.1. Sector Classification of Coal Consumption

Sectors using coal resource in Anhui Province can be divided into two major branches and several specific sectors, as shown in Figure 1. The first branch is transformation sectors. The further processing of coal resources can transform them into secondary energy that can be used clean and efficiently. The second branch includes sectors using coal directly as the terminal energy. In terms of transformation, according to the different processing methods, coal consumption is divided into three industries: the thermal power and heating supply, coal washing, coking and briquettes. In terms of terminal consumption, coal consumption is subdivided into three industrial sectors and residential sector. The primary industry is agriculture. The secondary industry includes building materials, steel, chemical industry, and other manufacturing industries according to the characteristics of products. The tertiary industry consists mainly of transport and storage, hotel and restaurant, and other service industries. The residential sector consists of urban and rural residents.

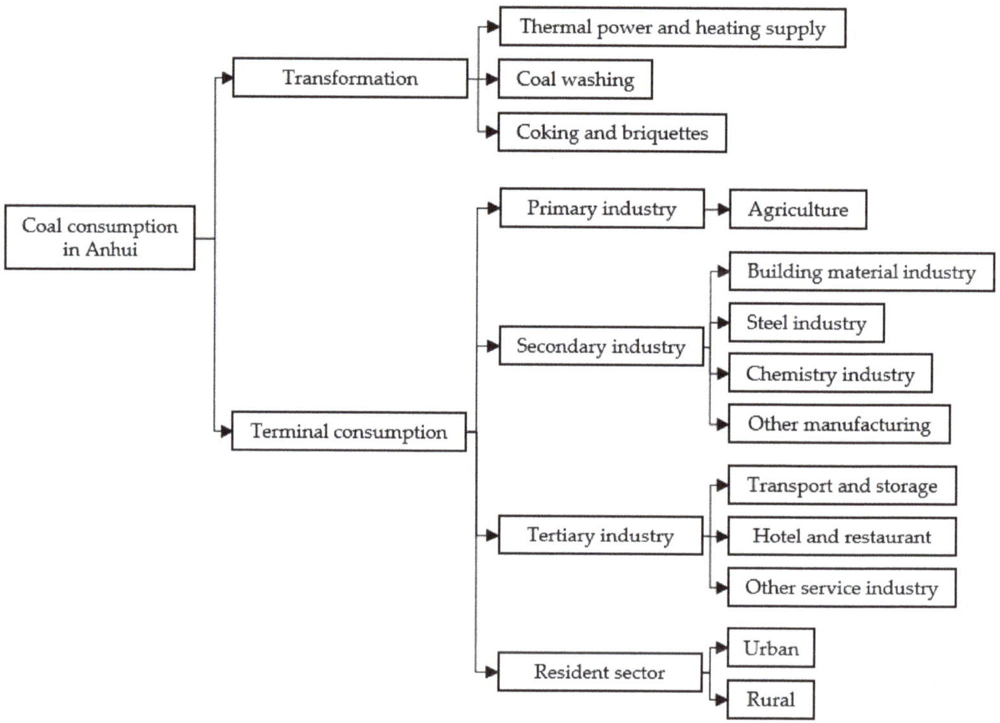

Figure 1. Coal consumption structure in Anhui.

2.2.2. Data Sources

The historical data of coal consumption in various industries (physical quantity) in Anhui are taken from the China Energy Statistical Yearbook and the Anhui Statistical Yearbook [41,42]. The emission factors of coal consumption in different industries are mainly derived from literatures and shown in Table 1. We assume that the emission factors will decrease by 5% every five years due to technological advance and pollution treatment in the future.

Table 1. Emission factors of coal consumption in different industries (Unit: kg/t).

	Sector/Industry		SO_2	NO_x	PM_{10}	$PM_{2.5}$
	Thermal power and heating supply		0.65	0.62	0.13	0.11
Transformation	Coal washing		0.71	0.92	2.26	1.61
	Coking and briquettes		0.91	1.23	1.45	1.20
	Primary industry	Agriculture	8.04	1.20	7.16	5.13
		Building material industry	8.40	2.70	1.59	0.76
	Secondary industry	Steel industry	6.72	2.16	1.27	0.61
		Chemistry industry	5.38	2.06	1.02	0.49
Terminal consumption		Other manufacturing	7.56	2.43	1.43	0.68
		Transport and storage	6.75	2.30	2.96	2.43
	Tertiary industry	Hotel and restaurant	8.04	1.20	7.16	5.13
		Other service industry	8.04	1.20	7.16	5.13
	Residential sector	Urban	8.04	1.20	13.13	9.41
		Rural	8.04	1.20	13.13	9.41

Note: The emission factors of thermal power and heating supply industry come from the "China Power Industry Annual Development Report 2020" [43], while other sectors/industries mainly refer to Gao and Guo's researches [44,45]. Missing data in some industries are adjusted according to the "Practical Manual for Discharge Declaration and Registration" and the "The Second National Pollution Source Census Production and Discharge Accounting Coefficient Manual" [35,46].

In terms of health benefits, the original IF_j were collected from Liu et al. [47]. and adjusted based on Anhui's population density. The IF_j of SO_2, NO_x, PM_{10}, and $PM_{2.5}$ are 3.6×10^{-6}, 25.3×10^{-6}, 29.2×10^{-6}, and 28.9×10^{-6}, respectively, and will continue to be adjusted with the change of population density in the future. The concentration-response coefficient (CR_{hj}) and morbidity/mortality incidence rates (f_{hj}) used in this study were taken from different studies [48,49]. The unit loss of various health risk in the Yangtze River Delta of China was obtained [50] and adjusted according to GDP per capita in 2020, as shown in Table 2.

Table 2. Unit value of different health risks (USD/case).

Health Risk	Unit Loss (Mean and 95% CI)
Mortality	127,591 (119,937, 135,354)
Respiratory hospital admission	835 (765, 896)
Cardiovascular hospital admission	1225 (1052, 1322)
Asthma attack	6 (2, 9)

Note: Adjusted according to GDP per capita in 2020.

2.2.3. Scenario Design

If there is no technological revolution, China's total carbon sink is expected to be about 1.5 billion tons. In 2019, China's carbon dioxide emissions were around 10 billion tons. If the target is to achieve carbon neutrality by 2060, China's fossil energy must be cut to at least one-sixth of the present amount. During 2015–2020, the Anhui government has formulated a series of measures to control coal consumption, including the structural optimization of coal consumption and gross coal consumption control. The coal regulation measures will be implemented sequentially in the future, and are also the main sources to design future coal consumption scenarios.

Based on the government's targets and analysis, we constructed the following four scenarios to evaluate coal consumption in Anhui from 2020 to 2050. The scenarios are (1) Business as Usual (BU), (2) Structure Optimization (SO), (3) Gross Consumption Control (GC), and (4) Comprehensive Measures (CM). The division into four scenarios is shown in Figure 2, and the parameter settings in these four scenarios are shown in Table 3.

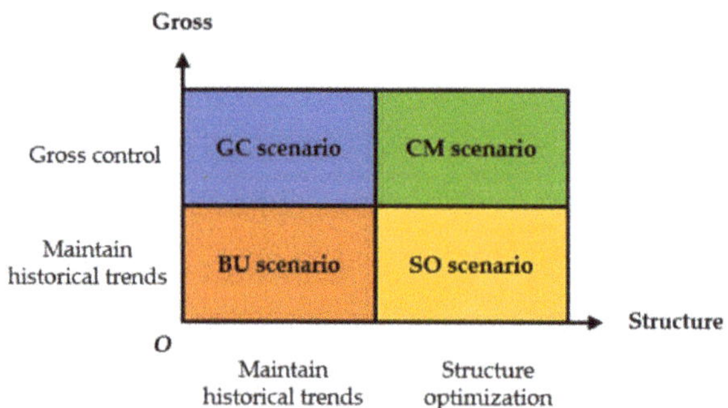

Figure 2. Scenario setting of future coal consumption in Anhui.

Table 3. Parameter setting in the four scenarios.

Gross Coal Consumption	BU Scenario	SO Scenario	GC Scenario	CM Scenario
2020	170 Mt	170 Mt	170 Mt	170 Mt
2025	180 Mt	175 Mt	180 Mt	175 Mt
2030	185 Mt	170 Mt	185 Mt	170 Mt
2040	160 Mt	120 Mt	160 Mt	120 Mt
2050	100 Mt	60 Mt	100 Mt	60 Mt
2060	30 Mt	1 Mt	30 Mt	1 Mt

(1) BU scenario: Anhui will be bound by the commitments made by China in Association of Paris. By developing new primary energy sources, such as hydropower and photovoltaic power generation, and expanding the linkage scale of the "Project of Natural Gas Transmission from West to East China", it will reduce the dependence on coal resources. Coal consumption will reach a peak in 2030 and drop to one-sixth of the current level in 2060. The coal consumption structure will maintain the historical trend.

(2) SO scenario: The trend of total coal consumption in Anhui was consistent with the BU scenario. In the future, the newly introduced coal regulation policy will focus on the control of the consumption of scattered coal. In the terminal coal consumption industry, it will further increase the promotion of the technology for replacing coal with electricity, such as electric boilers, electric kilns, and electric heating, to reduce the proportion of scattered coal consumption [3]. The decline rate of the coal consumption ratio in each terminal industry will be twice that of the BU scenario, and the optimization of the coal consumption structure will be accelerated.

(3) GC scenario: In the future, the coal consumption structure will continue to maintain the historical trend. The newly introduced coal regulation policy will focus on controlling the gross coal consumption. In addition to developing hydropower, photovoltaic power generation, and expanding natural gas consumption, new energy technologies such as hydrogen energy utilization and the co-firing of coal and biomass will be promoted in the future [51]. The province will strive to reach the peak of carbon moderately ahead, leaving a little room to avoid the passive peak situation in 2030. Coal consumption will reach the peak in 2025 and decrease to one-sixth of the current total coal consumption in 2055, realizing carbon neutrality in advance.

(4) CM scenario: It is the combination of the SO scenario and the GC scenario. In the future, the newly introduced coal regulation policies will focus not only on the optimization of the coal consumption structure, but also on the control of gross coal consumption. The decline rate of coal consumption in each terminal industry is over that of BU scenario. In addition, the total coal consumption will peak in 2025 and drop to one-sixth of the current level in 2055, which will guarantee that the national carbon neutrality target can be realized before 2060.

3. Results

3.1. Coal Consumption in Different Scenarios

Based on the simulation of the implementation intensity of the coal regulation policy in Anhui in the future, the coal consumption in the four scenarios established in this study is shown in Figure 3. In the scenarios of BU and SO, the annual coal consumption in Anhui will continue to increase to the peak in 2030 and decrease to 30 Mt in 2060, achieving carbon neutrality and basically realizing the emission reduction commitment made by China. In the GC and CM scenarios, gross coal consumption control will be further strengthened in the future, and China's target of carbon peak and carbon neutrality will be achieved 5 years ahead of schedule, respectively to avoid "passive peak" and "passive neutralization".

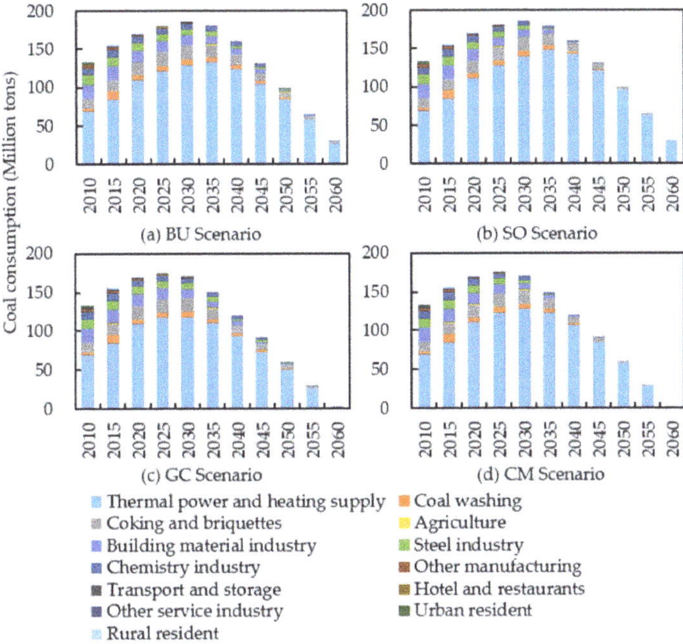

Figure 3. Coal consumption in Anhui in the four scenarios in the future. (**a**) BU Scenario; (**b**) SO Scenario; (**c**) GC Scenario; (**d**) CM Scenario.

The coal consumption in different industries in different scenarios was further analyzed. In the scenarios of BU and TC, the proportion of coal in each industry will maintain the historical change trend. The residential sector will achieve "zero coal consumption" by 2024 and 2028, respectively. Transport and storage and other manufacturing will achieve "zero coal consumption" by 2038. Hotel and restaurant and other service industries are expected to achieve "zero coal consumption" by 2045 and 2046, respectively. The steel industry will achieve "zero coal consumption" by 2059. Building materials and chemical industries with high energy consumption will not achieve "zero coal consumption" by 2060, which indicates that coal regulation policies such as electricity substitution still have great potential for implementation in the secondary industry. In the thermal power and heating supply and coking and briquettes industry, the future annual coal consumption will further increase, indicating that the use of coal will be cleaner and more efficient.

In the SO and CM scenarios, the coal consumption ratio of each terminal industry declines twice as fast as the historical trend. The residential sector will achieve "zero coal consumption" by 2021 and 2023, respectively. The sectors of transport and storage and other manufacturing will achieve "zero coal consumption" by 2028. The hotels and restaurants and other service industries will achieve "zero coal consumption" by 2032. Agriculture will achieve "zero coal consumption" by 2045. The steel, building materials and chemical industries will achieve "zero coal consumption" by 2038, 2041, and 2045, respectively. After 2045, all coal consumption will be used in the transformation sector, and the scattered coal consumption will be effectively eliminated.

3.2. Environmental Benefit Assessment

As can be seen from Figure 4, in all scenarios, the emissions of SO_2, NO_x, PM_{10}, and $PM_{2.5}$ caused by coal consumption will continue to decline from 2020 to 2060. In the BU scenario, even if coal consumption continues to grow during 2020–2030, coal consumption structure can completely offset the increase in pollutant emission caused by the growth of coal consumption.

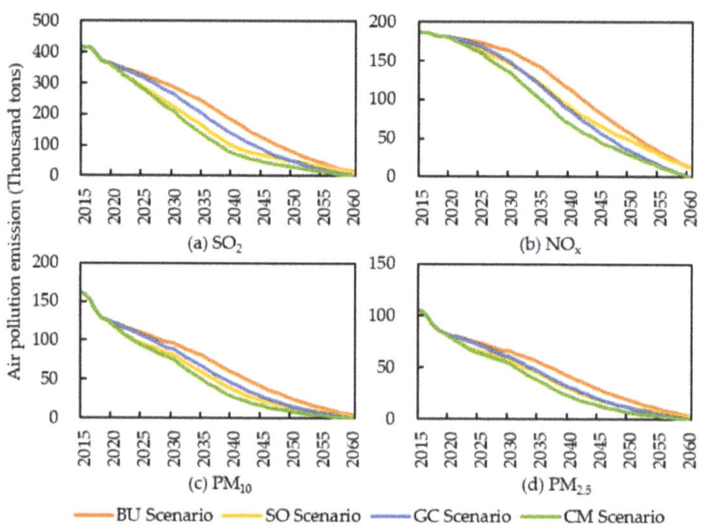

Figure 4. Air pollutants emission in the future four scenarios. (**a**) SO_2; (**b**) NO_X; (**c**) PM_{10}; (**d**) $PM_{2.5}$.

3.2.1. SO_2 Emission Abatement

The variation of SO_2 emissions in the four scenarios are shown in Figure 4a. From the change trend of annual SO_2 emission, the CM scenario has the lowest emission in 2060. During 2020–2050, the annual emission in the SO scenario was lower than that in the GC scenario, and during 2050–2060, the annual emission in GC scenario was lower than that in the SO scenario. The annual emission of SO_2 in the BU scenario is the highest in future. From the perspective of cumulative emissions during 2020–2060, SO_2 emissions in the BU, SO, GC, and CM scenarios are 7.48 Mt, 5.66 Mt, 6.38 Mt, and 4.94 Mt, respectively. The differences indicate that coal regulation can contribute to SO_2 emission reduction. Compared with BU scenario, the cumulative emission reduction of SO_2 in SO scenario is 66.19%. It can be seen that the main factor affecting SO_2 emissions is the coal consumption structure rather than the gross coal consumption. Accelerating optimization of coal consumption structure is one of the effective measures to reduce SO2 emissions in the future.

The SO_2 emissions were further segmented by industry. In 2020, the SO_2 emissions of the building material industry are the highest, adding up to 124.19 kt and accounting for 34.58% of the total annual emission. The thermal power and heating supply industry, which consumes the largest proportion of coal (64.63%), emitted only 71.41 kt. This is because of the implementation of an ultra-low emission reform of thermal power units, resulting in a low SO_2 emission coefficient. The building material, steel, and chemical industries are the key industries to reduce SO_2 emissions in the future.

3.2.2. NO_x Emission Abatement

In the next 40 years, the change trend of NO_x emission caused by coal consumption in the four scenarios is shown in Figure 4b. Similarly to SO_2 emissions, annual NO_x emissions are the highest in the BU scenario and the lowest in the CM scenario. Before 2032, the annual NO_x emissions in the GC scenario are higher than that in the SO scenario, and the emissions in the GC scenario after 2032 are lower than that in the SO scenario. From the perspective of cumulative emissions, NO_x emissions in the BU, SO, GC, and CM scenarios are 4.40 Mt, 3.89 Mt, 3.68 Mt, and 3.28 Mt, respectively. It can be found that gross coal consumption control has a better effect on NO_x emission reduction than coal consumption structure optimization, different from SO_2 emission. To reduce NO_x emission, we must start from the source and control the gross amount of coal consumption. The NO_x emission were segmented by industry. In 2020, NO_x emission are mainly concentrated in the thermal power and heating supply industry, which consume a large amount of coal, accounting for 37.80% of the total annual NO_x emission.

3.2.3. Abatement of PM_{10} and $PM_{2.5}$ Emissions

The emission trends of PM_{10} and $PM_{2.5}$ in the four scenarios in the future are shown in Figure 4c,d. During 2020–2060, the cumulative emissions of PM_{10} in the BU, SO, GC, and CM scenarios are 2.50 Mt, 1.95 Mt, 2.14 Mt, and 1.71 Mt, respectively. The cumulative emissions of $PM_{2.5}$ were 1.72 Mt, 1.41 Mt, 1.46 Mt, and 1.22 Mt, respectively. It can be seen that the emission trend of PM in the four scenarios is similar to that of SO_2, and the emission of PM in the CM scenario is the lowest. NO_x emissions are the highest in the BU scenario. Compared with the BU scenario, the cumulative emission reduction of PM_{10} and $PM_{2.5}$ in the SO scenario was 51.71% and 20.40% higher than that in the GC scenario, respectively. The optimization of the coal consumption structure is also an effective measure to reduce particulate emissions. PM emissions are further divided into industries. In 2020, PM emissions are mainly concentrated in coking and briquettes and the building materials industry. Although the consumption of coal in the residential sector is relatively low, accounting for only 0.62%, the emissions of PM_{10} and $PM_{2.5}$ are 11.18% and 12.09% of the total, respectively. This indicates that the control of PM emissions in the future needs to further strengthen the supervision of the consumption of scattered coal.

3.3. Health Benefit Assessment

From 2020 to 2060, the cumulative number of cases with various diseases caused by coal consumption pollutant emission in the four scenarios is shown in Figure 5. The emission of pollutants from coal consumption leads to the largest number of asthma patients. In the BU scenario, the cumulative number of asthma patients in the next 40 years is up to 6.70 million cases, while in the CM scenario it is 4.75 million cases, with a reduction of 29.03%. The second is respiratory hospital admission. In the BU scenario, the cumulative number of patients will be 0.88 million cases by 2060, which will be reduced by 18.97%, 15.21%, and 29.93% in the SO, GC, and CM scenarios, respectively. The third is cardiovascular hospital admission. The cumulative number of patients in the BU scenario will be 0.40 million cases in the next 40 years, while that in the SO, GC, and CM scenarios will decrease by 16.48%, 15.71%, and 28.54%, respectively. Finally, in terms of mortality, in the BU scenario, the total number of deaths will be 0.23 million cases by 2060, which will be reduced by 14.80%, 16.01%, and 27.52%, respectively, in the SO, GC, and CM scenarios.

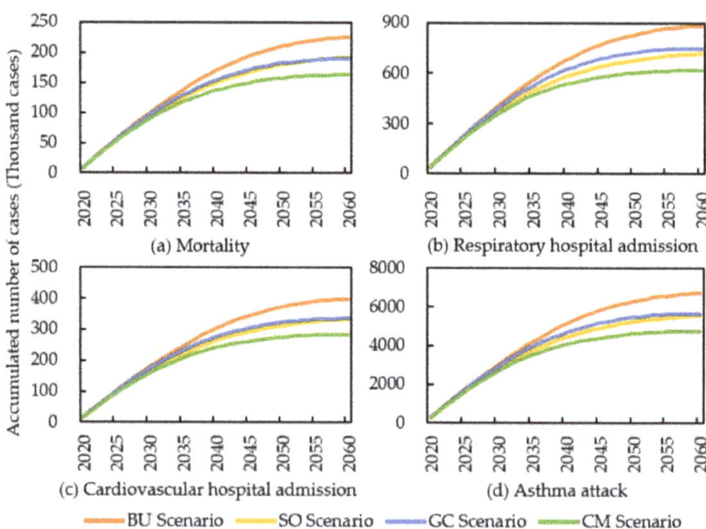

Figure 5. Cumulative number of various diseases in the four scenarios. (**a**) Mortality; (**b**) Respiratory hospital admission; (**c**) Cardiovascular hospital admission; (**d**) Asthma attack.

From the economic perspective of health benefits, compared with the BU scenario, the residents' health benefits of pollutant emission reduction during 2020–2060 in the three coal regulation scenarios are shown in Figure 6. By 2060, the cumulative health benefits of residents in the CM scenario will be the highest, with 8.296 billion USD, followed by the GC scenario, with 4.802 billion USD, and finally the SO scenario, with 4.486 billion USD. It can be further seen that, before 2056, the optimization of the coal consumption structure in the SO scenario has a significant effect on pollutant emission reduction, and the cumulative health benefits of residents are always higher than in the GC scenario. After 2056, the optimization of the coal consumption structure in the SO scenario is basically completed, and the effect of gross coal consumption control on the pollutant emission reduction in the GC scenario begins to be prominent, as the cumulative health benefits of residents exceed those of the SO scenario. In general, coal regulation can achieve significant health benefits for residents.

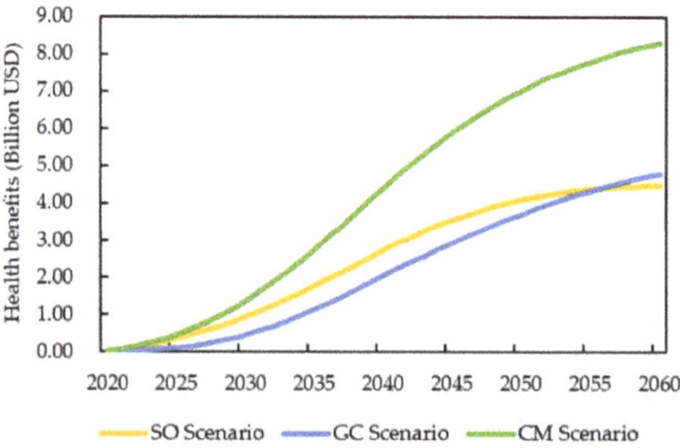

Figure 6. Health benefits of residents in different scenarios of coal regulation.

4. Discussion

4.1. Uncertainty and Sensitivity Analysis

In this study, the uncertainty of parameters will affect the results of the health benefit assessment. The overall uncertainty mainly comes from the following two aspects: (1) estimation of air pollutant emissions; (2) assessment of residents' health benefits.

In the process of pollutant emission calculation, the uncertainty of parameters mainly comes from the selection of the emission coefficients of SO_2, NO_x, PM_{10}, and $PM_{2.5}$ of coal consumption in various industries. However, the different sulfur, nitrogen, and ash contents in coal from different regions, as well as the different installation efficiency of sulfur, nitrogen, and dust removal equipment in the coal utilization equipment of different industries, all have a certain influence on the emission coefficient. At present, there is no authoritative, unified emission factor standard covering all fields needed for this study in the data published in China. We compared and referred to a large number of relevant studies to minimize the error. In addition, in order to be more consistent with the future development of Anhui, we assume that the emission factors of SO_2, NO_x, and PM will decrease by five percentage points every five years.

As for the assessment of the residents' health benefits, the uncertainty of the calculation results mainly involves the selection of the following four parameters: air pollutants in the population intake fraction IF_j; concentration-response coefficient CR_{hj}; mortality/morbidity of related health outcomes f_{hj}; unit economic value of related health outcomes UV_h. The selection of CR_{hj} and f_{hj} is mainly based on the relevant literature research data, and IF_j and UV_h are adjusted according to the number of permanent residents and the level of economic development in Anhui on the basis of relevant studies. As for the assessment of the residents' health benefits, the pollutant concentration (mg/m^3) (non-pollutant emissions (kt)) is usually used to calculate health risks. There is a mismatch between pollutant concentration and pollutant emission, which will be affected by climate, atmospheric environmental factors such as temperature, humidity, and wind direction, therefore the residents' health risk assessment will also have a certain deviation.

To further analyze the uncertainty, the sensitivity analysis of eight parameters involved in the two uncertain aspects of the calculation process is carried out. During each calculation, only one input parameter is changed, and then the residents' health benefits calculated from this parameter are compared with the results calculated without changing the parameters. The sensitivity analysis results of each input parameter are shown in Table 4.

Table 4. Sensitivity analysis of variable parameters on the health benefit.

Parameter	Change Range	Health Benefit Change		
		SO	GC	CM
	Estimation of air pollutant emissions			
SO_2 emission factors	−10%	−1.55%	−0.87%	−1.17%
	10%	1.55%	0.87%	1.17%
NO_x emission factors	−10%	−5.29%	−7.01%	−6.29%
	10%	5.29%	7.01%	6.29%
PM_{10} emission factors	−10%	−2.09%	−1.29%	−1.62%
	10%	2.09%	1.29%	1.62%
$PM_{2.5}$ emission factors	−10%	−1.07%	−0.83%	−0.93%
	10%	1.07%	0.83%	0.93%
	Assessment of residents' health benefits			
Intake fraction IF_j	−10%	−10.00%	−10.00%	−10.00%
	10%	10.00%	10.00%	10.00%
Concentration-response coefficient CR_{hj}	Lower values of 95% confidence	−75.00%	−77.03%	−76.13%
	Higher values of 95% confidence	68.55%	72.45%	70.79%
Mortality/Morbidity f_{hj}	−10%	−10.00%	−10.00%	−10.00%
	10%	10.00%	10.00%	10.00%
Unit economic value UV_h	Lower values of 95% confidence	−6.24%	−6.21%	−6.23%
	Higher values of 95% confidence	6.31%	6.26%	6.28%

When estimating air pollutant emissions, the emission factors of SO_2, NO_x, PM_{10}, and $PM_{2.5}$ of coal consumption in each industry are set to be 10% higher or lower than the original value. The variation of cumulative health benefits of residents in next 40 years in the three coal regulation scenarios are shown in Table 4. It can be seen that when the value of various emission factors is 10% higher than the original value, the residents' health benefits in all coal regulation scenarios will increase. Compared with various emission factors of air pollutants, it is obvious that the NO_x emission factor has the greatest impact on the residents' health benefits. A further analysis of the impact of the uncertainty of pollutant emission factors on the health benefits of residents in various scenarios shows that the uncertainty of the emission factors of SO_2, PM_{10}, and $PM_{2.5}$ are more sensitive to the health benefits of residents in the SO scenario, while the uncertainty of the emission factors of NO_x is more sensitive to the health benefits of residents in the GC scenario.

In terms of the assessment of the residents' health benefits, the concentration-response coefficient CR_{hj} and the unit economic value of related health outcomes UV_h adopted the data with a 95% confidence interval, indicating that the CR_{hj} coefficient had the greatest impact on the residents' health benefits. In addition, the value of the population intake fraction IF_j and the mortality/morbidity f_{hj} of related health results will directly affect the result of residents' health benefits.

4.2. Policy Implications

Although Anhui formulated strict coal consumption control targets, the gross coal consumption is still growing, indicating that it is a tough task. The main reason is that the supporting policies and measures are insufficient. Based on the consideration of environmental and health benefits, the following suggestions are provided for the coal consumption control strategy in the future. Firstly, the government should continue to promote the development of cleaner energies, such as wind energy and solar energy, to reduce the dependence on coal resources. Secondly, it will be useful to allocate annual coal consumption quotas to key enterprises, give priority to enterprises with high energy efficiency and low emissions, and shut down enterprises that do not meet the energy efficiency and emission standards. Thirdly, it is necessary to accelerate the coal consumption structure optimization so as to reduce emissions of SO_2, NO_x, and PM. For enterprises that need energy substitution, fiscal subsidies for equipment replacement are necessary. At present, the building materials, steel, and chemical industries still have a great potential to implement "coal to electricity or gas". Therefore, it is necessary to promote the implementation of "coal to electricity or gas" projects by enterprises. Finally, the government needs to continuously strengthen the monitoring of air pollutant emissions from enterprises and further improve the installation of pollution treatment equipment.

5. Conclusions

Based on historical data analysis and related research on coal regulation policies, this study constructs four scenarios of coal consumption in Anhui from 2020 to 2060. Furthermore, the coal consumption, air pollutant emissions, and effects on residents' health in each scenario are estimated. The main conclusions are summarized as follows.

In the scenarios of BU and SO, coal consumption will continue to increase to the peak in 2030 and decrease to 30 Mt in 2060 to achieve carbon neutrality. In the GC and CM scenarios, the carbon peak and carbon neutrality will be achieved five years in advance, respectively. In the CM scenario, after 2045, all terminal coal consumption industries will completely achieve "zero coal consumption", and coal resources will only be used for transformation.

Air pollutants emissions due to coal consumption are estimated in different scenarios, which will change in different paths. The cumulative emissions of SO_2, NO_x, PM_{10}, and $PM_{2.5}$ during 2020–2060 in the BU scenario are 7.48 Mt, 4.40 Mt, 2.50 Mt, and 1.72 Mt, respectively. Compared with the BU scenario, four types of air pollutants in the SO scenario are reduced by 24.33%, 11.62%, 22.02%, and 18.11%, respectively; and in the GC scenario, they are reduced by 14.64%, 16.50%, 14.51%, and 15.04%, respectively. In the CM scenario, the cumulative emissions of various air pollutants are the lowest in all scenarios, which are 33.92%, 25.54%, 31.56%, and 28.91% lower than those in the BU scenario, respectively. Structure optimization of coal consumption has an obvious effect on the emission reduction of SO_2 and PM, and gross coal consumption control has a better effect on the emission reduction of NO_x.

Residents' health will benefit from reduction of air pollutants emissions due to coal regulation as the number of patients and deaths will be decreased. In the BU scenario, the cumulative number of asthma patients will reach 6.70 million from 2020 to 2060; in the SO, GC, and CM scenarios, it can be effectively reduced by 17.39%, 15.52%, and 29.03%, respectively. From the economic perspective of the residents' health benefits, in the next 40 years, the CM scenario has the highest cumulative health benefits, with 8.30 billion USD, followed by the GC scenario, with 4.80 billion USD, and finally the SO scenario, with 4.49 billion USD. The estimation indicates that coal regulation can achieve significant health benefits for residents.

In summary, reducing coal consumption in China and some key provinces is not only essential for realizing carbon emission peak before 2030 and carbon neutrality before 2060, but also conducive to air quality and health improvement. Technological advance and energy transition in coal-related sectors should be encouraged by government with a strong effort. There are some limitations in this study. We considered the effects of SO_2, NO_x, and PM emissions, but overlooked the impact of other pollutants on residents' health. The intake fraction method was used to evaluate the health benefits of residents, and the results were uncertain to some extent. In addition, the carbon emission permit trading policy that will be promoted in the future will also have a certain impact on coal consumption in Anhui, which should be further explored in future research.

Author Contributions: Conceptualization, W.G. and F.L.; methodology, W.X. and W.G.; software, W.G.; validation, W.G. and W.S.; formal analysis, W.G. and W.S.; investigation, W.G. and F.L.; resources, W.G. and F.L.; data curation, W.G. and W.S.; writing—original draft preparation, W.X. and W.G.; writing—review and editing, W.S. and Z.T.; visualization, W.G.; supervision, W.X. and Z.T.; project administration, F.L.; funding acquisition, F.L. and Z.T. All authors have read and agreed to the published version of the manuscript.

Funding: This research was funded by the National Key Research and Development Program of China (Grant no. 2016YFA0602804) and the Fundamental Research Funds for the Central Universities of China (Grant no. JZ2021HGTB0069).

Institutional Review Board Statement: Not applicable.

Informed Consent Statement: Not applicable.

Conflicts of Interest: The authors declare no conflict of interest.

References

1. NBSC. *China Statistical Yearbook*; China Statistics Press: Beijing, China, 2020.
2. BP. *BP Statistical Review of World Energy 2020*; British Petroleum: London, UK, 2020.
3. Niu, D.; Song, Z.; Xiao, X. Electric power substitution for coal in China: Status quo and SWOT analysis. *Renew. Sustain. Energy Rev.* **2017**, *70*, 610–622. [CrossRef]
4. Yorifuji, T.; Kashima, S.; Doi, H. Acute exposure to fine and coarse particulate matter and infant mortality in Tokyo, Japan (2002–2013). *Sci. Total Environ.* **2016**, *551–552*, 66–72. [CrossRef] [PubMed]
5. Tao, Y.; Mi, S.; Zhou, S.; Wang, S.; Xie, X. Air pollution and hospital admissions for respiratory diseases in Lanzhou, China. *Environ. Pollut.* **2014**, *185*, 196–201. [CrossRef]
6. Sun, J.; Wang, J.; Wei, Y.; Li, Y.; Liu, M. The Haze Nightmare Following the Economic Boom in China: Dilemma and Tradeoffs. *Int. J. Environ. Res. Public Health* **2016**, *13*, 402. [CrossRef] [PubMed]
7. Tran, B.; Chang, C.; Hsu, C.; Chen, C.; Tseng, W.; Hsu, S. Threshold Effects of PM2.5 Exposure on Particle-Related Mortality in China. *Int. J. Environ. Res. Public Health* **2019**, *16*, 3549. [CrossRef]
8. Weschenfelder, F.; de Novaes Pires Leite, G.; Araújo da Costa, A.C.; de Castro Vilela, O.; Ribeiro, C.M.; Villa Ochoa, A.A.; Araújo, A.M. A review on the complementarity between grid-connected solar and wind power systems. *J. Clean. Prod.* **2020**, *257*, 120617. [CrossRef]

9.   Li, J.; Tian, Y.; Deng, Y.; Zhang, Y.; Xie, K. Improving the estimation of greenhouse gas emissions from the Chinese coal-to-electricity chain by a bottom-up approach. *Resour. Conserv. Recycl.* **2021**, *167*, 105237. [CrossRef]

10.   Shin, H.; Kim, T.H.; Kim, H.; Lee, S.; Kim, W. Environmental shutdown of coal-fired generators for greenhouse gas reduction: A case study of South Korea. *Appl. Energy* **2019**, *252*, 113453. [CrossRef]

11.   Squalli, J. Renewable energy, coal as a baseload power source, and greenhouse gas emissions: Evidence from U.S. state-level data. *Energy* **2017**, *127*, 479–488. [CrossRef]

12.   Kopas, J.; York, E.; Jin, X.; Harish, S.P.; Kennedy, R.; Shen, S.V.; Urpelainen, J. Environmental Justice in India: Incidence of Air Pollution from Coal-Fired Power Plants. *Ecolog. Econ.* **2020**, *176*, 106711. [CrossRef]

13.   Zhou, Y.; Zi, T.; Lang, J.; Huang, D.; Wei, P.; Chen, D.; Cheng, S. Impact of rural residential coal combustion on air pollution in Shandong, China. *Chemosphere* **2020**, *260*, 127517. [CrossRef] [PubMed]

14.   Du, X.; Jin, X.; Zucker, N.; Kennedy, R.; Urpelainen, J. Transboundary air pollution from coal-fired power generation. *J. Environ. Manag.* **2020**, *270*, 110862. [CrossRef]

15.   Yu, C.; Kang, J.; Teng, J.; Long, H.; Fu, Y. Does coal-to-gas policy reduce air pollution? Evidence from a quasi-natural experiment in China. *Sci. Total Environ.* **2021**, *773*, 144645. [CrossRef]

16.   Liu, C.; Zhu, B.; Ni, J.; Wei, C. Residential coal-switch policy in China: Development, achievement, and challenge. *Energy Policy* **2021**, *151*, 112165. [CrossRef]

17.   Sun, J.; Du, T.; Sun, W.; Na, H.; He, J.; Qiu, Z.; Yuan, Y.; Li, Y. An evaluation of greenhouse gas emission efficiency in China's industry based on SFA. *Sci. Total Environ.* **2019**, *690*, 1190–1202. [CrossRef]

18.   Yang, X.; Teng, F. The air quality co-benefit of coal control strategy in China. *Resour. Conserv. Recycl.* **2018**, *129*, 373–382. [CrossRef]

19.   Ai, H.; Guan, M.; Feng, W.; Li, K. Influence of classified coal consumption on PM2.5 pollution: Analysis based on the panel cointegration and error-correction model. *Energy* **2021**, *215*, 119108. [CrossRef]

20.   Cao, Y.; Zhao, Y.; Wen, L.; Li, Y.; Li, H.; Wang, S.; Liu, Y.; Shi, Q.; Weng, J. System dynamics simulation for CO2 emission mitigation in green electric-coal supply chain. *J. Clean. Prod.* **2019**, *232*, 759–773. [CrossRef]

21.   Sun, L.; Xiang, M.; Shen, Q. A comparative study on the volatility of EU and China's carbon emission permits trading markets. *Phys. A Stat. Mech. Appl.* **2020**, *560*, 125037. [CrossRef]

22.   Wang, G.; Zhang, Q.; Su, B.; Shen, B.; Li, Y.; Li, Z. Coordination of tradable carbon emission permits market and renewable electricity certificates market in China. *Energy Econ.* **2021**, *93*, 105038. [CrossRef]

23.   Bashir, M.F.; Ma, B.; Shahbaz, M.; Shahzad, U.; Vo, X.V. Unveiling the heterogeneous impacts of environmental taxes on energy consumption and energy intensity: Empirical evidence from OECD countries. *Energy* **2021**, *226*, 120366. [CrossRef]

24.   Yue, H.; Worrell, E.; Crijns-Graus, W.; Zhang, S. The potential of industrial electricity savings to reduce air pollution from coal-fired power generation in China. *J. Clean. Prod.* **2021**, *301*, 126978. [CrossRef]

25.   Chen, H.; Chen, W. Potential impact of shifting coal to gas and electricity for building sectors in 28 major northern cities of China. *Appl. Energy* **2019**, *236*, 1049–1061. [CrossRef]

26.   Yang, H.; Tao, W.; Liu, Y.; Qiu, M.; Liu, J.; Jiang, K.; Yi, K.; Xiao, Y.; Tao, S. The contribution of the Beijing, Tianjin and Hebei region's iron and steel industry to local air pollution in winter. *Environ. Pollut.* **2019**, *245*, 1095–1106. [CrossRef] [PubMed]

27.   Zhang, Y.; Song, Y.; Zou, H. Transformation of pollution control and green development: Evidence from China's chemical industry. *J. Environ. Manag.* **2020**, *275*, 111246. [CrossRef] [PubMed]

28.   Zhu, X.; Li, H.; Chen, J.; Jiang, F. Pollution control efficiency of China's iron and steel industry: Evidence from different manufacturing processes. *J. Clean. Prod.* **2019**, *240*, 118184. [CrossRef]

29.   Xu, S.; Ge, J. Sustainable shifting from coal to gas in North China: An analysis of resident satisfaction. *Energy Policy* **2020**, *138*, 111296. [CrossRef]

30.   Wang, K.; Tong, Y.; Yue, T.; Gao, J.; Wang, C.; Zuo, P.; Liu, J. Measure-specific environmental benefits of air pollution control for coal-fired industrial boilers in China from 2015 to 2017. *Environ. Pollut.* **2021**, *273*, 116470. [CrossRef] [PubMed]

31.   Xiao, K.; Li, F.; Dong, C.; Cai, Y.; Li, Y.; Ye, P.; Zhang, J. Unraveling effects of coal output cut policy on air pollution abatement in China using a CGE model. *J. Clean. Prod.* **2020**, *269*, 122369. [CrossRef]

32.   Wang, Y.; Wang, D.; Shi, X. Exploring the dilemma of overcapacity governance in China's coal industry: A tripartite evolutionary game model. *Resour. Pol.* **2021**, *71*, 102000. [CrossRef]

33.   Yang, J.; Zhao, Y.; Cao, J.; Nielsen, C.P. Co-benefits of carbon and pollution control policies on air quality and health till 2030 in China. *Environ. Int.* **2021**, *152*, 106482. [CrossRef] [PubMed]

34.   Chen, H.; Li, L.; Lei, Y.; Wu, S.; Yan, D.; Dong, Z. Public health effect and its economics loss of PM2.5 pollution from coal consumption in China. *Sci. Total Environ.* **2020**, *732*, 138973. [CrossRef] [PubMed]

35.   SEPA. *Practical Manual for Discharge Declaration and Registration*; China Environmental Science Press: Beijing, China, 2004.

36.   Li, F.; Xiao, X.; Xie, W.; Ma, D.; Song, Z.; Liu, K. Estimating air pollution transfer by interprovincial electricity transmissions: The case study of the Yangtze River Delta Region of China. *J. Clean. Prod.* **2018**, *183*, 56–66. [CrossRef]

37.   Li, F.; Cai, B.; Ye, Z.; Wang, Z.; Zhang, W.; Zhou, P.; Chen, J. Changing patterns and determinants of transportation carbon emissions in Chinese cities. *Energy* **2019**, *174*, 562–575. [CrossRef]

38.   Chen, S.-M.; He, L.-Y. Welfare loss of China's air pollution: How to make personal vehicle transportation policy. *China Econ. Rev.* **2014**, *31*, 106–118. [CrossRef]

39. Yang, S.; He, L.-Y. Fuel demand, road transport pollution emissions and residents' health losses in the transitional China. *Transp. Res. Part. D Transp. Environ.* **2016**, *42*, 45–59. [CrossRef]

40. Bennett, D.H.; McKone, T.E.; Evans, J.S.; Nazaroff, W.W.; Margni, M.D.; Jolliet, O.; Smith, K.R. Defining intake fraction. *Environ. Sci. Technol.* **2002**, *36*, 207A–211A. [CrossRef]

41. NBSC. *China Energy Statistical Yearbook*; China Statistics Press: Beijing, China, 2020.

42. NBSC. *Anhui Statistical Yearbook*; China Statistics Press: Beijing, China, 2021.

43. CEC. *China Power Industry Annual Development Report 2020*; China Market Press: Beijing, China, 2020.

44. Gao, T.; Zhou, F.; Yan, Q.; Zhang, Y. Comparison of main air pollutant emission in different ways of coal utilization. *China Min. Mag.* **2017**, *26*, 74–80, 95.

45. Guo, X.; Zhao, L.; Chen, D.; Jia, Y.; Chen, D.; Zhou, Y.; Cheng, S. Prediction of reduction potential of pollutant emissions under the coal cap policy in BTH region, China. *J. Environ. Manag.* **2018**, *225*, 25–31. [CrossRef]

46. MEEC. *The Second National Pollution Source Census Production and Discharge Accounting Coefficient Manual*; China Environmental Science Press: Beijing, China, 2020.

47. Liu, L.; Wang, K.; Wang, S.; Zhang, R.; Tang, X. Assessing energy consumption, CO2 and pollutant emissions and health benefits from China's transport sector through 2050. *Energy Policy* **2018**, *116*, 382–396. [CrossRef]

48. Chen, H.; Wang, Z.; Xu, S.; Zhao, Y.; Cheng, Q.; Zhang, B. Energy demand, emission reduction and health co-benefits evaluated in transitional China in a 2 °C warming world. *J. Clean. Prod.* **2020**, *264*, 121773. [CrossRef]

49. Wang, K.; Wang, S.; Liu, L.; Yue, H.; Zhang, R.; Tang, X. Environmental co-benefits of energy efficiency improvement in coal-fired power sector: A case study of Henan Province, China. *Appl. Energy* **2016**, *184*, 810–819. [CrossRef]

50. Wang, P.; Shen, J.; Zhu, S.; Gao, M.; Ma, J.; Liu, J.; Gao, J.; Zhang, H. The aggravated short-term PM 2.5-related health risk due to atmospheric transport in the Yangtze River Delta. *Environ. Pollut.* **2021**, *275*, 116672. [CrossRef] [PubMed]

51. Liu, Q.; Shi, Y.; Zhong, W.; Yu, A. Co-firing of coal and biomass in oxy-fuel fluidized bed for CO2 capture: A review of recent advances. *Chin. J. Chem. Eng.* **2019**, *27*, 2261–2272. [CrossRef]

 **sustainability**

Article

Comparison of Driving Forces to Increasing Traffic Flow and Transport Emissions in Philippine Regions: A Spatial Decomposition Study

Geoffrey Udoka Nnadiri [1], Anthony S. F. Chiu [2], Jose Bienvenido Manuel Biona [1,3] and Neil Stephen Lopez [1,*]

[1] Mechanical Engineering Department, De La Salle University, Manila 0922, Philippines; geoffrey_nnadiri@dlsu.edu.ph (G.U.N.); jose.bienvenido.biona@dlsu.edu.ph (J.B.M.B.)

[2] Industrial & Systems Engineering Department, De La Salle University, Manila 0922, Philippines; anthony.chiu@dlsu.edu.ph

[3] Enrique K. Razon Logistics Institute, De La Salle University, Manila 0922, Philippines

* Correspondence: neil.lopez@dlsu.edu.ph

Abstract: The warming of the climate system has raised a lot of concerns for decades, and this is traceable to human activities and energy use. Conspicuously, the transportation sector is a great contributor to global emissions. This is largely due to increasing dependence on private vehicles and a poorly planned public transportation system. In addition to economic impacts, this also has significant environmental and sustainability implications. This study demonstrates a novel approach using spatial logarithmic mean Divisia index (LMDI) to analyze drivers of traffic flow and its corresponding CO_2 emissions in regions through an illustrative case study in the Philippines. Population growth is revealed as the main driver to traffic flow in most regions with the exception of a few regions and the national capital which are driven by economic activity. The economic activity effect shows positive trends contributing positively to traffic flow which is greatly linked to income level rise and increase in vehicle ownership. Concerning the impacts, results revealed that an increase in economic activity generally causes traffic intensity to decrease, and switching to more sustainable modes is not a guarantee to reduce carbon emissions. The authors recommend increasing equity on the appropriation of transport infrastructure projects across regions, quality improvement of public transport services and promoting mixed-use development.

Keywords: transport; spatial LMDI; emissions; Philippines

Citation: Nnadiri, G.U.; Chiu, A.S.F.; Biona, J.B.M.; Lopez, N.S. Comparison of Driving Forces to Increasing Traffic Flow and Transport Emissions in Philippine Regions: A Spatial Decomposition Study. *Sustainability* **2021**, *13*, 6500. https://doi.org/10.3390/su13116500

Academic Editors: Wei-Hsin Chen, Alvin B. Culaba, Aristotle T. Ubando and Steven Lim

Received: 17 May 2021
Accepted: 6 June 2021
Published: 7 June 2021

Publisher's Note: MDPI stays neutral with regard to jurisdictional claims in published maps and institutional affiliations.

1. Introduction

Scientific assessments disclosed by the Philippine Atmospheric, Geophysical and Astronomical Services Administration (PAGASA) revealed that the warming of the climate system is likely due to human activities such as the combustion of fossil fuels and land use [1]. Human influence on the climate system is clear, and climate change is among the most critical global issues that must be addressed. From a global report, the most important drivers of increases in CO_2 emissions from fossil fuel combustion had been traced to economic and population growth [2]. According to a report from the International Energy Agency [3], the annual CO_2 emissions from fossil fuels have increased from an estimate of 23.6 billion tons of carbon dioxide ($GtCO_2$) to around 32.4 $GtCO_2$ from 1990 to 2014, and in 2018, the global energy consumption grew almost twice the average growth rate since 2010 [4]. Coincidentally, transportation is noted to be the fastest consumer of fossil fuels and source of carbon emissions. In 2012, the transportation sector accounted for 23% of global CO_2 emissions due to fossil fuel combustion. With rapid urbanization in developing countries, energy consumption and CO_2 emissions by metropolitan vehicles are expanding quickly as the transport sector continues to experience challenges in reducing CO_2 emissions [5]. In line with this, carbon emissions from the transport sector are gradually increasing each year, responsible for approximately 24% of global carbon emissions in 2018 [6]. On the other hand, GDP growth remains to be the leading factor in energy consumption growth [7].

To further reveal transportation problems and gaps, a large body of literature has examined the drivers of increasing energy use and emissions from transport. Chen and Yang [8] further disclosed that energy-related emissions are linked to economic growth and the population's standard of living and suggested a shift in the economic structure towards sectors with lower energy intensity. Dangay and Gately [9] carried out an investigation and projection of income's impact on global vehicle ownership utilizing the Gompertz function and inferred that there exists a solid connection between the growth of per capita income and the growth of vehicle ownership per capita. As per capita income increases, so will the number of vehicles until saturation is reached. To address increasing vehicle use, Martin et al. [10] recommended carsharing for non-work-related travel. In support of alternatives, Ubando et al. [11] proposed an optimization methodology for electric vehicle use, considering the capital expense and power costs. Papagiannaki and Diakoulaki [12] focused on changes in transportation profiles to explain the rapid growth of the transport sector, while Jiang [13] found a surge in car ownership, with the modal share effect being consistently positive, suggesting that more people are preferring less efficient modes. In conjunction with this, Ding et al. [14] noted that the growth in transport CO_2 emissions is primarily due to modal shifts. Wang et al. [15] introduced the avoid, shift and improve approaches focusing on passenger transport. Guo et al. [16] further highlighted the need to approach each region or country exclusively to further understand the main drivers. In previous studies, multi-regional studies have been conducted through independent decomposition analyses and then comparing the different indices and major drivers among regions. While previous studies have unveiled the drivers of increasing transport activity, energy use and emissions using country- or city-specific temporal data, there exists a gap in the literature with regards to the spatial comparison of these drivers. As different cities, countries and regions can adopt different growth trajectories, they can experience different futures, may it be sustainable or unsustainable.

In response to the above-identified gaps in transport issues, researchers find index decomposition as a suitable technique to investigate transport-related problems. Index decomposition analysis is one of the techniques widely used in analyzing the change of energy consumption over a period. Plenty of related studies have used the logarithmic mean Divisia index (LMDI) method to analyze transport sector problems after it was first introduced by Ang et al. [17]. The two primarily utilized decomposition analysis techniques are the Laspeyres and Divisia related methods in Ang [18]. The LMDI technique is favored over other approaches because of its ideal decomposition and capacity to deal with cases with zero quantities. LMDI has been demonstrated to be multidisciplinary, such as in Lopez et al. [19], dissecting the impacts of various drivers to fuel combustion and electricity generation in the Philippines, and in Lopez et al. [6], which further broke down the drivers to CO_2 emissions from the Philippine transportation sector. Ang [20], in addition, provided a practical guide and case studies in decomposing industrial energy consumption and environmental emissions. Due to the gap in comparing regions, cities, or countries, Ang [21] introduced a novel methodology for using LMDI in spatial studies. This approach provides a simple but informative way to benchmark between cities, regions, or countries compared to the traditional temporal-based LMDI.

In this present work, the authors utilize a novel application of the spatial LMDI methodology by Ang [21] to road transportation. As different regions are currently on different stages and trajectories of development, each one can be taken as a representative of a particular stage. Thus, the hypothesis of this study is that data from each region can be used to predict what can happen to other regions as they adopt the same development trajectory. This approach promises interesting insights which can be used for reviewing existing sustainable transport policies and recommending new ones.

The paper proceeds as follows: Section 2 presents a literature review; Section 3 discusses the methodology and data sources for the study; Section 4 discusses the results and Section 5 concludes the study.

2. Literature Review

In this section, key drivers and findings from existing studies using decomposition analysis to address transportation issues are summarized (see Table 1). This affirms that decomposition analysis can be a reliable tool in analyzing transportation problems.

Table 1. Summary of the literature review using decomposition analysis to solve transportation problems.

Author/s	Drivers Identified	Field of Study	Method
Feng et al. [22]	Scale impact (largest); production innovation; energy-saving innovation; and energy structure.	Transport sector (China)	LMDI * (temporal) and production theoretical decomposition analysis (PDA)
Sorrell et al. [23]	Value of domestically manufactured goods to GDP.	Road freight energy use (UK)	LMDI (temporal)
Papagiannaki and Diakoulaki [12]	Passenger car usage.	Road transport CO_2 emissions (Greece and Denmark)	LMDI (temporal)
Lu et al. [24]	Economic activity and vehicle ownership (contributor); population intensity (inhibitor).	Transport CO_2 emissions (Germany, Japan, South Korea and Taiwan)	Divisia index decomposition analysis (DIDA)
Kwon [25]	Vehicle driving distance per individual.	Vehicle CO_2 emissions (Great Britain)	Index decomposition analysis (IDA) (temporal)
Liu and Feng [26]	Per capita service output (contributor); urbanization (inhibitor).	Transport energy and CO_2 emissions (China)	LMDI (temporal)
Andreoni and Galmarini [27]	Economic growth	Water and aviation transport sector (14 European Countries)	LMDI (temporal)
Timilsina and Shrestha [28]	Economic growth and transportation energy intensity.	Transport sector (20 Latin American countries)	LMDI (temporal)
Zhang, Liu and Yao [29]	Income (largest), energy intensity and transportation structure.	Transport sector (China)	LMDI (temporal and Spatial decomposition SD)
Tu et al. [30]	Mode structure and motorization	Urban transport sectors (London, Paris, New York and Tokyo)	LMDI (temporal)
Cao et al. [31]	Aviation transport	Regional transport modes CO_2 emissions (Pearl River Delta)	LMDI (temporal)
Feng, Xia and Sun [32]	Transportation demand and urbanization (contributor); energy intensity and industrial structure (inhibitor).	Transport CO_2 emissions (China)	LMDI (temporal)
Li et al. [33]	Income (contributor) and energy intensity (inhibitor).	Transport sector (China)	LMDI (temporal)

Table 1. *Cont.*

Author/s	Drivers Identified	Field of Study	Method
Lian et al. [34]	Total output (intermediate use effect, domestic final demand, import substitution effect and export extension effect) and the energy intensity	Transport sector (China)	Structural decomposition approach
Román-Collado and Morales-Carrión [35]	Activity and population effect (contributor) and intensity effect (inhibitor).	Energy CO_2 emissions (Latin America)	Multiregional spatial decomposition analysis
Timilsina and Shrestha ([36]	Per capita GDP, Population growth, per capita economic growth and transportation energy intensity.	Transport sector (11 Asian countries)	LMDI (temporal)
Mendiluce and Schipper [37]	Transport activity	Transport sector (Spain)	LMDI (temporal)
Shi et al. [38]	Household income, number of household, energy intensity, energy structure and carbon emission coefficient.	Houshold energy CO_2 emissions (China)	LMDI (temporal and SD)
Wang et al. [15]	Economic activity and transportation modal shifting effect (contributor); transportation intensity and transportation services share effect (inhibitor).	Transport sector (China)	LMDI (temporal)
Guo et al. [16]	Economic activity and population effect	Transport sector (China)	LMDI (temporal)
Sumabat et al. [39]	Economic growth and better quality of living (inhibitor).	Energy use CO_2 emissions (Philippines)	LMDI (temporal)

* Logarithmic mean Divisia index (LMDI) method.

Based on this review, the most common decomposition method used is temporal analysis due to its ability to analyze time-series data over a long period of time. However, there are still gaps when it comes to comparing and understanding the interrelationship between different regions, and this can be covered using spatial analysis. Common drivers suggested were economic growth, income effect and population effect. Nevertheless, the performance of various driving factors varies in each study. Similar recommendations include: (a) the establishment and implementation of policies to encourage the use of green transport modes and fuel switching; (b) interregional collaborations to encourage the reduction of CO_2 emissions from transport; (c) cities with high population densities should focus on improving public and non-motorized transport; and (d) control private vehicle ownership through vehicle stock restriction and higher tax rates when purchasing a second vehicle. Activity reduction, as recommended, can be achieved through regulations, restrictions and mobility plans, by increasing availability and accessibility of high-speed rail transport and by improving intermodal transport networks.

3. Methods and Data

This study uses the spatial LMDI decomposition methodology to compare the drivers (i.e., contributors and inhibitors) of increasing traffic flow, transport energy use and emissions across Philippine regions. As different regions are currently on different stages and trajectories of development, each one can be taken as a representative of a particular stage. Thus, the hypothesis of this study is that data from each region can be used to predict what can happen to other regions as they adopt the same development trajectory (see Figure 1).

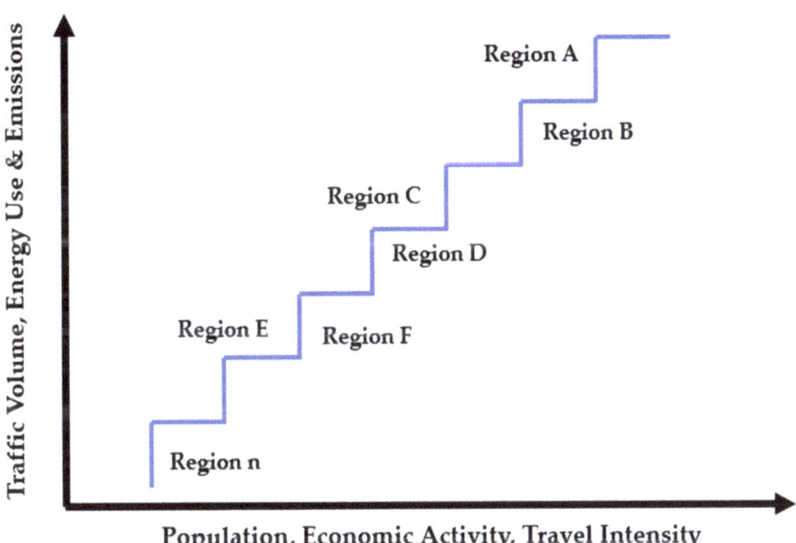

Figure 1. Conceptual framework of the study.

The Philippines is divided into 17 regions, as illustrated in Figure 2 below. There are three main island groups—Luzon, Visayas and Mindanao. Within the island group of Luzon are Regions 1 to 5, the Cordillera Administrative Region (CAR) and the National Capital Region (NCR); in Visayas are Regions 6 to 8; and in Mindanao are Regions 9 to 13, the Autonomous Region in Muslim Mindanao (ARMM) and the Caraga Administrative Region (CARAGA). It is important to note that NCR is the economic capital of the Philippines, contributing ~36% of the national GDP and about ~10% of the national population. Consequently, these economic activities and prosperity in the NCR spillover into the adjacent Regions III and IV-A. In addition, 122 urban communities make up the Philippines, of which 33 are delegated as "exceptionally urbanized" and 5 as "autonomous segments", while the rest are urban communities of the regions in which they are topographically found. Other key highly urbanized cities in the Philippines outside Luzon include Cebu City (Visayas, Region VII) and Davao City (Mindanao, Region XI).

Two angles are analyzed in this study. The first investigates the drivers of changing traffic flow. The identity function used for the decomposition of this angle is shown in Equation (1).

$$\text{VKM} = \sum_i \left(\text{pop} \times \frac{\text{GDP}}{\text{pop}} \times \frac{\text{VKM}}{\text{GDP}} \times \frac{\text{VKM}_i}{\text{VKM}} \right) = \sum_i (\text{pop} \times \text{act} \times \text{int} \times \text{str}_i) \qquad (1)$$

where VKM refers to vehicle-kilometers traveled (km), pop refers to the population (no. of persons), GDP refers to the gross domestic product (PhP) and i refers to the transport mode (e.g., private car, rail, bus, etc.). Moreover, the equivalent indices are on the righthand side of the identity function: pop refers to the population effect, act refers to the economic activity effect, int refers to the travel intensity effect and str refers to the modal structure effect.

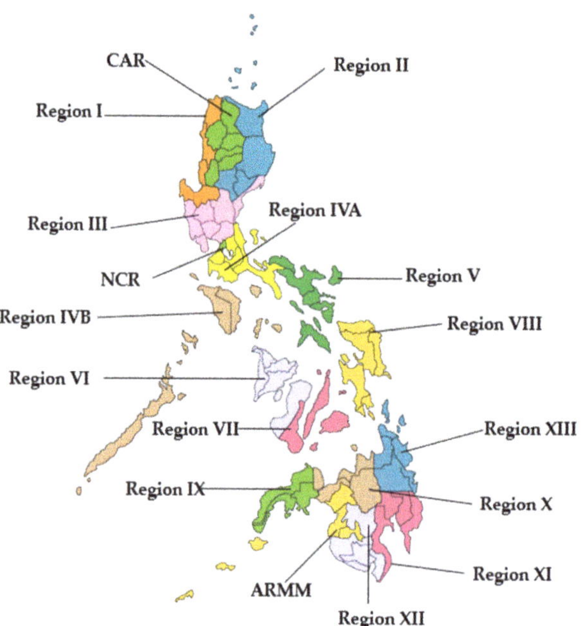

Figure 2. Map of Philippines with Regions.

The second angle investigates the drivers of changes in CO_2 emissions from transport. The identity function for decomposition used for this angle is shown in Equation (2).

$$CO_2 = \sum_i \left(pop \times \frac{GDP}{pop} \times \frac{VKM}{GDP} \times \frac{VKM_i}{VKM} \times \frac{CO_{2i}}{VKM_i} \right) = \sum_i (pop \times act \times int \times str_i \times emf_i) \tag{2}$$

where CO_2 refers to carbon dioxide emitted. The definitions of the other variables also present in Equation (1) remain the same. Regarding the equivalent indices on the righthand side, emf refers to the emission factor effect. However, it was assumed in this study that the emission factors are constant between regions, so this can be neglected.

The drivers to transport energy consumption are no longer analyzed separately, as a majority of the transportation in the Philippines is fossil powered. The drivers for energy consumption are most likely going to be exactly the same as that of emissions.

The calculation of the contribution of each effect (i.e., pop, act, int and str) to traffic flow and emissions was done using the procedure described in Ang [21]. The logarithmic functions shown from Equations (3)–(6) are used to calculate each effect per region, where R refers to any particular region, and μ refers to the benchmark region. It is important to note that in spatial decomposition, the indices of each region are compared to the benchmark region. As recommended in Ang [21], the benchmark region for this study is taken to be the average of all regions. To illustrate this, the population of the benchmark region is the average population of all regions; the GDP is the average GDP of all regions; and so on.

$$\Delta VKM_{pop} = \frac{VKM^R - VKM^\mu}{\ln VKM^R - \ln VKM^\mu} \ln \left(\frac{pop^R}{pop^\mu} \right) \tag{3}$$

$$\Delta VKM_{act} = \frac{VKM^R - VKM^\mu}{\ln VKM^R - \ln VKM^\mu} \ln \left(\frac{act^R}{act^\mu} \right) \tag{4}$$

$$\Delta VKM_{int} = \frac{VKM^R - VKM^\mu}{\ln VKM^R - \ln VKM^\mu} \ln \left(\frac{int^R}{int^\mu} \right) \tag{5}$$

$$\Delta VKM_{str} = \sum_i \frac{VKM_i^R - VKM_i^\mu}{\ln VKM_i^R - \ln VKM_i^\mu} \ln\left(\frac{str_i^R}{str^\mu}\right) \tag{6}$$

where ΔVKM refers to the contribution of a particular index to traffic flow.

$$\Delta CO_{2pop} = \frac{CO_2^R - CO_2^\mu}{\ln CO_2^R - \ln CO_2^\mu} \ln\left(\frac{pop^R}{pop^\mu}\right) \tag{7}$$

$$\Delta CO_{2act} = \frac{CO_2^R - CO_2^\mu}{\ln CO_2^R - \ln CO_2^\mu} \ln\left(\frac{act^R}{act^\mu}\right) \tag{8}$$

$$\Delta CO_{2int} = \frac{CO_2^R - CO_2^\mu}{\ln CO_2^R - \ln CO_2^\mu} \ln\left(\frac{int^R}{int^\mu}\right) \tag{9}$$

$$\Delta CO_{2str} = \sum_i \frac{CO_{2i}^R - CO_{2i}^\mu}{\ln CO_{2i}^R - \ln CO_{2i}^\mu} \ln\left(\frac{str_i^R}{str^\mu}\right) \tag{10}$$

where ΔCO_2 refers to the contribution of a particular index to transport emissions.

Furthermore, the authors opted to analyze vehicle-kilometers (VKM) instead of passenger-kilometers (PKM) because, from an environmental perspective, transport emissions are more correlated to VKM than PKM. For instance, rail transport and private vehicle transport can accrue the same amount of PKM, but not the same amount of emissions because rail can carry more passengers at a time, becoming more efficient with its energy use.

The data in this study were obtained from the Family Income and Expenditure Survey of the Philippine Statistics Authority [40] and the Gross Regional Domestic Product data from [41]. To obtain travel activity data, household expenditure data were converted using the estimated fuel prices, fuel economy of private vehicles and average fare prices per kilometer of public transport modes from the Department of Energy and Land Transportation Franchising and Regulatory board [42,43]. The emission factors used in this study were from Fabian and Gita [44].

4. Results and Discussion

4.1. Drivers to Traffic Flow

The results of the spatial LMDI calculations are illustrated in Figure 3. Each effect (e.g., ΔVKM_{pop}) is represented by a column. The total difference (ΔVKM) between a particular region's VKM and the national average VKM is also plotted using a dotted line. On this chart, a positive value means that the additional VKM due to the specific effect (e.g., population growth for ΔVKM_{pop}) is higher than the national average and vice versa. Similarly, a negative value for the total difference (ΔVKM) means that the total VKM in that region is less than the national average. As can be seen in the figure, only Regions 3, 4A, NCR and 7 (marginally) produce a VKM higher than the national average. This demonstrates one of the issues brought up in Ang [21] on the creation of a hypothetical benchmark region using the average of all other regions. If a few regions dominate the other regions, they can skew the characteristics of the benchmark region towards them.

Figure 3. Effect estimates by region, as compared with the national average, μ. ΔVKM_{pop} refers to the population effect, ΔVKM_{act} refers to the activity effect, ΔVKM_{int} refers to the intensity effect and ΔVKM_{str} refers to the structural effect.

Regions 3, 4A and NCR stand out among others in terms of vehicle-kilometers traveled per year. NCR is the economic capital of the country, while Regions 3 and 4A are directly adjacent to the north and south, respectively. Thus, it is understandable how the transport activity from NCR easily dissipates to them. It is quite common for Region 3 and 4A residents to find employment inside NCR.

With regards to the effects, it is interesting to note that population growth is the primary driver in most regions, with exception of Regions 1, 5, 6, 8, 12, ARMM and NCR, which are driven by economic activity. In Figure 4, the intensity and structural effects are presented separately. It is seen that structural effects are negligible compared to all other effects, suggesting that the difference in modal structure across regions does not significantly vary. On the other hand, it is interesting to note that NCR has a very low transport intensity. This can be explained by the fact that the majority of the major mass transport projects have been funneled into the capital and the booming high-rise residential building development in it, which created multiple mixed-use districts [45,46]. Mixed-use development can lead to reduced traveling since persons can live near their places of employment [47].

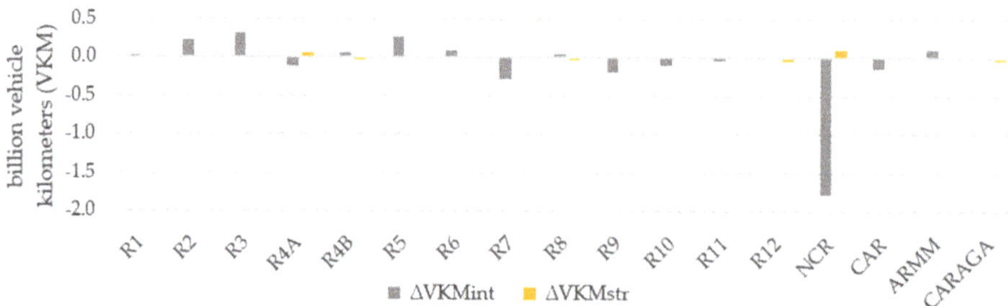

Figure 4. Effect estimates by region for traffic intensity (ΔVKM_{int}) and modal structure (ΔVKM_{str}), as compared with the national average, μ.

Figure 5 breaks down the total VKM by mode. While the gasoline private car dominates in most regions, it is only second to the diesel private car in NCR. This can be because of various reasons, which include the NCR residents' higher capacity to pay, since diesel variants are usually more expensive upfront and to maintain than their gasoline counterparts and regional differences in pump oil prices. The total VKM traveled by the Jeepney (i.e., traditional open-air, minibus-type public transport vehicle) is most significant in Regions 4A and NCR. Across all regions, private vehicles and jeepneys dominate land transport.

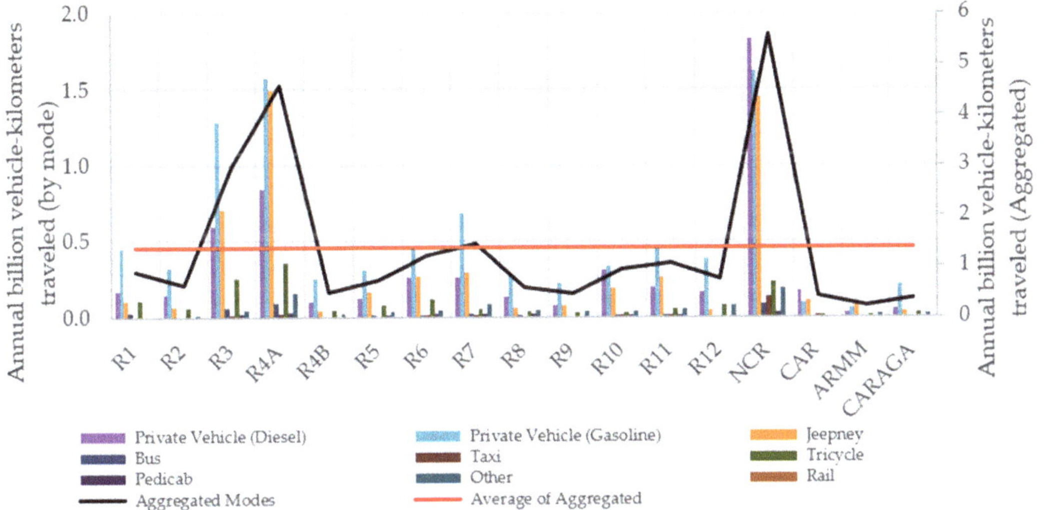

Figure 5. Breakdown of modal share by region.

Insights from Polar Plots: Traffic Flow

The activity effect was plotted against other effects and returned interesting trends. First, the economic activity effect (ΔVKM_{act}) was plotted against the population effect (ΔVKM_{pop}) in Figure 6. The scatterplot reveals an increasing trend suggesting that both contribute positively to traffic flow. Understandably, as income levels rise, vehicle ownership also rises [48] and as the population increases, the demand for additional transport also increases.

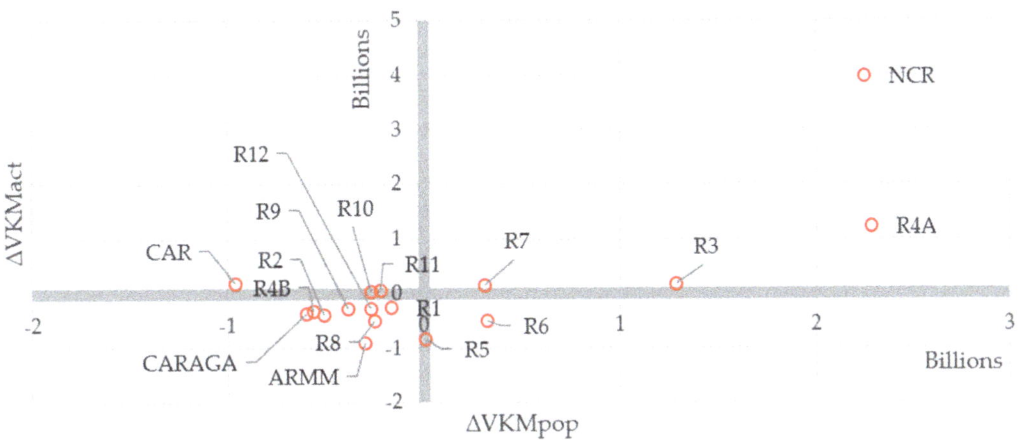

Figure 6. Polar plot between economic activity and population. ΔVKM_{act} and ΔVKM_{pop} refer to the activity and population effects, respectively.

However, if a trendline is fitted with the data, an asymptotic curve can be imagined, suggesting that a saturation point can be reached. This is when further population growth in a region can no longer contribute to additional traffic flow.

Economic activity (ΔVKM_{act}) is plotted against traffic intensity (ΔVKM_{int}) in Figure 7. As the additional traffic flow from economic activity increases, it is observed that the traffic intensity decreases. Notably, NCR is on the far left of this scatterplot, and as explained in the earlier sub-section, this is possibly because of the fact that most major land transport infrastructure projects are concentrated in NCR. Furthermore, a variety of transport options are more abundant in NCR than in other regions. For example, there are city buses, shuttle vehicles (i.e., UV Express) and taxi services in NCR, while there are none in most regions. On the other hand, in general, the trend suggests that as the economies of each region grow, they become less traffic-intensive. Another possible explanation is that the people have a minimum and maximum need to travel. At the minimum, people need to go to work, or to school and to go home. The maximum travel need is limited by the hours of the day (i.e., you cannot travel forever in the day, there is a certain limit). Coincidentally, the high traffic intensity regions seem to be the lower-income regions (e.g., Regions 2 and 5) and the low traffic intensity regions are the high-income regions (e.g., NCR, Regions 7 and 4A). As traffic intensity is the ratio of traffic flow to GDP, a significant increase in GDP can make this quantity very small and vice versa. However, it is quite alarming that Region 3 is both on the positive axes of economic activity and traffic intensity. This suggests that if not controlled, the region is heading in an unsustainable direction.

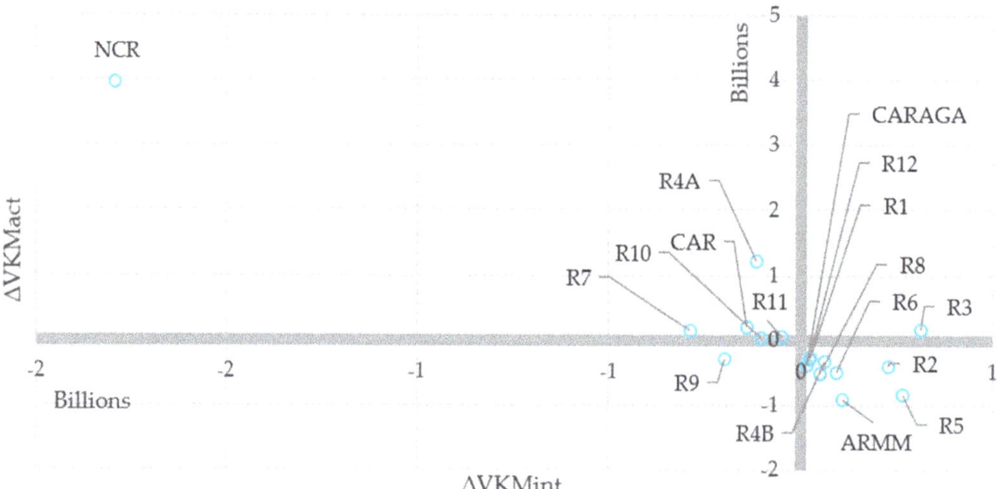

Figure 7. Polar plot between economic activity and traffic intensity. ΔVKM_{act} and ΔVKM_{int} refer to the activity and intensity effects, respectively.

The economic activity effect (ΔVKM_{act}) is plotted against the structural effect (ΔVKM_{str}) in Figure 8. While the traffic intensity tends to decrease as the economic activity increases (see Figure 7), it results in a shift to less efficient modes. It has to be noted though that the additional traffic flow due to structural effects is quite negligible compared to other effects.

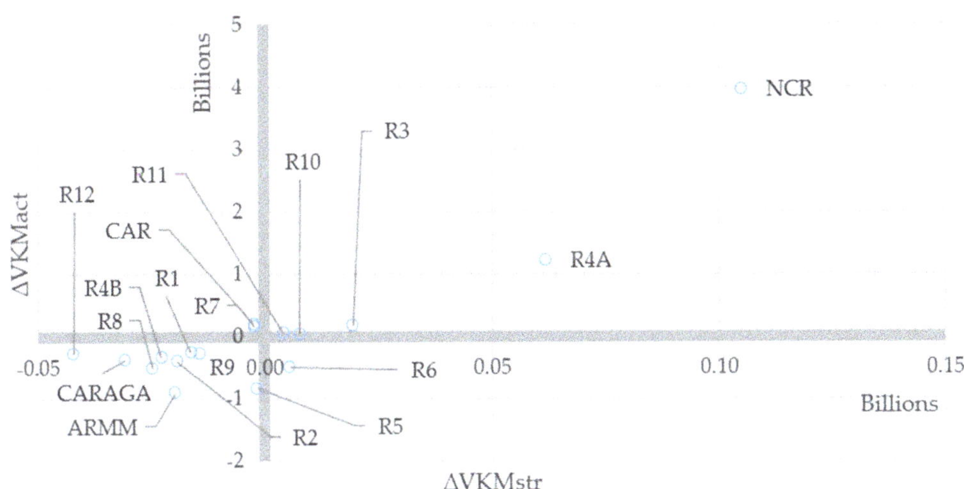

Figure 8. Polar plot between economic activity and modal structure. ΔVKM_{act} and ΔVKM_{str} refer to the activity and structural effects, respectively.

The economic activity effect (ΔVKM_{act}) is plotted against the structural effect due to private diesel vehicles (ΔVKM_{str} (PV, Diesel)) in Figure 9. The trend shows that as the income levels increase, the population tends to shift to private diesel vehicles. This can be due to a variety of factors, which include the higher capacity to pay. This can also be an indication of the high sensitivity of the population to pump oil prices. However, it is important to note that this is not true for Region 4A. Region 4A used to enjoy lower gasoline pump prices due to the presence of a refinery in the region.

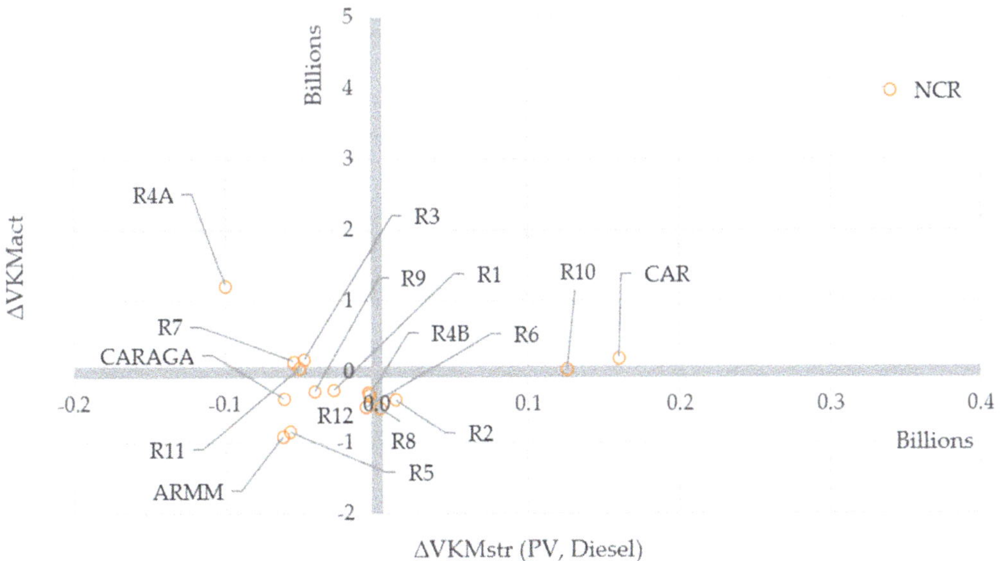

Figure 9. Polar plot between economic activity and modal share of private diesel vehicles. ΔVKM_{act} and ΔVKM_{str} (PV, Diesel) refer to the activity effect and the structural effect of private diesel vehicles, respectively.

Finally, the structural effect due to jeepneys (ΔVKM_{str} (Jeepney)) was plotted against the structural effect due to private gasoline vehicles (ΔVKM_{str} (PV, Gasoline)) in Figure 10. The plot shows an interesting trend wherein as the percentage share of jeepney traffic increases, the percentage share of private gasoline vehicle traffic decreases. This suggests that private gasoline vehicle trips are the ones substituted for jeepney trips and vice versa.

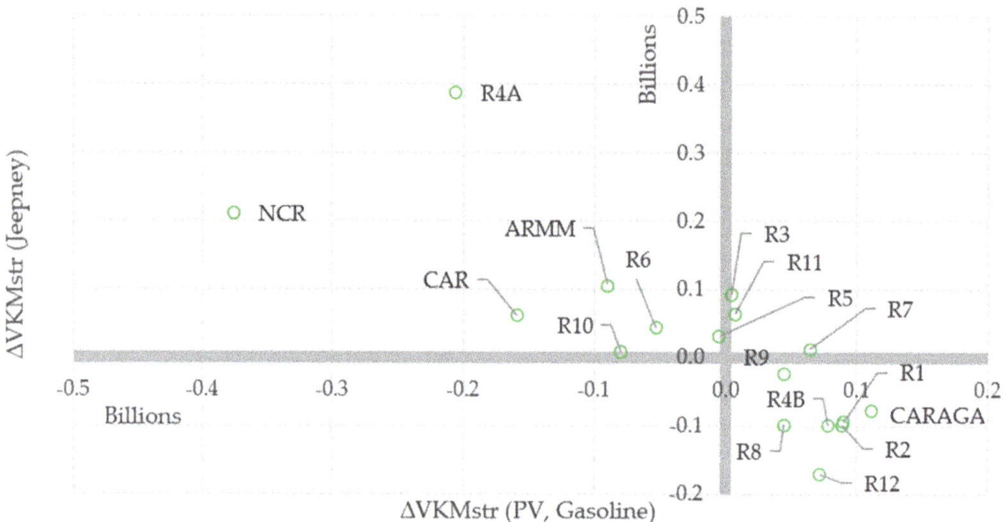

Figure 10. Polar plot between modal shares of jeepneys and private gasoline vehicles. ΔVKM_{str} (Jeepney) and ΔVKM_{str} (PV, Gasoline) refer to the structural effect of public jeepney vehicles and the structural effect of private gasoline vehicles, respectively.

4.2. Drivers of Transport Emissions

Comparing the scatterplots for both traffic flow and transport emissions, the trends are generally similar (i.e., the same drivers that cause the increase in traffic flow are responsible for the increase in emissions). This is not surprising, because a majority of the transportation in the Philippines is fossil powered. As the authors reflected on this, unique drivers can only be observed if a significant portion of the transport activity is active (e.g., biking) or from renewables (e.g., electric vehicles charged using renewables). However, this does not mean there are no other interesting results from this portion of the study. Other notable findings are discussed below.

In the Philippines, as the results suggest, switching to public transport does not ultimately reduce emissions. As shown in Figure 11, increasing public transport share does not show a significant decrease in transport emissions. What this suggests is that the increasing public transport share is just a spillover from private transport, that is, if people can travel by private modes, they will. The need is to reduce the total VKM or to switch to cleaner energy sources for transport. However, switching to cleaner energy sources does not guarantee that emissions will be reduced because of the concept of induced demand. As shown in previous literature, the concept of using cleaner, more sustainable sources can encourage the public to travel and to consume more. While promoting alternatively powered vehicles, it is also important to take note that electric vehicles would also depend on the electricity grid mix. An electricity grid reliant on fossil power will not enable electric vehicles to meet their full potential with regards to mitigating carbon emissions.

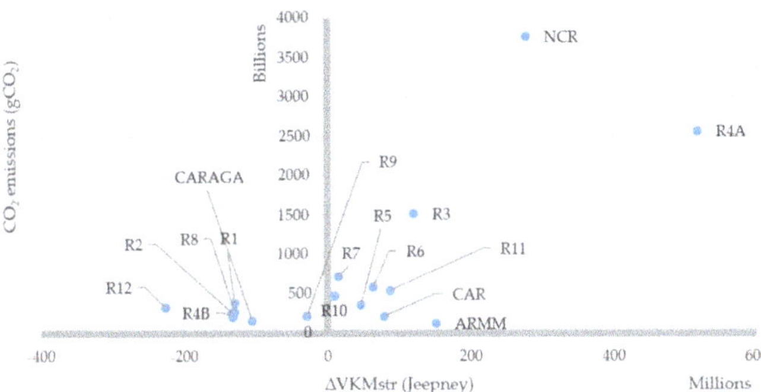

Figure 11. Plot between CO_2 emissions and effect of increasing modal share of jeepneys, ΔVKM_{str} (Jeepney).

Another interesting observation is that regions with high transport emissions tend to have very low active transport activity (see Figure 12). However, increasing active transport does not seem to reverse this, or affect emissions significantly. Reflecting on this, this can be because there are practical limits to active transport, that is, not everyone can bike or walk to work since it entails a level of physical fitness. Also, regions with high transport emissions tend to be more urbanized and spread out, further discouraging the practicality of biking or walking. Moreover, the weather is a big factor especially in a tropical country like the Philippines. With that in mind, policymakers should temper the promotion of active transport with regards to reducing carbon emissions. While it is helpful to some extent, from a larger perspective, it is not the solution that is needed. On the other hand, Banister [47] argues the important role played by mixed-use development in the sustainable mobility paradigm. In the national capital region of the Philippines, mixed-use development areas are on the rise. These are districts that combine residential, commercial and business areas in one place, for example, Bonifacio Global City, Eastwood City and Filinvest City. This connects to active transport, as one key feature of mixed-use development is the reduction of the daily commute distance, making biking or walking more practical to everybody.

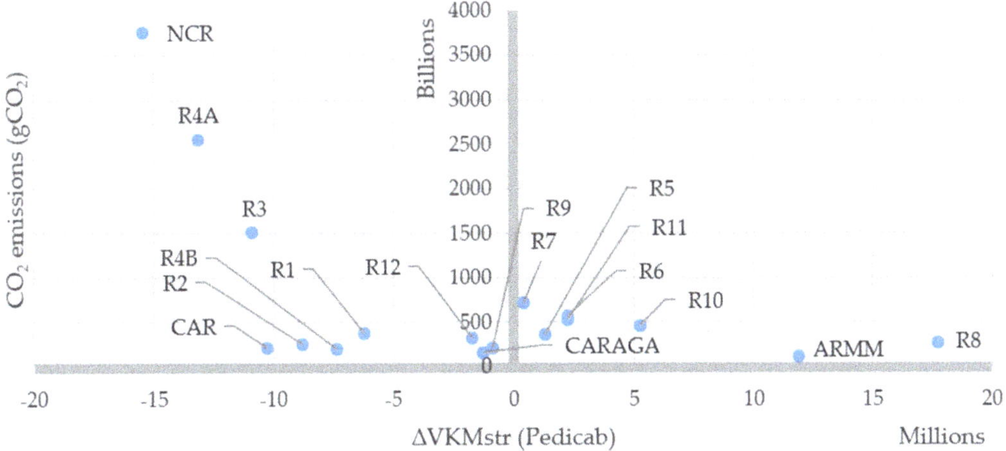

Figure 12. Plot between CO_2 emissions and effect of increasing modal share of pedicabs (i.e., pedaled tricycles), ΔVKM_{str} (Pedicab).

Analyzing a stacked column of transport emission drivers across regions (see Figure 13), some key observations were also found. For Regions 3 and 4A, the main driver is population growth, while for NCR, the main driver is economic activity. This shows the developmental evolution of a region in the Philippines. Since Regions 3 and 4A are directly adjacent to NCR, it can be assumed that the development is more of a spillover from the national capital, NCR. To address this, more aggressive decentralization should be strategized by the government, so as not to sacrifice the sustainability of Regions 3 and 4A. In a few years, both might become as unsustainable as NCR, if not controlled. Regions way up north and down south need to be developed rapidly to control the unsustainable growth of Regions 3, 4A and NCR. Furthermore, it can be observed that while NCR has a low travel intensity, the sheer size of its economic activity significantly drives its transport emissions upwards. Thus, in the national capital, it is more of an over-demand problem than a technology problem. Also, it is alarming that Region 3 is showing an unsustainable trend, with both a positive transport intensity effect and economic activity effect. This suggests a strong need to improve public transport infrastructure in the region, as recommended in [6].

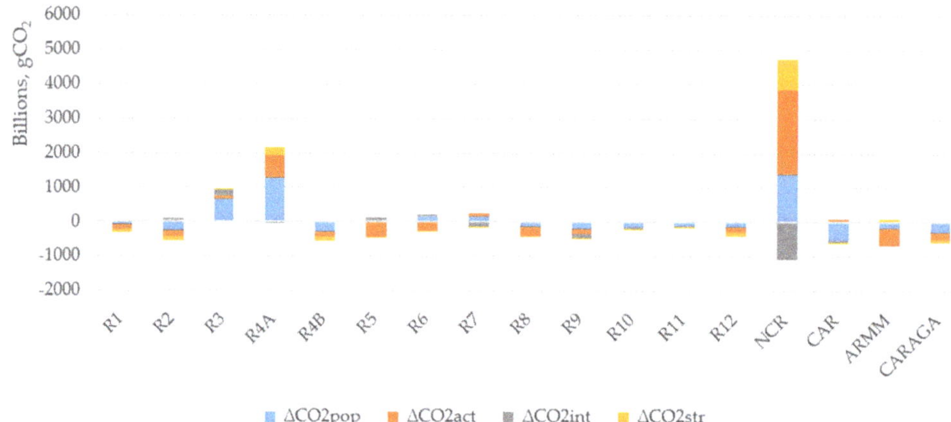

Figure 13. Transport emission drivers across regions.

5. Conclusions and Way Forward

The bottom line of this study is to reduce the need to travel. This is highly relevant considering how the recent pandemic has affected mobility. In an instant, the world had learned to adapt to telecommuting, as global lockdowns were put in place. In April 2020, up to 90% of cities worldwide were in lockdown at a particular point in time [49]. With regards to increasing traffic flow and emissions from transport, it deserves to be contemplated that technological innovation might not be the key but behavioral change. In a way, sound urban planning can also reduce the need to travel.

Transportation is responsible for almost 30% of global CO_2 emissions, and this growth is largely because of increasing dependence on private vehicles. This study demonstrated the potential of using spatial decomposition in the early diagnosis of traffic flow problems in regions. By comparing regions with each other, the cause for increasing or decreasing traffic flow and transport emissions can be determined and ideas can be tested. For example, as shown in the paper, increasing public transport share did not result in reduced transport emissions, nor did increasing active transport share. Using spatial decomposition analysis, potential driving factors are not merely determined. Instead, their individual contributions to traffic flow and emissions are compared with one another to show the potential effect of growth in a certain region. A new tool is explored by this paper that can help in the mitigation of worsening traffic flows and emissions in regions. However, in future studies, a better way to define the hypothetical benchmark region needs to be explored.

The following policy recommendations arising from the results of this paper can be further investigated:

1. *Equity on the distribution of transport infrastructure projects.* The government needs to invest in mass transport services outside NCR to reduce the dependence of other regions on private vehicle modes. This can accelerate the decentralization of growth to other regions up north and down south. While these regions' shares in traffic flow are not yet significant, the current situation could be leading them toward an unsustainable and irreversible route. The results of this study support the increased taxation of new private vehicle purchases; however, the authors advocate that revenues from these new taxes have to be reinvested in the development of quality public transport services.

2. *Improve the quality of public transport services.* The data showed that jeepney trips are substituted for private gasoline vehicle trips and vice versa. As income levels rise, people grab the opportunity to own private vehicles. This is understandable especially if the quality of public transport services is unbearable. The population deserves a decent means to go to work, to school, etc., and it should be a priority of the government. In fact, the national government of the Philippines had been pushing for the jeepney modernization program, but however, social and cultural issues have delayed its implementation.

3. *Support telecommuting.* The improvement of public transport services goes hand in hand with the reduction of the demand for it. As shown by the data in this study, increasing public transport share does not ultimately result in lower transport emissions. At the end of the day, reducing transport demand is the key. The government should find ways to support and encourage employers who successfully adapted telecommuting in their operations to significantly reduce transport demand.

4. *Promote mixed-use development.* As shown in this study, active transport does not help the reduction of transport emissions significantly, perhaps due to practical reasons. Mixed-use development decreases the commute distance and thus can encourage more biking and walking. The pursuance of active transport projects (e.g., bike lanes) has to be connected to mixed-use areas to maximize their effectiveness.

Author Contributions: Conceptualization, G.U.N. and N.S.L.; Data curation, G.U.N. and J.B.M.B.; Formal analysis, N.S.L.; Funding acquisition, G.U.N.; Methodology, G.U.N., A.S.F.C. and N.S.L.; Project administration, J.B.M.B.; Visualization, G.U.N.; Writing—original draft, G.U.N. and N.S.L.; Writing—review and editing, A.S.F.C. and J.B.M.B. All authors have read and agreed to the published version of the manuscript.

Funding: De La Salle University, through the Office of the Vice Chancellor of Research and Innovation (OVCRI), funded Geoffrey Udoka Nnadiri during the initial development of this manuscript through the international conference subsidy grant. The Commission on Higher Education (CHED), Philippines funded Neil Stephen Lopez and Jose Bienvenido Manuel Biona during the development of this study, through the CHED Grants-in-Aid Program.

Institutional Review Board Statement: Not applicable.

Informed Consent Statement: Not applicable.

Data Availability Statement: The primary data used in this study can be accessed on request from the Philippine Statistics Authority.

Conflicts of Interest: The authors declare no conflict of interest or competing interests in the conduct of this study.

References

1. Philippine Atmospheric, Geophysical Astronomical Services Administration (PAGASA). Climate Projections in the Philippines. Climate Change in the Philippines. Available online: https://www1.pagasa.dost.gov.ph/index.php/93-cad1/472-climate-projections (accessed on 10 May 2021).
2. IPCC. Climate Change 2014: Synthesis Report. In *Contribution of Working Groups I, II and III to the Fifth Assessment Report of the Intergovernmental Panel on Climate Change (IPCC)*; IPCC: Geneva, Switzerland, 2014.
3. IEA. CO_2 emissions from fuel combustion. *Int. Energy Agency* **2016**, *533*. [CrossRef]
4. International Energy Agency Global Energy & CO_2 Status Report; Paris. 2019. Available online: https://www.iea.org/reports/global-energy-co2-status-report-2019 (accessed on 13 May 2021).
5. The World Bank. Urban Transport and Climate Change. Available online: https://www.worldbank.org/en/news/feature/2012/08/14/urban-transport-and-climate-change (accessed on 5 May 2021).
6. Lopez, N.S.; Chiu, A.S.F.; Biona, J.B.M. Decomposing drivers of transportation energy consumption and carbon dioxide emissions for the Philippines: The case of developing countries. *Front. Energy* **2018**, *12*, 389–399. [CrossRef]
7. Wu, H.M.; Xu, W. Cargo transport energy consumption factors analysis: Based on LMDI decomposition technique. *IERI Procedia* **2014**, *9*, 168–175. [CrossRef]
8. Chen, L.; Yang, Z. A spatio-temporal decomposition analysis of energy-related CO_2 emission growth in China. *J. Clean. Prod.* **2015**, *103*, 49–60. [CrossRef]
9. Dargay, J.; Gately, D. Income's effect on car and vehicle ownership, worldwide: 1960–2015. *Transp. Res. Part A Policy Pract.* **1999**, *33*, 101–138. [CrossRef]
10. Martin, E.; Shaheen, S.A.; Lidicker, J. Impact of carsharing on household vehicle holdings: Results from North American shared-use vehicle survey. *Transp. Res. Rec.* **2010**, *2143*, 150–158. [CrossRef]
11. Ubando, A.T.; Gue, I.H.; Rith, M.; Gonzaga, J.; Lopez, N.S.; Biona, J.B. A systematic approach to the optimal planning of energy mix for electric vehicle policy. *Chem. Eng.* **2019**, *76*. [CrossRef]
12. Papagiannaki, K.; Diakoulaki, D. Decomposition analysis of CO_2 emissions from passenger cars: The cases of Greece and Denmark. *Energy Policy* **2009**, *37*, 3259–3267. [CrossRef]
13. Jiang, J. A factor decomposition analysis of transportation energy consumption and related policy implications. *IATSS Res.* **2015**, *38*, 142–148. [CrossRef]
14. Ding, J.; Jin, F.; Li, Y.; Wang, J.E. Analysis of transportation carbon emissions and its potential for reduction in China. *Chin. J. Popul. Resources Environ.* **2013**, *11*, 17–25. [CrossRef]
15. Wang, Y.; Hayashi, Y.; Kato, H.; Liu, C. Decomposition analysis of CO_2 emissions increase from the passenger transport sector in Shanghai, China. *Int. J. Urban Sci.* **2011**, *15*, 121–136. [CrossRef]
16. Guo, B.; Geng, Y.; Franke, B.; Hao, H.; Liu, Y.; Chiu, A. Uncovering China's transport CO_2 emission patterns at the regional level. *Energy Policy* **2014**, *74*, 134–146. [CrossRef]
17. Ang, B.W.; Zhang, F.Q.; Choi, K.H. Factorizing changes in energy and environmental indicators through decomposition. *Energy* **1998**, *23*, 489–495. [CrossRef]
18. Ang, B.W. Decomposition analysis for policymaking in energy: Which is the preferred method? *Energy Policy* **2004**, *32*, 1131–1139. [CrossRef]
19. Lopez, N.S.; Mousavi, B.; Biona, J.B.; Chiu, A.S. Decomposition Analysis of CO_2 Emissions with Emphasis on Electricity Imports and Exports: EU as a Model for ASEAN Integration. *Chem. Eng. Trans.* **2017**, *61*, 739–744. [CrossRef]
20. Ang, B.W. The LMDI approach to decomposition analysis: A practical guide. *Energy Policy* **2005**, *33*, 867–871. [CrossRef]
21. Ang, B.W. LMDI decomposition approach: A guide for implementation. *Energy Policy* **2015**, *86*, 233–238. [CrossRef]
22. Feng, C.; Sun, L.X.; Xia, Y.S. Clarifying the "gains" and "losses" of transport climate mitigation in China from technology and efficiency perspectives. *J. Clean. Prod.* **2020**, *263*, 121545. [CrossRef]
23. Sorrell, S.; Lehtonen, M.; Stapleton, L.; Pujol, J.; Champion, T. Decomposing road freight energy use in the United Kingdom. *Energy Policy* **2009**, *37*, 3115–3129. [CrossRef]
24. Lu, I.J.; Lin, S.J.; Lewis, C. Decomposition and decoupling effects of carbon dioxide emission from highway transportation in Taiwan, Germany, Japan and South Korea. *Energy Policy* **2007**, *35*, 3226–3235. [CrossRef]
25. Kwon, T.H. Decomposition of factors determining the trend of CO_2 emissions from car travel in Great Britain (1970–2000). *Ecol. Econ.* **2005**, *53*, 261–275. [CrossRef]
26. Liu, Y.; Feng, C. Decouple transport CO_2 emissions from China's economic expansion: A temporal-spatial analysis. *Transp. Res. Part D Transp. Environ.* **2020**, *79*, 102225. [CrossRef]
27. Andreoni, V. Galmarini, S. European CO_2 emission trends: A decomposition analysis for water and aviation transport sectors. *Energy* **2012**, *45*, 595–602. [CrossRef]
28. Timilsina, G.R.; Shrestha, A. Factors affecting transport sector CO_2 emissions growth in Latin American and Caribbean countries: An LMDI decomposition analysis. *Int. J. Energy Res.* **2009**, *33*, 396–414. [CrossRef]
29. Zhang, K.; Liu, X.; Yao, J. Identifying the driving forces of CO_2 emissions of China's transport sector from temporal and spatial decomposition perspectives. *Environ. Sci. Pollut. Res.* **2019**, *26*, 17383–17406. [CrossRef]
30. Tu, M.; Li, Y.; Bao, L.; Wei, Y.; Orfila, O.; Li, W.; Gruyer, D. Logarithmic Mean Divisia Index Decomposition of CO_2 Emissions from Urban Passenger Transport: An Empirical Study of Global Cities from 1960–2001. *Sustainability* **2019**, *11*, 4310. [CrossRef]

31. Cao, X.; OuYang, S.; Liu, D.; Yang, W. Spatiotemporal Patterns and Decomposition Analysis of CO_2 Emissions from Transportation in the Pearl River Delta. *Energies* **2019**, *12*, 2171. [CrossRef]
32. Feng, C.; Xia, Y.S.; Sun, L.X. Structural and social-economic determinants of China's transport low-carbon development under the background of aging and industrial migration. *Environ. Res.* **2020**, *188*, 109701. [CrossRef]
33. Li, W.; Li, H.; Zhang, H.; Sun, S. The analysis of CO_2 emissions and reduction potential in China's transport sector. *Math. Probl. Eng.* **2016**, *2016*. [CrossRef]
34. Lian, L.; Lin, J.; Yao, R.; Tian, W. The CO 2 emission changes in China's transportation sector during 1992 2015: A structural decomposition analysis. *Environ. Sci. Pollut. Res.* **2020**, 1–4. [CrossRef]
35. Román-Collado, R.; Morales-Carrión, A.V. Towards a sustainable growth in Latin America: A multiregional spatial decomposition analysis of the driving forces behind CO_2 emissions changes. *Energy Policy* **2018**, *115*, 273–280. [CrossRef]
36. Timilsina, G.R.; Shrestha, A. Transport sector CO_2 emissions growth in Asia: Underlying factors and policy options. *Energy policy* **2009**, *37*, 4523–4539. [CrossRef]
37. Mendiluce, M.; Schipper, L. Trends in passenger transport and freight energy use in Spain. *Energy Policy* **2011**, *39*, 6466–6475. [CrossRef]
38. Shi, Y.; Han, B.; Han, L.; Wei, Z. Uncovering the national and regional household carbon emissions in China using temporal and spatial decomposition analysis models. *J. Clean. Prod.* **2019**, *232*, 966–979. [CrossRef]
39. Sumabat, A.K.; Lopez, N.S.; Yu, K.D.; Hao, H.; Li, R.; Geng, Y.; Chiu, A.S. Decomposition analysis of Philippine CO_2 emissions from fuel combustion and electricity generation. *Appl. Energy.* **2016**, *164*, 795–804. [CrossRef]
40. Philippine Statistics Authority. Family Income and Expenditure Survey (FIES). 2015. Available online: https://psa.gov.ph/content/statistical-tables-2015-family-income-and-expenditure-survey (accessed on 9 March 2018).
41. Philippine Statistics Authority. The Gross Regional Domestic Product (GRDP). 2016. Available online: Psa.gov.ph/sites/default/files/2015-2017grdp%20publication.pdf (accessed on 11 March 2020).
42. Department of Energy. NCR/Metro Manila Prevailing Retail Pump Prices. Available online: https://www.doe.gov.ph/retail-pump-prices-metro-manila?ckattempt=2&fbclid=IwAR1an3gkTR8pOHqNZKb4K2JCUdJ57t1YieAWUANnx3arrnOzg0GrSFYxK5Q (accessed on 17 May 2021).
43. Land Transportation Franchising and Regulatory Board. Fare Rates. Available online: https://ltfrb.gov.ph/fare-rates/?fbclid=IwAR3dqRxhhdhGenOMr8Li4fGlMLSbUgxPJIxfVKW4BHP3zDWoZXJcKMwCfVY (accessed on 17 May 2021).
44. Fabian, H.; Gota, S. CO_2 emissions from the land transport sector in the Philippines: Estimates and policy implications. In Proceedings of the 17th Annual Conference of the Transportation Science Society of the Philippines, Pasay, Philippines, 4 September 2009.
45. Tomeldan, M.V.; Antonio, M.; Arcenas, J.; Beltran, K.M.; Cacalda, P.A. "Shared Growth" Urban Renewal Initiatives in Makati City, Metro Manila, Philippines. *J. Urban Manag.* **2014**, *3*, 45–65. [CrossRef]
46. Philippine Institute for Development Studies. Land Use Planning in Metro Manila and the Urban Fringe: Implications on the Land and Real Estate Market. 2000. Available online: https://dirp3.pids.gov.ph/ris/dps/pidsdps0020.pdf (accessed on 5 January 2021).
47. Banister, D. The sustainable mobility paradigm. *Transp. Policy* **2008**, *15*, 73–80. [CrossRef]
48. Rith, M.; Roquel, K.I.; Lopez, N.S.; Fillone, A.M.; Biona, J.B. Towards more sustainable transport in Metro Manila: A case study of household vehicle ownership and energy consumption. *Transp. Res. Interdiscip. Perspect.* **2020**, *6*, 100163. [CrossRef]
49. Devi, S. Travel restrictions hampering COVID-19 response. *Lancet* **2020**, *395*, 1331–1332. [CrossRef]

 *sustainability*

Article

Modeling Traffic Flow, Energy Use, and Emissions Using Google Maps and Google Street View: The Case of EDSA, Philippines

Joshua Ezekiel Rito [1], Neil Stephen Lopez [1,*] and Jose Bienvenido Manuel Biona [1,2]

[1]   Mechanical Engineering Department, De La Salle University, Manila 0922, Philippines; joshua_rito@dlsu.edu.ph (J.E.R.); jose.bienvenido.biona@dlsu.edu.ph (J.B.M.B.)
[2]   Enrique Razon Logistics Institute, De La Salle University, Manila 0922, Philippines
*   Correspondence: neil.lopez@dlsu.edu.ph

Abstract: The general framework of the bottom-up approach for modeling mobile emissions and energy use involves the following major components: (1) quantifying traffic flow and (2) calculating emission and energy consumption factors. In most cases, researchers deal with complex and arduous tasks, especially when conducting actual surveys in order to calculate traffic flow. In this regard, the authors are introducing a novel method in estimating mobile emissions and energy use from road traffic flow utilizing crowdsourced data from Google Maps. The method was applied on a major highway in the Philippines commonly known as EDSA. Results showed that a total of 370,855 vehicles traveled along EDSA on average per day in June 2019. In comparison to a government survey, only an 8.63% error was found with respect to the total vehicle count. However, the approximation error can be further reduced to 4.63% if cars and utility vehicles are combined into one vehicle category. The study concludes by providing the limitations and opportunities for future work of the proposed methodology.

Keywords: Google Maps; transportation; energy use; emissions; modeling; vehicle flow

Citation: Rito, J.E.; Lopez, N.S.; Biona, J.B.M. Modeling Traffic Flow, Energy Use, and Emissions Using Google Maps and Google Street View: The Case of EDSA, Philippines. *Sustainability* **2021**, *13*, 6682. https://doi.org/10.3390/su13126682

Academic Editors: Wei-Hsin Chen, Alvin B. Culaba, Aristotle T. Ubando and Steven Lim

Received: 9 May 2021
Accepted: 8 June 2021
Published: 11 June 2021

Publisher's Note: MDPI stays neutral with regard to jurisdictional claims in published maps and institutional affiliations.

1. Introduction

Climate change, the effects of global warming, and air pollution are some of the contemporary crucial issues faced by the global community [1]. Countries, especially developing economies, are susceptible to risks of climates change as well as to the adverse impacts of increasing air pollution and energy intensity caused by rapid industrialization in recent years [2]. The rising global energy consumption demand brought about by increased manufacturing outputs and transportation of goods and services has resulted in a record high of 33.1 Gt CO_2 in 2018 [3]. The latest 2.3% increase in CO_2 emissions is almost twice the average annual growth rate of CO_2 emissions since 2010. The increase in CO_2 emissions is alarming despite the implementation of the Paris Climate Agreement, which aims to limit global warming to a temperature increase of 2 °C in a 10-year span by reducing greenhouse gas (GHG) emissions. All 196 countries who are signatories of the Paris Climate Accord must develop individual long-term plans called nationally determined contributions in lowering GHG emissions. In connection to that, conferences are held every five years by all stakeholders to evaluate the progress in reducing GHG emissions based on the best available research [4]. Air pollution can be directly linked to climate change: it not only intensifies the effects of global warming but also contributes dreadful repercussions to human health which result in increasing morbidity and mortality [5]. Approximately 60% of the total air pollution globally is attributable to outdoor particulate matter [6]. Particulate matter was responsible for 2.42 million deaths worldwide in 2007, while the mortality cases increased by more than half a million in 2017 or about 21.6% [7].

The transportation sector accounts for about 21% of the total carbon emissions globally due to its high fossil fuel consumption [8]. Carbon emissions by the transportation sector are steadily increasing. In 2018, it was estimated to be responsible for 24% of the global carbon emissions, wherein road transportation was one of the main contributors [8–10]. Road transportation accounts for as high as 77% of the total GHG emissions from transportation globally [10]. With regard to air pollution and its health impacts, the share of transportation-related $PM_{2.5}$-associated mortality at the global scale was estimated to be at 11.6% in 2015, while the health costs due to $PM_{2.5}$ transportation-attributable deaths were $891 billion [11]. These estimates are significant and there is an exigency to regularly quantify, monitor, and evaluate whether the goals in curbing carbon emissions and improving air quality are being met. A yearly cross-sectional study is highly recommended to objectively meet this requirement.

The increase in CO_2 emissions is often correlated to increases in energy consumption [12,13]. According to IEA [14], the transport sector had the highest share (35%) in final energy consumption globally in 2018. Furthermore, 89% of the total transport energy consumption was from road transport. However, a high percentage share does not imply high utility. Generally, the efficiency of energy use in the transport sector is low [15]. On the bright side, the efficiency is gradually increasing, and a 30% reduction in fuel consumption is expected by 2050 due to improvements in the technologies of the internal combustion engine [16]. That being said, much more needs to be done in order to have better efficiency in road transport energy use.

Emissions and energy use from road transport are quantified based on the number of vehicles traveling along a road. The traditional methods of traffic flow data collection fall under two main categories: (1) the manual counting method and (2) the automatic counting method [17]. The manual counting method can be conducted by individuals standing by the roadside collecting and recording their observations on tally sheets [18]. Alternatively, manual counting can also be done by people in a moving vehicle [19,20], where the observers categorically count the number of vehicles moving in the opposite direction. This may also be done along the same direction of the moving vehicle, which will take into account the number of vehicles that are overtaking and are overtaken. The automatic counting method, on the other hand, uses detectors based on electromagnetics and wireless communication to observe vehicular presence on or proximate to the road [21]. Examples of the most common detectors used in automatic counting are: (1) pneumatic tubes, (2) inductive loops, (3) weigh-in-motion sensors, (4) micro-millimeter wave radar, and (5) video camera [17]. The majority of previous studies in traffic flow analysis utilized the manual counting and automatic counting methods as the gold standard. However, these methods require considerable amounts of resources and manpower, which can be considered impediments to some developing countries in conducting such studies. Thus, there is a need for an alternative methodology that is less resource-intensive but also compendious.

Previous researchers have developed alternative approaches with regard to quantifying traffic flow and eventually converting these into emissions and energy use estimates. Zhao et al. [22] recommended an approach to obtain traffic flow characteristic parameters such as traffic volume, average travel speed, and traffic density using electronic toll collection (ETC) data. Seo and Kusakabe [23] introduced a method of estimating traffic flow dynamics on the basis of spacing and positioning data of probe vehicles. Aksoy et al. [24] integrated a fuel emission and consumption calculation model in logistics to develop strategies to increase economic benefits and reduce the environmental impacts of pollution and energy use. Jabali et al. [25] analyzed CO_2 emissions and fuel consumption in time-dependent vehicle routing by assigning and scheduling predetermined destinations of vehicles. Bharadwaj et al. [26] analyzed the fuel consumption and greenhouse gas emissions from road transport using vehicle kilometers traveled, obtained from an actual traffic survey. Nesamani et al. [27] analyzed air pollutant emissions and energy use of vehicles in relation to various operating conditions of vehicles such as geometric design elements, traffic characteristics, the roadside environment, weather conditions, and driving style. Maes et al. [28] proposed a methodology that evaluates road transport emission inventories in high resolution. The authors employed a probabilistic bottom-up approach in their transportation model while utilizing fleet data such as engine size, model-year, and fuel types of vehicles. Iqbal et al. [29] introduced an innovative approach through a mesoscopic model where the magnitude of emission variations as well as the underlying factors that affect emissions were evaluated. Zhang et al. [30] performed a microscopic analysis by characterizing toxic emissions from gasoline and diesel vehicles on a busy mountain road. The authors acquired real-world measurements through the use of a portable emissions measurement system. Iqbal et al. [31] introduced a technique involving Monte Carlo simulation for probabilistic health risk assessment in an urban area with respect to vehicular emissions. While these studies and other related research have been extremely useful in evaluating traffic flow, emissions, and energy use, the fact remains that conducting research in this context is a challenge for developing countries, especially considering the time and costs of doing it.

On this note, crowdsourcing is an emerging tool that can provide cost-effective solutions to traditional problems [32,33]. In transportation planning, crowdsourcing is an effective tool that can consolidate data from a large group of individuals on the same platform in order to address a shared problem among its members [34]. Key areas in which crowdsourcing proved to be very useful in transportation are the following: (1) development of strategies for managing traffic flow in an urban environment; (2) efficient detour routing of vehicles to avoid traffic congestion; (3) monitoring and assessment of road conditions; (4) data analysis on traffic accident and road crime; and many more [35–39]. Transport modeling provides insights for the possible future development of a particular area [40]. In the road transport sector, transportation modeling is an essential tool used in estimating emissions and energy use.

To assess the potential of using publicly available crowdsourced data for transport planning, the authors introduce in this paper a novel method for modeling road transport emissions and energy use. In particular, crowdsourced data from Google Maps and Google Street View are utilized to estimate vehicle and traffic data. Furthermore, emission and energy consumption factors are derived from various references to calculate the total emission load and energy usage, respectively, with respect to the estimated vehicle count. For the purpose of demonstrating the methodology, greenhouse gas emissions in terms of CO_2eq and air pollution in terms of $PM_{2.5}$ emissions are also presented. The proposed approach aims to provide an alternative to resource-constrained transport planners for traffic data gathering. While data uncertainty from Google Maps and Google Street View is not yet tackled in this manuscript, the results presented here are validated with an actual government-commissioned annual average daily traffic survey (AADT), and recommendations on how to tackle data uncertainty are provided towards the end of the manuscript.

The paper proceeds as follows: Section 2 describes the proposed methodology in detail; Section 3 provides an illustrative case study in Epifanio de los Santos Avenue (EDSA), Philippines; Section 4 discusses limitations and opportunities for future work of the proposed methodology; and Section 5 concludes the study.

2. Methods and Data

2.1. Research Methodology

The proposed general procedure in estimating road transport emissions and energy use by means of utilizing crowdsourced data from Google Maps is presented in Figure 1. The average travel time and segment length for each road segment are collected in Google Maps to estimate the bulk speed or the average speed of all vehicles. The computed bulk speed is subsequently used as an independent variable to estimate hourly vehicle flows in terms of passenger car units (PCU) for the full 24 h of the day from a speed-flow curve. A speed-flow curve is empirically generated and plots bulk speed as a function of the flow rate of PCUs. With the help of a speed-flow curve which refers to a roadside friction index (RSFI), the PCU count is estimated by either manually plotting bulk speed values on the curve or using it as an input to a fitted regression equation. The hourly PCU counts for one whole day must be consolidated into a daily PCU count and then translated into an overall PCU count based on the intended periodicity of the study (e.g., monthly, annually, etc.). The PCU count is eventually broken down into different vehicle categories (i.e., modal share) through classified vehicle counting using the Street View feature of Google Maps and using passenger car equivalence factors (PCEF). This will be further elaborated on the succeeding sections below. Finally, the total road transport energy use and emission load are estimated by multiplying the overall vehicle count with the assumed energy economy and emission factors, respectively.

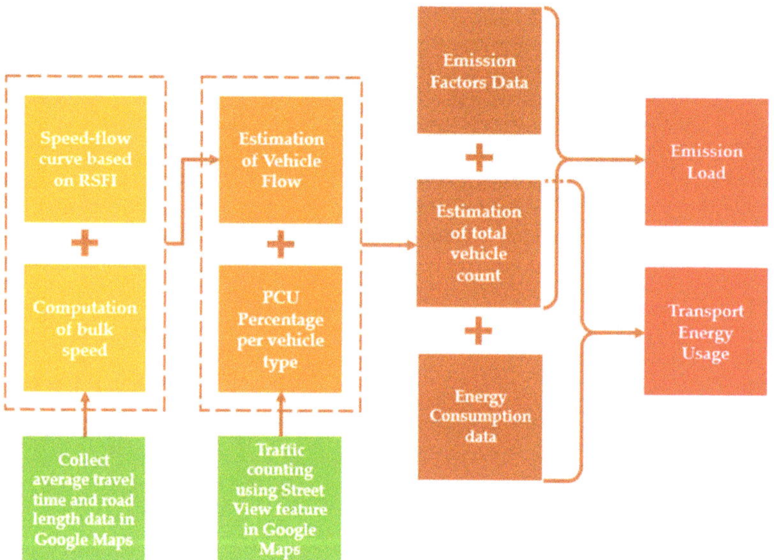

Figure 1. The general framework of the proposed method for estimating road transport energy use and emissions using crowdsourced data in Google Maps and Google Street View.

The proposed detailed process for estimating energy consumption and emissions from road transport using Google Maps and Google Street View data is outlined below:

- To begin, identify the specific road or highway that will be studied and determine its starting and end points in Google Maps. The road must be divided into a preferred number of segments. The segment length in kilometers (km) and the actual number of lanes on each side of the road (e.g., northbound and southbound) shall be recorded;
- Decide on the periodicity (e.g., weekly, monthly, and yearly) of the study and select the days of the week from which data collection will be made. Days can be classified according to the identicality of their traffic situations. For example, the modeler can opt to assume that Tuesday, Wednesday, and Thursday have similar traffic conditions;
- Across each road segment, collect and tabulate the average travel time, T_{Ave}, from Google Maps. By default, this is provided by Google in terms of minutes (min). Convert the unit of time from minutes to hours (hr). Do this at least 24 times, getting at least one data point per hour of the day (i.e., from 0:00 to 23:00). Note that this will be repeated for all the days covered in the study. This tabulation was done manually in the illustrative case study below, but the modeler has the option to automate this using Google Maps' API service.

- Calculate the bulk speed, V_B, in terms of kilometers per hour (km/hr) by dividing each segment length, L_S, by the hourly average travel time, T_{Ave} [41] (see Equation (1) below);

$$V_B = \frac{L_S}{T_{Ave}} \tag{1}$$

- Through the use of a speed-flow curve, the calculated bulk speed, V_B, from Google Maps is converted to passenger car units per hour (PCU/hr). The unit PCU is used to convert the heterogeneous characteristics of vehicle flow due to the presence of different vehicle types on the road into an equivalent homogenous quantity, using relative weightage factors (i.e., PCEF) [42]. For example, the road space occupied by a bus is equivalent to approximately two passenger cars;
- Assuming that the available speed-flow curve only applies to particular hours of the day, correction factors can be used to adjust the rate of vehicle flow in hours when a significant drop in traffic volume is expected, such as from midnight to the earliest hours of the morning;
- Consolidate the estimated hourly PCU counts for the full 24 h into a total daily PCU count;
- Within each road segment, assign point coordinates having roughly equidistant spacing with one another as shown in Figure 2 [43]. The modeler has the discretion to determine the distance/spacing between these points, with consideration of road structures, such as the presence of an underpass, flyover, road intersection, etc. The modeler must avoid potential duplication in the counting of vehicles. These points will be used to estimate the modal share;
- Utilizing the Street View feature of Google Maps, perform a classified vehicle count on all of the points identified in the previous step. Count the number of vehicles per variant/category (i.e., motorcycle, tricycle, car, taxi, utility vehicle, jeepney, bus, truck, etc.).
- Multiply the total vehicle count of each category (from Street View counting in Google Maps) with its PCEF, and then divide it by the sum-product of vehicle counts and PCEFs across all categories. This shall generate the PCU mix (i.e., modal share);
- To convert the PCU count into vehicle counts by category, VC, break down the PCU count using the PCU mix obtained in the previous step, and then divide it by the corresponding PCEF for each vehicle category;
- Derive mobile emission factors for each vehicle type with respect to greenhouse gas and air pollutant emissions in terms of grams per kilometer ($g_{emissions}$/km), taking into account the variants, fuel type, local emission standards, fuel economy in grams of fuel per kilometer (g_{fuel}/km), and specific emission factors in grams of emissions per gram of fuel ($g_{emissions}$/g_{fuel}). The emission factors used in this study are shown in the illustrative case study in Section 3. An aggregated emission factor, EF, can be estimated using the PCU mix obtained above;
- Multiply the emission factors, EF, ($g_{emissions}$/km) to the segment length, L_S, (km) and to the total vehicle count, VC, in order to obtain the total emissions load, EL. See Equation (2);

$$EL = EF \times L_S \times VC \tag{2}$$

- Derive energy consumption for each vehicle type, EC, in terms of megajoules per kilometers (MJ/km) by taking into account the variants, fuel type, local emission standards, fuel economy in grams of fuel per kilometer (g_{fuel}/km), and calorific value of fuels, specifically the lower heating values, in terms of megajoules per grams of fuel (MJ/g_{fuel}).
- Multiply energy consumption, EC, (MJ/km) to the segment length, L_S, (km) and to the total vehicle count, VC, in order to get the total energy use, EU. See Equation (3).

$$EU = EC \times L_S \times VC \tag{3}$$

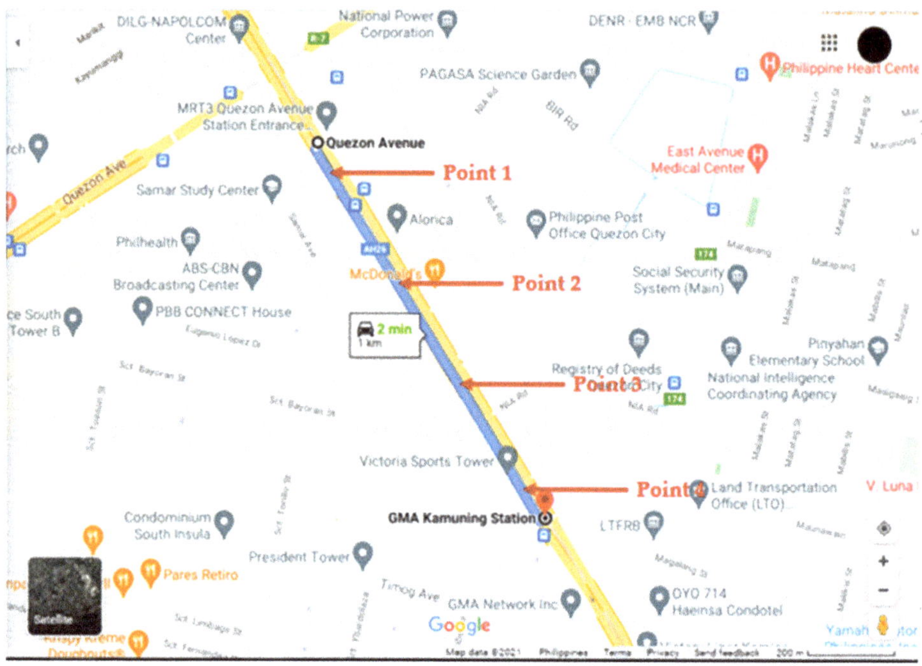

Figure 2. Assigning of point coordinates in Google Maps. Data Source: Google Maps.

2.2. Data Collection for the Illustrative Case Study

Epifanio de los Santos Avenue (EDSA) was the subject of the illustrative case study in this paper. It interlinks seven major cities in Metro Manila, making it the main thoroughfare of the economic capital of the Philippines [44]. EDSA will be divided into 12 road segments; furthermore, the authors decided to use the Metro Rail Transit Line 3 (MRT-3) stations as the basis for dividing the avenue into segments. Figure 3 illustrates the whole range of MRT-3 as well as the approximate locations of each station starting from Taft Avenue Station and all the way to North Avenue Station [45]. Currently, MRT-3 has a total of 13 stations spread across five cities: North Avenue Station, Quezon Avenue Station, GMA Kamuning Station, Araneta Center-Cubao Station, and Santolan-Annapolis Station are located in Quezon City; Ortigas Avenue station is located in Pasig City; Shaw Boulevard Station and Boni Avenue Station are located in Mandaluyong City; Guadalupe Station, Buendia Station, Ayala Station, and Magallanes Station are located in Makati City; furthermore, Taft Avenue Station is located in Pasay City.

In the case study, the data collected by the authors was for June 2019. June is the start of the rainy season in the Philippines, thus creating heavier traffic flow [46], while most schools in the country typically start their academic year at this time of the year as well [47]. In the month of June, data were collected for Mondays, Weekdays (i.e., Tuesday, Wednesday, and Thursday), Fridays, Saturdays, and Sundays, assuming that the traffic situations on Tuesdays, Wednesdays, and Thursdays are identical. However, during the actual data collection in Google Maps, it had been observed that the data can have marginal variations despite inputting the same date and time when collected on different occasions. For this reason, the authors collected data on five separate occasions and utilized the average. The dates looked up in Google Maps were 1 June, 2 June, 3 June, 4 June, and 7 June in the year 2019.

As discussed in the previous subsection, the bulk speed can be calculated by dividing average travel time with its segment length. By making use of speed-flow curves, the average speed of all vehicles for a given range can be translated into a rate of vehicle flow.

A speed-flow curve is created using what is referred to as the roadside friction index (RSFI), which describes side friction factors or occurrences on the side of the road that affect traffic flow [48]. For example, road segments with schools and wet markets tend to have a higher roadside friction. The speed-flow curve illustrated in Figure 4 was adopted for this case study, with reference to the paper of Pal and Roy [49].

Figure 3. The approximate locations of the 13 MRT-3 stations which will be used as the basis for dividing EDSA into road segments in this paper. Data Source: Google Maps.

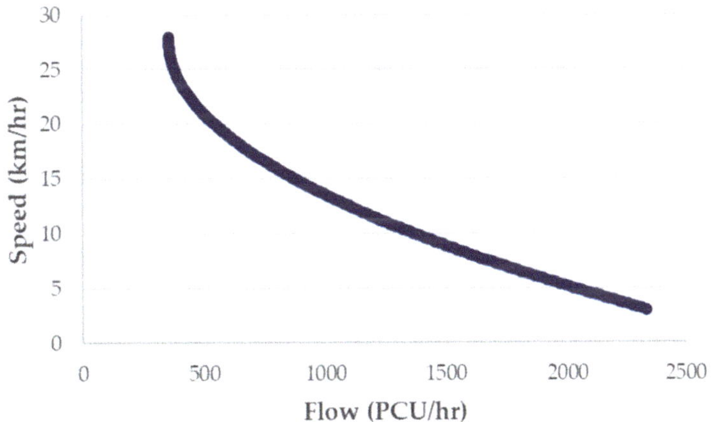

Figure 4. Adopted speed-flow curve for EDSA.

A regression equation was fitted to the speed-flow curve shown on Figure 4. This is a relatively more accurate and convenient approach to obtain vehicle flow rates using the curve [50]. Equation (4) shows the fitted quadratic equation. This will be used to estimate the hourly vehicle flow or PCU_{Count}, as a function of bulk speed. In an actual study, the authors recommend that speed flow curves for specific road links be utilized to ensure accuracy.

$$PCU_{Count} = 3.26V_B{}^2 - 180.09V_B + 2843 \tag{4}$$

The outcome was then multiplied to the average number of lanes assumed for both southbound and northbound directions of the road. An average number of lanes equal to 5 was assumed for EDSA (each way) in this study [51]. However, it is worth noting that the second-degree polynomial equation pertaining to the best fit curve was only applied from 5:00 to 22:00, considering that the traffic is normally heavier during this period. In fact, the average daily traffic volume in terms of PCU according to a transport and environmental survey for some roads in the Philippines had been observed to comparatively drop from 23:00 to 4:00 [52]. Thus, correction factors (or proportionality constants) were introduced for the hours 23:00, 1:00, 2:00, 3:00, and 4:00. Each correction factor was multiplied to the average PCU count value from 5:00 to 22:00. These factors are determined and calibrated by the modeler based on actual on-road observations. The proportionality constants used in the illustrative case study are presented in Table 1.

Table 1. Correction factors used as constants of proportionality during hours with anticipated significant drops in traffic volume.

Hours with Anticipated Significant Vehicle Volume Drops	Correction Factor
23:00	0.03
0:00	0.02
1:00	0.01
2:00	0.01
3:00	0.01
4:00	0.02

Vehicle volume counts need to be normalized as different vehicle categories have different footprints or road space requirements. Expressing volume counts in terms of PCU considers the approximate footprint of each vehicle category by estimating the number of passenger cars occupying the same footprint [53]. These estimates, considered as weightage factors, are called passenger car equivalence factors (PCEFs). Shown in Table 2 are the passenger car equivalence factors used in this study. The PCEF values for a car, jeepney, tricycle, bus, light-duty truck, and heavy-duty truck were adopted from a feasibility study relevant to the Philippine road setting [54] while the PCEF values for van, taxi, and motorcycles were adopted from the study of Adnan [55]. Moreover, utility vehicle (UV) as a vehicle type was assigned by the authors to have a PCEF equal to 1. Dividing the PCU count per vehicle type with their respective PCEF generates the estimated vehicle count per category, which will then be used to estimate energy use and emissions. It is worth noting that the jeepney is a unique vehicle category for the Philippines. The Philippine jeepney is the most common mode of public transportation in the country because of its accessibility and affordability [56]. Jeepneys are diesel-powered vehicles that were originally refurbished American military utility vehicles left after the Second World War [57]. Due to most jeepneys being old, the specific engine technology and standards used are not documented, but they are most likely all pre-Euro engines. Fuel-wise, as of writing, all fuel sold in the Philippines is at least Euro 4 quality.

For the classified vehicle count using the Street View feature of Google Maps, four points were identified per road segment, and points were approximately one kilometer from each other.

In the process of estimating emissions, transport activity data is multiplied with emission factors [58,59]. In the context of the transport sector, the greenhouse gas emissions arise from the combustion and evaporation of fuels from diverse forms of transport activities [60]. In particular, the calculation of greenhouse gas emissions on the basis of transport activities and emission factors is referred to as the bottom-up approach in which the transport activities are measured per vehicle kilometer traveled (VKT), and the emission factors are in the units of grams per kilometer [61]. In relation to the current study, the total estimated vehicle count of each road segment together with the respective road lengths determine transport activity.

Table 2. Passenger car equivalence factors for various vehicle types.

Type of Vehicle	PCEF
Tricycle	1
Motorcycle	0.25
Jeepney	1.5
Taxi	1
Car	1
Utility Vehicle	1
Bus	2
Light Truck	2
Heavy Truck	2.2

Emission factors represent the ratio between the quantities of pollutants, expressed in unit weight, that are being discharged to the atmosphere and the activity, expressed in unit weight, volume, distance or time, in connection to the release of these pollutants [62]. However, deriving emission factors relevant to a particular study is an arduous task. With regard to road transport, one needs to take into account the emission standards (i.e., Euro 4 in the Philippines), shares in use of fuel types (diesel or petrol), fuel economies (km/l_{fuel} or $g_{emissions}/km$), fuel properties, specific emission factors ($g_{emissions}/g_{fuels}$), and vehicle variants. In this study, the fuel economy, fuel properties, and specific emission factors were obtained from the databases of Argonne National Laboratory. Specifically, these are from the Greenhouse Gases, Regulated Emissions, and Energy use in Transportation (GREET) Model [63] and the Alternative Fuel Life-Cycle Environmental and Economic Transportation (AFLEET) Tool [64]. The energy use in road transport can also be estimated in the same manner, but the calorific value of the fuel is used as the multiplier instead of the specific emission factor. The calorific value of gasoline and diesel were obtained from the *Mechanical Engineer's Data Handbook* authored by James Carvill [65]. For emissions and energy use, the Metro Manila drive cycle utilized in [2] was used.

3. Illustrative Case Study

3.1. Estimation of Monthly Vehicle Count

The estimated vehicle counts in June 2019 are summarized in Table 3. The data reflects that there were no heavy-duty trucks and tricycles traveling along EDSA both in the southbound and northbound directions, except for the Magallanes Station to Taft Avenue Station road segment, which had an estimated total count of 3546 tricycles. Jeepneys plying EDSA were also limited in number relative to other vehicle types. This is because of the prohibitions by the government for these vehicle types to use most parts of the circumferential road, to accelerate traffic flow [51].

Another assumption made by the authors that can affect the classified vehicle counts involved the assumed variants belonging to each vehicle type. Car as a vehicle type includes the following variants: mini-compact, subcompact, compact, and full-size sedan. On the other hand, utility vehicle considers all variants of sport utility vehicles (SUV), pick-up trucks, and vans such as multi-purpose vehicles (MPV), crossover utility vehicles (CUV), mid-size SUVs, and full-size SUVs. Thus, it is understandable why the majority of the road segments counted more utility vehicles than cars.

Table 3. Estimated Vehicle Count in June 2019.

Monthly Vehicle Count (Southbound EDSA)									
Road Segment	**Tricycle**	**Motorcycle**	**Jeepney**	**Taxi**	**Car**	**Utility Vehicle**	**Bus**	**Light Truck**	**Heavy Truck**
North A-Quezon A	0	51,933	0	81,610	59,352	126,124	44,514	3710	0
Quezon A-G Kamuning	0	81832	0	37768	44,063	176,253	0	0	0
G Kamuning-A C Cubao	0	339,941	0	59,837	60,704	105,798	64,173	0	0
A C Cubao-Santolan A	0	45,296	2831	49,543	117,487	134,473	84,931	11,324	0
Santolan A-Ortigas A	0	55,772	2145	23,596	98,674	178,042	39,684	1073	0
Ortigas A-Shaw B	0	16,773	0	16,773	122994	178,900	36,339	2795	0
Shaw B-Boni	0	46,081	0	31,902	106341	233,951	21,268	7089	0
Boni-Guadalupe	0	54,241	0	27,121	57631	149,164	30,511	0	0
Guadalupe-Buendia	0	29,357	1957	25,443	111556	187,884	45,014	0	0
Buendia-Ayala	0	78,073	7435	33,460	74355	66,919	29,742	7435	0
Ayala-Magallanes	0	82,409	0	12,361	61807	86,530	127,735	0	0
Magallanes-Taft A	0	56,876	23,334	30,626	56876	107,919	18,959	16,042	0

Monthly Vehicle Count (Northbound EDSA)									
Road Segment	**Tricycle**	**Motorcycle**	**Jeepney**	**Taxi**	**Car**	**Utility Vehicle**	**Bus**	**Light Truck**	**Heavy Truck**
North A-Quezon A	0	86,341	13,283	43,170	149,436	209,211	36,529	6642	0
Quezon A-G Kamuning	0	63,742	0	93,162	112,775	93,162	24,516	4903	0
G Kamuning-A C Cubao	0	275,358	0	76,707	64,906	94,408	51,138	983	0
A C Cubao-Santolan A	0	74,151	0	28,837	148,303	177,139	74,151	8239	0
Santolan A-Ortigas A	0	63,442	1322	26,434	158,605	229,977	34,364	5287	0
Ortigas A-Shaw B	0	71,129	0	35,565	138,308	209,436	59,274	3953	0
Shaw B-Boni	0	93,141	0	37,256	193,732	230,989	67,061	0	0
Boni-Guadalupe	0	95,722	0	0	245,288	233,324	41,879	0	0
Guadalupe-Buendia	0	102,646	0	10,265	164,233	243,784	38,492	0	0
Buendia-Ayala	0	40,424	5775	21,174	94,322	115,496	32,724	0	0
Ayala-Magallanes	0	82,708	0	9543	89,071	213,133	22,268	9543	0
Magallanes-Taft A	3546	35,457	56,732	46,094	109,917	120,555	24,820	0	0

The largest shares of vehicle type by road segment are as follows. For southbound EDSA, GMA Kamuning Station to Araneta Center-Cubao Station had a 36.2% share for motorcycles, Magallanes Station to Taft Avenue Station had a 61.9% share for jeepneys; North Avenue Station to Quezon Avenue Station had a 19% share for taxis; Ortigas Avenue Station to Shaw Boulevard Station had a 12.7% share for cars; Shaw Boulevard Station to Boni Avenue Station had a 13.5% share for utility vehicles; Ayala Station to Magallanes Station had a 23.5% share for buses; and lastly, Magallanes Station to Taft Avenue Station had a 32.4% share for light-duty trucks. For northbound EDSA, Magallanes Station to Taft Avenue Station had the highest share of jeepneys at 73.6%. Moreover, it was the only road segment that had tricycles. GMA Kamuning Station to Araneta Center-Cubao Station had the highest share of motorcycles at 25.4%; Quezon Avenue Station to GMA Kamuning Station had a 21.8% share for taxis; Boni Station to Guadalupe Station had a 14.7% share for cars; Guadalupe Station to Buendia Station had an 11.2% share for utility vehicles; Araneta Center-Cubao Station to Santolan-Annapolis Station had a 14.6% share for buses; and finally, Ayala Station to Magallanes Station had a 24.1% share for light-duty trucks.

3.2. Estimation of Transport Emissions and Energy Use

The warming effect of greenhouse gases is based on a unit called carbon dioxide equivalent (CO_2eq) [66,67]. Therefore, the authors estimated global warming potential (GWP) using the CO_2eq of each road segment in EDSA for both northbound and southbound directions. The different greenhouse gas emissions are converted to their CO_2eq to measure their heat trapping ability [68]. Shown in Figure 5 are the estimates of CO_2eq emissions in terms of tonnes for the month of June 2019 through the use of clustered heatmaps. In consideration of the effects of air pollutant emissions to human health and environment [5], estimations of fine particulate matter, particularly $PM_{2.5}$ [69,70], were also done in the present study. $PM_{2.5}$ refers to the infinitesimal particles emitted in the air which measure two and half microns or less in average diameter. Figure 6 shows $PM_{2.5}$ emissions in terms of kilograms per road segment and vehicle type. The transport energy use in terms of tonnes of oil equivalent (toe) is shown in Figure 7.

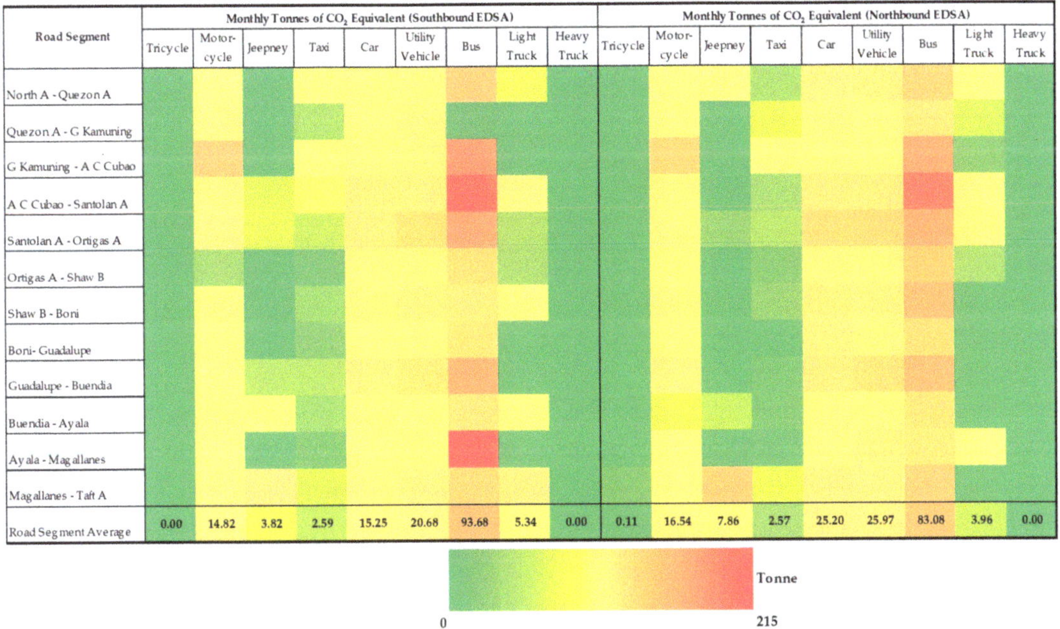

Figure 5. Clustered heatmap of the monthly tonnes of CO_2eq in EDSA.

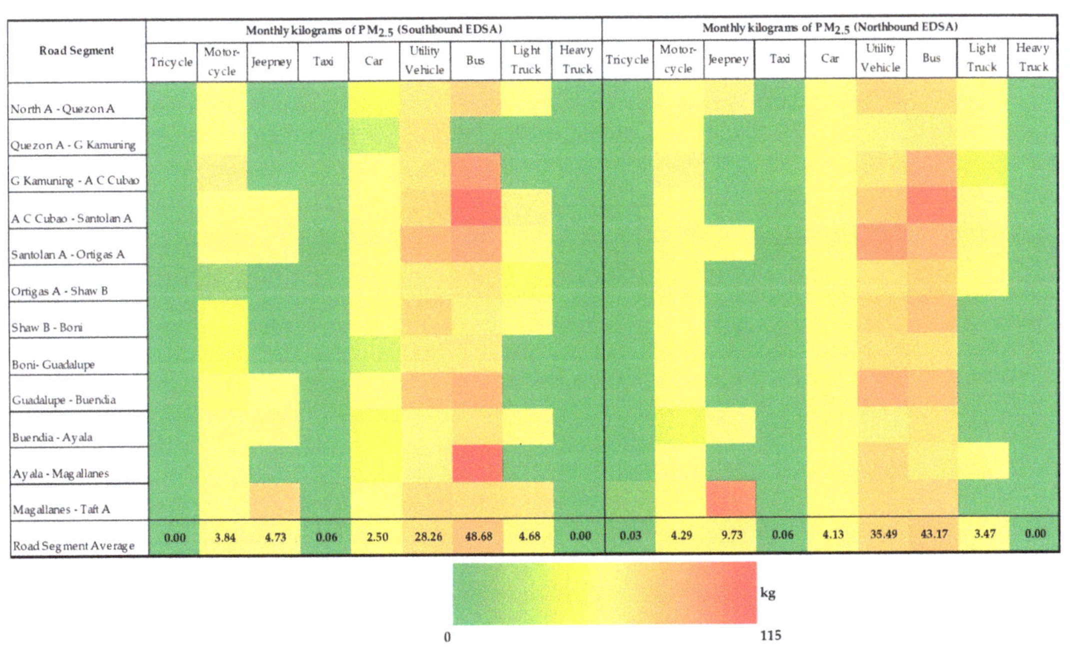

Figure 6. Clustered heatmap of the monthly kg of $PM_{2.5}$ in EDSA.

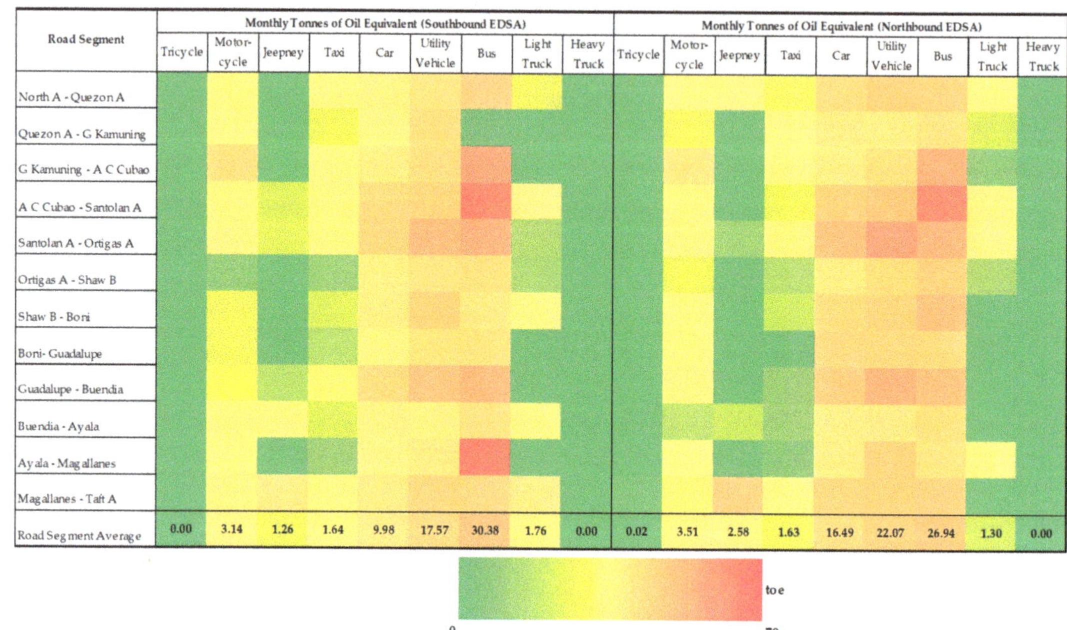

Figure 7. Clustered heatmap of monthly transport energy use in EDSA in terms of toe.

Summarized in Table 4 are the emission factors and energy consumption data which were used in the case study. Firstly, the columns for tricycle and heavy truck in Figures 5–7 had the coolest color (i.e., with respect to the heat map). This is because none of these vehicles pass through the majority of the road segments in EDSA, as indicated in Table 3. For some road segments, jeepneys and light trucks also had none, due to a relatively small number of vehicle count, thus resulting in plenty of green tiles on the heatmap. Conversely, the majority of the motorcycle, car, utility vehicle, and bus tiles showed warm colors, indicating substantial contributions to the emission load or energy usage of the road segments.

Table 4. Summary of emission factors and energy consumption per vehicle type.

	Emission Factors Data ($g_{emissions}$/km)									CO_2 Equivalence (g_{CO2eq}/$g_{emissions}$)
Types of Emission	Tricycle	Motorcycle	Jeepney	Taxi	Car	Utility Vehicle	Bus	Light Truck	Heavy Truck	
$PM_{2.5}$	0.0562	0.0336	0.8466	0.0011	0.0221	0.1430	0.7539	0.7519	0.6731	-
CH_4	4.0906	2.3022	0.2357	0.3000	0.7408	0.3538	1.2873	0.3648	1.0238	30.0000
N_2O	0.0021	0.0015	0.0316	0.0039	0.0099	0.0063	0.0222	0.0226	0.0247	265.0000
CO_2	66.9747	60.0983	668.7415	41.9204	109.8958	92.4039	1406.2301	842.0852	1672.4363	1.0000

	Energy Consumption Data (MJ/km)								
	Tricycle	Motorcycle	Jeepney	Taxi	Car	Utility Vehicle	Bus	Light Truck	Heavy Truck
Energy Consumption	1.5285	1.1504	9.4130	1.3812	3.6924	3.7241	19.6944	11.8412	23.3813	-

In Figures 5–7, the columns or tiles for buses generally had the warmest colors in comparison to the other vehicle types. However, the vehicle counts for buses based on Table 3 are comparatively smaller compared to motorcycles, cars, and utility vehicles. This observation has something to do with the fact that buses, as indicated in Table 4, have a significantly higher emission factor and energy consumption in comparison to the other vehicle types. For comparison, the emission factors and energy consumption for utility vehicles are 0.14, 0.01, 0.35, 0.01, 92.40 $g_{emissions}$/km and 3.72 MJ/km for $PM_{2.5}$, CH_4, N_2O, CO_2, and energy consumption, respectively. For buses, these are about 0.75, 0.05, 1.29, 0.02, 1406.23 $g_{emissions}$/km and 19.69 MJ/km. The same effect was observed in taxis—despite being greater in vehicle count than jeepneys and light trucks, the estimates of CO_2eq, $PM_{2.5}$, and energy use were relatively lower. This observation was also due to the fact that the emission factors and energy consumption for taxis were comparatively lower than jeepneys and light trucks. It is also interesting to note that the road segments with the warmest tiles under the bus column in Figures 5–7, namely Araneta Center-Cubao to Santolan-Annapolis and Ayala-Magallanes, are where most provincial bus terminals and major bus stops are located.

3.3. Data Validation

Table 5 shows the comparison between the annual average daily traffic (AADT) reported by the Metropolitan Manila Development Authority (MMDA) in 2019 [71] and the average daily vehicle count estimated using crowdsourced data from Google Maps. The total number of vehicles from both sources were within close range. The MMDA reported a total of 405,882 vehicles while the authors estimated 370,855 vehicles in total (i.e., 8.63% error). Having said that, considerable discrepancies on the individual percentage share for each vehicle type were observed especially in cars, utility vehicles, and buses. One possible cause for the discrepancies in car and utility vehicle percentage shares is the assignment of vehicle variants belonging to each type. This assumption might have been different with how the MMDA did it in their actual survey. Note that if the vehicle shares of cars and utility vehicles were to be aggregated in both reports, the percentage error will only be 4.63%. It is important to take note also that the PCEF values for cars and utility vehicles (see Table 2) are the same. In fact, the two can theoretically be classified into one vehicle type in as much as PCU count is concerned. Another possible reason for the discrepancies in the percentage shares is the timing when data collection took place. It is important to note that the Google Maps Street View feature does not specify the time of data collection. In this regard, the authors assumed a constant PCU percentage share (modal share) throughout the whole day. The same is true with the survey performed by MMDA—there was also no information regarding the time of data collection in the survey report.

Table 5. Comparison of average daily vehicle counts from MMDA and the proposed method using Google Maps data.

Vehicle Type	MMDA	Percentage Share	Google Maps	Percentage Share
Tricycle	9	0%	118	0%
Motorcycle	110,167	27%	70,299	19%
Jeepney	2166	1%	3827	1%
Taxi	18,913	5%	29,745	8%
Car	255,732	63%	91,817	25%
Utility Vehicle	6285	2%	136,514	37%
Bus	11,313	3%	35,496	10%
Light Truck	1297	0%	3038	1%
Heavy Truck	0	0%	0	0%
Total	405,882	100%	370,854	100%

4. Limitations and Future Work

The study demonstrated a convenient and cost-effective way to estimate emission load and energy use in road transport. One major advantage of using the proposed method is the availability and accessibility of the data. Furthermore, it makes multi-regional or multi-locality studies for benchmarking purposes easier to do. For instance, researchers can easily conduct studies on a city aside from their own. Most importantly, the gathering of data according to the said method can be used as an alternative to conducting an actual survey, which is more laborious and resource-intensive. It is important to take note that the proposed methodology in this work provides an alternative only for vehicle data and traffic flow collection. It is not meant to replace the whole emissions modeling approach, as it is rather more complex than the simplified approach used in this study. If desired by the modeler, the proposed approach to collect vehicle and traffic data in Google Maps can be coupled to other more sophisticated emissions models for more accurate estimates.

Despite the advantages, there are also some disadvantages. Aside from the Google Maps data, the methodology is also highly dependent on data from various references and assumptions. Therefore, these can become sources of inaccuracies in the results. In some countries, data concerning the transportation sector can be more accessible. Another major assumption used in the current case study is the speed-flow curve adopted. In vehicle flow analysis, speed-flow curves and PCEFs are derived based on the dynamic characteristics and positions of vehicles for specific roads. The illustrative case study used a speed-flow curve that had been modeled from a specific highway in India. On the other hand, the PCEFs used, although relevant to Philippine roads, were taken from a feasibility study published in 1987.

To address these limitations for the Philippines, the authors propose the following for future work. First, research on speed-flow curve modeling, particularly for the major thoroughfare, EDSA, is needed. Relevant literature in developing speed-flow curves includes [49,72,73]. Second, a future research study should develop updated values of PCEFs considering the current traffic characteristics of Philippine highways. Relevant literature in measuring PCEFs includes [55,74,75]. To reduce uncertainty in data, a probabilistic analysis could be used to improve the robustness of the results. For example, a Monte Carlo approach can be used to explore the robustness of the data collected from Google Maps and Google Street View.

5. Conclusions

A novel method for estimating emissions and energy use in road transport using crowdsourced data from Google Maps was introduced and demonstrated by the authors. Two particular sets of data are utilized from Google Maps. The first is the travel time and road length data which will be converted into bulk speed or the average speed of all vehicles. Subsequently, the rate of vehicle flow in terms of PCU/hr is determined using the calculated bulk speed. The second dataset collected from Google Maps is the classified vehicle count using Google Street View. After establishing the rate of vehicle flows and PCU mix, PCEF values are utilized to break down the PCU into different types of vehicles. Finally, emissions and energy use in road transport are estimated by multiplying the vehicle count with the assumed emission factors and energy consumption data.

An illustrative case study was performed to demonstrate the actual use of this methodology, which was also validated using government-reported daily traffic data (AADT). The average daily vehicle count estimated using the proposed method had about an 8.63% error when compared to the AADT survey conducted by MMDA in the same year. However, this can be further reduced to 4.63% if cars and utility vehicles were combined. This means that the proposed method can accurately estimate average traffic flow, given proper assumptions on the data.

Furthermore, CO_2eq, $PM_{2.5}$, and energy use were estimated in the case study. The results were illustrated through clustered heatmaps showing the share of each road segment and vehicle type. Results indicated that buses, despite having less vehicle counts, had the largest contributions to CO_2eq, $PM_{2.5}$, and energy use. On the contrary, taxis, despite being greater in vehicle count than jeepneys and light-duty trucks, had one of the least contributions to CO_2eq, $PM_{2.5}$, and energy use. Based on the average values in southbound and northbound EDSA, buses produced 88.38 tonnes of CO_2eq and 43.93 kg of $PM_{2.5}$

monthly per road segment while consuming 28.66 toe. Furthermore, taxis discharged 2.58 tonnes of CO_2eq. and 0.06 kg of $PM_{2.5}$ monthly per road segment while consuming 1.64 toe.

Since the novelty of this study is using crowdsourced data in Google Maps, this study can give rise to future works which can improve the methodology in such a way that possible inaccuracies and uncertainties can be mitigated and considered better. While the proposed methodology demonstrated promising results in the illustrative case study, this research is an evolving work that shows great potential to streamline methodologies in estimating emission load and energy use in road transport. While much of the data collection work was done manually in this paper, future research can easily automate plenty of these steps.

Author Contributions: Conceptualization, N.S.L. and J.B.M.B.; Methodology, N.S.L., J.E.R. and J.B.M.B.; software, N.S.L., J.E.R. and J.B.M.B.; formal analysis, N.S.L. and J.E.R.; writing—original draft preparation, N.S.L. and J.E.R.; writing—review and editing, J.B.M.B.; visualization, J.E.R.; funding acquisition, J.B.M.B. All authors have read and agreed to the published version of the manuscript.

Funding: At the time of writing, Joshua Ezekiel Rito was under full scholarship from the St. La Salle Financial Assistance Grant for Graduate Students of De La Salle University, Manila, Philippines for his MS in Mechanical Engineering degree. The APC was funded by the Office of the Vice Chancellor for Research and Innovation of De La Salle University, Manila, Philippines.

Institutional Review Board Statement: Not applicable.

Informed Consent Statement: Not applicable.

Data Availability Statement: Publicly available datasets were analyzed in this study. This data can be found here: https://www.google.com/maps (accessed on 12 December 2020).

Conflicts of Interest: The authors declare no conflict of interest.

References

1. Stern, N. The Economics of Climate Change. *Am. Econ. Rev.* **2008**, *98*, 1–37. [CrossRef]
2. Lopez, N.S.; Soliman, J.; Biona, J.B.M.; Fulton, L. Cost-benefit analysis of alternative vehicles in the Philippines using immediate and distant future scenarios. *Transp. Res. Part D Transp. Environ.* **2020**, *82*, 102308. [CrossRef]
3. International Energy Agency. *Global Energy & CO$_2$ Status Report*; International Energy Agency: Paris, France, 2019.
4. United Nations. *Paris Agreement*; United Nations: Paris, France, 2015.
5. Manisalidis, I.; Stavropoulou, E.; Stavropoulos, A.; Bezirtzoglou, E. Environmental and Health Impacts of Air Pollution: A Review. *Front. Public Health* **2020**, *8*, 1–13. [CrossRef] [PubMed]
6. Ritchie, H.; Roser, M. Outdoor Air Pollution. Available online: https://ourworldindata.org/outdoor-air-pollution (accessed on 20 May 2021).
7. Stanaway, J.D.; Afshin, A.; Gakidou, E.; Lim, S.S.; Abate, D.; Abate, K.H.; Abbafati, C.; Abbasi, N.; Abbastabar, H.; Abd-Allah, F.; et al. Global, regional, and national comparative risk assessment of 84 behavioural, environmental and occupational, and metabolic risks or clusters of risks for 195 countries and territories, 1990–2017: A systematic analysis for the Global Burden of Disease Stu. *Lancet* **2018**, *392*, 1923–1994. [CrossRef]
8. Lopez, N.S.; Chiu, A.S.F.; Biona, J.B.M. Decomposing drivers of transportation energy consumption and carbon dioxide emissions for the Philippines: The case of developing countries. *Front. Energy* **2018**, *12*, 389–399. [CrossRef]
9. International Energy Agency. *Clean Energy Transitions Programme*; International Energy Agency: Paris, France, 2021.
10. Hall, D.; Lutsey, N. Estimating the infrastructure needs and costs for the launch of zero-emission trucks. *Int. Counc. Clean Transp.* **2019**, 1–31. [CrossRef]
11. Anenberg, S.C.; Miller, J.; Henze, D.K.; Minjares, R.; Achakulwisut, P. The global burden of transportation tailpipe emissions on air pollution-related mortality in 2010 and 2015. *Environ. Res. Lett.* **2019**, *14*, 094012. [CrossRef]
12. Acaravci, A.; Ozturk, I. On the relationship between energy consumption, CO2 emissions and economic growth in Europe. *Energy* **2010**, *35*, 5412–5420. [CrossRef]
13. Zou, S.; Zhang, T. CO$_2$ Emissions, Energy Consumption, and Economic Growth Nexus: Evidence from 30 Provinces in China. *Math. Probl. Eng.* **2020**, *2020*, 1–10. [CrossRef]
14. International Energy Agency. *Energy Efficiency Indicators Highlights*; International Energy Agency: Paris, France, 2020.
15. Song, M.; Wu, N.; Wu, K. Energy Consumption and Energy Efficiency of the Transportation Sector in Shanghai. *Sustainability* **2014**, *6*, 702–717. [CrossRef]

16. Furfari, S. Energy efficiency of engines and appliances for transport on land, water, and in air. *Ambio* **2016**, *45*, 63–68. [CrossRef] [PubMed]
17. Engeset, P.; Keitheile, O.B. *Traffic Data Collection and Analysis*; Ministry of Works and Transport: Gaborone, Botswana, 2004.
18. Paľo, J.; Caban, J.; Kiktová, M.; Černický, Ľ. The comparison of automatic traffic counting and manual traffic counting. *IOP Conf. Ser. Mater. Sci. Eng.* **2019**, *710*, 012041. [CrossRef]
19. Wardrop, J.G.; Charlesworth, G. A Method of Estimating Speed and Flow of Traffic from a Moving Vehicle. *Proc. Inst. Civ. Eng.* **1954**, *3*, 158–171. [CrossRef]
20. Mortimer, W.J. Moving Vehicle Method of Estimating Traffic Volumes and Speeds. *Highw. Res. Board Bull.* **1957**, *156*, 1–13.
21. Venkatcharyulu, S.; Mallikarjunareddy, V. Traffic volume Analysis of Newly Developing semi-urban Road. *E3S Web Conf.* **2020**, *184*, 01116. [CrossRef]
22. Zhao, N.; Qi, T.; Yu, L.; Zhang, J.; Jiang, P. A Practical Method for Estimating Traffic Flow Characteristic Parameters of Tolled Expressway Using Toll Data. *Procedia Soc. Behav. Sci.* **2014**, *138*, 632–640. [CrossRef]
23. Seo, T.; Kusakabe, T. Probe vehicle-based traffic flow estimation method without fundamental diagram. *Transp. Res. Procedia* **2015**, *9*, 149–163. [CrossRef]
24. Aksoy, A.; Küçükoğlu, İ.; Ene, S.; Öztürk, N. Integrated Emission and Fuel Consumption Calculation Model for Green Supply Chain Management. *Procedia Soc. Behav. Sci.* **2014**, *109*, 1106–1109. [CrossRef]
25. Jabali, O.; Van Woensel, T.; de Kok, A.G. Analysis of Travel Times and CO_2 Emissions in Time-Dependent Vehicle Routing. *Prod. Oper. Manag.* **2012**, *21*, 1060–1074. [CrossRef]
26. Bharadwaj, S.; Ballare, S.; Rohit Chandel, M.K. Impact of congestion on greenhouse gas emissions for road transport in Mumbai metropolitan region. *Transp. Res. Procedia* **2017**, *25*, 3538–3551. [CrossRef]
27. Nesamani, K.S.; Saphores, J.; McNally, M.G.; Jayakrishnan, R. Estimating impacts of emission specific characteristics on vehicle operation for quantifying air pollutant emissions and energy use. *J. Traffic Transp. Eng.* **2017**, *4*, 215–229. [CrossRef]
28. de Maes, A.S.; Hoinaski, L.; Meirelles, T.B.; Carlson, R.C. A methodology for high resolution vehicular emissions inventories in metropolitan areas: Evaluating the effect of automotive technologies improvement. *Transp. Res. Part D Transp. Environ.* **2019**, *77*, 303–319. [CrossRef]
29. Iqbal, A.; Allan, A.; Zito, R. Meso-scale on-road vehicle emission inventory approach: A study on Dhaka City of Bangladesh supporting the 'cause-effect' analysis of the transport system. *Environ. Monit. Assess.* **2016**, *188*, 149. [CrossRef] [PubMed]
30. Zhang, L.; Lin, J.; Qiu, R. Characterizing the toxic gaseous emissions of gasoline and diesel vehicles based on a real-world on-road investigation. *J. Clean. Prod.* **2021**, *286*, 124957. [CrossRef]
31. Iqbal, A.; Afroze, S.; Rahman, M.M. Probabilistic Health Risk Assessment of Vehicular Emissions as an Urban Health Indicator in Dhaka City. *Sustainability* **2019**, *11*, 6427. [CrossRef]
32. Chatzimilioudis, G.; Zeinalipour-Yazti, D. Crowdsourcing for Mobile Data Management. In Proceedings of the 2013 IEEE 14th International Conference on Mobile Data Management, Milan, Italy, 3–6 June 2013; Volume 2, pp. 3–4.
33. Brabham, D.C. Crowdsourcing as a Model for Problem Solving. *Converg. Int. J. Res. N. Media Technol.* **2008**, *14*, 75–90. [CrossRef]
34. Misra, A.; Gooze, A.; Watkins, K.; Asad, M.; Le Dantec, C.A. Crowdsourcing and Its Application to Transportation Data Collection and Management. *Transp. Res. Rec. J. Transp. Res. Board* **2014**, *2414*, 1–8. [CrossRef]
35. Chandra, S.; Naik, R.T.; Jimenez, J. A Framework for Smart Freight Mobility with Crowdsourcing. *Transp. Res. Procedia* **2020**, *48*, 494–502. [CrossRef]
36. Nair, D.J.; Gilles, F.; Chand, S.; Saxena, N.; Dixit, V. Characterizing multicity urban traffic conditions using crowdsourced data. *PLoS ONE* **2019**, *14*, e0212845. [CrossRef]
37. Ferster, C.; Nelson, T.; Laberee, K.; Vanlaar, W.; Winters, M. Promoting Crowdsourcing for Urban Research: Cycling Safety Citizen Science in Four Cities. *Urban Sci.* **2017**, *1*, 21. [CrossRef]
38. Wang, X.; Zheng, X.; Zhang, Q.; Wang, T.; Shen, D. Crowdsourcing in ITS: The State of the Work and the Networking. *IEEE Trans. Intell. Transp. Syst.* **2016**, *17*, 1596–1605. [CrossRef]
39. Zheng, X.; Chen, W.; Wang, P.; Shen, D.; Chen, S.; Wang, X.; Zhang, Q.; Yang, L. Big Data for Social Transportation. *IEEE Trans. Intell. Transp. Syst.* **2016**, *17*, 620–630. [CrossRef]
40. Linton, C.; Grant-Muller, S.; Gale, W.F. Approaches and Techniques for Modelling CO_2 Emissions from Road Transport. *Transp. Rev.* **2015**, *35*, 533–553. [CrossRef]
41. Jiménez-Meza, A.; Arámburo-Lizárraga, J.; de la Fuente, E. Framework for Estimating Travel Time, Distance, Speed, and Street Segment Level of Service (LOS), based on GPS Data. *Procedia Technol.* **2013**, *7*, 61–70. [CrossRef]
42. Sharma, M.; Biswas, S. Estimation of Passenger Car Unit on urban roads: A literature review. *Int. J. Transp. Sci. Technol.* **2020**. [CrossRef]
43. Google Maps. Available online: https://www.google.com/maps/dir/Quezon+Avenue/GMA+Kamuning+Station/@14.6389009,-121.0371689,16z/data=!4m18!4m17!1m5!1m1!1s0x3397b7aa16a4f333:0x5eefaaf26ee44220!2m2!1d121.0385078!2d14.6427595!1m5!1m1!1s0x3397b7af0a4f1251:0x867338649729026b!2m2!1d121.0433426 (accessed on 30 April 2021).
44. Ganiron, T.U.J. Exploring the Emerging Impact of Metro Rail Transit (MRT-3) in Metro Manila. *Int. J. Adv. Sci. Technol.* **2015**, *74*, 11–24. [CrossRef]

45. Google Maps. Available online: https://www.google.com/maps/dir/Taft+Avenue/MRT+North+Ave.+Station,+Bagong+Pag-asa,+Quezon+City,+Metro+Manila/@14.5950342,120.9740679,12z/data=!4m18!4m17!1m5!1m1!1s0x3397c945008b5cb5:0x23dd98e8b1d43815!2m2!1d121.0021747!2d14.537669!1m5!1m1!1s0x3397b6fd9e1 (accessed on 20 April 2021).

46. Cruz, F.T.; Narisma, G.T.; Villafuerte, M.Q.; Cheng Chua, K.U.; Olaguera, L.M. A climatological analysis of the southwest monsoon rainfall in the Philippines. *Atmos. Res.* **2013**, *122*, 609–616. [CrossRef]

47. Villafuerte, M.Q.; Juanillo, E.L.; Hilario, F.D. Climatic insights on academic calendar shift in the Philippines. *Philipp. J. Sci.* **2017**, *146*, 267–276.

48. Salini, S.; George, S.; Ashalatha, R. Effect of Side Frictions on Traffic Characteristics of Urban Arterials. *Transp. Res. Procedia* **2016**, *17*, 636–643. [CrossRef]

49. Pal, S.; Roy, S.K. Impact of Roadside Friction on Travel Speed and LOS of Rural Highways in India. *Transp. Dev. Econ.* **2016**, *2*, 9. [CrossRef]

50. Srikanth, S.; Mehar, A. Estimation of Equivalency Units for Vehicle Types under Mixed Traffic Conditions: Multiple Non-Linear Regression Approach. *Int. J. Technol.* **2017**, *8*, 820. [CrossRef]

51. Boquet, Y. Battling Congestion in Manila: The EDSA Problem. *Transp. Commun. Bull. Asia Pacific* **2013**, 45–59.

52. Japan International Cooperation Agency; Department of Public Works and Highways. Transport and Environmental Surveys. In *The Feasibility Study and Implementation Support for Cavite-Laguna East-West National Road Project*; Japan International Cooperation Agency: Tokyo, Japan, 2006; pp. 1–84.

53. Al-Kaisy, A.; Jung, Y.; Rakha, H. Developing Passenger Car Equivalency Factors for Heavy Vehicles during Congestion. *J. Transp. Eng.* **2005**, *131*, 514–523. [CrossRef]

54. Japan International Cooperation Agency; Department of Public Works and Highways. *Feasibility Study of the Road Improvement Project on the Pan-Philippine Highway (Philippine-Japan Friendship Highway)*; Japan International Cooperation Agency: Tokyo, Japan, 1987.

55. Adnan, M. Passenger Car Equivalent Factors in Heterogenous Traffic Environment-are We Using the Right Numbers? *Procedia Eng.* **2014**, *77*, 106–113. [CrossRef]

56. Coz, M.C.; Flores, P.J.; Hernandez, K.L.; Portus, A.J. An Ergonomic Study on the UP-Diliman Jeepney Driver's Workspace and Driving Conditions. *Procedia Manuf.* **2015**, *3*, 2597–2604. [CrossRef]

57. Agaton, C.B.; Guno, C.S.; Villanueva, R.O.; Villanueva, R.O. Diesel or Electric Jeepney? A Case Study of Transport Investment in the Philippines Using the Real Options Approach. *World Electr. Veh. J.* **2019**, *10*, 51. [CrossRef]

58. Kim, H.; Tae, S.; Yang, J. Calculation Methods of Emission Factors and Emissions of Fugitive Particulate Matter in South Korean Construction Sites. *Sustainability* **2020**, *12*, 9802. [CrossRef]

59. Olaguer, E.P. Emission Inventories. In *Atmospheric Impacts of the Oil and Gas Industry*; Elsevier: Amsterdam, The Netherlands, 2017; pp. 67–77; ISBN 9780128018835.

60. Waldron, C.D.; Harnisch, J.; Lucon, O.; Mckibbon, R.S.; Saile, S.B.; Wagner, F.; Walsh, M.P.; Kapshe, M. Mobile Combustion. In *2006 IPCC Guidelines for National Greenhouse Gas Inventories Volume 2 Energy*; Institute for Global Environmental Strategies: Kanagawa, Japan, 2006; pp. 1–78.

61. Bongardt, D.; Eichhorst, U.; Dünnebeil, F.; Reinhard, C. *Monitoring Greenhouse Gas Emissions of Transport Activities in Chinese Cities*; Deutsche Gesellschaft für Internationale Zusammenarbeit: Bonn, Germany, 2016.

62. Cheremisinoff, N.P. Pollution Management and Responsible Care. In *Waste*; Elsevier: Cambridge, UK, 2011; pp. 487–502; ISBN 9780123814753.

63. Argonne National Laboratory The Greenhouse gases, Regulated Emissions, and Energy use in Technologies (GREET) Model. Available online: https://greet.es.anl.gov/greet.models (accessed on 6 January 2021).

64. Argonne National Laboratory Alternative Fuel Life-Cycle Environmental and Economic Transportation (AFLEET) Tool. Available online: https://greet.es.anl.gov/afleet (accessed on 6 January 2021).

65. Carvill, J. *Mechanical Engineer's Data Handbook*, 1st ed.; Elsevier: Oxford, UK, 1993; ISBN 9780080511351.

66. Yoro, K.O.; Daramola, M.O. CO2 emission sources, greenhouse gases, and the global warming effect. In *Advances in Carbon Capture*; Elsevier: Cambridge, UK, 2020; pp. 3–28; ISBN 9780128196571.

67. Dincer, I.; Abu-Rayash, A. Sustainability modeling. In *Energy Sustainability*; Elsevier: Amsterdam, The Netherlands, 2020; pp. 119–164; ISBN 9780128195567.

68. Vallero, D.A. Air pollution biogeochemistry. In *Air Pollution Calculations*; Elsevier: Amsterdam, The Netherlands, 2019; pp. 175–206; ISBN 9780128149348.

69. Myong, J.-P. Health Effects of Particulate Matter. *Korean J. Med.* **2016**, *91*, 106–113. [CrossRef]

70. Xing, Y.F.; Xu, Y.H.; Shi, M.H.; Lian, Y.X. The impact of PM2.5 on the human respiratory system. *J. Thorac. Dis.* **2016**, *8*, E69–E74. [CrossRef]

71. Metropolitan Manila Developement Authority Metropolitan Manila Annual Average Daily Traffic (AADT) 2019. Available online: https://mmda.gov.ph/2-uncategorised/3345-freedom-of-information-foi.html (accessed on 30 April 2021).

72. Bains, M.S.; Ponnu, B.; Arkatkar, S.S. Modeling of Traffic Flow on Indian Expressways using Simulation Technique. *Procedia Soc. Behav. Sci.* **2012**, *43*, 475–493. [CrossRef]

73. Bharadwaj, N.; Kumar, P.; Arkatkar, S.S.; Joshi, G. Deriving capacity and level-of-service thresholds for intercity expressways in India. *Transp. Lett.* **2020**, *12*, 182–196. [CrossRef]

74. Ahmed, U. *Passenger Car Equivalent Factors for Level Freeway Segments Operating under Moderate and Congested Conditions*; Marquette University: Milwaukee, WI, USA, 2010.
75. Lu, P.; Zheng, Z.; Tolliver, D.; Pan, D. Measuring Passenger Car Equivalents (PCE) for Heavy Vehicle on Two Lane Highway Segments Operating Under Various Traffic Conditions. *J. Adv. Transp.* **2020**, *2020*, 1–9. [CrossRef]

Article

Effects of Organic Solvents on the Organosolv Pretreatment of Degraded Empty Fruit Bunch for Fractionation and Lignin Removal

Danny Wei Kit Chin [1,2], Steven Lim [1,2,*], Yean Ling Pang [1,2], Chun Hsion Lim [1,2], Siew Hoong Shuit [1,2], Kiat Moon Lee [3] and Cheng Tung Chong [4]

[1] Department of Chemical Engineering, Lee Kong Chian Faculty of Engineering and Science, Universiti Tunku Abdul Rahman (UTAR), Jalan Sungai Long, Kajang 43000, Selangor, Malaysia; dannycwk@1utar.my (D.W.K.C.); pangyl@utar.edu.my (Y.L.P.); chlim@utar.edu.my (C.H.L.); shuitsh@utar.edu.my (S.H.S.)

[2] Centre for Photonics and Advanced Materials Research, Universiti Tunku Abdul Rahman (UTAR), Kampar 31900, Malaysia

[3] Department of Chemical & Petroleum Engineering, Faculty of Engineering, Technology and Built Environment, UCSI University, Cheras 53000, Malaysia; leekm@ucsiuniversity.edu.my

[4] China-UK Low Carbon College, Shanghai Jiao Tong University, Lingang, Shanghai 201306, China; ctchong@sjtu.edu.cn

* Correspondence: stevenlim@utar.edu.my or stevenlim_my2000@hotmail.com

Citation: Chin, D.W.K.; Lim, S.; Pang, Y.L.; Lim, C.H.; Shuit, S.H.; Lee, K.M.; Chong, C.T. Effects of Organic Solvents on the Organosolv Pretreatment of Degraded Empty Fruit Bunch for Fractionation and Lignin Removal. *Sustainability* **2021**, *13*, 6757. https://doi.org/10.3390/su13126757

Academic Editor: Changhyun Roh

Received: 2 May 2021
Accepted: 29 May 2021
Published: 15 June 2021

Publisher's Note: MDPI stays neutral with regard to jurisdictional claims in published maps and institutional affiliations.

Abstract: Empty fruit bunch (EFB), which is one of the primary agricultural wastes generated from the palm oil plantation, is generally discharged into the open environment or ends up in landfills. The utilization of this EFB waste for other value-added applications such as activated carbon and biofuels remain low, despite extensive research efforts. One of the reasons is that the EFB is highly vulnerable to microbial and fungi degradation under natural environment owning to its inherent characteristic of high organic matter and moisture content. This can rapidly deteriorate its quality and results in poor performance when processed into other products. However, the lignocellulosic components in degraded EFB (DEFB) still largely remain intact. Consequently, it could become a promising feedstock for production of bio-products after suitable pretreatment with organic solvents. In this study, DEFB was subjected to five different types of organic solvents for the pretreatment, including ethanol, ethylene glycol, 2-propanol, acetic acid and acetone. The effects of temperature and residence time were also investigated during the pretreatment. Organosolv pretreatment in ethylene glycol (50 v/v%) with the addition of NaOH (3 v/v%) as an alkaline catalyst successfully detached 81.5 wt.% hemicellulose and 75.1 wt.% lignin. As high as 90.4 wt.% cellulose was also successfully retrieved at mild temperature (80 °C) and short duration (45 min), while the purity of cellulose in treated DEFB was recorded at 84.3%. High-purity lignin was successfully recovered from the pretreatment liquor by using sulfuric acid for precipitation. The amount of recovered lignin from alkaline ethylene glycol liquor was 74.6% at pH 2.0. The high recovery of cellulose and lignin in DEFB by using organosolv pretreatment rendered it as one of the suitable feedstocks to be applied in downstream biorefinery processes. This can be further investigated in more detailed studies in the future.

Keywords: organosolv pretreatment; delignification; fractionation; organic solvent; degraded empty fruit bunch

1. Introduction

Depletion of finite natural energy resources such as fossil fuel is one of the most actively debated topics recently. Lignocellulosic biomass as one of the most abundantly available resources has attracted intensive research over the years since 2008. Numerous lignocellulosic biomass, such as napier grass [1], corn stover [2], sugarcane bagasse, rice straw [3], sunflower stalk [4] and cotton stalk [5], have been constantly investigated by researchers for production of higher value bio-products such as biofuels. Indonesia and Malaysia are the major palm oil producers in the world, producing more than 85% of the global palm oil supply. During the palm oil production, the oil palm fruits are stripped from the fresh fruit bunch and thus enormous amounts of empty fruit bunch (EFB) will be generated as waste. It is conservatively estimated that the production of 1 ton of palm oil can generate as much as 1.05 tons of EFB [6]. Every year, both of the countries generate more than 60 million tons of EFB, which presents a huge burden to the environment [7]. Under optimal conditions, this EFB waste is reutilized as building blocks or composite for wooden furniture and bedding [8]. Another potential usage of EFB is being treated as solid fuel after energy densification processes such as torrefaction. Palm oil mills with substantial capacity usually resorted to combust EFB as solid fuel to generate energy for their own usage, which could partially offset the exorbitant energy cost.

There are several drawbacks related to electricity generation with EFB, such as considerable investment capital for smallholders to set up the necessary infrastructure and the high moisture content in the biomass, which renders it difficult to be stored for a pro-longed period. As a result, EFB is often discharged into the open environment or ends up in the landfills. Furthermore, EFB is an organic matter which is highly vulnerable to microbial and fungal attacks. This will cause its quality to degrade rapidly, especially in terms of physical appearances and chemical composition. Consequently, degraded EFB (DEFB) was seldom recycled and eventually disposed of into the environment, and left to decompose naturally. As a matter of fact, lignocellulosic content in DEFB has high potential for reutilization after undergoing appropriate pretreatment. These lignocellulosic materials can be fractionated and synthesized into higher value-added products such as biofuels, activated carbon or clean electricity generation from biohydrogen [9,10]. These can effectively improve the commercial value of EFB and diversify the sources of income for the palm oil industry in addition to the principal income derived from the conventional production of palm oil.

Generally, pretreatment for biomass is applied to destroy the barrier or reduce the recalcitrance of lignocellulosic biomass towards the subsequent conversion reaction [11]. Organosolv pretreatment, which applied organic solvents to pretreat lignocellulosic biomass, is one of the promising pretreatments in the presence or absence of catalyst at elevated temperature (80 to 200 °C) for short duration (30 to 120 min). Organosolv pretreatment has several advantages in pretreating the lignocellulosic biomass compared to acidic or alkaline hydrothermal pretreatment. It is able to obtain a very distinctive separation of high-purity cellulose content from the rest of the lignocellulosic components, such as lignin and hemicellulose. Furthermore, it can produce high-quality lignin which can be recovered from its pretreatment liquor. The lignin can be further processed into other bio-products which helps to minimize the waste generation and fulfills the circular economy concept [12]. Several prominent solvents used in the organosolv pretreatment are ethanol, acetic acid, acetone, glycerol and γ-valerolactone (GVL) [13]. Smit and Huijgen fractionated corn stover by using 50 wt.% acetone in the presence of sulfuric acid [2]. The acidified acetone pretreatment was conducted at 140 °C for 120 min. The treated corn stover was delignified at 81.5% with high hemicellulose removal of 91.5%. The treated corn stover also achieved high cellulose recoverability at 89.3%. Wu et al. successfully performed the organosolv pretreatment for cotton stalk using GVL as a solvent [5]. They discovered that the lignin obtained after the pretreatment contained higher yields and purity due to the higher degree of repolymerization and degradation.

From the literature, there were a lot of studies being conducted using fresh EFB [6,8–10]. However, relevant scientific studies which used DEFB for pretreatment are still very rare, despite its existence in large quantity in the industry, to the best of our knowledge. Therefore, this study aims to apply suitable pretreatment to the DEFB produced from the palm oil industry to uncover its potential as feedstock for the production of other bio-products from the recovered lignin and cellulose content. As an extension of our previous works [6,14], five different organic solvents were applied in this study, including ethanol, ethylene glycol, 2-propanol, acetic acid and acetone. Each of these solvents was chosen based on their different characteristics, which might provide different solvation effects during the pretreatment. The effects of temperature and residence time were also investigated for the most effective organosolv pretreatment against delignification, hemicellulose removal and cellulose recovery. The dissolved lignin was also isolated and recovered by acidification of the pretreatment liquor [15,16]. In order to understand the influences of pH to the amount of condensed lignin from the pretreatment liquor, sulfuric acid was used to acidify the pretreatment liquor. The effect of this organosolv pretreatment towards DEFB is still relatively lacking in the literature and thus the results presented in this study would be useful to provide the essential scientific information for future organosolv pretreatment studies on DEFB.

2. Materials and Methods

2.1. Preparation of Feedstock

DEFB and fresh EFB were obtained from an oil palm plantation estate in Segamat, Malaysia. Prior to the collection, fresh EFB was left in the open field for two weeks to allow natural degradation to take place in order to obtain DEFB. The average humidity is estimated at about 80%, which helps to promote the microbial and fungal attacks from the oil palm plantation through a series of enzymatic and oxidation reactions. DEFB and fresh EFB were cut into pieces with length less than 5 cm and washed thoroughly with distilled water several times in order to remove the dirt and contaminants. Subsequently, the cleaned fresh EFB and DEFB were dried in a hot air oven (Mermert, Schwabach, Germany) at 105 °C overnight until constant weight was obtained. The dried fresh EFB and DEFB were ground with a mechanical grinder to reduce the particle size to powder form. The ground fresh EFB and DEFB in powder form were then sieved with No. 20 mesh size in order to maintain uniform particle size. Finally, the samples were kept in resealable containers inside a desiccator prior to the pretreatment. The lignocellulosic compositions for both the DEFB and fresh EFB were characterized and compared in Figure 1. The schematic experimental flowchart for the entire experimental work in this study is shown in Figure 2.

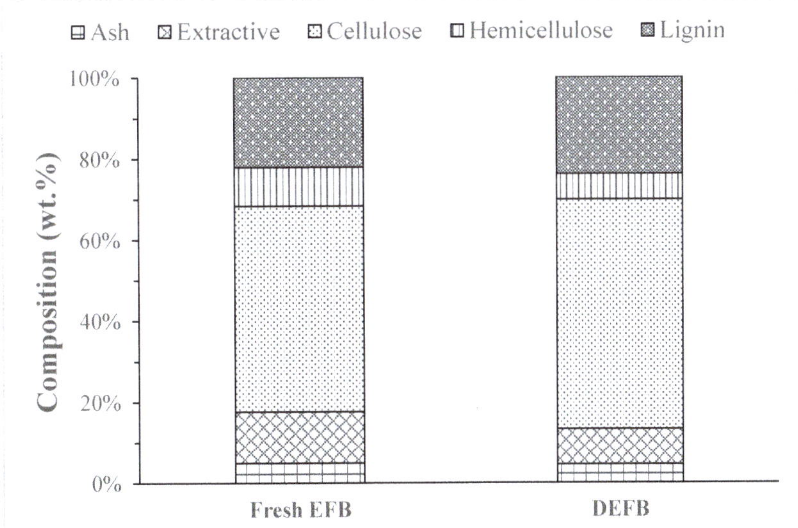

Figure 1. Lignocellulosic composition of fresh EFB and DEFB.

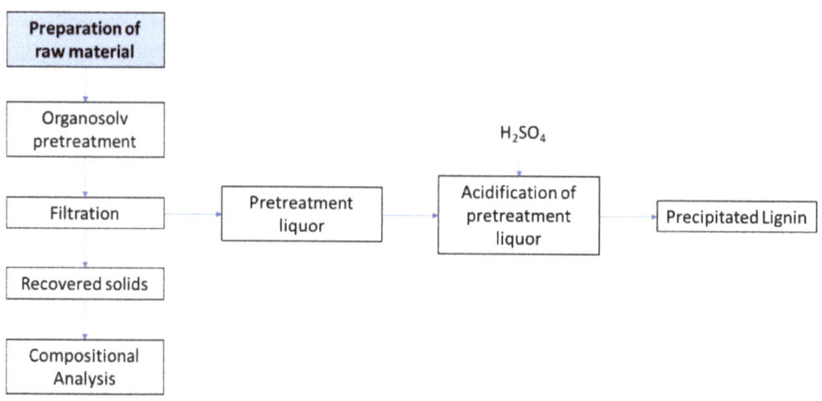

Figure 2. Schematic experimental flow of organosolv pretreatment.

2.2. Pretreatment with Different Organic Solvents

A predetermined amount of powdered fresh EFB and DEFB (10 $w/v\%$) was treated with a total of five different organic solvents, including ethanol, ethylene glycol, 2-propanol, acetic acid and acetone. For all organic solvents, 30 $v/v\%$ solvent concentration was used to pretreat the DEFB at 80 °C for 30 min based on the parameters from the previous works [6,14]. The dried and treated DEFB was repeatedly filtered and cleaned with distilled water until no more solvent was detected in the effluent. Subsequently, the biomass was dried in the hot air oven (Mermert, Schwabach, Germany) at 105 °C overnight. The dried and treated DEFB was then subjected to compositional analysis and characterization tests. Each of the experimental runs were repeated thrice to determine the variability of the data.

2.3. Pretreatment of DEFB with Ethylene Glycol Assisted by NaOH

Two primary factors including reaction temperature and pretreatment duration were studied for alkaline pretreatment of DEFB using ethylene glycol based on the best results obtained from Section 2.2. For the effect of reaction temperature, 10 $w/v\%$ of a predetermined amount DEFB sample was subjected to pretreatment with the concentration of ethylene glycol at 50 $v/v\%$ and addition of NaOH as the alkaline catalyst (3 $v/v\%$) at temperature from 60 to 140 °C for 30 min. Treated DEFB was washed with distilled water and filtered repeatedly to remove all the pretreatment liquor. Afterwards, it was dried at 105 °C overnight to remove all the excess water and solvent in the sample. The pretreatment and washing steps were performed repeatedly with different pretreatment durations from 15 to 75 min under the same solvent and catalyst concentrations. For the temperature and residence time parameter study, the dried DEFB was then subjected to compositional analysis to evaluate the pretreatment performances in terms of cellulose recovery, delignification and hemicellulose removal. Each of the experimental runs were repeated thrice to determine the variability of the data.

2.4. Precipitation of Lignin from the Pretreatment Liquor by Using Acidification

Alkaline ethylene glycol pretreatment was conducted at optimum conditions according to Section 2.3 to produce sufficient amounts of pretreatment liquor in order to study the effects of pH toward the amount of precipitated lignin. All pretreatment liquors were collected immediately upon the completion of the pretreatment. The collected pretreatment liquors were kept in the cold storage at 4 °C to preserve the samples before their pH values were measured from the produced pretreatment liquors. The initial pH of pretreatment liquors from the alkaline ethylene glycol pretreatment was in the range of 12.3 to 12.5. Sulfuric acid with 5.0 mol/L concentration was slowly added in order to acidify the pretreatment liquor until the designated pH values from 6.5 to 2.0 were obtained. The acidified pretreatment liquor was subsequently washed thoroughly and filtered under vacuum several times in order to recover all the precipitated lignin. The precipitated lignin was then dried in the hot air oven (Mermert, Germany) at 105 °C overnight until constant weight was obtained.

2.5. Compositional Analysis of Fresh EFB and DEFB

Untreated and treated fresh EFB and DEFB were subjected to compositional analysis. For composition of cellulose and hemicellulose in the substrate, they were determined according to TAPPI T203 cm-09. For total lignin content, the Klason lignin was quantified by using the Klason method while the acid soluble lignin from the Klason lignin test and the depolymerized lignin in pretreatment liquor were measured with a UV-Vis spectrophotometer (Agilent, Cary 100, Selangor, Malaysia) at 215 and 280 nm wavelength [14]. The ash and extractives content in the substrate were determined with TAPPI T211 om-02 and TAPPI T204 cm-97, respectively.

2.6. Equipment Characterizations

Scanning electron microscopy (SEM) was conducted using the model Hitachi, S-3400N, Japan, to obtain the surface morphology of the treated, untreated DEFB and fresh EFB, with $500\times$ magnifications, non-destructively. FTIR (Thermo-Fisher, Nicolet IS-10, Selangor, Malaysia) was used to analyze the chemical bonding of the related substrate by using the wavelength range from 500 to 4000 cm^{-1}.

3. Results and Discussion

3.1. Effects of Different Organic Solvents on Fresh EFB and DEFB

Both the fresh EFB and DEFB were used as substrates and subjected to organosolv pretreatment in order to study the substrate effects on the pretreatment performances. Five different organic solvents, including acetic acid, 2-propanol, ethylene glycol, glycerol and acetone, were investigated, as shown in Figure 3. For fair comparison, the organosolv pretreatment was conducted with different organic solvents in fixed conditions based on the previous works: 30 $v/v\%$ concentration at 80 °C for 30 min under 10 $w/v\%$ mass loading.

As shown in Figure 3, the pretreatment with DEFB was able to achieve moderately higher delignification but lower hemicellulose removal compared to fresh EFB for all the organic solvents investigated. Furthermore, DEFB also achieved slightly lower cellulose recovery. The degradation process by the microbial action on the DEFB when left in open environment had helped to loosen up the lignocellulosic structure. This would increase its surface area which was exposed to the pretreatment environment. Higher surface area of DEFB also resulted in it being more susceptible to chemical attacks, which enhanced the delignification and degradation of cellulose.

In terms of delignification, 2-propanol was able to achieve the highest delignification at 50.6% for DEFB, while acetone recorded the highest delignification at 44.4% for fresh EFB. The mechanism of lignin depolymerization through the action of major organic solvents was mainly through the cleavage of aryl ether linkages [17]. The highest delignification achieved by 2-propanol could be attributed to its closer Hildebrand solubility parameter (23.8 MPa^{-1}) with lignin, which was estimated to have a value of 22.5 MPa^{-1} [13]. Hildebrand solubility parameter is a square root of the cohesive energy density and is often used to gauge the solvency behavior of the solvents. It was advocated that the smaller difference in Hildebrand solubility parameter between organic solvent and lignin would result in the largest possible solubility [18,19]. It could be evidently observed that the decrement of delignification was in accordance with the increasing differences between the Hildebrand solubility of organic solvents and lignin, except for acetic acid (acetone 20.5 MPa^{-1}, ethylene glycol 32.8 Mpa^{-1}, glycerol 33.8 MPa^{-1}). Low delignification was reported for acetic acid despite its close Hildebrand solubility parameter with lignin. The possible reasons could be due to the lower concentration and milder temperature applied, which led to insufficient cleavage of the lignin. Tang et al. studied the organosolv pretreatment with different acetic acid concentrations in the presence of 0.7 wt.% sulfuric acid for tea oil fruit hull under the same operating conditions (135 °C for 30 min) [20]. The delignification was improved by more than 5-fold, from 11.9% to 67.1%, when the acetic acid concentration was increased from 20 to 60 wt.%. This indicated that the delignification performance of acetic acid was strongly affected by its concentration.

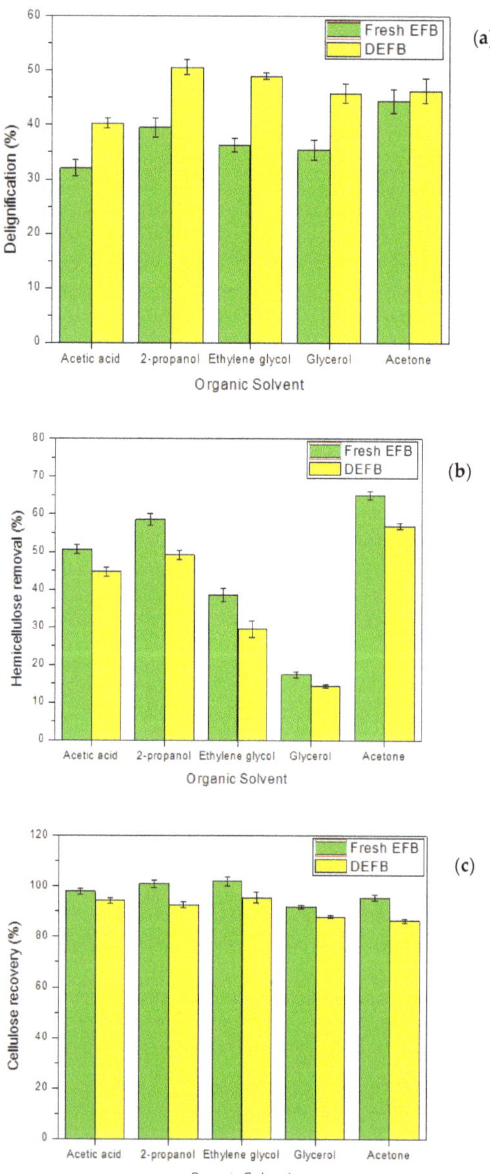

Figure 3. Pretreatment performances, including (**a**) delignification, (**b**) hemicellulose removal and (**c**) cellulose recovery of different organic solvents (acetic acid, 2-propanol, ethylene glycol, glycerol and acetone).

On the other hand, the percentage of hemicellulose removal for different organic solvents was recorded in the following sequence (from highest to lowest), regardless of fresh EFB and DEFB: acetone > 2-propanol > acetic acid > ethylene glycol > glycerol. Increasing hydroxyl functional groups in the organic solvent did not favor the hemicellulose hydrolysis since 2-propanol recorded 50% while glycerol only recorded 15% hemicellulose removal. Generally, acid catalyst was able to facilitate the hemicellulose hydrolysis more efficiently and achieved higher removal of hemicellulose [21]. It was interesting to report that acetic acid merely recorded 44% hemicellulose hydrolysis, which was lower as compared to 2-propanol and acetone. This could be due to the lower concentration of acetic acid being applied in this study, which could result in a smaller amount of H+ ions being dissociated. The fact that acetic acid is a weak organic acid compared to sulfuric acid further strengthens this theory. Similar to delignification, Tang et al. reported that the hemicellulose was only effectively removed under higher acetic acid concentrations [20]. Hemicellulose hydrolysis was substantially improved from 9.0% to 81.2% when the concentration of acetic acid was increased from 20 to 60 wt.% under the same operating conditions. On the other hand, glycerol recorded the lowest hemicellulose hydrolysis as compared to other organic solvents. It was reported that glycerol was ineffective in hemicellulose hydrolysis, where large amounts of hemicellulose would be retained in the treated pulp after the pretreatment [22,23]. Lv et al. performed the pretreatment for sugarcane bagasse with 80 wt.% glycerol at 130 °C for 60 min. Their study merely recorded 8.7% of hemicellulose removal [23]. In another similar work, Phi Trinh et al. observed that the continuous increment of 70 wt.% glycerol pretreatment temperature (130–210 °C) was not effective in removing hemicellulose, which only recorded the removal in the range of 6.1–14.0% [22].

The surface morphology of treated DEFB by different organic solvents is shown in Figure 4. The SEM images agree with the pretreatment performances reported in Figure 3. Large opened and deepened line structures were observed for treated DEFB by 2-propanol and acetone, which indicated that substantial lignin and hemicellulose content had been removed. Numerous opened structures were also prevalently observed on the surface of DEFB treated with acetic acid. However, the surface was relatively smooth, which implied that the lignin was removed to a relatively smaller extent. Similar surface morphology was also observed for ethylene glycol, where numerous pores were formed and silica bodies were removed. The enlargement of the pores showed that the delignification process took place through the pores' enlargement and swelling as the delignification intensified. DEFB treated with glycerol had a relatively smooth surface as compared to DEFB treated with ethylene glycol, which could be due to its lower hemicellulose removal. Despite that similar opened surface structures were observed for DEFB treated with glycerol, the opened structures were relatively less obvious since glycerol only removed a smaller amount of lignin and hemicellulose.

The FTIR spectrum of DEFB treated with different organic solvents in the range of 500 to 4000 cm^{-1} is shown in Supplementary Figure S1. As compared to FTIR spectrum of untreated DEFB, the reduction of intensity of the absorbance peak at 2850 cm^{-1} was observed for all organic solvents. The absorbance peak at 2850 cm^{-1} represents the acetyl group of hemicellulose or the methyl oxide group of lignin [24], which indicated that all the tested organic solvents were able to remove considerable amounts of lignin or hemicellulose. Since organic solvents including 2-propanol and acetone were able to remove significant amounts of lignin and hemicellulose, the absorbance peak at 3330 cm^{-1}, which corresponded to the hydrogen bond or hydroxyl of cellulose [25], was shown to be intensified. The intensification of the peak indicated the removal of lignin and hemicellulose content, which ramified the exposure of cellulose.

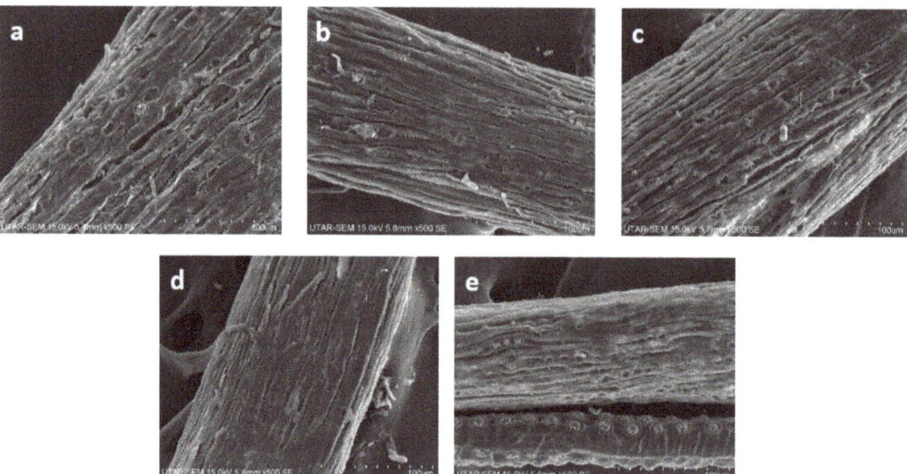

Figure 4. Surface morphology of DEFB treated with (**a**) 2-propanol, (**b**) acetic acid, (**c**) acetone, (**d**) ethylene glycol and (**e**) glycerol under 500× magnifications.

3.2. Effects of Reaction Temperature for Organosolv Pretreatment

The alkaline pretreatment of DEFB was conducted with 50 $v/v\%$ ethylene glycol and addition of 3 $v/v\%$ sodium hydroxide as the catalyst. The operating temperature of the pretreatment varied from 60 to 140 °C for 30 min. From Figure 5, it was evident that the pretreatment temperature significantly affected the degree of delignification. As the pretreatment temperature increased from 60 to 140 °C, the delignification recorded improvement from 56.1% to 71.5%. Higher pretreatment temperature could provide more energy to facilitate the cleavage of lignin bonds, which led to higher delignification. On the other hand, when the temperature was increased from 60 to 140 °C, the hemicellulose removal from DEFB also increased, from 52% to 78.0%, which was a similar trend to the delignification. The addition of alkaline catalyst was able to promote the cleavage of the glycosidic linkages between the hemicellulose and benzyl ester [22]. The cellulose recovery was diminished to 78.5%, down from the previous 92%, when the pretreatment temperature was increased from 60 to 140 °C Higher pretreatment temperature could lead to more intensive cleavage of the glycosidic bonds in cellulose. Lv et al. demonstrated that the addition of a suitable alkaline catalyst was able to significantly improve the delignification [23]. They conducted two types of glycerol pretreatments to treat sugarcane bagasse, with one scenario using 0.8 wt.% sodium methoxide and another without the addition of any catalyst [23]. The glycerol pretreatment with sodium methoxide achieved 70.4% delignification, while the glycerol pretreatment without any catalyst merely recorded 19.1% delignification.

It was generally hypothesized that the increasing pretreatment temperature would encourage the destruction of glycosidic linkages in the biomass structure connecting the cellulose and hemicellulose content. Nevertheless, it was interesting to take note that the degree of delignification at higher temperature did not improve significantly based on the comparison of the initial composition of the DEFB shown in Table 1 and the results obtained in Figure 5. A closer observation revealed that there was no significant reduction of lignin content for the composition of treated pulp when the temperature was varied from 100 to 140 °C. Similarly, the cellulose content in the treated DEFB did not exhibit any improvement either. By observing the lignocellulosic composition in Table 1, composition of all components remained almost similar in the temperature range investigated, which insinuated that all the compositions were being detached simultaneously at the same rate. It could also be due to the high exposure of treated DEFB to the pretreatment environment after significant lignin content was removed. Ideally, satisfactory pretreatment performances by organosolv pretreatment should preferentially remove the lignin and hemicellulose content only, while recovering as much cellulose content as possible with high purity. Consequently, the optimum pretreatment temperature at 80 °C was fixed for the subsequent experimental investigations based on the best pretreatment results obtained, which recorded the lignin composition at 11.9%, 90.8% cellulose recovery and 70.4% hemicellulose removal. By comparing to the acidified ethylene glycol pretreatment, ethylene glycol pretreatment assisted by NaOH catalyst in this study successfully obtained high delignification percentage at lower temperature (80 °C), albeit it required a longer pretreatment period. The longer pretreatment duration could be attributed to the DEFB, which had higher lignin content as compared to newspaper in a similar study conducted by Lee and co-workers, where they applied the ethylene glycol with sulfuric acid to pretreat the newspaper. The acidified ethylene glycol pretreatment was conducted at 150 °C for short duration (15 min) and the delignification was recorded at 75% [26].

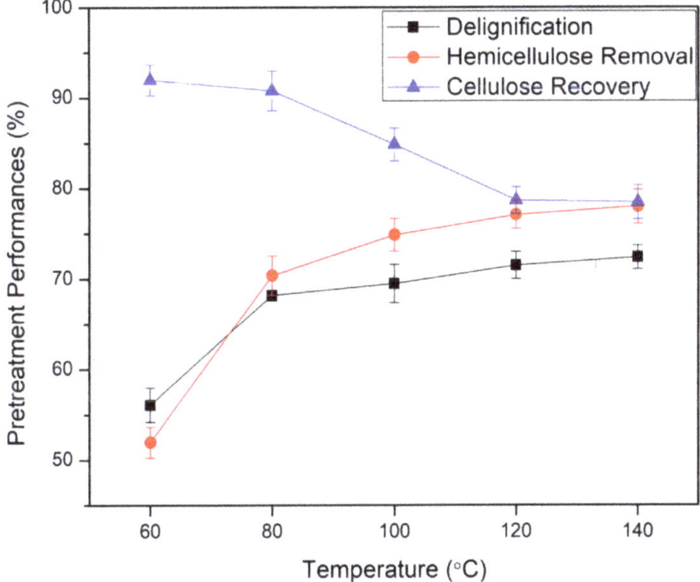

Figure 5. Pretreatment performances of alkaline ethylene glycol for DEFB from 60 to 140 °C.

Table 1. The compositional analysis of treated pulp with variations in pretreatment temperatures.

Temperature (°C)	Pulp Yield (g)	Composition (wt.%)				
		Lignin	Hemicellulose	Cellulose	Ash	Extractive
Untreated	-	23.9	6.10	55.9	4.45	8.80
60	6.87	15.3	3.93	75.4	2.33	3.06
80	6.29	11.9	2.70	81.2	1.91	2.23
100	5.89	12.4	2.38	81.2	2.72	1.36
120	5.55	12.1	2.34	79.8	4.32	1.44
140	5.42	12.0	2.40	81.5	3.14	0.92

Figure 6a–c demonstrates the surface characteristics of the treated DEFB at 60, 80 and 120 °C. From the figures shown, the surface structures were observed to be severely cracked open with prevalence of rough surfaces at higher temperatures, which proved the

successful detachment of lignin and hemicellulose from DEFB. However, excessive high pretreatment temperature can be detrimental to the pretreatment efficiency since it will cause higher loss of cellulose during the pretreatment. As shown in Figure 6c, at 120 °C, the surface morphology was significantly disrupted, and the internal structures were widely exposed. This could explain the reasons behind the low cellulose recovery after 120 °C. The larger exposure of the internal structure to the pretreatment environment could induce more severe bond cleavage.

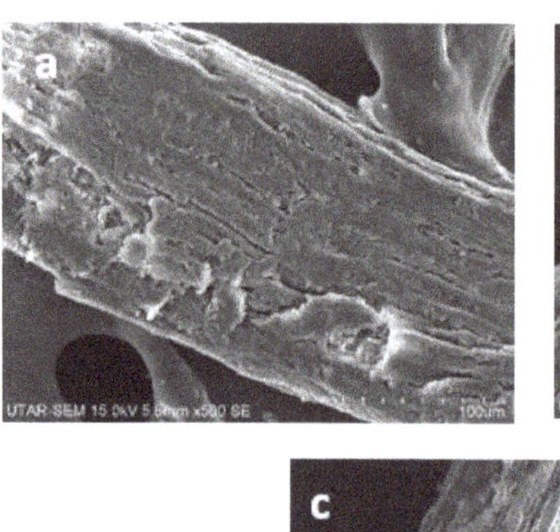
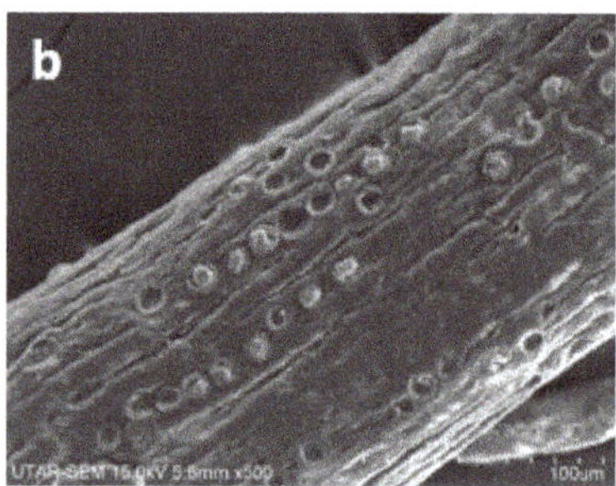
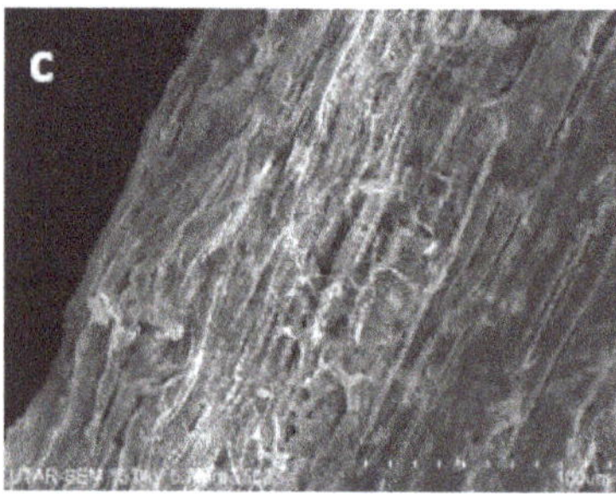

Figure 6. Surface morphology of treated DEFB by alkaline ethylene glycol at (**a**) 60 °C, (**b**) 80 °C and (**c**) 120 °C.

Supplementary Figure S2 shows the FTIR of treated DEFB at 60, 80 and 120 °C, as well as untreated DEFB. FTIR results show that the reduction in absorbance peak intensity at 1243 cm^{-1} (C-O-C aryl alkyl ether of lignin), 1630 cm^{-1} (C=C aromatic group of lignin) [27] and 2850 cm^{-1} (OCH$_3$ group of lignin) indicated that a higher amount of lignin was removed with increasing temperature. This was in agreement with the higher delignification results being reported in the previous pretreatment performance. The reduction of 1243 and 1735 cm^{-1} peak intensities coupled with the reduction in absorbance peak at 2850 cm^{-1} represented the removal of hemicellulose during the pretreatment. The increment in absorbance of peak intensity at 3330 cm^{-1} indicated higher exposure of cellulose with higher pretreatment temperature, which was in agreement with the surface morphology observed with SEM characterization. This could also be explained as the reason for lower cellulose recovery being reported at 100–120 °C.

3.3. Effects of Duration Time for Organosolv Pretreatment

The effects of residence time are shown in Figure 7, while the composition of treated DEFB at different residence times is shown in Table 2. Initially, the increased residence time of pretreatment from 15 to 45 min improved the delignification from 63.1% to 75.1%. However, the delignification reduced slightly to 72.8% as the pretreatment residence time was further increased to 75 min. The removal of hemicellulose was discovered to increase significantly from 53.2% to 85.2% when the residence time of pretreatment was increased from 15 to 75 min. This resulted in the decrement of hemicellulose content from 6.10% to 1.39%. On the other hand, the cellulose recovery was only affected minimally, even though the residence time was further prolonged from 15 to 75 min. The cellulose recovery was recorded at 91.8% for 15 min, while the recovery was only 88.0% for 75 min. By comparing the results from Sections 3.2 and 3.3, it seems that prolonged pretreatment time had a milder influential effect on cellulose recovery as compared to pretreatment temperature. However, low reduction of cellulose recovery with extended residence time could be due to the lower operating temperature (80 °C) applied to the pretreatment. This was in agreement with a similar work in the literature, where the cellulose recovery was found to reduce from 93.1% to 85.6%, even though the residence time for the pretreatment was prolonged from 30 to 90 min at the same temperature as in this work. However, when the pretreatment was conducted at 140 °C with the same residence time range, the reduction of recoverable cellulose content became very significant. At this high temperature, it reported a cellulose recovery reduction of 46.1% to 35.7% [28]. Therefore, it can be concluded that the operating temperature for the pretreatment played a dominant role in determining the cellulose yield. Consequently, this study chose the residence time of 45 min since it recorded the highest cellulose purity of treated DEFB at 84.5%, lignin removal efficiency at 75.1%, hemicellulose removal up to 81.5% and recovery of cellulose at 90.4%.

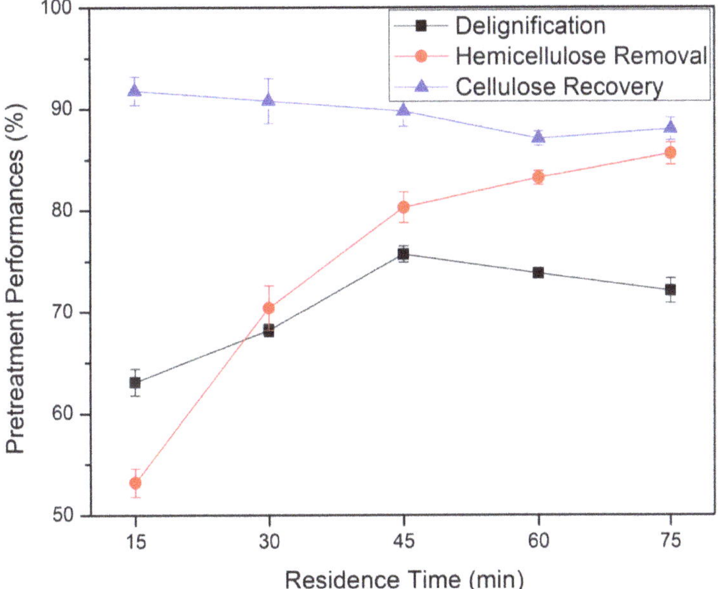

Figure 7. Pretreatment performances of alkaline ethylene glycol for DEFB from 15 to 75 min.

Table 2. The compositional analysis of treated pulp with variations in residence times.

Time (min)	Pulp Yield (g)	Composition (wt.%)				
		Lignin	Hemicellulose	Cellulose	Ash	Extractive
Untreated	10.0	23.9	6.10	55.9	4.45	8.80
15	6.71	13.6	3.56	77.0	3.27	2.10
30	6.29	11.9	2.70	81.2	1.91	2.23
45	5.98	10.0	1.89	84.4	1.73	1.98
60	5.85	11.3	1.50	83.2	1.76	2.19
75	5.84	11.6	1.39	84.3	1.86	0.92

Figure 8a–c shows the SEM images of treated DEFB, which revealed its surface morphology after the designated pretreatment duration. Increased pretreatment residence

time led to higher uneven and opened surface structures, as shown on treated DEFB. This showed that both the lignin and hemicellulose content were removed or cleaved after the pretreatment. Supplementary Figure S3 shows the FTIR spectrum of treated DEFB at different pretreatment residence times (15, 45 and 75 min) and untreated DEFB for comparison. FTIR analysis indicated that increasing pretreatment residence time successfully improved the delignification due to the reduction in intensity of the absorbance peaks observed at 1243, 1630 and 2850 cm^{-1}. Similarly, peaks at 1735 and 2850 cm^{-1}, which represented hemicellulose bonding, were reduced. In addition, the increased intensity of absorbance peaks at 3330 and 1157 cm^{-1} signified that increased pretreatment time enhanced the exposure of cellulose. The continuous removal of lignin and hemicellulose resulted in the higher exposure of cellulose. However, similar to temperature, increasing the pretreatment residence time would result in a higher degradation of the cellulose.

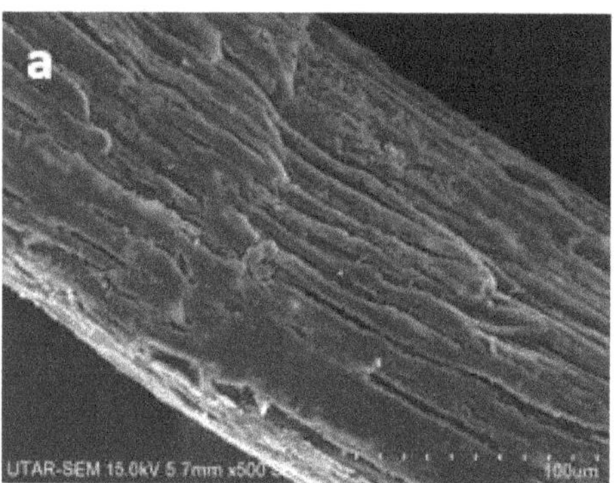
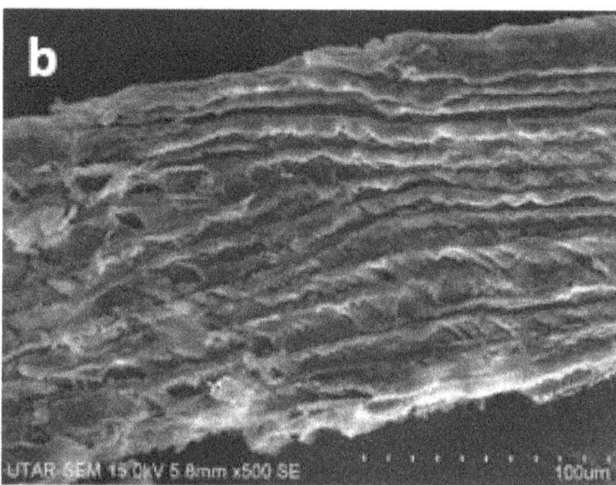
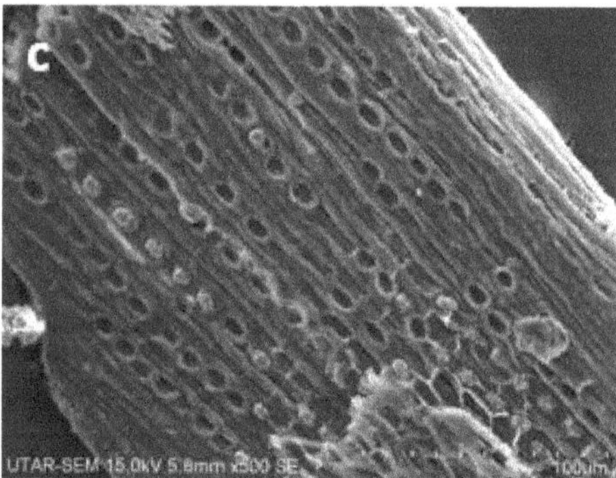

Figure 8. Surface morphology of DEFB treated with alkaline ethylene glycol at (**a**) 15, (**b**) 45 and (**c**) 75 min.

3.4. Comparison of Different Pretreatments with That in This Study

Table 3 shows the comparison of pretreatment performances of different pretreatments in the literature with that in this study. As compared to other pretreatments, the alkaline ethylene glycol pretreatment that was applied in this study was capable to achieve comparable and effective pretreatment performances with lower pretreatment temperatures and shorter duration. The comparable pretreatment performances indicated the effectiveness of alkaline organosolv pretreatment to remove the lignin and hemicellulose but retain significant amounts of cellulose. Furthermore, the pretreatment in this study was performed below 100 °C, which removed the requirement of a pressurized reactor which can minimize the initial capital cost for industrial-scale production.

Table 3. Comparisons of different pretreatments with DEFB.

Biomass	Pretreatment Conditions	Delignification (%)	Hemicellulose Removal (%)	Cellulose Recovery (%)	Reference
DEFB	50 v/v% ethylene glycol + 3 v/v% NaOH at 80 °C for 45 min	75.1	81.5	90.4	This Study
EFB	50 wt.% ethanol + 0.07 wt.% H_2SO_4 at 210 °C for 90 min	90.0	84.6	80.0	[29]
EFB	3–7 wt.% Acetic acid at 170–190 °C for 10–20 min followed by: 5–20 wt.% Ammonium hydroxide at 140–220 °C for 5–25 min	53.9	53.6	98.3	[30]
EFB	8 v/v% H_2SO_4 at 121 °C for 1 h followed by: 2.5 mol/dm^3 NaOH at 90 °C for 20 min	72.9	90.7	N.D. *	[31]
Extractive-free EFB	76.3 wt.% lactic acid + 23.7 wt.% choline chloride at 120 °C for 8 h	88.0	100	100	[32]

* Not Determined.

3.5. Lignin Precipitation from Spent Pretreatment Liquor after Organosolv Pretreatment

Acidification of pretreatment liquor is commonly applied to precipitate and recover the lignin which dissolved in the pretreatment liquor. The amount of precipitated lignin and its corresponding purity at different pH values is shown in Figure 9. Decreasing pH was found to improve the amount of precipitated lignin. From Figure 9, it shows that the lignin recovery increased significantly from 47.7% to 74.6% at pH 6.5 and pH 2.0, respectively. The purity of the precipitated lignin was measured between 94.5% and 97.2%, while the rest consisted of a hemicellulose component which was still attached together with the lignin due to the lignin–hemicellulose bonding, which remained intact [33]. Mussatto et al., who performed a similar study in the literature, investigated the delignified brewer's spent grain as the feedstock with sodium hydroxide as the alkaline catalyst to recover its precipitated lignin [15]. They successfully precipitated up to 81.4% of lignin content at pH 2.15, compared to this study, which recorded 74.6% recoverable lignin content at pH 2.0. This difference might be caused by the intrinsic phenolic composition in the different feedstock which resulted in different lignin precipitation rates. Furthermore, Supplementary Figure S4 shows the color of the pretreatment liquor after the removal of precipitated lignin from the acidified pretreatment liquor. It could be observed that the color changed significantly from dark brown to light brown when the pH was reduced from pH 6.5 to pH 2.0. The color change advocated that the dark brown color was derived from the chromophore of lignin, such as phenolic hydroxyl groups, carbonyl groups and quinones, which are generally soluble in alkaline medium [34]. Therefore, the color change corresponded to the higher amount of precipitated lignin being recorded. Depolymerized lignin from the pretreatment was able to be recovered in huge amounts from the pretreatment liquor, which signified the sustainability of organosolv pretreatment and reduced the wastage from the production. This was also in line with the findings from Lauwaert et al. [16]. The color of black liquor was reduced upon acidification and yielded light brown precipitate, which represented the recovered lignin at pH 5.2. Generally, the lignin that is recovered from organosolv liquor has a very high reutilization potential due to the negligible sulfur content, which will reduce the environmental pollution. The recovered lignin from organosolv pretreatment liquor can be incorporated or synthesized into composites, bioplastics, nanoparticles, dispersants, adsorbents and carbon fibers [35].

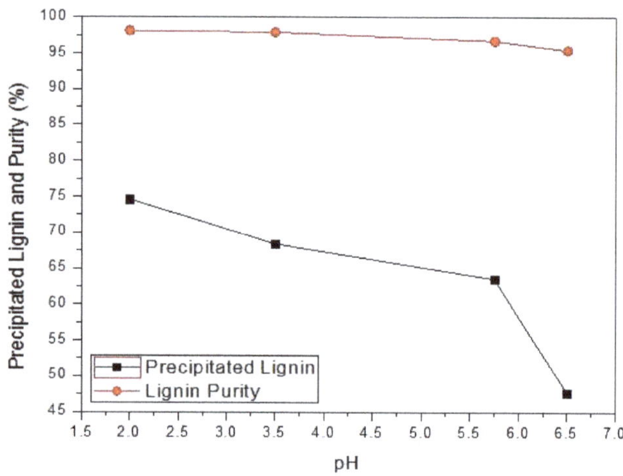

Figure 9. Precipitated lignin from alkaline ethylene glycol pretreatment liquor at different pH values.

4. Conclusions

This study successfully demonstrated that the DEFB could be a promising low-cost feedstock for various applications due to its high lignocellulosic content, which was still preserved even after microbial degradation. In fact, the DEFB possessed a more open and deepened surface structure, as evidenced from the SEM characterizations. Its high cellulose content, which was recoverable up to 55.9% by weight, could be a potential source for biofuel production. The comparison between different organic solvents during the pretreatment proved that ethylene glycol was the best solvent for DEFB. Under the designated ethylene glycol pretreatment with addition of sodium hydroxide as the alkaline catalyst, 75.1% of lignin and 81.5% of hemicellulose contents were successfully removed, while at the same time, 90.4% of cellulose content was recovered. Its cellulose purity was enhanced from 55.9% in fresh EFB to 84.5% in treated DEFB, which rendered the DEFB to be suitable for utilization as a substrate for the subsequent hydrolysis process. The next phase of research will be to evaluate the conversion of the extracted cellulose into sugar before being subjected to enzymatic processes to produce bioethanol. Last but not least, addition of sulfuric acid was able to precipitate the lignin from the pretreatment liquor of DEFB, which enhanced the practicability of organosolv pretreatment. In this study, the optimum lignin recovery was obtained at 74.6% at pH 2.0. Further analysis could be performed in the future to upgrade the recovered lignin into more usable bio-products such as bioplastics, adsorbents and other specialty chemicals through a consolidated bioprocess.

Supplementary Materials: The following are available online at https://www.mdpi.com/article/10.3390/su13126757/s1.

Author Contributions: Conceptualization, D.W.K.C. and S.L.; methodology, D.W.K.C. and S.L.; validation, Y.L.P. and C.H.L.; formal analysis, D.W.K.C. and S.L.; investigation, D.W.K.C. and S.H.S.; resources, K.M.L. and C.T.C.; writing—original draft preparation, D.W.K.C., S.L. and Y.L.P.; writing—review and editing, C.H.L., S.H.S., K.M.L. and C.T.C.; supervision, S.L. and Y.L.P.; funding acquisition, S.L. and Y.L.P. All authors have read and agreed to the published version of the manuscript.

Funding: This research was funded by the Fundamental Research Grant Scheme (FRGS/1/2018/TK10/UTAR/02/1) from the Ministry of Education, Malaysia, and the Universiti Tunku Abdul Rahman Research Grant (IPSR/RMC/UTARRF/2020-C2/W01).

Institutional Review Board Statement: Not Applicable.

Informed Consent Statement: Not Applicable.

Data Availability Statement: All the data is available inside this paper and in the supplementary figures.

Conflicts of Interest: The authors declare no conflict of interest. The funders had no role in the design of the study; in the collection, analyses, or interpretation of data; in the writing of the manuscript, or in the decision to publish the results.

References

1. Chin, D.W.K.; Lim, S.; Pang, Y.L.; Wong, K.H. Application of organosolv pretreatment on Pennisetum Purpureum for lignin removal and cellulose recovery. In Proceedings of the 2017 4th International Conference on Biomedical and Bioinformatics Engineering, Seoul, Korea, 12–14 November 2017; pp. 84–89.
2. Smit, A.T.; Huijgen, W.J.J. Effective fractionation of lignocellulose in herbaceous biomass and hardwood using a mild acetone organosolv process. *Green Chem.* **2017**, *22*, 5505–5514. [CrossRef]
3. Morone, A.; Pandey, R.A.; Chakrabarti, T. Evaluation of OrganoCat process as a pretreatment during bioconversion of rice straw. *Ind. Crops Prod.* **2017**, *77*, 7–18. [CrossRef]
4. Hesami, S.M.; Zilouei, H.; Karimi, K.; Asadineshad, A. Enhanced biogas production from sunflower stalks using hydrothermal and organosolv pretreatment. *Ind. Crops Prod.* **2015**, *76*, 449–455. [CrossRef]
5. Wu, M.; Liu, J.K.; Yan, Z.Y.; Wang, B.; Zhang, X.M.; Feng, X.; Sun, R.C. Efficient recovery and structural characterization of lignin from cotton stalk based on a biorefinery process using a γ-valerolactone/water system. *RSC Adv.* **2016**, *6*, 6196–6204. [CrossRef]
6. Chin, D.W.K.; Lim, S.; Pang, Y.L.; Leong, L.K.; Lim, C.H. Investigation of organosolv pretreatment to natural microbial-degraded empty fruit bunch for sugar based substrate recovery. *Energy Procedia* **2019**, *158*, 1065–1071. [CrossRef]
7. Wong, W.Y.; Lim, S.; Pang, Y.L.; Shuit, S.H.; Chen, W.H.; Lee, K.T. Synthesis of renewable heterogeneous acid catalyst from oil palm empty fruit bunch for glycerol-free biodiesel production. *Sci. Total Environ.* **2020**, *727*, 138534. [CrossRef] [PubMed]
8. Zhang, K.; Pei, Z.J.; Wang, D.H. Organic solvent pretreatment of lignocellulosic biomass for biofuels and biochemicals: A review. *Bioresour. Technol.* **2016**, *199*, 21–33. [CrossRef] [PubMed]
9. Mahmood, H.; Moniruzzaman, M.; Igbal, T.; Khan, M.J. Recent advances in the pretreatment of lignocellulosic biomass for biofuels and value-added products. *Curr. Opin. Green Sustain. Chem.* **2019**, *20*, 18–24. [CrossRef]
10. Thoresen, P.P.; Matsakas, L.; Rova, U.; Christakopoulos, P. Recent advances in organosolv fractionation: Towards biomass fractionation technology of the future. *Bioresour. Technol.* **2020**, *306*, 123189. [CrossRef] [PubMed]
11. Guo, Y.; Zhou, J.; Wen, J.; Sun, G.; Sun, Y. Structural transformations of triploid of Populus tomentosa Carr. lignin during auto-catalyzed ethanol organosolv pretreatment. *Ind. Crops Prod.* **2015**, *75*, 522–529. [CrossRef]
12. Karnaouri, A.; Asimakopoulou, G.; Kalogiannis, K.G.; Lappas, A.; Topakas, E. Efficient d-lactic acid production by Lactobacillus delbrueckii subsp. bulgaricus through conversion of organosolv pretreated lignocellulosic biomass. *Biomass Bioenergy* **2020**, *140*, 105672. [CrossRef]
13. Chin, D.W.K.; Lim, S.; Pang, Y.L.; Lam, M.K. Fundamental review of organosolv pretreatment and its challenges in emerging consolidated bioprocessing. *Biofuel Bioprod. Biorefin.* **2020**, *14*, 808–829. [CrossRef]
14. Chin, D.W.K.; Lim, S.; Pang, Y.L.; Lim, C.H.; Lee, K.M. Two-staged acid hydrolysis on ethylene glycol pretreated degraded oil palm empty fruit bunch for sugar based substrate recovery. *Bioresour. Technol.* **2019**, *292*, 121967. [CrossRef]
15. Mussatto, S.I.; Fernandes, M.; Roberto, I.C. Lignin recovery from brewer's spent grain black liquor. *Carbohydr. Polym.* **2007**, *70*, 218–223. [CrossRef]
16. Lauwaert, J.; Stals, I.; Lancefield, C.S.; Deschaumes, W.; Depuydt, D.; Vanlerberghe, B.; Devlamynck, T.; Bruijnincx, P.C.A.; Verberckmoes, A. Pilot scale recovery of lignin from black liquor and advanced characterization of the final product. *Sep. Purif. Technol.* **2019**, *221*, 226–235. [CrossRef]
17. Zhou, X.; Ding, D.; You, T.; Zhang, X.; Takabe, K.; Xu, F. Synergetic dissolution of branched xylan and lignin opens the way for enzymatic hydrolysis of poplar cell wall. *J. Agric. Food Chem.* **2018**, *66*, 3449–3456. [CrossRef]
18. Zhang, Z.; Harrison, M.D.; Rackemann, D.W.; Doherty, W.O.S.; O'Hara, I.M. Organosolv pretreatment of plant biomass for enhanced enzymatic saccharification. *Green Chem.* **2016**, *20*, 360–381. [CrossRef]
19. Sameni, J.; Krigstin, S.; Sain, M. Acetylation & lignin solubility. *Bioresources* **2017**, *12*, 1548–1565.
20. Tang, S.; Liu, R.; Sun, F.F.; Dong, C.; Wang, R.; Gao, Z.; Zhang, Z.; Xiao, Z.; Li, C.; Li, H. Bioprocessing of tea oil fruit hull with acetic acid organosolv pretreatment in combination with alkaline H2O2. *Biotechnol. Biofuels* **2017**, *10*, 1–12. [CrossRef]
21. Oliva, A.; Tan, L.C.; Papirio, S.; Esposito, G.; Lens, P.N.L. Effect of methanol-organosolv pretreatment on anaerobic digestion of lignocellulosic materials. *Renew. Energy* **2021**, *169*, 1000–1012. [CrossRef]
22. Phi Trinh, L.T.; Lee, J.W.; Lee, H.J. Acidified glycerol pretreatment for enhanced ethanol production from rice straw. *Biomass Bioenergy* **2016**, *94*, 39–45. [CrossRef]
23. Lv, X.; Lin, J.; Luo, L.; Zhang, D.; Lei, S.; Xiao, W.; Xu, Y.; Gong, Y.; Liu, Z. Enhanced enzymatic saccharification of sugarcane bagasse pretreated by sodium methoxide with glycerol. *Bioresour. Technol.* **2018**, *249*, 226–233. [CrossRef]
24. Banerjee, D.; Mukherjee, S.; Pal, S.; Khowala, S. Enhanced saccharification efficiency of lignocellulosic biomass of mustard stalk and straw by salt pretreatment. *Ind. Crops Prod.* **2016**, *80*, 42–49. [CrossRef]
25. Avanthi, A.; Banerjee, D. A strategic laccase mediated lignin degradation of lignocellulosic feedstocks for ethanol production. *Ind. Crops Prod.* **2016**, *92*, 174–185. [CrossRef]

26. Lee, D.H.; Cho, E.Y.; Kim, C.J.; Kim, S.B. Pretreatment of waste newspaper using ethylene glycol for bioethanol production. *Biotechnol. Bioprocess Eng.* **2010**, *15*, 1094–1101. [CrossRef]
27. Patowary, D.; Baruah, D.C. Effect of combined chemical and thermal pretreatments on biogas production from lignocellulosic biomasses. *Ind. Crops Prod.* **2018**, *124*, 735–746. [CrossRef]
28. Sun, S.N.; Chen, X.; Tao, Y.H.; Cao, F.X.; Li, M.F.; Wen, J.L.; Nie, S.X.; Sun, R.C. Pretreatment of Eucalyptus urophylla in γ-valerolactone/dilute acid system for removal of non-cellulosic components and acceleration of enzymatic hydrolysis. *Ind. Crops Prod.* **2019**, *132*, 21–28. [CrossRef]
29. Mondylaksita, K.; Ferreira, J.A.; Millati, R.; Budhijanto, W.; Niklasson, C.; Taherzadeh, M.J. Recovery of high purity lignin and digestible cellulose from oil palm empty fruit bunch using low acid-catalyzed organosolv pretreatment. *Agronomy* **2020**, *10*, 674. [CrossRef]
30. Kim, D.Y.; Kim, Y.S.; Kim, T.H.; Oh, K.K. Two-stage, acetic acid-aqueous ammonia, fractionation of empty fruit bunches for increased lignocellulosic biomass utilization. *Bioresour. Technol.* **2016**, *199*, 121–127. [CrossRef]
31. Akhtar, J.; Idris, A. Oil palm empty fruit bunches a promising substrate for succinic acid production via simultaneous saccharification and fermentation. *Renew. Energy* **2017**, *114*, 917–923. [CrossRef]
32. Tan, Y.T.; Ngoh, G.C.; Chua, A.S.M. Evaluation of fractionation and delignification efficiencies of deep eutectic solvents on oil palm empty fruit bunch. *Ind. Crops Prod.* **2018**, *123*, 271–277. [CrossRef]
33. Giummarella, N.; Pu, Y.Q.; Ragauskas, A.J.; Lawoko, M. A critical review on the analysis of lignin carbohydrate bonds. *Green Chem.* **2019**, *20*, 1573–1595. [CrossRef]
34. Fengel, D.; Wegener, G. Wood: Chemistry, ultrastructure, reactions. *J. Polym. Sci.* **1985**, *23*, 601–602.
35. Norgen, M.; Edlund, H. Lignin: Recent advances and emerging applications. *Curr. Opin. Colloid Interface Sci.* **2014**, *19*, 409–416. [CrossRef]

Review

The Influence of COVID-19 on Global CO_2 Emissions and Climate Change: A Perspective from Malaysia

Chung Hong Tan [1], Mei Yin Ong [1], Saifuddin M. Nomanbhay [1,*], Abd Halim Shamsuddin [1] and Pau Loke Show [2]

[1] Institute of Sustainable Energy (ISE), Universiti Tenaga Nasional (UNITEN), Jalan IKRAM-UNITEN, Kajang 43000, Malaysia; chtan4996@gmail.com (C.H.T.); meiyin993@gmail.com (M.Y.O.); abdhalim@uniten.edu.my (A.H.S.)

[2] Department of Chemical and Environmental Engineering, Faculty of Science and Engineering, University of Nottingham Malaysia, Semenyih 43500, Malaysia; PauLoke.Show@nottingham.edu.my

* Correspondence: Saifuddin@uniten.edu.my

Abstract: The rapid spread of coronavirus disease 2019 (COVID-19) in early 2020 prompted a global lockdown from March to July 2020. Due to strict lockdown measures, many countries experienced economic downturns, negatively affecting many industries including energy, manufacturing, agriculture, finance, healthcare, food, education, tourism, and sports. Despite this, the COVID-19 pandemic provided a rare opportunity to observe the impacts of worldwide lockdown on global carbon dioxide (CO_2) emissions and climate change. Being the main greenhouse gas responsible for rising global surface temperature, CO_2 is released to the atmosphere primarily by burning fossil fuels. Compared to 2019, CO_2 emissions for the world and Malaysia decreased significantly by 4.02% (-1365.83 MtCO$_2$) and 9.7% (-225.97 MtCO$_2$) in 2020. However, this is insufficient to cause long-term impacts on global CO_2 levels and climate change. Therefore, in this review, we explored the effects of worldwide lockdown on global CO_2 levels, the impacts of national lockdown on Malaysia's CO_2 emissions, and the influence of climate change in Malaysia.

Keywords: climate change; COVID-19; CO_2; fossil fuel; Malaysia

Citation: Tan, C.H.; Ong, M.Y.; Nomanbhay, S.M.; Shamsuddin, A.H.; Show, P.L. The Influence of COVID-19 on Global CO_2 Emissions and Climate Change: A Perspective from Malaysia. *Sustainability* **2021**, *13*, 8461. https://doi.org/ 10.3390/su13158461

Academic Editor: Antonio Caggiano

Received: 10 June 2021
Accepted: 23 July 2021
Published: 29 July 2021

Publisher's Note: MDPI stays neutral with regard to jurisdictional claims in published maps and institutional affiliations.

1. Introduction

Coronavirus disease 2019 (COVID-19) is the first worldwide pandemic caused by a coronavirus. Since the first confirmed case was found in China, the novel coronavirus has evolved and spread rapidly to many countries around the world [1]. This onset of COVID-19 health pandemic in early 2020 has seen a new global recession bringing about adverse impacts on economies and labour markets. The global gross domestic product (GDP) in 2020 is expected to decline by 3.9%, whereas China's GDP loss may hit 4.3% [2]. The COVID-19 pandemic has negatively affected many industries such as energy, manufacturing, agriculture, finance, healthcare, food, education, tourism, and sports [3]. The drastic lockdown measures implemented to curb the rapid spread of COVID-19, which included social distancing, minimum travelling, banning of gatherings, and border closures, have decreased the oil demand in many countries. These measures, coupled with the Saudi–Russia oil price war, have lowered the global greenhouse gas (GHG) emissions, but this phenomenon is only temporary [3]. In March to July 2020, strict lockdown measures were enforced by many countries around the world, but after the lockdown was lifted, the global CO_2 emissions increased quickly due to the partial startup of the economy, especially in China and some European countries [4].

The primary cause of rising global temperatures is the release of carbon dioxide (CO_2) from burning coal, oil, and natural gas [5]. There are two primary factors affecting CO_2 concentration in the atmosphere: (i) the natural carbon cycle whereby carbon is exchanged between the atmosphere and the Earth, and (ii) anthropogenic CO_2 emissions to the atmosphere via the combustion of fossil fuels [6,7]. In 2020, natural gas and coal remain the top two energy sources used for electricity generation in the Organisation for Economic Co-operation and Development (OECD) countries, having shares of 29.5% and 19.2%, respectively [8]. Given that fossil fuels are currently still the main energy source for many countries and fossil fuel power plants will continue operation until net zero in the global energy sector is achieved by 2050 [9], it is important to fund research in raising the efficiencies and dropping CO_2 emissions of existing modern gas- and coal-fired power plants [10–12]. At the same time, it is also crucial to fund research in renewable energy as electricity production from renewables in the OECD increased by 7.5% (228.8 TWh) in 2020 compared to 2019. Solar, wind, and hydropower recorded significant increments of 20.2% (73.0 TWh), 11.5% (95.0 TWh), and 3.5% (52.3 TWh), generating 4.2%, 8.9%, and 15.0% of total electricity in 2020, respectively [8]. Following the COVID-19 pandemic, the global natural gas consumption and the global demand for coal to generate electricity declined by 4% and 3.3% in 2020 compared to 2019 [13,14]. A mild winter in the northern hemisphere in early 2020 and the global lockdown during the first half of 2020 negatively affected all major gas markets, except for China where growth was sluggish at best [13]. The greatest reduction in coal-fired electricity production was in the European Union (−19%, −111 TWh) and the United States of America (U.S.A.) (−14%, −196 TWh) [14]. The reduction in demand for fossil fuels due to the global lockdown, together with the increase in utilization of renewables, 2020 saw an overall decrease in CO_2 emissions of 4.02% (−1365.83 $MtCO_2$) [15]. According to a June 2020 report by the United Nations Environment Program (UNEP), the slowdown caused by COVID-19 pandemic in the fossil fuel industry has made investments in renewable energy more cost-effective than before. The reduction in cost for renewable energy is mainly attributed to better technologies, improved production at scale, and more experienced renewable developers [16].

In the search for alternatives to fossil fuels, the concept of bioeconomy was proposed. A bioeconomy, or biobased economy, can be described as an economy where the raw materials for manufacturing, chemicals, and energy are obtained from renewable biological sources [17]. Not only is the transition from fossil-based to biobased economy essential for stopping climate change, but also important for resolving current issues surrounding health, food security, industrial restructuring, and energy security [18,19]. So far, the driving force for the development of bioeconomy worldwide has been the societal transition towards circular and low-carbon economies, which are based on the potential of providing green biofuels. The continuous backing of low-carbon technologies should be coupled with policies to gradually phase out fossil fuels. The COVID-19 pandemic, however, revealed a much wider role that bioeconomy can have to diversify supplies for food, feed, and raw materials. Views and new knowledge on a fresh perspective for future sustainable development and operation of biobased value chains are needed. The integration of renewable energy (solar, wind, hydro, geothermal, and biomass) with CO_2 utilization and carbon recycling (C-recycling) is the way forward for the circular carbon economy. C-recycling is the construction of an artificial carbon cycle, which can be combined with the natural cycle, to enhance the rate of CO_2 sequestration and increase the selectivity of end-products that are currently produced by fossil sources [20]. Liu et al. (2020) and Sikarwar et al. (2021) published comprehensive articles detailing the impacts of global lockdown due to the COVID-19 pandemic on global CO_2 emissions and the different sectors (such as power, transportation, industry, aviation, commercial, and residential) in major countries [4,21]. Naderipour et al. compiled the impacts of COVID-19 on solar energy generation and the relation between energy consumption (1973–2019) and GHG emissions (2008–2019) in Malaysia [22]. Yusup et al. analysed and compared the monthly differences in atmospheric CO_2 concentration for the world and Malaysia from 2018–2020 (from November to May of each year) [23]. To the best of our knowledge, no journal article has described the effects of COVID-19-related lockdown on daily CO_2 emissions in Malaysia.

Therefore, in this review, we discuss how worldwide lockdown due to the COVID-19 pandemic significantly reduced global CO_2 emissions. We also discuss the impacts of national lockdown in Malaysia on its daily CO_2 emissions in 2020 together with the influence of climate change on various sectors in Malaysia.

2. Overview of Greenhouse Gases

In 1896, Svante Arrhenius (1859–1927), a Swedish scientist, was the first to relate the effect of atmospheric CO_2 levels on Earth's temperature. He observed that the infrared absorption ability of water vapour and CO_2 had kept Earth's mean surface temperature at a comfortable 15 °C. This phenomenon is known as the natural greenhouse effect. Arrhenius calculated that a two-fold increase in CO_2 level would increase temperature by 5 °C [24]. Around the same time, Thomas Chamberlin (1843–1928), an American geologist, suggested that large alterations on Earth's climate could be caused by variations in CO_2 levels. Both Arrhenius and Chamberlin concluded that the combustion of fossil fuels released CO_2 to the atmosphere, thereby warming the Earth [24].

Now, scientists have concluded with high confidence that human activities in the last five decades have warmed the Earth. The industrial activities that support our technologically advanced world have elevated atmospheric CO_2 concentrations from 280 ppm to 414 ppm in the past 150 years, and concentrations continue to rise [25]. This marks a 47% rise in CO_2 level compared to the start of the Industrial Revolution (about 280 ppm) and an 11% rise compared to the year 2000 (about 370 ppm) [26]. CO_2 and water vapour, together with methane (CH_4) and nitrous oxide (N_2O), are the primary GHGs that heat the Earth's atmosphere. GHGs function by absorbing energy from sunlight, which heats the atmosphere and Earth's surface. Water vapour and clouds contribute the most to the greenhouse effect (about 75%). Noncondensing GHGs make up the remainder, with CO_2 and CH_4 accounting for about 24%, whereas N_2O and ozone (O_3) cover the remaining 1%. At temperatures and pressures similar to Earth's surface, water vapour and clouds behave as feedbacks of the greenhouse effect, whereas noncondensing GHGs are the drivers that influence the intensity of the greenhouse effect [27,28]. The noncondensable nature of the smaller group of GHGs (CO_2, CH_4, N_2O, and O_3) means that larger amounts of these gases can be accumulated in the atmosphere, thereby increasing the strength of the greenhouse effect [29]. Although there are other greenhouse gases, CO_2 receives the most attention because of two main reasons: (1) All other noncondensing GHGs together contribute only around one third of the total global warming potential, as their concentration in the atmosphere is very low compared to CO_2, and (2) CO_2 remains in the atmosphere longer than other GHGs. The time needed for CH_4 to leave the atmosphere is one decade, while N_2O is around one century. In contrast, 70% of emitted CO_2 will stay in the atmosphere for 100 years, 20% will stay for 1000 years, and the remaining 10% will stay for 10,000 years [30]. This long lifespan of CO_2 in the atmosphere makes it the most critical gas when tackling the issues of climate change.

Naturally, CO_2 is also generated in the atmosphere in the natural carbon cycle, but the carbon cycle maintains the balance of carbon in the atmosphere, on land, and in the ocean. In other words, the concentration of carbon in the atmosphere, land, plants, and oceans are kept relatively constant. Despite this, the carbon cycle has been altered following changes in climate. Shifts in Earth's orbit occur continuously and in predictable cycles, varying the amount of energy Earth acquires from the Sun. This results in a cycle of warm periods and ice ages on Earth. When summers in the Northern Hemisphere become colder, ice accumulates on land and the carbon cycle slows. At the same time, several factors such as lower temperature and higher phytoplankton growth rate likely enhance the absorption of atmospheric carbon into the ocean. The reduction in atmospheric carbon further cools the Earth. Likewise, when the last Ice Age ended about 10,000 years ago, CO_2 concentrations in the atmosphere climbed rapidly as Earth's temperatures rose [31–33].

Detailed analyses have been performed on air trapped in ice cores taken from Antarctica, and results have shown that for the past 800,000 years, CO_2 levels in the atmosphere had only fluctuated between 170 and 210 ppm [34]. Presently, the CO_2 level in the atmosphere is approaching 414 ppm, double that of the past 800,000 years. This rise in atmospheric CO_2 levels is due to the combustion of fossil fuels. The carbon released from combusting fossil fuels contains a different ratio of heavy-to-light carbon atoms. This produces a distinct "fingerprint" that can be quantified as (i) burning fossil fuel causes a relative decrease in the concentration of heavy carbon-13 atoms (^{13}C) in the atmosphere, and (ii) burning fossil fuels reduces the ratio of oxygen to nitrogen in the atmosphere [26]. There are three different isotopes for carbon, which are carbon-12 (^{12}C), carbon-13 (^{13}C), and carbon-14 (^{14}C). The most common is ^{12}C, while ^{13}C makes up around 1% of the total and ^{14}C only accounts for around one in one trillion carbon atoms. Plants and fossil fuels (ultimately derived from ancient plants) all contain a similar $^{13}C/^{12}C$ ratio. The $^{13}C/^{12}C$ ratio in the atmosphere is 2% higher than in plants (and hence fossil fuels). When the carbon from fossil fuels is liberated into the atmosphere, the average atmospheric $^{13}C/^{12}C$ ratio decreases [26].

One way to reduce CO_2 emissions is the substitution of fossil fuels with biomass-derived biofuels. Although utilizing biomass-derived biofuels also emit CO_2, the benefit of biomass-derived biofuels is that the combustion of biofuels discharges carbon which is part of the natural carbon cycle. In contrast, the combustion of fossil fuels discharges carbon which was previously kept in the ground for millions of years, adding more carbon into Earth's biosphere [35]. The additional carbon from burning fossil fuels is primarily responsible for global warming. To mitigate this, it is imperative that the global energy industry switches to biofuels and renewable energies, and reduces the dependence on fossil fuels. To this end, the implementation of renewable energy has been growing globally. For instance, the global electricity production from renewables was 27% in 2019 and 29% in 2020. In the midst of the COVID-19 pandemic in 2020, renewable energy usage increased by 3% while demand for fossil fuels dropped [36].

3. Global CO_2 Emissions during COVID-19 Pandemic

When compared to 2019, there was a significant reduction in global CO_2 emissions in 2020 due to the COVID-19 pandemic. From Figure 1 [15], the total global CO_2 emissions in 2020 showed a decline of 4.02% (-1365.83 $MtCO_2$), with the greatest decrease in emissions occurring in the U.S.A. (-477.37 $MtCO_2$, -9.44%), followed by EU27–UK (-230.58 $MtCO_2$, -7.48%), and India (-198.44 $MtCO_2$, -8.07%). Lesser decrease in emissions were observed in Japan (-54.92 $MtCO_2$, -5.04%), Russia (-45.16 $MtCO_2$, -2.92%), and Brazil (-41.95 $MtCO_2$, -9.80%). When compared to the first half of 2019 (January to June 2019), the first half of 2020 observed a decline in global CO_2 emissions by roughly 8.8% (-1551 $MtCO_2$). The period when emissions declined coincided with the global spread of COVID-19 and implementation of lockdown measures in many countries worldwide between March and July 2020, with the greatest decline recorded in April 2020. The absolute reduction in CO_2 emissions over the first half of 2020 was the greatest reduction in history, surpassing the drop in emissions during the Great Recession in 2008–2009 and the end of World War II (annual decrease of 790 Mt CO_2). However, since July 2020, the global CO_2 emissions started to increase due to the relaxation of lockdown measures and partial restart of the economy, particularly in China and some European countries [4].

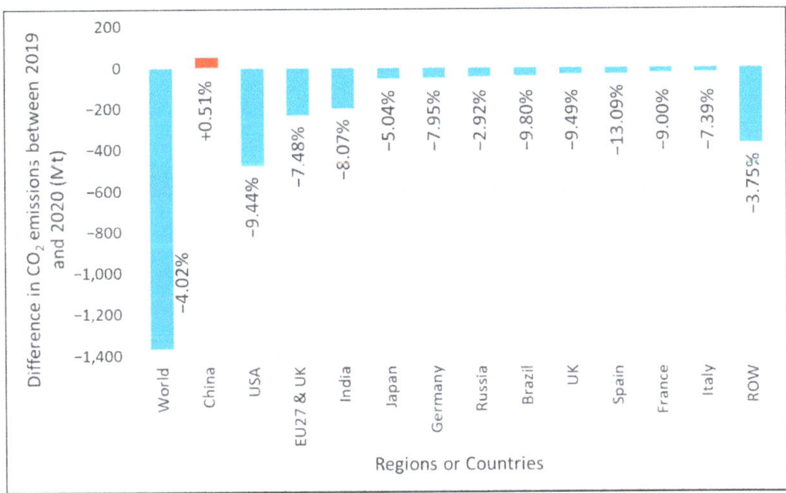

Figure 1. Changes in CO_2 emissions (million tonnes (Mt)) during 2019–2020 (ROW = rest of the world).

As shown in Figure 1, China recorded overall a slight increase in total emissions in 2020 compared to 2019 (+52.96 $MtCO_2$, +0.51%). At the start of 2020, there was a sharp decline in China's emissions due to the outbreak of COVID-19 and stringent lockdown measures, but these restrictions were gradually lifted starting in March. After this, CO_2 emissions in China quickly recovered in April and May. When compared to 2019, China in 2020 experienced a reduction in emissions of −18.4% in February and −9.2% in March, followed by an increase of 0.6% in April and 5.4% in May [4].

3.1. Power Sector

From Figure 2 [15], global CO_2 emissions from the power sector in 2020 decreased by 2.54% (−343.40 $MtCO_2$). The largest reduction was found in U.S.A. (−166.06 $MtCO_2$, −10.35%) and EU27–UK (−107.46 $MtCO_2$, −11.51%). During the lockdown period, global electricity consumption was reduced to weekend levels. Electricity demand was substantially reduced in services and industries, and this reduction was partially offset by increased residential usage [37]. Typically, weekends record lower electricity consumption than weekdays. This is because most commercial offices are closed, reducing the need to power lighting, air-conditioning, and computers [38]. When the lockdowns in Italy and Germany were lifted in April, electricity consumption gradually increased and rebounded to near 2019 levels in August. This trend was also observed in other countries (France, Spain, Great Britain, and India), which lifted their lockdowns in May and saw electricity demand rebound to near 2019 levels in August [37].

In India, electricity consumption increased 3.4% in September 2020 compared to 2019 levels due to greater demand from commercial and industrial sectors, as well as increased need for irrigation compared to 2019. In October 2020, India's recovering economy and relaxed restrictions drove up the consumption of electricity by more than 10% compared to 2019, which was similar to pre-COVID-19 levels. In the last two weeks of November, electricity consumption dropped to 2019 levels due to the Diwali festival and strikes in the agriculture industry. In December, electricity demand again rose more than 8% compared to 2019 [37].

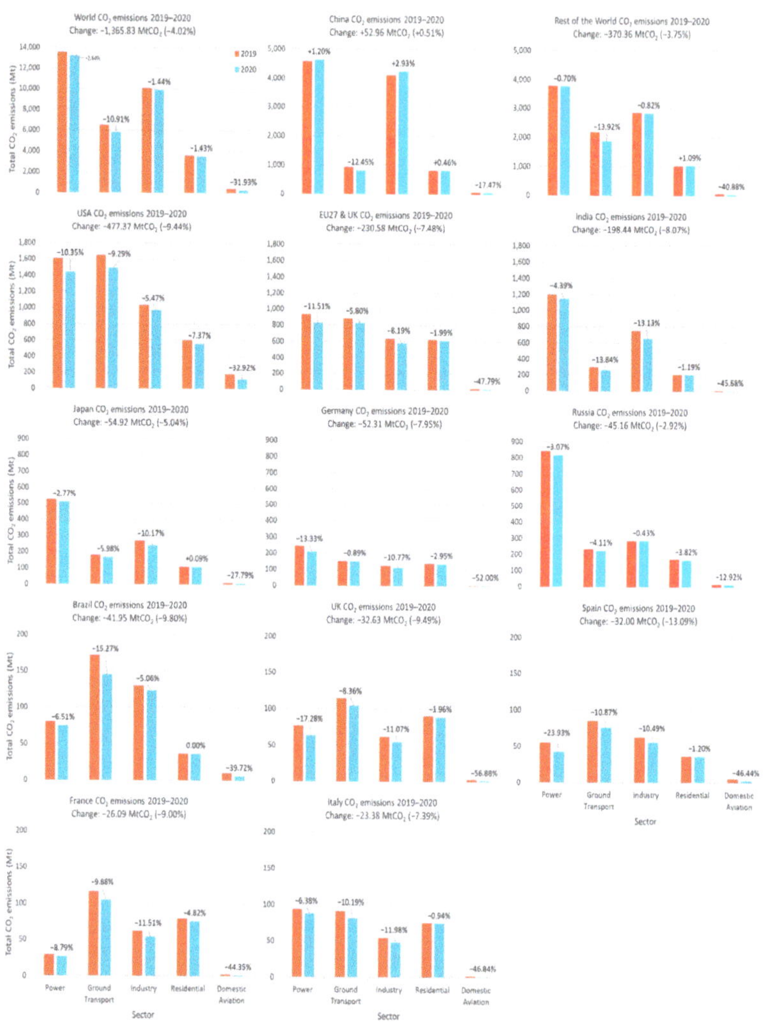

Figure 2. Changes in CO$_2$ emissions (million tonnes (Mt)) in different sectors for various regions and countries.

After China enforced lockdown in January 2020, electricity consumption began to drop and a significant reduction of 11% was recorded in February 2020 compared to 2019 levels. One of the reasons for lower electricity usage is the warmer temperatures recorded in February 2020 compared to February 2019. After China eased lockdown measures, electricity consumption quickly rebounded to pre-COVID-19 levels in April, and increased to a consistent 6% higher from August to November compared to 2019 levels [4,37]. Overall, electricity usage in China showed a slight growth of 1.20% (+54.83 MtCO$_2$).

3.2. Ground Transportation

The global CO$_2$ emissions from ground transportation in 2020 were reduced by 10.91% (−710.06 MtCO$_2$) (Figure 2). The largest contributors to this reduction were U.S.A. (−153.02 MtCO$_2$, −9.29%) and China (−115.05 MtCO$_2$, −12.45%). Moderate reductions were observed in EU27–UK (−51.03 MtCO$_2$, −5.80%) and India (−41.47 MtCO$_2$, −13.84%).

In China, the mean emissions from transportation in January 2020 rose by 7.4% compared to January 2019. This occurred because the Spring Festival travel rush began earlier in 2020 (10 January) before the onset of lockdown (last week of January). However, after lockdown was initiated, transport emissions declined sharply showing -75.1% in February and -34.2% in March [39]. In contrast, although transportation in U.S.A., Brazil, and Japan were not restricted, emissions from ground transportation in these three countries still declined, with Brazil registering -15.27% (-26.14 MtCO$_2$) and Japan -5.98% (-10.66 MtCO$_2$) [4].

3.3. Industry and Cement Production

On average, 29% of the worldwide annual CO$_2$ emissions are released from industries such as steel, chemicals, fossil-fuel-derived goods, and cement production. Developing countries, such as China and India, typically have a greater share in these emissions [40]. However, due to the COVID-19 pandemic, the worldwide industrial emissions in 2020 declined by 1.44% (-145.05 MtCO$_2$) (Figure 2). This included significant reductions in India (-98.63 MtCO$_2$, -13.13%), U.S.A. (-56.40 MtCO$_2$, -5.47%), and EU27–UK (-51.82 MtCO$_2$, -8.19%). In China's case, its industrial emissions recorded a small increase of 2.93% ($+120.36$ MtCO$_2$) in 2020 compared to 2019. This increase can be attributed to two of China's major industries, steel and cement, which accounted for roughly 42% and 22% of China's industrial emissions from fuel consumption in the past several years. Despite the lockdown measures, China's steel industry recorded a 2.2% increase in emissions in the first half of 2020 compared to 2019. For China's cement industry, a significant decline in emissions was observed in the first three months of 2020 (-29.5% in January and February combined, and -18.3% in March). After the lockdown was relaxed in March, however, cement emissions surged to +3.8%, +8.6%, and +8.4% in April, May, and June compared to 2019 [4,41].

3.4. Residential and Commercial Premises

The rapid spread of COVID-19 disease caused more people to work from home and minimize travelling. The winter months in the northern hemisphere were unusually warm in 2020, reducing energy consumption for heating. As a result, CO$_2$ emissions from global heating demand fell 1.43% (-50.83 MtCO$_2$) in 2020 compared to 2019 (Figure 2). The estimation of global heating demand assumes that most of the energy usage in residential and commercial premises was for temperature control, with no considerable change in their intrinsic energy consumption throughout the lockdown phase [4].

3.5. Aviation Industry

CO$_2$ emissions from global domestic aviation were reduced by 31.93% (-116.49 MtCO$_2$) in 2020 compared to 2019 (Figure 2). There were two occurrences where global aviation emissions were considerably reduced in 2020, one in Asia close to the end of January and another during the implementation of lockdown measures worldwide in mid-March. Global international aviation was greatly affected by the worldwide lockdowns, which showed a reduction in emissions of 72% in July 2020 compared to July 2019 [4]. According to the International Air Transport Association (IATA), the aviation industry's revenue passenger kilometres (RPKs) declined by 94.3% year-on-year in April 2020, following a huge drop of 55% year-on-year in March [42]. The industry-wide RPKs still showed reductions in July (-79.5%) and August 2020 (-75.3%) compared to 2019 levels. However, the easing of lockdowns globally has increased domestic market demands, allowing the aviation industry to gradually rebound [43]. Apart from RPKs, industry-wide cargo tonne kilometres (CTKs) also saw year-on-year reductions from -14.7% in March to -27.7% in April. This resulted from a reduced demand for cargo shipping and decreased manufacturing activities due to lockdowns [44]. In August, although industry-wide CTKs still displayed a reduction of 12.6%, international cargo traffic increased in many countries [45].

3.6. Comparison of CO$_2$ Emissions in 2019, 2020, and 2021

Figure 3 shows the comparison of CO$_2$ emissions for the world and major contributing countries or regions in the first five months of 2019, 2020, and 2021 [15,46]. Data from the first five months sufficiently covers the lockdown periods for the major countries and regions [4]. From Figure 3, global CO$_2$ emissions recovered back to 2019 levels in 2021 despite a sharp decrease of 8.19% ($-$1160.03 MtCO$_2$) in 2020. After experiencing CO$_2$ reductions in 2020, emissions in 2021 quickly rose back to 2019 levels for EU27–UK, India, and Russia. In the case of China, its CO$_2$ emissions rose 9.4% above 2019 levels, observed the highest recovery of 15.44% (+610.06 MtCO$_2$) in 2021 compared to 2020. In contrast, emissions from U.S.A., Japan, and rest of the world (ROW) in 2021 were 7.5%, 4.1%, and 5.4% lower than their 2019 levels, despite all three registering positive emissions growth in 2021 compared to 2020. Even though there is a net reduction in emissions in 2020, the mean atmospheric CO$_2$ level still increased by 0.6% in 2020 (412 ppm) compared to 2019 (410 ppm). This annual rate of increase is similar to 2019, which also showed a 0.6% increase compared to 2018 (408 ppm) [47]. After global emissions in 2021 returned to 2019 levels, the mean atmospheric CO$_2$ level for May 2021 reached 419 ppm as measured by National Oceanic and Atmospheric Administration's (NOAA) Mauna Loa Observatory [48]. These data showed that the CO$_2$ emissions savings from drastic lockdown measures on a global scale was only temporary and showed no significant effect on the annual growth of CO$_2$ levels [3]. This agrees with historical observations where brief decline in carbon emissions did not lead to significant climate change. To achieve long-term impacts, a systemic shift is required. Even if the lower GHG emissions due to COVID-19 were to become normal, we must still cut emissions by 50% by 2030 to prevent global mean temperature from rising more than 1.5 °C compared to pre-industrial level. This will allow us to mitigate some of the worst effects of climate change [49]. This has been the goal of the international Paris Agreement on climate change, which aims to control the rise of the global surface mean air temperature well below 2 °C compared to pre-industrial levels [50].

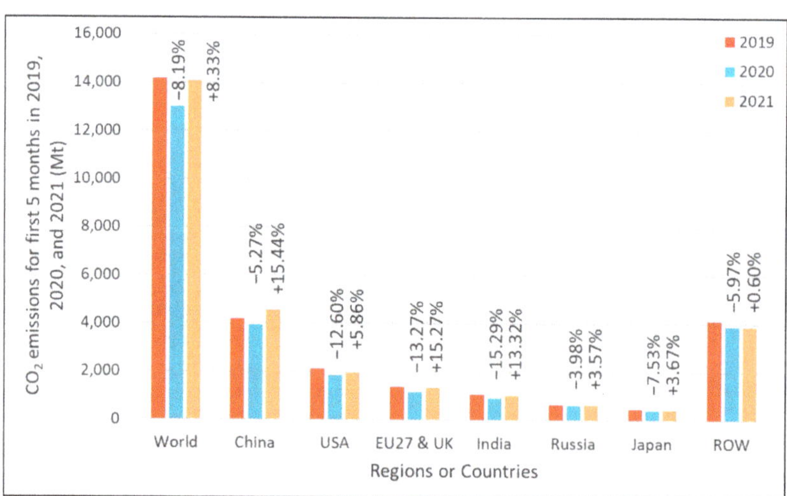

Figure 3. Comparison of CO$_2$ emissions from the first five months in 2019, 2020, and 2021 for major contributing countries/regions (ROW = rest of the world). The percentage differences in emissions are based on the previous year.

4. CO$_2$ Emissions in Malaysia during COVID-19 Pandemic

Figure 4 [51] shows a decline in Malaysia's CO$_2$ emissions during 2020 in six different areas, namely the power sector, industrial sector, ground transportation, public and residential sectors, and domestic aviation. The values in Figure 4 represent the estimated difference between daily CO_2 emissions in 2020 compared to the mean 2019 levels. From Figure 4, total CO_2 emissions showed the highest reduction (-0.202 $MtCO_2$/d) in the middle of March 2020, which coincided with the implementation of lockdown measures by the government, known as Movement Control Order (MCO), to curb the rapid spread of COVID-19. The spike in positive cases from 29 cases on 1 March to 553 cases on 16 March, due to mass gathering at a religious event, prompted the government to enforce first phase MCO from 18 March to 31 March [52,53]. The MCO enforced rules such as banning of movement and assembly, banning of overseas and interstate travel, closure of all educational institutions, closure of all nonessential public and private premises, social distancing, and allowing only one person per household to leave home for purchasing essential items [54,55]. The government then implemented three more phases of MCO, extending the lockdown period from April to early May. These restrictions were reflected by significant reductions in CO_2 emissions in power generation (-0.052 $MtCO_2$/d, 26% share of total reductions during MCO), industrial sector (-0.047 $MtCO_2$/d, 23% share), and ground transportation (-0.093 $MtCO_2$/d, 46% share) from March to May as all nonessential sectors were closed (Figure 4). The public sector and aviation industry saw a lesser decrease in emissions of -0.006 $MtCO_2$/d (28% share) and -0.005 $MtCO_2$/d (23% share), respectively. In the same period, CO_2 emissions from the residential sector saw a slight increase of 0.0005 $MtCO_2$/d as all residents stayed at home. Outside of MCO, residential emissions stayed the same as their mean 2019 levels.

The government gradually relaxed the lockdown measures by establishing Conditional Movement Control Order (CMCO) in two phases from 4 May to 9 June. Under the CMCO, several sectors of the economy were permitted to resume provided that all employers and employees complied to the Standard Operating Procedures (SOPs) specified by the government [56,57]. The partial restart of the economy increased the daily CO_2 emissions during CMCO to -0.133 $MtCO_2$/d (34% recovery) compared to the MCO period (Figure 4). Emissions from power generation recorded the highest recovery of 67% (-0.017 $MtCO_2$/d) compared to the MCO period. This was followed by industrial and ground transportation emissions, registering recoveries of 29% (-0.033 $MtCO_2$/d) and 20% (-0.075 $MtCO_2$/d).

As the rate of infection was largely under control due to the MCO and CMCO measures, the government introduced Recovery Movement Control Order (RMCO) after CMCO ended. The first two phases of RMCO lasted from 10 June to 31 December 2020 [58,59], while the third phase continued from 1 January to 31 March 2021 [60]. Under RMCO, restrictions were further relaxed while still adhering to the SOPs. RMCO allowed interstate travel (but not international travel), normal business operations, small-scale gatherings for recreation and sports (except sporting events and contact sports), gradual reopening of schools, and small congregations of worshippers at mosques and suraus [58]. The RMCO initiated another increase in daily CO_2 emissions to -0.043 $MtCO_2$/d (79% recovery compared to the MCO period), which stabilized until the end of 2020 (Figure 4). Upon the start of RMCO, CO_2 emissions from power generation completely recovered to 2019 levels. Compared to the MCO period, emissions from ground transportation and public sector showed high recoveries of 80% (-0.019 $MtCO_2$/d) and 78% (-0.0009 $MtCO_2$/d), respectively. Meanwhile, industrial and aviation emissions showed lower recoveries of 57% (-0.020 $MtCO_2$/d) and 34% (-0.003 $MtCO_2$/d). By the end of 2020, Malaysia's total CO_2 emissions were 225.97 $MtCO_2$, a notable decline of 9.7% from 250.09 $MtCO_2$ in 2019 (Figure 5) [51].

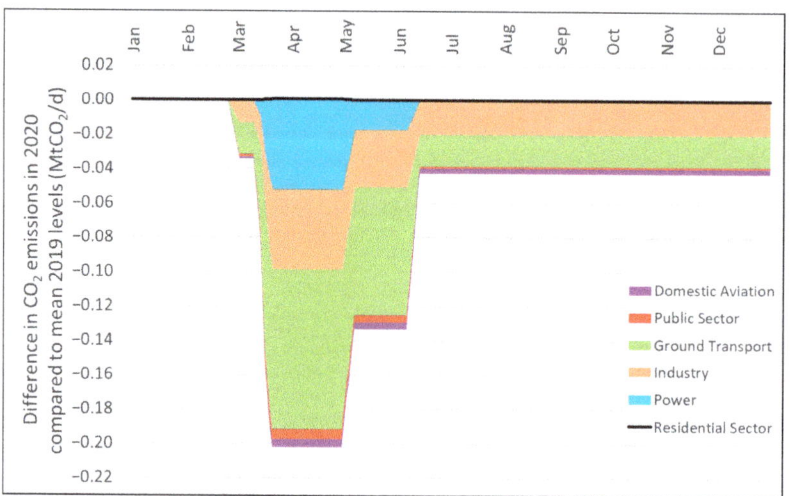

Figure 4. Malaysia's daily CO_2 emissions in 2020, represented as the difference in daily emissions compared to the mean 2019 levels. Values are in million tonnes of CO_2 per day ($MtCO_2/d$).

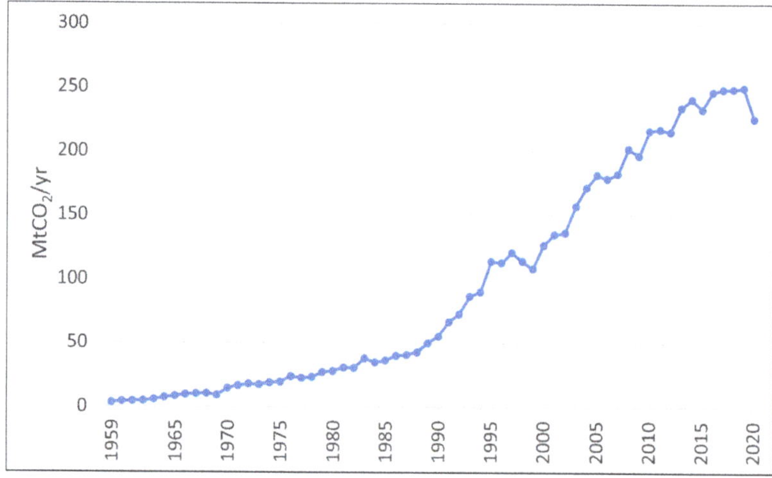

Figure 5. Malaysia's CO_2 emissions from 1959–2020 in million tonnes of CO_2 per year ($MtCO_2/year$).

5. Climate Change in Malaysia

Despite the considerable drop in global and Malaysia's CO_2 emissions in 2020 due to the COVID-19 pandemic, it was not enough to cause a long-term impact on global CO_2 levels [3]. Given the direct link between rising CO_2 levels and increased global warming, it is prudent to look into the effects of climate change [30]. This is because climate change not only poses a grave threat to the planet and health of the human race, but it is also endangering the world economy. To solve this problem, we need to change our technologies of commodity production to modern technologies that guarantee and drive the growth of sustainable economies. Climate change, increasing energy consumption, and sustainable development have become the most critical challenges in the 21st century. The climbing atmospheric CO_2 concentrations (currently >400 ppm) coupled with the rising average global temperatures (increase >2 °C by 2100) will severely affect the human population worldwide, resulting in extreme weather events (higher frequency of intense heat waves and cyclones) and melting of glaciers (causing rapid sea-level rise) [61]. If measures to halt global warming are not implemented swiftly, coastal cities may become flooded and unsuitable for living in the near future. Apart from this, the growing world population, projected to reach 9.7 billion in 2050 [62], together with the expected rise in global energy consumption of 30–50% higher in 2050 [63,64], are great challenges that must be solved by the entire human society.

Being a global phenomenon, the impacts of climate change on a global and regional scale, including in Malaysia, must be studied. Malaysia, located in Southeast Asia, is a country that comprises two regions, namely Peninsular Malaysia (West Malaysia) and Malaysian Borneo (East Malaysia). The whole country has a land area of 330,803 km^2 and an estimated population of 32.7 million in 2020 [65,66]. Malaysia is located near the equator and has hot and humid climate throughout the year. The variabilities of Malaysia's climate are closely linked to the Northeast and Southwest Monsoons. The Northeast Monsoon happens from October to March and brings higher precipitation, whereas the Southwest Monsoon occurs from April to September and displays drier climate with decreased rainfall [67].

5.1. Mean Monthly Temperature in Malaysia

Figure 6 [68,69] shows the mean monthly temperature in five different towns/cities in Malaysia, two from East Malaysia (Kota Kinabalu and Kuching) and three from West Malaysia (Malacca, Subang, and Kuantan). These five towns/cities were selected because they are suitably located in different regions of Malaysia, thereby providing a good overview of climate conditions throughout the country. In Figure 6, there is a clear upward trend in temperatures at all five locations. The rate of increase in mean monthly temperature, as shown by the slope of the linear trendline (the red dotted straight lines in Figure 6), was smallest for Kuching, followed by Kota Kinabalu, Malacca, Kuantan, and Subang, in increasing order. These data coincide with observations from the Malaysian Meteorological Department, wherein Kuching recorded the lowest temperature rise due to its slower growth in development [70]. However, the slope of the trendlines does not give the actual rate of mean temperature rise. Based on a 2015 report by the Ministry of Natural Resources and Environment Malaysia, the estimated rate of mean temperature rise was 0.25 °C every decade for Peninsular Malaysia, 0.20 °C for Sabah, and 0.14 °C for Sarawak [71].

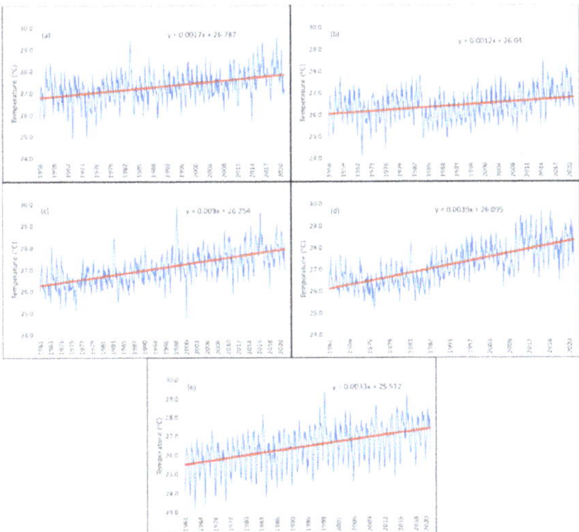

Figure 6. Mean monthly temperature in (**a**) Kota Kinabalu (1956–2020), (**b**) Kuching (1956–2020), (**c**) Malacca (1961–2020), (**d**) Subang (1961–2020), and (**e**) Kuantan (1961–2020).

5.2. Extreme Weather

Malaysia has increasingly experienced extreme climate events in the last several decades, where days or weeks of high temperature, high precipitation, dry weather, thunderstorms, and strong winds are becoming more frequent. Since the 1980s, days with very high precipitation levels have been rising. In addition, the number of days with extreme thunderstorm and wind conditions has also increased. These factors have promoted more frequent flooding in many regions of Malaysia. The most extreme flooding event was documented in the southern region of Peninsular Malaysia during the Northeast Monsoon in 2006–2007 [72]. The rising temperature trends shown in Figure 6a–e depict progressively hotter weather in Malaysia. This change is influenced by El Nino events [70]. The co-occurrence of high precipitations and dry spells within the same year is becoming an apparent weather pattern in Malaysia [73]. Table 1 shows a list of major flooding events in Malaysia for the past 50 years.

Table 1. Major floods in Malaysia [74,75].

Date	Location	Event
Jan 1971	Kuala Lumpur, Malaysia	Most extreme flood in Malaysia since 1926. Caused by heavy monsoon rains. 32 deaths and 180,000 people affected.
Dec 2006–Jan 2007	West Malaysia	Johor was severely affected. Malacca, Pahang, and Negeri Sembilan were affected to a lesser extent. Caused by heavy rain following Typhoon Utor.
Oct–Nov 2010	Thailand and Northern Peninsular Malaysia	Flooding caused by abnormally late monsoon rains from the La Nina event and the Bay of Bengal. Killed 232 people in Thailand and 4 in Malaysia.
Jan–Feb 2014	Sabah, Malaysia	Heavy rainfall and flash flooding in various regions of Sabah.
Dec 2014–Jan 2015	Southeast Asia and South Asia	Northeast monsoon caused flooding in South Thailand, West Malaysia, Indonesia, and Sri Lanka. More than 417,000 people were affected.
Jan–Feb 2015	East Malaysia	Stronger northeast monsoon causing floods in many regions of Sarawak and Sabah, affecting roughly 13,878 people.
Feb–Mar 2016	Malaysia	Heavy rainfall causing floods in Sarawak, Johor, Malacca, and part of Negeri Sembilan.
Dec 2016–early 2017	Southern Thailand and Northern Peninsular Malaysia	Northeast monsoon caused floods in southern Thailand and in Kelantan and Terengganu, northern Peninsular Malaysia. Resulted in a loss of roughly USD 4 billion due to the infrastructure damage and disruption of agriculture and tourism.
Nov 2017	Penang and Kedah, Malaysia	Flash floods due to hours of torrential rain in Penang killed a minimum of 7 people. The Malaysian military personnel assisted in evacuating more than 3500 people. In Kedah, floods displaced more than 2000 people.
2017	Kundang, Malaysia	More than 6000 people were affected by flash floods and landslides. A few stretches of roads, infrastructure, and properties were severely damaged.
Jan 2018	Malaysia	Downpour due to annual monsoon created severe flooding in six states. About 5000 people were displaced in Pahang. The hardest hit state, Kuantan, saw 3931 people displaced by floods and stayed in 22 evacuation centers.
Feb 2018	Sarawak, Malaysia	Flooding due to heavy rainfall displaced 4859 people. 25 evacuation centers were set up for the flood victims.

5.3. Rainfall

The large variability found in Malaysia's historical precipitation data coincides with observations from other reports [70,72,76]. From Figure 7 [68,69], there is a lack of precipitation data between 1973–1978 in all five locations. Despite this, a rising trend can be seen in Figure 7a–e, with the greatest increase in precipitation occurring in Kuantan. This is followed by Kuching and Subang, indicating moderate increase in precipitation. Kota Kinabalu and Malacca share the lowest increase in precipitation in this selected group. The yearly precipitation increment in Malaysia is attributed to the Northeast Monsoon, which usually takes place between October to March. The strong northeasterly and easterly winds carry moisture from the western Pacific to the South China Sea, increasing the amount of rain in the Southeast Asian region [67,77]. The rising mean precipitation in Malaysia is linked to a greater number of rainy days, and rainfalls during the Northeast Monsoon season are heavier than at other times of the year. Although the mean precipitation varies throughout the year, one study reported that the yearly precipitation and Northeast Monsoon rainfall have increased significantly between 1971 and 2010, with 95% and 90% confidence levels, respectively. During the same period, the number of days with greater than 20 mm precipitation also increased by 1.5 every decade [78].

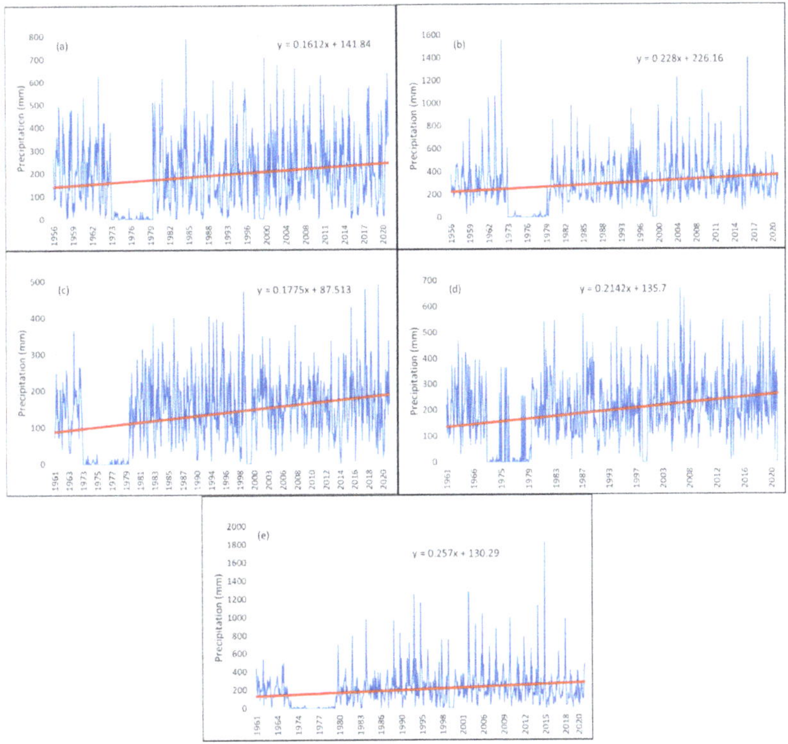

Figure 7. Mean monthly precipitation levels in (**a**) Kota Kinabalu (1956–2020), (**b**) Kuching (1956–2020), (**c**) Malacca (1961–2020), (**d**) Subang (1961–2020), and (**e**) Kuantan (1961–2020).

5.4. Impacts of Climate Change in Malaysia

There are seven sectors in Malaysia that are susceptible to changes in climate, namely agriculture, biodiversity, coastal and marine resources, forestry, water resources, energy, and public health. Despite this, most of the research in Malaysia centres around investigating the effects of climate change on agriculture, especially the effects on rice cultivation [75]. It was reported that agricultural yields and growth rates were adversely but weakly affected by climate change. However, there was a notable positive correlation between crop production index and per capita CO_2 emissions, indicating that the agriculture sector is a CO_2 source [79]. Although the considerable CO_2 emissions from Malaysia's agricultural practices were not highlighted, this phenomenon still raised concerns regarding the sustainability of the agriculture sector. While higher atmospheric CO_2 concentration can promote the rate of photosynthesis in C3 plants (where the first product of photosynthesis is 3-phosphoglycerate which comprises three carbon atoms, for instance paddy), the increasing temperature instead inhibits photosynthesis and promotes respiration when temperature rises above the tolerance limit of 26 °C. The final result is a lower rate of grain filling in paddy [80]. The future weather conditions in Malaysia were projected to have higher temperatures and unanticipated rainfall patterns, and these conditions do not favour the production of grains [75].

It is predicted that between 2025 and 2050, climate change will significantly increase flow in the watersheds of Terengganu, Kelantan, Perak, and Pahang during the Northeast Monsoon, while considerably reduce flowing in the watersheds of Klang and Selangor during the Southwest Monsoon [81]. This change in flow has the potential to cause heavy precipitation, floods, and rainfall-induced landslides, which can damage private properties and public infrastructure [82]. In the fishing industry, a study revealed that unstable weather decreases the days for fishing and amplifies the risks to fishing activities [83]. Even though statistical data are insufficient, another study showed that unstable weather resulted in an overall reduction in earnings of 9–32% for fishermen living along the east coast of Peninsular Malaysia [84].

In terms of energy consumption, higher temperatures resulting from climate change promotes the usage of air-conditioning in commercial and residential buildings. Compared to the year 2000, the cooling load in Malaysia is projected to rise by 2.96%, 8.08%, and 11.7% in 2020, 2050, and 2080, respectively. Since the maximum cooling load in 2000 was 297,000 KJ/h, this projection translates into maximum energy consumptions of 305,000 KJ/h in 2020, 321,000 KJ/h in 2050, and 330,000 KJ/h in 2080. The greater outdoor temperatures causes heat to enter buildings through walls, thereby increasing the indoor space cooling demand [85]. Apart from varying temperatures and rainfall patterns, climate change also raises the frequency of fire and haze in the tropical peatlands. This will disrupt the capacity of the peatlands to store carbon, instead turning them from carbon sinks into carbon sources, which in turn aggravates climate change [86].

In the public health sector, climate change is recognized for enhancing the spread of water-borne (such as cholera, typhoid fever, hepatitis A, and dysentery) and vector-borne diseases (or mosquito-related diseases such as malaria, dengue, and chikungunya) [87,88]. Global warming causes an increase in rainfall, which enhances the risk of flooding and the formation of ground pools where drinking water can be contaminated by untreated river or sewage water. Increase in precipitation also enhances the formation of stagnant waters that encourage breeding of mosquitoes, thereby raising the occurrence of mosquito-related diseases and expanding the affected areas [87]. Although the occurrences of water-borne diseases have been largely reduced in the last century, there were still incidences of typhoid and cholera outbreaks due to low coverage of clean water supply and lack of water supply during droughts, respectively. The best practices to minimize the incidences of typhoid and cholera are to extend clean water supply infrastructure to cover remote rural areas and to prepare and ration water supply for dry spells. On the other hand, the higher ambient temperature and increased rainfall in Malaysia provide the optimum conditions for mosquitoes to multiply and spread diseases. Warmer temperature reduces the time mosquito larvae reach maturity, and given favourable conditions, mosquito eggs can hatch within a day and mature as adults within a week. Even under dry conditions, the eggs can last about nine months. The policies and programmes to control the spread of malaria, dengue, and chikungunya include public health awareness, active detection of outbreaks, regular fogging, and reduction of breeding grounds [88]. In spite of this, little research has been carried out in Malaysia to examine the impacts of climate change on the occurrences of these diseases and conditions of the areas affected. From a review article, in the year 2050, the number of malarial cases could potentially increase by 15% when temperatures rise by 1.5 °C. This projection considered factors such as increased rainfall and intrusion of saline water. The review also described a positive correlation between precipitation and dengue, showing that greater frequency of rainy days and higher temperatures were favourable conditions for the propagation of dengue viruses [88].

6. Conclusions

There is no doubt that the COVID-19 pandemic is a serious global matter and has claimed the lives of many people worldwide. Many countries also faced economic downturns due to stringent lockdown measures implemented to curb the rapid global spread of COVID-19. However, there is a silver lining in this scenario, and this pandemic provided a once-in-a-lifetime opportunity to study the global atmospheric CO_2 levels during the global lockdown. Although the concentration of CO_2 in the atmosphere is relatively low, the high heat absorbing ability and long atmospheric lifespan of CO_2 makes it the most important GHG to study and mitigate. If CO_2 levels are not lowered quickly, we will be faced with global warming and more frequent and severe weather events. Thus, it has become a global mission to reduce CO_2 emissions to prevent the global mean temperature from rising above 1.5 °C. There is strong evidence that the additional CO_2 in the atmosphere comes from human activities, primarily the burning of fossil fuels. In order to reduce our reliance on fossil fuels, it is imperative that more research, funding, and political considerations are put into developing renewable sources of energy.

Author Contributions: C.H.T.: data curation, visualization, writing—original draft, and writing—review and editing; M.Y.O.: writing—review and editing; S.M.N.: conceptualization, supervision, writing—original draft, and writing—review and editing; A.H.S.: funding acquisition; P.L.S.: writing—review and editing. All authors have read and agreed to the published version of the manuscript.

Funding: This research was funded by AAIBE Chair of Renewable Energy (ChRE) (C.H. Tan under postdoctoral fellowship programme, and M.Y. Ong under project code 202006KETTHA). In addition, M.Y. Ong would also like to thank UNITEN for the UNITEN Postgraduate Excellence Scholarship 2019 (YCU-COGS). The APC was funded by BOLD Refresh Publication Fund 2021 under iRMC, UNITEN (grant: J5100D4103-BOLDREFRESH2025-CENTER OF EXCELLENCE).

Institutional Review Board Statement: Not applicable.

Informed Consent Statement: Not applicable.

Data Availability Statement: The CO_2 emissions data for the world and major countries from 2019–2021 are publicly available online from https://carbonmonitor.org/ (accessed on 12 April 2021 and 8 July 2021). Malaysia's daily CO_2 emissions data in 2020 and annual CO_2 emissions from 1959–2020 are publicly available online from https://www.icos-cp.eu/gcp-COVID19 (accessed on 10 July 2021) and https://www.icos-cp.eu/gcp-COVID19-v12 (accessed on 10 July 2021). The mean monthly temperature and rainfall data for Malaysia are publicly available online from https://www.ncdc.noaa.gov/cdo-web/datasets/GHCNDMS/locations/FIPS:MY/detail#stationlist (accessed on 30 April 2021) and https://www.worldweatheronline.com/malaysia-weather.aspx (accessed on 30 April 2021).

Acknowledgments: The authors would like to acknowledge UNITEN for the research facilities and funding.

Conflicts of Interest: The authors declare no conflict of interest.

Abbreviations

The following abbreviations are used in the manuscript:

^{12}C	Carbon-12 atom
^{13}C	Carbon-13 atom
^{14}C	Carbon-14 atom
CH_4	Methane
CMCO	Conditional Movement Control Order
CO_2	Carbon dioxide
COVID-19	Coronavirus disease 2019

C-recycling	Carbon recycling
CTK	Cargo tonne kilometre
EJ	Exajoules
EU27	The 27 European Union countries after the UK left the EU
GDP	Gross domestic product
GHG	Greenhouse gas
Gt	Giga tonnes
IATA	International Air Transport Association
KJ/h	Kilojoules per hour
MCO	Movement Control Order
mm	Millimetre
Mt	Million tonnes
N_2O	Nitrous oxide
O_3	Ozone
ppm	Parts per million
RMCO	Recovery Movement Control Order
ROW	Rest of the world
RPK	Revenue passenger kilometre
SOP	Standard Operating Procedures
TWh	Terawatt-hour
U.S.A.	United States of America
UK	United Kingdom
UNEP	United Nations Environment Program

References

1. Liu, Y.-C.; Kuo, R.-L.; Shih, S.-R. COVID-19: The first documented coronavirus pandemic in history. *Biomed. J.* **2020**, *43*, 328–333. [CrossRef]
2. Maliszewska, M.; Mattoo, A.; Van der Mensbrugghe, D. *The Potential Impact of COVID-19 on GDP and Trade: A Preliminary Assessment*; The World Bank: Washington, DC, USA, 2020.
3. Nicola, M.; Alsafi, Z.; Sohrabi, C.; Kerwan, A.; Al-Jabir, A.; Iosifidis, C.; Agha, M.; Agha, R. The socio-economic implications of the coronavirus and COVID-19 pandemic: A review. *Int. J. Surg.* **2020**, *78*, 185–193. [CrossRef]
4. Liu, Z.; Ciais, P.; Deng, Z.; Lei, R.; Davis, S.J.; Feng, S.; Zheng, B.; Cui, D.; Dou, X.; Zhu, B. Near-real-time monitoring of global CO_2 emissions reveals the effects of the COVID-19 pandemic. *Nat. Commun.* **2020**, *11*, 5172. [CrossRef]
5. Dehghani-Sanij, A.; Bahadori, M.N. Energy consumption and environmental consequences. In *Ice-Houses: Energy, Architecture, and Sustainability*; Dehghani-Sanij, A., Bahadori, M.N., Eds.; Academic Press: Cambridge, MA, USA, 2021; pp. 1–55.
6. Jackson, R.B.; Le Quéré, C.; Andrew, R.; Canadell, J.G.; Korsbakken, J.I.; Liu, Z.; Peters, G.P.; Zheng, B. Global energy growth is outpacing decarbonization. *Environ. Res. Lett.* **2018**, *13*, 120401. [CrossRef]
7. Peters, G.P.; Andrew, R.M.; Canadell, J.G.; Friedlingstein, P.; Jackson, R.B.; Korsbakken, J.I.; Le Quéré, C.; Peregon, A. Carbon dioxide emissions continue to grow amidst slowly emerging climate policies. *Nat. Clim. Chang.* **2020**, *10*, 3–6. [CrossRef]
8. International Energy Agency (IEA). Key Electricity Trends 2020: Annnual Trends from OECD Countries. Available online: https://www.iea.org/articles/key-electricity-trends-2020 (accessed on 21 July 2021).
9. International Energy Agency (IEA). *Net Zero by 2050: A Roadmap for the Global Energy Sector*; IEA: Paris, France, 2021.
10. Khallaghi, N.; Hanak, D.P.; Manovic, V. Techno-economic evaluation of near-zero CO_2 emission gas-fired power generation technologies: A review. *J. Nat. Gas. Sci. Eng.* **2020**, *74*, 103095. [CrossRef]
11. Tumanovskii, A.G.; Shvarts, A.L.; Somova, E.V.; Verbovetskii, E.K.; Avrutskii, G.D.; Ermakova, S.V.; Kalugin, R.N.; Lazarev, M.V. Review of the coal-fired, over-supercritical and ultra-supercritical steam power plants. *Therm. Eng.* **2017**, *64*, 83–96. [CrossRef]
12. Hoseinzadeh, S.; Stephan Heyns, P. Advanced energy, exergy, and environmental (3E) analyses and optimization of a coal-fired 400 MW thermal power plant. *J. Energy Resour Technol.* **2021**, *143*, 082106. [CrossRef]
13. International Energy Agency (IEA). *Gas 2020: Analysing the Impact of the COVID-19 Pandemic on Global Natural Gas Markets*; IEA: Paris, France, 2020.
14. International Energy Agency (IEA). *Coal 2020: Analysis and Forecast to 2025*; IEA: Paris, France, 2020.
15. Carbon Monitor. CO2 Emissions Variation (%): January 1st to December 31st, 2020 vs. January 1st to December 31st, 2019. Available online: https://carbonmonitor.org/ (accessed on 12 April 2021).
16. Frankfurt School-UNEP Centre/BNEF. *Global Trends in Renewable Energy Investment 2020*; FS-UNEP Centre: Frankfurt, Germany, 2020.
17. McCormick, K.; Kautto, N. The Bioeconomy in Europe: An Overview. *Sustainability* **2013**, *5*, 2589–2608. [CrossRef]
18. Ollikainen, M. Forestry in bioeconomy—Smart green growth for the humankind. *Scand. J. For. Res.* **2014**, *29*, 360–366. [CrossRef]
19. Pülzl, H.; Kleinschmit, D.; Arts, B. Bioeconomy—An emerging meta-discourse affecting forest discourses? *Scand. J. For. Res.* **2014**, *29*, 386–393. [CrossRef]

20. Aresta, M.; Dibenedetto, A. Carbon Recycling Through CO_2-Conversion for Stepping Toward a Cyclic-C Economy. A Perspective. *Front. Energy Res.* **2020**, *8*, 159. [CrossRef]
21. Sikarwar, V.S.; Reichert, A.; Jeremias, M.; Manovic, V. COVID-19 pandemic and global carbon dioxide emissions: A first assessment. *Sci. Total Environ.* **2021**, *794*, 148770. [CrossRef] [PubMed]
22. Naderipour, A.; Abdul-Malek, Z.; Ahmad, N.A.; Kamyab, H.; Ashokkumar, V.; Ngamcharussrivichai, C.; Chelliapan, S. Effect of COVID-19 virus on reducing GHG emission and increasing energy generated by renewable energy sources: A brief study in Malaysian context. *Environ. Technol. Innov.* **2020**, *20*, 101151. [CrossRef]
23. Yusup, Y.; Kayode, J.S.; Ahmad, M.I.; Yin, C.S.; Hisham, M.S.M.N.; Isa, H.M. Atmospheric CO_2 and total electricity production before and during the nation-wide restriction of activities as a consequence of the COVID-19 pandemic. *arXiv* **2020**, arXiv:2006.04407.
24. Maslin, M. *Global Warming: A Very Short Introduction*, 2nd ed.; Oxford University Press: Oxford, UK, 2008. [CrossRef]
25. NASA—Global Climate Change. The Causes of Climate Change. Available online: https://climate.nasa.gov/causes/ (accessed on 9 May 2021).
26. NASA—Global Climate Change. The Atmosphere: Getting a Handle on Carbon Dioxide. Available online: https://climate.nasa.gov/news/2915/the-atmosphere-getting-a-handle-on-carbon-dioxide/ (accessed on 15 October 2020).
27. Lacis, A.A.; Schmidt, G.A.; Rind, D.; Ruedy, R.A. Atmospheric CO_2: Principal control knob governing Earth's temperature. *Science* **2010**, *330*, 356–359. [CrossRef]
28. Hansen, J.; Sato, M.; Hearty, P.; Ruedy, R.; Kelley, M.; Masson-Delmotte, V.; Russell, G.; Tselioudis, G.; Cao, J.; Rignot, E. Ice melt, sea level rise and superstorms: Evidence from paleoclimate data, climate modeling, and modern observations that 2 C global warming could be dangerous. *Atmos. Chem. Phys.* **2016**, *16*, 3761–3812. [CrossRef]
29. Lacis, A.A.; Hansen, J.E.; Russell, G.L.; Oinas, V.; Jonas, J. The role of long-lived greenhouse gases as principal LW control knob that governs the global surface temperature for past and future climate change. *Tellus B Chem. Phys. Meteorol.* **2013**, *65*, 19734. [CrossRef]
30. Archer, D.; Eby, M.; Brovkin, V.; Ridgwell, A.; Cao, L.; Mikolajewicz, U.; Caldeira, K.; Matsumoto, K.; Munhoven, G.; Montenegro, A. Atmospheric lifetime of fossil fuel carbon dioxide. *Annu. Rev. Earth Planet. Sci* **2009**, *37*, 117–134. [CrossRef]
31. Lüthi, D.; Le Floch, M.; Bereiter, B.; Blunier, T.; Barnola, J.-M.; Siegenthaler, U.; Raynaud, D.; Jouzel, J.; Fischer, H.; Kawamura, K. High-resolution carbon dioxide concentration record 650,000–800,000 years before present. *Nature* **2008**, *453*, 379–382. [CrossRef]
32. Jouzel, J.; Masson-Delmotte, V.; Cattani, O.; Dreyfus, G.; Falourd, S.; Hoffmann, G.; Minster, B.; Nouet, J.; Barnola, J.-M.; Chappellaz, J. Orbital and millennial Antarctic climate variability over the past 800,000 years. *Science* **2007**, *317*, 793–796. [CrossRef]
33. NASA—Earth Observatory. Changes in the Carbon Cycle. Available online: https://earthobservatory.nasa.gov/features/CarbonCycle/page4.php (accessed on 9 May 2021).
34. Joos, F.; Spahni, R. Rates of change in natural and anthropogenic radiative forcing over the past 20,000 years. *Proc. Natl. Acad. Sci. USA* **2008**, *105*, 1425–1430. [CrossRef] [PubMed]
35. IEA Bioenergy. *Is Energy from Woody Biomass Positive for the Climate?* IEA Bioenergy: Paris, France, 2018.
36. International Energy Agency (IEA). *Global Energy Reviews 2021: Renewables*; IEA: Paris, France, 2021.
37. International Energy Agency (IEA). *COVID-19 Impact on Electricity*; IEA: Paris, France, 2021.
38. U.S. Energy Information Administration (EIA). Hourly Electricity Consumption Varies throughout the Day and Across Seasons. Available online: https://www.eia.gov/todayinenergy/detail.php?id=42915 (accessed on 10 May 2021).
39. Liu, Z.; Deng, Z.; Ciais, P.; Lei, R.; Davis, S.J.; Feng, S.; Zheng, B.; Cui, D.; Dou, X.; He, P. COVID-19 Causes Record Decline in Global CO2 Emissions. Available online: https://escholarship.org/uc/item/2fv7n055 (accessed on 26 May 2021).
40. Oliver, J.; Peters, J. *Trends in Global CO_2 and Total Greenhouse Gas Emissions: 2019 Report*; PBL Netherlands Environmental Assessment Agency: The Hague, The Netherlands, 2019.
41. Liu, Z.; Ciais, P.; Deng, Z.; Davis, S.J.; Zheng, B.; Wang, Y.; Cui, D.; Zhu, B.; Dou, X.; Ke, P. Carbon Monitor, a near-real-time daily dataset of global CO_2 emission from fossil fuel and cement production. *Sci. Data* **2020**, *7*, 392. [CrossRef] [PubMed]
42. IATA. *Air Passenger Market Analysis: April 2020*; IATA: Montreal, QC, Canada, 2020.
43. IATA. *Air Passenger Market Analysis: August 2020*; IATA: Montreal, QC, Canada, 2020.
44. IATA. *Air Cargo Market Analysis: April 2020*; IATA: Montreal, QC, Canada, 2020.
45. IATA. *Air Cargo Market Analysis: August 2020*; IATA: Montreal, QC, Canada, 2020.
46. Carbon Monitor. CO2 Emissions Variation (%): January 1st to May 31st, 2021 vs. January 1st to May 31st, 2020. Available online: https://carbonmonitor.org/ (accessed on 8 July 2021).
47. Dlugokencky, E.; Tans, P. Trends in Atmospheric Carbon Dioxide. Available online: https://www.esrl.noaa.gov/gmd/ccgg/trends/global.html (accessed on 9 May 2021).
48. NOAA. Carbon Dioxide Peaks Near 420 Parts per Million at Mauna Loa Observatory. Available online: https://research.noaa.gov/article/ArtMID/587/ArticleID/2764/Coronavirus-response-barely-slows-rising-carbon-dioxide (accessed on 8 July 2021).
49. Forster, P.M.; Forster, H.I.; Evans, M.J.; Gidden, M.J.; Jones, C.D.; Keller, C.A.; Lamboll, R.D.; Le Quéré, C.; Rogelj, J.; Rosen, D. Current and future global climate impacts resulting from COVID-19. *Nat. Clim. Chang.* **2020**, *10*, 913–919. [CrossRef]
50. Jackson, R.; Friedlingstein, P.; Andrew, R.; Canadell, J.; Le Quéré, C.; Peters, G. Persistent fossil fuel growth threatens the Paris Agreement and planetary health. *Environ. Res. Lett.* **2019**, *14*, 121001. [CrossRef]

51. Le Quéré, C.; Jackson, R.B.; Jones, M.W.; Smith, A.J.; Abernethy, S.; Andrew, R.M.; De-Gol, A.J.; Shan, Y.; Canadell, J.G.; Friedlingstein, P.; et al. Temporary reduction in daily global CO2 emissions during the COVID-19 forced confinement (Version 1.3). *Glob. Carbon Project* **2020**. [CrossRef]

52. DG of Health Malaysia. Press Statement 1 Mar 2020—Updates on the Coronavirus Disease 2019 (COVID-19): Situation in Malaysia. Available online: https://kpkesihatan.com/2020/03/01/kenyataan-akhbar-kpk-1-mac-2020-situasi-semasa-jangkitan-penyakit-coronavirus-2019-COVID-19-di-malaysia/ (accessed on 8 July 2021).

53. DG of Health Malaysia. Press Statement 16 Mar 2020—Updates on the Coronavirus Disease 2019 (COVID-19): Situation in Malaysia. Available online: https://kpkesihatan.com/2020/03/16/kenyataan-akhbar-ybmk-16-mac-2020-situasi-semasa-jangkitan-penyakit-coronavirus-2019-COVID-19-di-malaysia/ (accessed on 8 July 2021).

54. Prime Minister's Office of Malaysia. The Prime Minister's Special Message on COVID-19—16 March 2020. Available online: https://www.pmo.gov.my/2020/03/perutusan-khas-yab-perdana-menteri-mengenai-COVID-19-16-mac-2020/ (accessed on 8 July 2021).

55. Tang, K.H.D. Movement control as an effective measure against COVID-19 spread in Malaysia: An overview. *J. Public Health* **2020**, 1–4. [CrossRef]

56. DG of Health Malaysia. Press Statement 4 May 2020—Updates on the Coronavirus Disease 2019 (COVID-19): Situation in Malaysia. Available online: https://kpkesihatan.com/2020/05/04/kenyataan-akhbar-kpk-4-mei-2020-situasi-semasa-jangkitan-penyakit-coronavirus-2019-COVID-19-di-malaysia/ (accessed on 8 July 2021).

57. The Star. PM: CMCO Extended Till June 9. Available online: https://www.thestar.com.my/news/nation/2020/05/11/pm-cmco-extended-till-june-9 (accessed on 8 July 2021).

58. New Straits Times. CMCO to End, Replaced with RMCO Until Aug 31. Available online: https://www.nst.com.my/news/nation/2020/06/598700/cmco-end%C2%A0replaced-rmco-until-aug-31 (accessed on 8 July 2021).

59. New Straits Times. RMCO Extended Until Dec 31, Says PM Muhyiddin. Available online: https://www.nst.com.my/news/nation/2020/08/620307/rmco-extended-until-dec-31-says-pm-muhyiddin (accessed on 8 July 2021).

60. The Sun Daily. RMCO Extended Till March 31 (Updated). Available online: https://www.thesundaily.my/home/rmco-extended-till-march-31-updated-DJ5902252 (accessed on 8 July 2021).

61. NASA—Global Climate Change. A Degree of Concern: Why Global Temperatures Matter. Available online: https://climate.nasa.gov/news/2878/a-degree-of-concern-why-global-temperatures-matter/ (accessed on 19 April 2021).

62. United Nations. Growing at a Slower Pace, World Population Is Expected to Reach 9.7 Billion in 2050 and Could Peak at Nearly 11 Billion around 2100. Available online: https://www.un.org/development/desa/en/news/population/world-population-prospects-2019.html (accessed on 19 April 2021).

63. International Atomic Energy Agency (IAEA). *Energy, Electricity and Nuclear Power Estimates for the Period up to 2050*; IAEA: Vienna, Austria, 2020.

64. U.S. Energy Information Administration (IEA). EIA Projects Nearly 50% Increase in World Energy Usage by 2050, Led by Growth in Asia. Available online: https://www.eia.gov/todayinenergy/detail.php?id=41433 (accessed on 19 April 2021).

65. Ab Rahman, A.K.; Abdullah, R.; Balu, N.; Shariff, F.M. The impact of La Niña and El Niño events on crude palm oil prices: An econometric analysis. *Oil Palm Ind. Econ. J.* **2013**, *13*, 38–51.

66. Department of Statistics Malaysia. Current Population Estimates, Malaysia. 2020. Available online: https://www.dosm.gov.my/v1/index.php?r=column/cthemeByCat&cat=155&bul_id=OVByWjg5YkQ3MWFZRTN5bDJiaEVhZz09&menu_id=L0pheU43NWJwRWVSZklWdzQ4TlhUUT09 (accessed on 10 May 2021).

67. Kwan, M.S.; Tangang, F.T.; Juneng, L. Projected changes of future climate extremes in Malaysia. *Sains Malays.* **2013**, *42*, 1051–1059.

68. NOAA. Monthly Summaries Location Details: Malaysia. Available online: https://www.ncdc.noaa.gov/cdo-web/datasets/GHCNDMS/locations/FIPS:MY/detail#stationlist (accessed on 30 April 2021).

69. World Weather Online. World Weather: Malaysia. Available online: https://www.worldweatheronline.com/malaysia-weather.aspx (accessed on 30 April 2021).

70. Malaysia Meteorological Department. *Climate Change Scenarios for Malaysia 2001–2099*; Malaysia Meteorological Department: Petaling Jaya, Malaysia, 2009.

71. Ministry of Natural Resources and Environment Malaysia. *Malaysia: Biennial Update Report to the UNFCCC*; Ministry of Natural Resources and Environment Malaysia: Petaling Jaya, Malaysia, 2015.

72. Sammathuria, M.; Ling, L. Regional climate observation and simulation of extreme temperature and precipitation trends. In Proceedings of the 14th International Rainwater Catchment Systems Conference, Kuala Lumpur, Malaysia, 3–6 August 2009; pp. 3–6.

73. Khor, M. Lessons from the Great Floods. Available online: https://www.thestar.com.my/opinion/columnists/global-trends/2015/01/19/lessons-from-the-great-floods/ (accessed on 30 April 2021).

74. CFE-DMHA. Malaysia: Disaster Management Reference Handbook (June 2019). Available online: https://reliefweb.int/report/malaysia/malaysia-disaster-management-reference-handbook-june-2019 (accessed on 9 May 2021).

75. Tang, K.H.D. Climate change in Malaysia: Trends, contributors, impacts, mitigation and adaptations. *Sci. Total Environ.* **2019**, *650*, 1858–1871. [CrossRef] [PubMed]

76. Loh, J.L.; Tangang, F.; Juneng, L.; Hein, D.; Lee, D.-I. Projected rainfall and temperature changes over Malaysia at the end of the 21st century based on PRECIS modelling system. *Asia Pac. J. Atmos. Sci.* **2016**, *52*, 191–208. [CrossRef]

77. Chang, C.; Harr, P.A.; Chen, H.-J. Synoptic disturbances over the equatorial South China Sea and western Maritime Continent during boreal winter. *Mon. Weather Rev.* **2005**, *133*, 489–503. [CrossRef]
78. Mayowa, O.O.; Pour, S.H.; Shahid, S.; Mohsenipour, M.; Harun, S.B.; Heryansyah, A.; Ismail, T. Trends in rainfall and rainfall-related extremes in the east coast of peninsular Malaysia. *J. Earth Syst. Sci.* **2015**, *124*, 1609–1622. [CrossRef]
79. Murad, W.; Molla, R.I.; Mokhtar, M.B.; Raquib, A. Climate change and agricultural growth: An examination of the link in Malaysia. *Int. J. Clim. Chang. Strateg. Manag.* **2010**, *2*, 403–417. [CrossRef]
80. Matthews, R.B.; Kropff, M.J.; Bachelet, D.; van Laar, H. *Modeling the Impact of Climate Change on Rice Production in Asia*; CAB International and International Rice Research Institute: Wallingford, UK, 1995.
81. Shaaban, A.J.; Amin, M.; Chen, Z.; Ohara, N. Regional modeling of climate change impact on Peninsular Malaysia water resources. *J. Hydrol. Eng.* **2011**, *16*, 1040–1049. [CrossRef]
82. Shahid, S.; Pour, S.H.; Wang, X.; Shourav, S.A.; Minhans, A.; bin Ismail, T. Impacts and adaptation to climate change in Malaysian real estate. *Int. J. Clim. Chang. Strateg. Manag.* **2017**, *9*, 87–103. [CrossRef]
83. Shaffril, H.A.M.; Samah, B.A.; D'Silva, J.L. The process of social adaptation towards climate change among Malaysian fishermen. *Int. J. Clim. Chang. Strateg. Manag.* **2013**, *5*, 38–53. [CrossRef]
84. Yaakob, O.; Quah, P.C. Weather downtime and its effect on fishing operation in Peninsular Malaysia. *J. Teknol.* **2005**, *42*, 13–26. [CrossRef]
85. Yau, Y.H.; Hasbi, S. A Comprehensive Case Study of Climate Change Impacts on the Cooling Load in an Air-Conditioned Office Building in Malaysia. *Energy Procedia* **2017**, *143*, 295–300. [CrossRef]
86. Leng, L.Y.; Ahmed, O.H.; Jalloh, M.B. Brief review on climate change and tropical peatlands. *Geosci. Front.* **2019**, *10*, 373–380. [CrossRef]
87. Naish, S.; Dale, P.; Mackenzie, J.S.; McBride, J.; Mengersen, K.; Tong, S. Climate change and dengue: A critical and systematic review of quantitative modelling approaches. *BMC Infect. Dis.* **2014**, *14*, 167. [CrossRef] [PubMed]
88. Alhoot, M.A.; Tong, W.T.; Low, W.Y.; Sekaran, S.D. Climate Change and Health: The Malaysia Scenario. In *Climate Change and Human Health Scenario in South and Southeast Asia*; Akhtar, R., Ed.; Springer International Publishing: Cham, Switzerland, 2016; pp. 243–268.

 *sustainability*

Article

Solvent-Free Synthesis of MIL-101(Cr) for CO_2 Gas Adsorption: The Effect of Metal Precursor and Molar Ratio

Kok Chung Chong [1,2,*], Pui San Ho [1], Soon Onn Lai [1,2], Sze Shin Lee [1,2], Woei Jye Lau [3], Shih-Yuan Lu [4] and Boon Seng Ooi [5]

[1] Department of Chemical Engineering, Lee Kong Chian Faculty of Engineering and Science, Universiti Tunku Abdul Rahman (UTAR), Jalan Sungai Long, Kajang 43000, Selangor, Malaysia; puisan2@1utar.my (P.S.H.); laiso@utar.edu.my (S.O.L.); lsshin@utar.edu.my (S.S.L.)

[2] Centre for Photonics and Advanced Materials Research, Universiti Tunku Abdul Rahman (UTAR), Kampar 31900, Perak, Malaysia

[3] Advanced Membrane Technology Research Centre (AMTEC), School of Chemical and Energy Engineering, Universiti Teknologi Malaysia, Johor Bahru 81310, Johor, Malaysia; lwoeijye@utm.my

[4] Department of Chemical Engineering, National Tsing Hua University, Hsinchu 30013, Taiwan; sylu@mx.nthu.edu.tw

[5] School of Chemical Engineering, Engineering Campus, Universiti Sains Malaysia, Seri Ampangan, Nibong Tebal 14300, Pulau Pinang, Malaysia; chobs@usm.my

* Correspondence: chongkc@utar.edu.my

Citation: Chong, K.C.; Ho, P.S.; Lai, S.O.; Lee, S.S.; Lau, W.J.; Lu, S.-Y.; Ooi, B.S. Solvent-Free Synthesis of MIL-101(Cr) for CO_2 Gas Adsorption: The Effect of Metal Precursor and Molar Ratio. *Sustainability* **2022**, *14*, 1152. https://doi.org/10.3390/su14031152

Academic Editors: Wei-Hsin Chen, Alvin B. Culaba, Aristotle T. Ubando and Steven Lim

Received: 10 December 2021
Accepted: 9 January 2022
Published: 20 January 2022

Publisher's Note: MDPI stays neutral with regard to jurisdictional claims in published maps and institutional affiliations.

Abstract: MIL-101(Cr), a subclass of metal–organic frameworks (MOFs), is a promising adsorbent for carbon dioxide (CO_2) removal due to its large pore volume and high surface area. Solvent-free synthesis of MIL-101(Cr) was employed in this work to offer a green alternative to the current approach of synthesizing MIL-101(Cr) using a hazardous solvent. Characterization techniques including XRD, SEM, and FTIR were employed to confirm the formation of pure MIL-101(Cr) synthesized using a solvent-free method. The thermogravimetric analysis revealed that MIL-101(Cr) shows high thermal stability up to 350 °C. Among the materials synthesized, MIL-101(Cr) at the molar ratio of chromium precursor to terephthalic organic acid of 1:1 possesses the highest surface area and greatest pore volume. Its BET surface area and total pore volume are 1110 m^2/g and 0.5 cm^3/g, respectively. Correspondingly, its CO_2 adsorption capacity at room temperature is the highest (18.8 mmol/g), suggesting it is a superior adsorbent for CO_2 removal. The textural properties significantly affect the CO_2 adsorption capacity, in which large pore volume and high surface area are favorable for the adsorption mechanism.

Keywords: metal–organic framework; MIL-101; solvent free; adsorption; carbon dioxide

1. Introduction

Energy demand is rapidly increasing, and world demand is expected to rise by another 19% in the coming twenty years, as the global population is growing simultaneously [1]. Currently, industrial power usage heavily depends on non-renewable energy produced by power plants, which are a major human source of CO_2 emission into the atmosphere [2]. It was reported that the CO_2 level in the atmosphere had unprecedentedly approached 417 ppm on average in 2020 [2], and had increased by around 45% since the mid-1800s, the start of the Industrial Revolution.

Energy is now in transition from non-renewable energy to clean energy sources, such as biomass. Wind energy is thought to be a solution for the world's energy supply, and an excellent solution for cutting down the emission of CO_2 gas into the atmosphere in the future. However, realizing and applying this clean energy source in industrial usage is challenging and will require many more years to accomplish [3]. Hence, capturing the excessive CO_2 emissions from industrial activities is the immediate action that needs to be

Sustainability **2022**, *14*, 1152. https://doi.org/10.3390/su14031152   https://www.mdpi.com/journal/sustainability

taken in order to minimize the environmental impact while also developing clean energy sources.

Carbon dioxide gas accounts for 60% of the greenhouse effect by trapping heat in the atmosphere. The Earth's average surface temperature has increased 1.18 °C since the start of the Industrial Revolution, attributed to global warming [4]. By the 22nd century, global sea levels are forecast to rise 30.5 to 121.9 cm and flood low-lying areas, an event mainly caused by melting ice in Antarctica and Greenland [4]. Additionally, the rising temperature of the Earth will affect wildlife habitats, leading to extreme weather change by increasing precipitation in some areas, while some regions will experience hot weather and drought [5].

Carbon capture and sequestration (CSS) is a well-developed CO_2 elimination technology affixed to CO_2 emission sources. It comprises three stages: CO_2 is captured from flue gas, then shipped and stored underground [6]. Amine scrubbing is one of the most common CCS technologies commercialized and applied industrially. However, it suffers from several limitations, including amine chemical degradation, equipment corrosion, and high energy requirement for amine regeneration [3]. In this regard, physisorption-based CO_2 capture provides a potential replacement based on its simplicity, economic scale, and low energy requirement. Varieties of physical adsorbents, including zeolites and activated carbon, have been evaluated for CO_2 gas capture.

Compared to these traditional adsorbent materials, metal–organic frameworks (MOF) are distinguished by their extraordinarily high surface area and porosity, together with their flexibility in framework formation. They are an organic–inorganic hybrid formed from a metal precursor and an organic ligand [7]. In contrast to rigid zeolite and activated carbon materials, their adjustability and flexibility in pore structure and framework building have enabled them to replace traditional adsorbents. Moreover, their superior adsorption characteristic, rapid adsorption kinetics, and reversibility suggest they are a promising adsorbent material [8].

MIL-101(Cr) is an attractive type of MOF for serving as a CO_2 adsorbent due to its many outstanding features for adsorption, including high surface area, large pore volume, and excellent chemical and thermal strength [9,10]. There is an abundance of published studies relevant to the use of MIL-101(Cr) in CO_2 adsorption. Hong et al. [11] carried out a comparative study to determine the adsorptive properties of 13X zeolite monoliths and MIL-101(Cr) monoliths for CO_2 adsorption. The MIL-101(Cr) was fabricated using a hydrothermal reaction in this work, and the results showed that the CO_2 capture performance of MIL-101(Cr) monoliths was better than that of zeolite monoliths. Zhang et al. [12] successfully synthesized MIL-101(Cr) composites with a solvent-based hydrothermal method for CO_2 adsorption in 2019. In the same year, Yulia et al. [13] reported the fabrication of MIL-101(Cr) using a fluorine-free hydrothermal method, and achieved a CO_2 adsorption capacity of 2.28 mmol/g at a temperature of 298 K and pressure of 600 KPa.

Most of these studies adopted the hydrothermal method to synthesize the MIL-101(Cr) for CO_2 adsorption. To the best of our knowledge, none of the studies to date studying the CO_2 adsorption ability of MIL-101(Cr) used a solvent-free fabrication method. In addition, relevant scientific studies using solvent-free methods to synthesize MIL-101(Cr) are still lacking.

The most common MOF synthesis method applied in the laboratory is hydrothermal (solvothermal), but this method requires a massive amount of solvent, which is most probably toxic, during the organic linkage reaction. Furthermore, it can cause environmental issues and make the upscaling process challenging and expensive [14]. Therefore, in this study, we employed a solvent-free method for synthesizing MIL-101(Cr) to investigate its potential in CO_2 adsorption. Mechanochemical fabrication is a solvent-free MOF fabrication method that uses force to grind the mixture of starting materials in order to initiate the reaction [15]. It is the most simple, economical, and green technique for synthesizing MOF compared to the methods involving organic solvents.

In the present study, MIL-101(Cr) was fabricated using a solvent-free method in order to develop a clean and efficient way of synthesizing MOF. A total of seven MIL-101(Cr) samples with different Cr to organic linker molar ratios were prepared in order to investigate the influence of the molar ratio of starting materials on the MIL-101(Cr) MOF's CO_2 adsorption capacities.

2. Materials and Methods

2.1. Solvent-Free Preparation of MIL-101(Cr)

We purchased 99% pure chromium trinitrate nonahydrate ($Cr(NO_3)_3 \cdot 9H_2O$) and 98% pure 1,4-benzene dicarboxylate (BDC) from Sigma-Aldrich, St. Louis, MO, USA. VWR International supplied ethanol (EtOH, 99%). $Cr(NO_3)_3 \cdot 9H_2O$ and BDC were used as the metal precursor and organic linker, respectively, for the synthesis of MIL-101(Cr). The method we used solvent-free synthesis of chromium-based MIL-101 MOF is similar to the procedures reported by [16], with minor modifications. First, the weighed amounts of starting materials, $Cr(NO_3)_3 \cdot 9H_2O$ and BDC, were mixed and ground for around 30 min at room temperature. The resulting mixture was then transferred into a stainless-steel, Teflon-lined autoclave and heated at 220 °C for 4 h. After, it was washed with hot ethanol at 60 °C to eliminate any unreacted reactants. Then, the MIL-101(Cr) solid was recovered by 15 min centrifugation at 5500 rpm. Lastly, the obtained green MIL-101(Cr) powder was dried in an oven at 120 °C overnight. The color changes of MIL-101(Cr) that occurred along the solvent-free synthesis process are demonstrated in Figure S1.

2.2. Characterization Techniques

All the samples were scanned with a Shimadzu Model 6000 diffractometer (Shimadzu, Kyoto, Japan) over the 2θ angle ranging from 5° to 30° at a scan rate of 2°/min with Cu-Kα radiation to obtain their XRD patterns. The morphology and the elemental composition of the samples were obtained with Hitachi S-3400N scanning electron microscopy (SEM), coupled with energy-dispersive X-ray spectroscopy (EDX). The samples' Fourier transform infrared (FTIR) spectra were obtained using a Nicolet iS10 instrument. Each sample was scanned from wavenumbers in the region of 400–4000 cm^{-1}, with a 1 cm^{-1} interval. Thermal stability analysis of the samples was performed using a PerkinElmer STA 8000 simultaneous thermal analyzer. The samples were exposed to an increasing temperature condition from 50 to 800 °C at a heating rate of 10 °C/min, with the N_2 flow rate kept constant at 20 mL/min during the thermogravimetric analysis. The samples' surface textural properties, including the BET surface area, average pore size, and total pore volume, were obtained from N_2 adsorption isotherms at 350 °C on a Micromeritics 3Flex surface characterization analyzer. All the samples were degassed at 150 °C with N_2 gas before the analysis.

2.3. CO2 Adsorption Test Setup

Figure 1 illustrates the setup for determining the CO_2 gas adsorption capacity of the MIL-101(Cr) synthesized at room temperature using this procedure. Before each experiment, the sample was degassed at 120 °C overnight. After the degasification, a weighed sample was inserted into a porous bag and placed in a stainless-steel vessel. Subsequently, the gas valve was turned on to allow pure CO_2 to flow into the adsorption vessel and fill it up. The concentration of the purging CO_2 gas was measured by a CO_2 meter (AZ-001). After each experiment run (30 min), the sample was weighed to obtain its final weight after the adsorption. The CO_2 adsorption capacity for each sample was calculated by:

$$n_{ads} = \left(\frac{W_f - W_i}{W_i} \right) \div M_{CO2} \tag{1}$$

$$Q_{ads} = \frac{n_{ads}}{(W_i \div 1000)} \tag{2}$$

where n_{ads} is millimoles of adsorbed CO_2 (mmol)l; W_i is the weight of sample before the adsorption test (mg); W_f is the weight of sample after the adsorption test (mg); M_{CO2} is the molar mass of CO_2 (44.009 mg/mmol); Q_{ads} is the adsorption capacity (mmol/g).

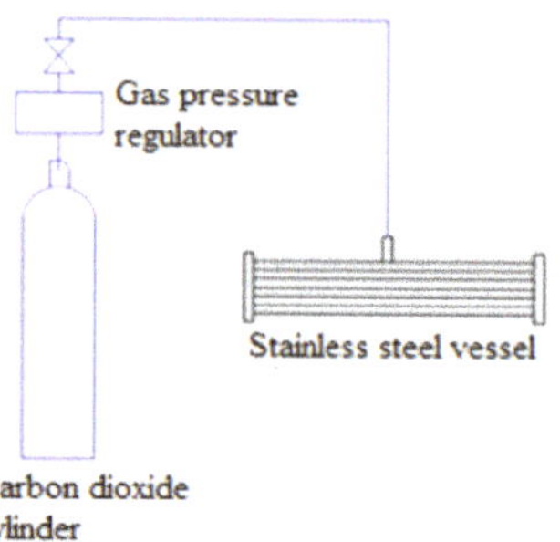

Figure 1. Setup for measuring CO_2 adsorption capacity.

3. Results and Discussion

3.1. Characterization of MIL-101(Cr)

The XRD patterns of the MIL-101(Cr) synthesized using the solvent-free method in Figure 2 are in agreement with those reported in the literature [16,17], confirming the successful formation of pure MIL-101(Cr). The diffraction peak became stronger as the Cr to BDC molar ratio increased. The breadth of the diffraction peak is related to the particle size, where a broader peak indicates a smaller particle [18]. Hence, the XRD patterns indicated that the size of MIL-101(Cr) solids decreased as the molar ratio between Cr and BDC linker increased. Therefore, the sample with a molar ratio of 2.5:1 exhibited the smallest particle size as it had a broader diffraction peak.

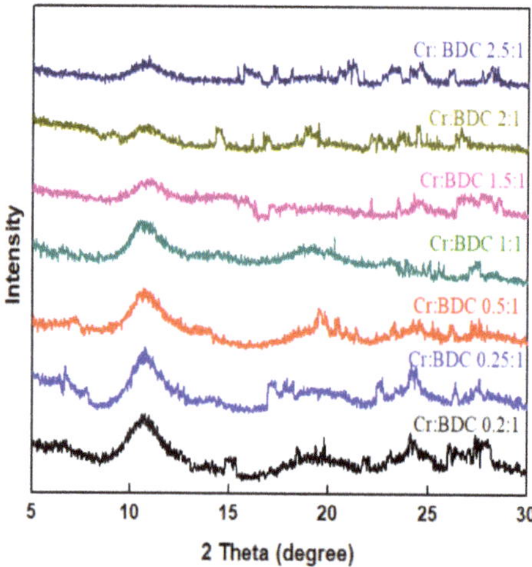

Figure 2. XRD patterns of the Cr-based MIL-101 MOF synthesized in this work.

The surface morphology of the synthesized MIL-101(Cr) is shown in Figure 3. As the molar ratio of Cr to BDC organic linker increased, one can notice the increase in the surface roughness of the MIL-101(Cr). The increased surface roughness is associated with reduced particle size. Based on the SEM images in Figure 3a–c, we observed that the particle size of MIL-101(Cr) decreased as a higher Cr to BDC molar ratio was employed. The SEM image of the sample with a Cr:BDC molar ratio of 1.5:1 (Figure 3e) shows that it consisted of small, irregularly shaped particles. When the molar ratio was decreased from 1.5:1 to 1:1, we observed fine growth of small, irregular, granular-shaped particles in an aggregation state (Figure 3d). The findings are similar to the MIL-101(Cr) morphology reported elsewhere [16,17], indicating that MIL-101(Cr) can grow well using the solvent-free method with an equal molar ratio of Cr to BDC organic linker. The surface of MIL-101(Cr) with 2:1 and 2.5:1 Cr to BDC molar ratio was found to have many regular round-shaped particles, as shown in Figure 3f,g, respectively. These round-shaped particles are thought to be the unreacted chromium particles [19] due to an excess in chromium metal reactant when a high ratio of Cr over BDC linker is utilized during the crystallization reaction. Proper manipulation of the amount of Cr and BDC organic linker used during the solvent-free reaction is vital for the growth of the desired morphology of Cr-based MIL-101 MOF.

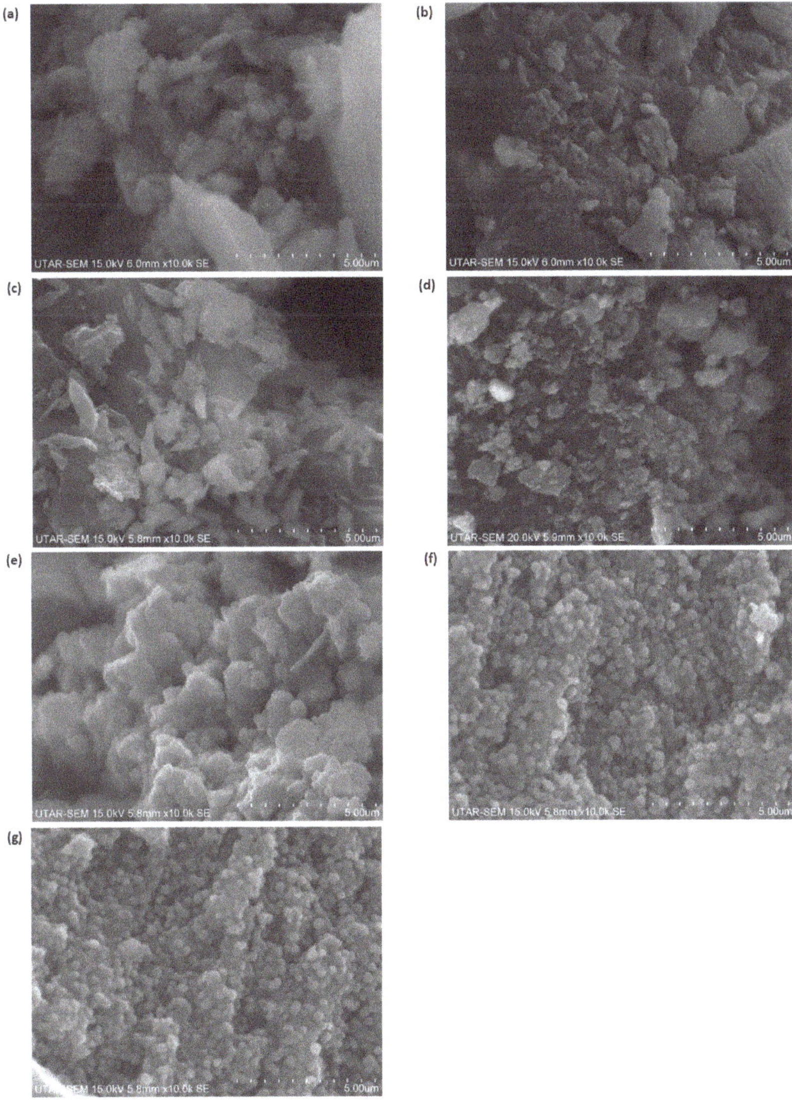

Figure 3. SEM images of MIL-101(Cr) with different Cr: BDC molar ratios: (**a**) 0.2:1, (**b**) 0.25:1, (**c**) 0.5:1, (**d**) 1:1, (**e**) 1.5:1, (**f**) 2:1, and (**g**) 2.5:1.

The elemental composition of each variant of MIL-101(Cr) synthesized in this work was determined using EDX analysis and is presented in Table 1. Cr was successfully incorporated into all samples regardless of the Cr-to-BDC molar ratio. The carbon and oxygen elements within the samples were sourced from BDC organic linkers that participated in the formation of MIL-101(Cr) [16]. A higher Cr-to-BDC molar ratio used during the organic linkage reaction is supposed to have a higher weight percentage of Cr element in the sample, but the EDX analysis showed that the sample with a Cr-to-BDC molar ratio of 0.5:1 had a lower Cr content than the sample with a Cr-to-BDC molar ratio of 0.25 to 1. This could be due to the improper substitution of Cr within the 0.5:1 molar ratio MIL-101(Cr). However, the overall EDX analysis result for the MIL-101(Cr) synthesized in this work is in line with the changes in the Cr to BDC molar ratio used.

Table 1. The weight percentage of chromium, carbon, and oxygen elements of MIL-101(Cr) synthesized in this work.

Cr to BDC Molar Ratio	Cr (wt %)	C (wt %)	O (wt %)
0.2:1	24.22	42.74	33.05
0.25:1	26.28	40.73	32.99
0.5:1	22.01	45.79	32.21
1:1	27.52	41.45	31.02
1.5:1	29.10	37.83	33.07
2:1	33.07	29.52	37.41
2.5:1	35.27	30.05	34.69

The results of the FTIR analysis conducted on the MIL-101(Cr) in Figure 4 are in accordance with the literature data [16–18], particularly in the fingerprint region between 400 and 1500 cm^{-1}. An intense absorption band around 520 cm^{-1} occurred due to the chromium-to-oxygen stretching vibration [18]. Most of the absorption bands at the region between 600 and 1600 cm^{-1} belonged to the functional groups and the benzene ring within the BDC organic linker that participated in the formation of MIL-101(Cr) [18]. The strong peak at 720 cm^{-1} was attributed to the mono-substituted benzene ring [17], while the peak at 750 cm^{-1} was the vibration of the C–H group [18]. In addition, the peaks around 1390 and 1450 cm^{-1} were due to the symmetrical vibration of the O–C–O of the benzene ring and the dicarboxylate functional group, respectively, within the BDC [20]. A peak at 1640 cm^{-1} was due to the presence of adsorbed water within the pores of the synthesized MIL-101(Cr) after exposure to air. Since the absorption band observed at 1715 cm^{-1} is was to the unreacted BDC [18], this peak was stronger in the samples that used a higher BDC content relative to the Cr.

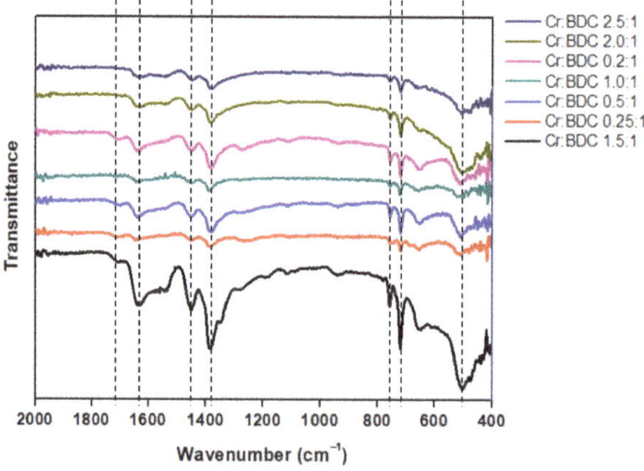

Figure 4. FTIR spectra of different MIL-101(Cr) variants synthesized in this work.

3.2. Textural Properties of MIL-101(Cr)

The N_2 adsorption and desorption isotherms at 350 °C for the Cr-based MIL-101 MOF synthesized in this work are displayed in Figure 5. It shows a Type I(b) isotherm according to IUPAC categorization [21]. The N_2 adsorption and desorption branches coincide well; no obvious hysteresis loop was observed, indicating there was no large pore structure within the sample. We observed that secondary uptake happened at a relative pressure (P/Po) of around 0.1 to 0.2 [22,23], as shown in Figure 6. This occurrence was mainly due to the existence of two kinds of mesoporous windows within the MIL-101(Cr). The two mesoporous structures were the pentagonal windows with a diameter of 2.9 nm, and the hexagonal windows with a diameter of 3.4 nm [24,25].

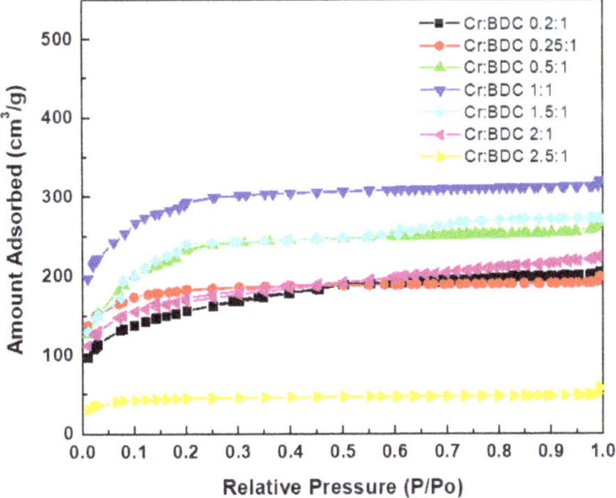

Figure 5. N_2 adsorption and desorption isotherms of chromium-based MIL-101 MOF synthesized in this work.

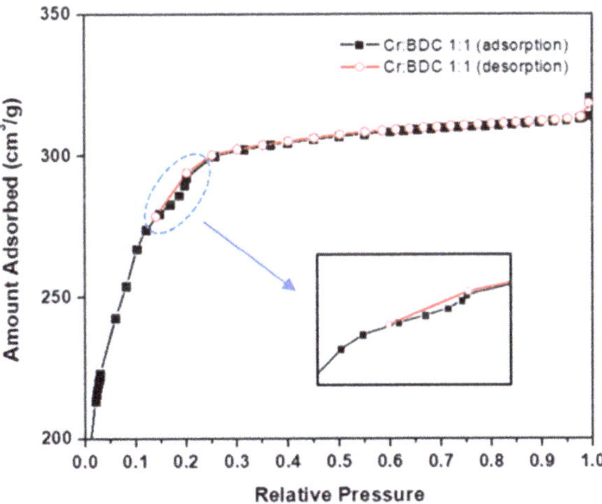

Figure 6. Nitrogen adsorption/desorption isotherm for the MIL-101(Cr) with a Cr:BDC molar ratio of 1:1 synthesized in this work. Inset: The secondary uptake portion is enlarged in the inset.

The detailed textural properties of the MIL-101(Cr) synthesized in this work are summarized in Table 2. The sample with a Cr to BDC molar ratio of 1:1 achieved the highest BET surface area of 1110.1 m^2/g. Although its average pore diameter (1.8 nm) was the smallest among the samples synthesized, its pore volume was the highest, attaining a 0.5 cm^3/g total pore volume. Its CO_2 adsorption capacity was the best compared to the other Cr o-BDC molar ratios. The lowest BET surface area and total pore volume were found in the MIL-101(Cr) with a Cr-to-BDC molar ratio of 2.5:1: only 178.6 m^2/g and 0.1 cm^3/g, respectively. Correspondingly, its adsorption capacity was the lowest compared to the other samples. High surface area and pore volume coupled with smaller pore diameter are favorable for the adsorption mechanism. Hence, the textural properties of the MIL-101(Cr) MOF synthesized in this work align with the CO_2 adsorption capacity result. The average pore diameter of MIL-101(Cr) ranged from 1.8 to 2.5 nm.

Table 2. Textural properties and CO_2 adsorption capacity of MIL-101(Cr) synthesized in this work.

Cr:BDC Molar Ratio	S_{BET} [1] (m^2/g)	Average Pore Diameter [2] (nm)	Pore Volume [3] (cm^3/g)	Q_{ads} [4] (mmol/g)
0.2:1	565.6	2.5	0.3	8.1
0.25:1	705.2	1.8	0.3	10.5
0.5:1	856.1	1.8	0.4	18
1:1	1110.1	1.8	0.5	18.8
1.5:1	728.2	2.3	0.4	18
2:1	636.8	2.2	0.3	9.4
2.5:1	178.6	2.0	0.1	1.2

[1] BET surface area obtained in a P/Po range of 0.05 to 0.15. [2] Average pore diameter calculated by the BET method. [3] Total pore volume at a P/Po of ~0.9. [4] Adsorption capacity of CO_2.

3.3. Thermogravimetric Analysis (TGA) of MIL-101(Cr)

The thermal stability test results of the synthesized MIL-101(Cr) are presented in Figure 7. As shown, all chromium-based MIL-101 MOFs experienced two weight losses during the thermal decomposition process, consistent with the reference works [7,16,26,27]. The first weight loss happened from room temperature to 200 °C, which could be ascribed to the loss of adsorbed water molecules, unreacted organic matter, and residual washing solvent [16,26,27]. Since all the samples were dried at 80 °C overnight before undergoing the TGA analysis, a small degree of weight loss occurred at the first stage. The second phase of weight loss occurred between 200 and 500 °C. At this stage, significant weight loss was observed due to the breakdown of the MIL-101(Cr) framework structure [16,26,27]. In short, the MIL-101(Cr) synthesized in this work generally exhibited high thermal stability for temperatures up to 350 °C.

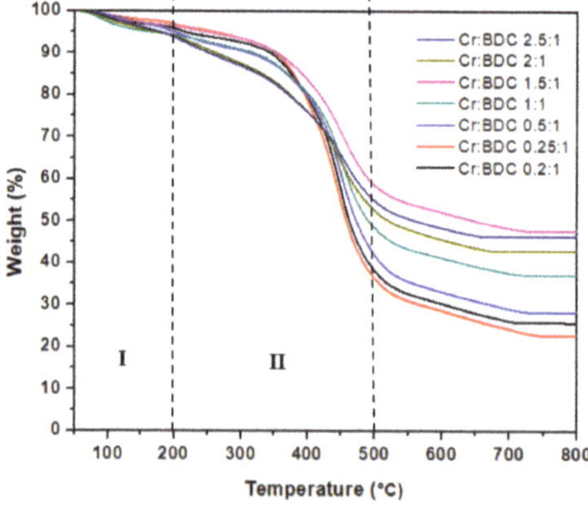

Figure 7. TGA curves of the MIL-101(Cr) synthesized in this work.

The total weight loss for the second phase of decomposition is presented in Table 3. As shown, the MIL-101(Cr) with a Cr-to-BDC molar ratio of 0.25 to 1 experienced the most severe weight loss during the second stage decomposition, up to 62.23%. In contrast, the weight loss percentage for MIL-101(Cr) with a Cr-to-BDC molar ratio of 1.5:1 was the lowest, indicating that it had the most robust structure of the samples considered.

Table 3. Total weight loss of MIL-101(Cr) during the second stage of decomposition.

Cr-to-BDC Molar Ratio	Initial Weight [1] (mg)	Final Weight [2] (mg)	Total Weight Loss (%)
0.2:1	14.88	5.92	60.22
0.25:1	14.51	5.48	62.23
0.5:1	19.79	8.75	55.79
1:1	17.96	9.20	48.78
1.5:1	23.58	14.34	39.19
2:1	17.81	9.98	43.96
2.5:1	15.92	9.34	41.33

[1] Initial weight during 2nd stage of weight loss at 200 °C. [2] Final weight during 2nd stage of weight loss at 500 °C.

3.4. CO_2 Adsorption Capacity

Table 4 shows the CO_2 gas adsorption test conducted at room temperature for the MIL-101(Cr) synthesized in this work. The CO_2 adsorption for the samples with the lowest Cr-to-BDC molar ratio (0.2:1) was found to be 8.1 mmol/g. The CO_2 adsorption capacity, however, improved as the molar ratio of metal precursor and organic ligand increased from 0.2:1 to 1:1. The MIL-101(Cr) with an equal Cr-to-BDC molar ratio was the best-performing sample, attaining 18.8 mmol/g, whereas the MIL-101(Cr) with 2.5:1 exhibited the lowest adsorption capacity of only 1.2 mmol/g.

Table 4. CO_2 gas adsorption result of the MIL-101(Cr) synthesized in this work.

Cr-to-BDC Molar Ratio	W_i [1] (mg)	W_f [2] (mg)	n_{ads} [3] (mmol)	Q_{ads} [4] (mmol/g)
0.2:1	1091.0 ± 57.8	1479.1 ± 81.4	8.8 ± 0.4	8.1 ± 0.5
0.25:1	907.5 ± 44.5	1325.2 ± 70.2	9.5 ± 0.5	10.5 ± 0.6
0.5:1	918.5 ± 52.4	1646.1 ± 79.0	16.5 ± 1.0	18.0 ± 0.9
1:1	903.9 ± 47.0	1650.7 ± 94.1	17.0 ± 1.0	18.8 ± 1.1
1.5:1	922.3 ± 52.6	1654.3 ± 97.6	16.6 ± 0.8	18.0 ± 1.0
2:1	1429.5 ± 77.2	2021.0 ± 97.0	13.4 ± 0.7	9.4 ± 0.6
2.5:1	1317.0 ± 63.2	1388.4 ± 75.0	1.6 ± 0.1	1.2 ± 0.1

[1] Initial weight before adsorption test. [2] Final weight after adsorption test. [3] Millimoles of adsorbed CO_2. [4] Adsorption capacity of CO_2.

It is well-known that the high surface area of MOFs leads to a high CO_2 adsorption capacity. Figure 8 illustrates the relationship between the textural properties of MIL-101(Cr) variants and their CO_2 adsorption capacity. It is clear that the CO_2 adsorption capacity was significantly affected by the textural properties of MIL-101(Cr). Owing to its highest surface area and pore volume, the MIL-101(Cr) with an equal Cr to BDC molar ratio demonstrated the highest CO_2 adsorption capacity amongst all the samples.

Table 5 compares the CO_2 adsorption capacity of MIL-101(Cr) synthesized in this work with the other data obtained from the literatures [1,10,22,23,28]. The CO_2 adsorption capacity of our MIL-101(Cr) is relatively higher than that reported in many studies. Although our CO_2 adsorption capacity is lower than that reported by Zhou et al. [28], our testing conditions are more economical as tests are performed at only 1 bar.

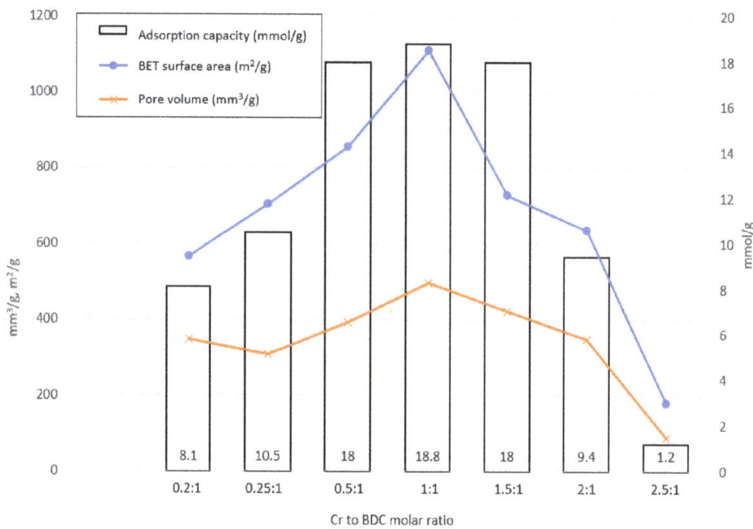

Figure 8. Relationship between CO_2 adsorption capacity, total pore volume, and BET specific surface area of the MIL-101(Cr) synthesized in this work.

Table 5. CO_2 gas adsorption capacities of MIL-101(Cr) reported in various works.

MOF Name	Temperature (K)	Pressure (bar)	Adsorption Capacity (mmol/g)	Reference
MIL-101(Cr)	298	6	2.28	[1]
MIL-101(Cr)	298	7	9.72	[10]
MIL-101(Cr)	273	1	5.77	[22]
MIL-101(Cr)	298	1	7.7	[23]
MIL-101(Cr)	298	25	22.4	[28]
MIL-101(Cr) (1:1)	298	1	18.8	This work

The reversibility of MIL-101(Cr) adsorption and its regeneration ability are key factors for evaluating it as an efficient CO_2 adsorbent. The regeneration of an adsorbent can be achieved by several methods, including temperature swing adsorption (TSA), pressure swing adsorption (PSA), vacuum swing adsorption (VSA), vacuum temperature swing adsorption (VTSA), and steam stripping [29].

The regenerative ability of MIL-101(Cr) has been previously investigated and optimized by other researchers. Five consecutive cycles of TSA adsorption–desorption were performed with MIL-101(Cr) by Ye et al. [30]. In this adsorption–desorption test, regeneration was carried out by increasing the temperature to 100 °C, with N_2 purging after the complete adsorption test. It was found that the adsorption capacity for each test cycle remained almost unchanged, indicating its adsorption was reversible, and MIL-101(Cr) could be easily regenerated. Soltanolkottabi et al. [23] also reported that MIL-101(Cr) could retain its CO_2 adsorption capacity even after more than 100 cycles of adsorption-desorption testing.

4. Conclusions

In this study, a green and solvent-free method was employed to synthesize chromium-based MIL-101 MOF for CO_2 adsorption. The molar ratio of the starting materials (Cr and BDC) and textural properties were found to play an essential role in determining the CO_2 adsorption capacity of MIL-101(Cr) at room temperature. Our results indicated that the MIL-101(Cr) made of an equal Cr-to-BDC molar ratio was the best sample and achieved an excellent CO_2 adsorption capacity of 18.8 mmol/g. The adsorption capacity is also higher compared to the values reported in the literature, indicating it is a superior adsorbent for CO_2 removal.

Supplementary Materials: The following are available online at https://www.mdpi.com/article/10.3390/su14031152/s1, Figure S1: Color changes along the synthesis of MIL-101(Cr).

Author Contributions: Conceptualization, K.C.C. and S.O.L.; methodology, S.-Y.L. and P.S.H.; validation, S.S.L., W.J.L. and B.S.O.; formal analysis, P.S.H.; investigation, P.S.H.; resources, K.C.C.; data curation, P.S.H.; writing—original draft preparation, P.S.H.; writing—review and editing, K.C.C., W.J.L. and S.O.L.; visualization, S.S.L.; supervision, K.C.C. and S.O.L.; project administration, K.C.C. and S.S.L.; funding acquisition, K.C.C. All authors have read and agreed to the published version of the manuscript.

Funding: This research was funded by Universiti Tunku Abdul Rahman under UTAR Research Fund (UTARRF), grant number IPSR/RMC/UTARRF/2020-C2/C05.

Institutional Review Board Statement: Not applicable.

Informed Consent Statement: Not applicable.

Data Availability Statement: All the data are available inside this paper and in the Supplementary Materials.

Conflicts of Interest: The authors declare no conflict of interest. The funders had no role in the design of the study; in the collection, analyses, or interpretation of data; in the writing of the manuscript; or in the decision to publish the results.

References

1. Yulia, F.; Nasruddin; Zulys, A.; Ruliandini, R. Metal-organic framework based chromium terephthalate (MIL-101 Cr) growth for carbon dioxide capture: A review. *J. Adv. Res. Fluid Mech. Therm. Sci.* **2019**, *57*, 158–174.
2. Yoro, K.O.; Daramola, M.O. CO_2 emission sources, greenhouse gases, and the global warming effect. In *Advances in Carbon Capture*; Woodhead Publishing: Pretoria, South Africa, 2020; pp. 3–28.
3. Hu, Z.; Wang, Y.; Shah, B.B.; Zhao, D. CO_2 capture in metal-organic framework adsorbents: An engineering perspective. *Adv. Sustain. Syst.* **2019**, *3*, 1800080. [CrossRef]
4. Gregory, J.M.; White, N.J.; Church, J.A.; Bierkens, M.F.P.; Box, J.E.; van den Broeke, M.R.; Cogley, J.G.; Fettweis, X.; Hanna, E.; Huybrechts, P.; et al. Twentieth-century global-mean sea level rise: Is the whole greater than the sum of the parts? *J. Clim.* **2013**, *26*, 4476–4499. [CrossRef]
5. Peterman, K.E. Climate change literacy and education: History and project overview. *ACS Symp. Series* **2017**, *1247*, 1–14.
6. Yu, J.; Xie, L.H.; Li, J.R.; Ma, Y.; Seminario, J.M.; Balbuena, P.B. CO_2 capture and separations using MOFs: Computational and experimental studies. *Chem. Rev.* **2017**, *117*, 9674–9754. [CrossRef]
7. Kooti, M.; Pourreza, A.; Rashidi, A. Preparation of MIL-101 nanoporous carbon as a new type of nanoadsorbent for HS removal from gas stream. *J. Nat. Gas Sci. Eng.* **2018**, *57*, 331–338. [CrossRef]
8. Bahri, M.; Haghighat, F.; Kazemian, H.; Rohani, S. A comparative study on metal organic frameworks for indoor environment application: Adsorption evaluation. *Chem. Eng. J.* **2017**, *313*, 711–723. [CrossRef]
9. Bhattacharjee, S.; Chen, C.; Ahn, W.S. Chromium terephthalate metal-organic framework MIL-101: Synthesis, functionalization, and applications for adsorption and catalysis. *RSC Adv.* **2014**, *4*, 52500–52525. [CrossRef]
10. Montazerolghaem, M.; Aghamiri, S.F.; Talaie, M.R.; Tangestaninejad, S. A comparative investigation of CO_2 adsorption on powder and pellet forms of MIL-101. *J. Taiwan Inst. Chem. Eng.* **2017**, *72*, 45–52. [CrossRef]
11. Hong, W.Y.; Perera, S.P.; Burrows, A.D. Comparison of MIL-101(Cr) metal-organic framework and 13X zeolite monoliths for CO_2 capture. *Micropor. Mesopor. Mat.* **2020**, *308*, 110525. [CrossRef]
12. Zhang, X.T.; Li, F.Q.; Ren, J.X.; Guan, Z.Z.; Zhang, L.J.; Feng, H.J.; Hou, X.; Ma, C. Preparation and CO_2 breakthrough adsorption of MIL-101(Cr)-D composites. *J. Nanopart. Res.* **2019**, *21*, 105. [CrossRef]
13. Yulia, F.; Utami, V.J.; Nasruddin; Zulys, A. Synthesis, characterizations, and adsorption isotherms of CO_2 on chromium terephthalate (MIL-101) metal-organic frameworks (MOF). *Int. J. Technol.* **2019**, *10*, 1427–1436. [CrossRef]
14. Bo, Q.B.; Pang, J.J.; Wang, H.Y.; Fan, C.H.; Zhang, Z.W. Hydrothermal synthesis, characterization and photoluminescent properties of the microporous metal organic frameworks with 1,3-propanediaminetetraacetate ligand and its auxiliary ligand. *Inorg. Chim. Acta* **2015**, *428*, 170–175. [CrossRef]
15. Pilloni, M.; Padella, F.; Ennas, G.; Lai, S.; Bellusci, M.; Rombi, E.; Sini, F.; Pentimalli, M.; Delitala, C.; Scano, A.; et al. Liquid-assisted mechanochemical synthesis of an iron carboxylate metal organic framework and its evaluation in diesel fuel desulfurization. *Micropor. Mesopor. Mat.* **2015**, *213*, 14–21. [CrossRef]

16. Leng, K.; Sun, Y.; Li, X.; Sun, S.; Xu, W. Rapid synthesis of metal-organic frameworks MIL-101(Cr) without the addition of solvent and hydrofluoric acid. *Cryst. Growth Des.* **2016**, *16*, 1168–1171. [CrossRef]
17. Utami, V.J.; Yulia, F.; Nasruddin; Budiyanto, M.A.; Zulys, A. CO_2/N_2 adsorption selectivity using metal-organic framework MIL-101(Cr) for marine engine exhaust model. *AIP Conf. Proc.* **2020**, *2255*, 060035.
18. Shafiei, M.; Alivand, M.S.; Rashidi, A.; Samimi, A.; Mohebbi-Kalhori, D. Synthesis and adsorption performance of a modified micro-mesoporous MIL-101(Cr) for VOCs removal at ambient conditions. *Chem. Eng. J.* **2018**, *341*, 164–174. [CrossRef]
19. Satgurunathan, T.; Bhavan, P.S.; Joy, R.D.S. Green synthesis of chromium nanoparticles and their effects on the growth of the prawn macrobrachium rosenbergii post-larvae. *Biol. Trace Elem. Res.* **2019**, *187*, 543–552. [CrossRef]
20. Huang, X.L.; Hu, Q.; Gao, L.; Hao, Q.R.; Wang, P.; Qin, D.L. Adsorption characteristics of metal-organic framework MIL-101(Cr) towards sulfamethoxazole and its persulfate oxidation regeneration. *RSC Adv.* **2018**, *8*, 27623–27630. [CrossRef]
21. Thommes, M.; Kaneko, K.; Neimark, A.V.; Olivier, J.P.; Rodriguez-Reinoso, F.; Rouquerol, J.; Sing, K.S.W. Physisorption of gases, with special reference to the evaluation of surface area and pore size distribution (IUPAC technical report). *Pure Appl. Chem.* **2015**, *87*, 1051–1069. [CrossRef]
22. Zhao, T.; Li, S.H.; Shen, L.; Wang, Y.; Yang, X.Y. The sized controlled synthesis of MIL-101(Cr) with enhanced CO_2 adsorption property. *Inorg. Chem. Commun.* **2018**, *96*, 47–51. [CrossRef]
23. Soltanolkottabi, F.; Talaie, M.R.; Aghamiri, S.; Tangestaninejad, S. Introducing a dual-step procedure comprising microwave and electrical heating stages for the morphology-controlled synthesis of chromium-benzene dicarboxylate, MIL-101(Cr), applicable for CO_2 adsorption. *J. Environ. Manag.* **2019**, *250*, 109416. [CrossRef] [PubMed]
24. Zhou, Z.; Mei, L.; Ma, C.; Xu, F.; Xiao, J.; Xia, Q.; Li, Z. A novel bimetallic MIL-101(Cr, Mg) with high CO_2 adsorption capacity and CO_2/N_2 selectivity. *Chem. Eng. Sci.* **2016**, *147*, 109–117. [CrossRef]
25. Ferey, G.; Mellot-Draznieks, C.; Serre, C.; Millange, F.; Dutour, J.; Surble, S.; Margiolaki, I. A chromium terephthalate-based solid with unusually large pore volumes and surface area. *Science* **2005**, *309*, 2040–2042. [CrossRef] [PubMed]
26. Zhao, H.; Li, Q.; Wang, Z.; Wu, T.; Zhang, M. Synthesis of MIL-101(Cr) and its water adsorption performance. *Micropor. Mesopor. Mat.* **2020**, *297*, 110044. [CrossRef]
27. Rallapalli, P.B.S.; Raj, M.C.; Senthikumar, S.; Somani, R.S.; Bajaj, H.C. HF-free synthesis of MIL-101(Cr) and its hydrogen adsorption studies. *Environ. Prog. Sustain. Energy* **2016**, *35*, 461–468. [CrossRef]
28. Zhou, X.; Huang, W.; Miao, J.; Xia, Q.; Zhang, Z.; Wang, H.; Li, Z. Enhanced separation performance of a novel composite material GrO@MIL-101 for CO_2/CH_4 binary mixture. *Chem. Eng. J.* **2014**, *266*, 339–344. [CrossRef]
29. Liu, Q.; Ning, L.; Zheng, S.; Tao, M.; Shi, Y.; He, Y. Adsorption of carbon dioxide by MIL-101(Cr): Regeneration conditions and influence of flue gas contaminants. *Sci. Rep.* **2013**, *3*, 2916. [CrossRef] [PubMed]
30. Ye, S.; Jiang, X.; Ruan, L.W.; Liu, B.; Wang, Y.M.; Zhu, J.F.; Qiu, L.G. Post-combustion CO_2 capture with the HKUST-1 and MIL-101(Cr) metal-organic frameworks: Adsorption, separation and regeneration investigations. *Micropor. Mesopor. Mat.* **2013**, *179*, 191–197. [CrossRef]

sustainability

Article

Thermal-Energy Performance of Bulk Insulation Coupled with High-Albedo Roof Tiles in Urban Pitched Residential Roof Assemblies in the Hot, Humid Climate

Mohammed Dahim [1], Syed Ahmad Farhan [2,*], Nasir Shafiq [2], Hashem Al-Mattarneh [3] and Rabah Ismail [4]

[1] Vice Presidency for Project Management, King Khalid University, Abha 61421, Saudi Arabia; madahim@kku.edu.sa

[2] Institute of Self-Sustainable Building for Smart Living, Universiti Teknologi PETRONAS, Seri Iskandar 32610, Malaysia; nasirshafiq@utp.edu.my

[3] Department of Civil Engineering, Yarmouk University, Irbid 21163, Jordan; hashem.mattarneh@yu.edu.jo

[4] Department of Civil Engineering, Jadara University, Irbid 21110, Jordan; r.ismail@jadara.edu.jo

* Correspondence: syed.af_g02626@utp.edu.my

Citation: Dahim, M.; Farhan, S.A.; Shafiq, N.; Al-Mattarneh, H.; Ismail, R. Thermal-Energy Performance of Bulk Insulation Coupled with High-Albedo Roof Tiles in Urban Pitched Residential Roof Assemblies in the Hot, Humid Climate. *Sustainability* **2022**, *14*, 2867. https://doi.org/10.3390/su14052867

Academic Editor: Wei-Hsin Chen

Received: 14 January 2022
Accepted: 26 February 2022
Published: 1 March 2022

Publisher's Note: MDPI stays neutral with regard to jurisdictional claims in published maps and institutional affiliations.

Abstract: The high rate of heat transfer through the residential roof assembly aggravates the condition of indoor thermal discomfort. Bulk insulation can be installed in the assembly to improve thermal performance. However, although it can efficiently reduce diurnal heat transfer from the outdoor environment into the indoor space through the roof assembly, it can also suppress nocturnal heat transfer in the opposite direction. Alternatively, high-albedo roof tiles employ cool colors to reflect heat at the roof surface, whereas bulk insulation hinders the conduction of heat through the roof assembly. In light of the potential of high-albedo roof tiles and bulk insulation in reducing heat transfer, thermal-energy performance of an urban pitched residential roof assembly, which adopted varying configurations of high-albedo roof tiles and bulk insulation under a hot, humid climate, was evaluated. Energy savings were generated, which were 15.13% when the change from a conventional to a high-albedo roof surface was performed, and 17.00% when the installation of bulk insulation was performed on the high-albedo roof assembly.

Keywords: air conditioner; cooling load; heat conduction; residential building; roof insulation; roof tile color; solar reflectance

1. Introduction

Buildings in countries that have hot, humid climates are exposed to intense solar radiation during the day, owing to the high altitude of the sun path [1,2]. In particular, the roof receives the highest amount of solar radiation in comparison to other components of the building envelope by virtue of the horizontal orientation and higher elevation of the roof [3,4].

Malaysia, which is located in Southeast Asia from 1° to 7° north of the equator [1], has a hot, humid climate throughout the year [1,5]. Its climate can be classified as a tropical rainforest climate, as per the Köppen–Geiger climate classification [6]. According to the annual moving averages reported in Tang [5] for selected urban areas, namely Kota Kinabalu, Kuantan, Kuching, Malacca and Subang Jaya, mean daily temperatures of Malaysia, from 1956 to 2016, ranged between 25.0 °C and 28.7 °C. Recently, the Malaysian Meteorological Department revealed that 38.6 °C was the highest peak daily temperature in Malaysia in 2020, which was recorded in Alor Setar [7]. Previously in 1998, a higher peak daily temperature of 40.1 °C was recorded in Chuping [7]. The average duration of exposure to sunshine throughout Malaysia ranged from six to eight hours per day [8,9].

Typical urban residential buildings in Malaysia are predominantly low-rise with pitched roof assemblies, where the heat transfer through the roof accounts for between 50% and 70% of the total heat gain in the indoor space beneath the roof [10]. The high rate of heat transfer through the residential roof assembly aggravates the condition of indoor thermal discomfort experienced by the occupants. Accordingly, dependence toward air conditioners increases, which is a huge concern, as air conditioners heavily consume energy [8,9].

Bulk insulation restricts the transfer of heat via conduction and convection by trapping air in millions of pockets within bulky materials that possess low density. Common bulk insulation materials employed for various applications, include, among others, cellulose, glass wool, mineral wool, polyester, polyisocyanurate, polystyrene and polyurethane. Bulk insulation products can be manufactured in various forms, which are, but not limited to, batts, loose-fills, rigid boards and rolls. Bulk insulation can be installed in the residential roof assembly to improve thermal performance owing to the presence of miniature air spaces that hinder heat conduction [11], which can potentially reduce the intensity and duration of the operation of air conditioners by the occupants [12]. Innovations pertaining to bulk insulation materials have been proposed by, among others, Husna et al. [13] and Ismail et al. [14], who adopted nano-materials that possess ultra-low thermal conductivity, as well as Nuruddin et al. [15], Farhan et al. [16] and Omar et al. [17], who adopted natural fibers, which are greener than synthetic fibers. Although bulk insulation has great potential in improving the thermal-energy performance of the roof assembly, its rate of adoption in Malaysia is still low [18]. Increases in cost related to the purchase, installation and maintenance of the insulation material, as well as lack of awareness and understanding of the long-term benefits of employing insulation, influence the decisions opted by homeowners [18]. Consequently, the omission of bulk insulation from the roof assembly may result in an increase in the rate of heat transfer through the roof assembly and into the indoor space. Hence, thermal-energy performance of the building during hours of high exposure to intense solar radiation will be negatively impacted.

Residential roof assemblies in Malaysia are typically lightweight and pitched. They comprise roof tiles, attic spaces and ceiling boards. For lightweight roofs, Malaysian Standard: Energy Efficiency and Use of Renewable Energy for Residential Buildings (MS 2680:2017) [19] recommends the installation of insulation within the roof assembly. Furthermore, a minimum thermal resistance (R-value) of 2.50 m^{2}K/W has been set as a mandatory compliance criterion for lightweight roofs as stated in MS 2680:2017 [19], Green Building Index Assessment Criteria for Residential New Construction [20], and Selangor Uniform Building By-Laws [21].

Previous studies on building insulation paid more attention to wall insulation in cold climates [22], and less emphasis was given to roof insulation in hot climates. In Malaysia, previous research pertaining to thermal performance evaluation of insulation materials installed in residential roof assemblies is limited to the studies of Farhan et al. [16], Halim et al. [23], Irwan et al. [24,25], Ismail et al. [14], Morris et al. [26], Nuruddin et al. [15,27], Puad et al. [28] and Zakaria et al. [29]. Findings indicated that installing roof insulation efficiently reduced diurnal heat transfer from the outdoor environment into the indoor space through the roof assembly. Conversely, findings also revealed that the presence of insulation suppressed nocturnal heat transfer through the roof assembly, which is in the opposite direction to that of the diurnal heat transfer. Consequently, the nocturnal energy consumption owing to the use of air conditioners will increase in view of the fact that indoor thermal comfort has to be sustained throughout the night in order to facilitate adequate rest and sleep among the occupants.

Alternatively, Al-Obaidi et al. [4], Al Yacouby et al. [30] and Farhan et al. [31] studied the effect of high-albedo roofs without insulation under the climate of Malaysia. High-albedo roof tiles reflect heat at the roof surface, whereas bulk insulation hinders the conduction of heat through the roof assembly. Adoption of high-albedo roofs has been reported in Synnefa et al. [32] to be effective at increasing thermal-energy performance for widely differing climate classes. Prevalently, previous studies have attempted to increase the albedo of roof tiles by applying high-albedo coatings. The coatings can be classified according to their binders, such as cementitious or elastomeric coatings. Alternatively, the coatings can also be categorized according to their carriers, such as solvent- or water-based coatings [33]. Essentially, for application on high-albedo roofs, the coatings are required to possess superficial thermal-optical properties that are appropriate for maintaining, under exposure to solar radiation, surface temperatures that are appreciably lower than those of conventional roofs. In general, high-albedo coatings that possess pre-eminent thermal-optical properties are those that are white in color. However, as aesthetics of buildings cannot be disregarded, studies have been conducted to develop innovative coatings, such as those that possess solar-reflective surfaces with non-white colors, those that are thermochromic, or those that are doped with phase-change materials [33].

Despite the potential for improving the thermal-energy performance, adoption of high-albedo roofs in countries that are exposed to the tropical rainforest climate is still low [9,30]. In particular, within the region of Southeast Asia, research on the effect of high-albedo roofs is currently deficient. Exclusive of the studies that were conducted in Malaysia, which are Al-Obaidi et al. [4], Al Yacouby et al. [30] and Farhan et al. [31], the research is limited to the studies of Syuhada and Maulana [34] in Indonesia, Zingre et al. [35] in Singapore and Thongkanluang et al. [36] in Thailand.

Although Al-Obaidi et al. [4] and Al Yacouby et al. [30] studied the effect of high-albedo roofs by varying the color of the roof surface, their methodologies employed test cells that did not comply with the clauses in Uniform Building By-Laws 1984 (UBBL 1984) [37] for habitable rooms of residential buildings in Malaysia. The methodology adopted in Farhan et al. [31] later addressed the shortcomings of the test cells employed in Al-Obaidi et al. [4] and Al Yacouby et al. [30] but focused solely on the effect of high-albedo roofs without considering its coupling with insulation. The scope of Syuhada and Maulana [34] zoomed in on zinc roofs and excluded the adoption of roof tiles. Zingre et al. [35] adopted a methodology that concentrated on flat roofs and did not consider pitched roof assemblies that have attic spaces and ceiling boards. Thongkanluang et al. [36] focused on synthesizing a coating material for potential application on the surface of high-albedo roofs, without performing any study on heat transfer through the roof assembly that endures exposure to solar radiation.

Hence, new studies are required to address the shortcomings of previous research on the effect of high-albedo roofs, in particular, those that zoom in on pitched roof assemblies that have roof tiles, attic spaces and ceiling boards, together with the adoption of insulation, under a hot, humid climate. In light of the potential of high-albedo roof tiles and bulk insulation in reducing heat transfer, thermal-energy performance of an urban pitched residential roof assembly, which adopted varying configurations of high-albedo roof tiles and bulk insulation under a hot, humid climate, was evaluated.

2. Materials and Methods

The thermal-energy performance of an urban pitched residential roof assembly was evaluated by developing a building information model using Integrated Environmental Solutions <Virtual Environment> (IESVE), which is a building information modeling (BIM) tool. Thermal-energy and computational fluid dynamics (CFD) analyses were performed on the model. Varying configurations of high-albedo roof tiles and bulk insulation within the roof assembly were adopted. The roof was exposed to the hot, humid climate of Shah Alam in Malaysia, which is an urban area.

The second law of thermodynamics states that the total entropy, which is a measure of the disorder of a system and its environment, will never decrease. Therefore, heat transfer through the residential roof assembly will occur from the hotter to the colder bodies, as the building and its environment attempt to gain entropy over time and reach its maximum, which is when thermal equilibrium is achieved [38]. Accordingly, as outdoor and sky conditions change over time throughout the day, magnitude and direction of heat transfer through the roof assembly will continually change in conformity with the second law of thermodynamics.

In the present study, evaluation of thermal-energy performance considered the conduction, convection and radiation modes of heat transfer through the roof assembly. Thermal properties of materials that constitute the assembly were also taken into account. The rate of heat transfer by conduction ($Q_{conduction}$), convection ($Q_{convection}$) and radiation ($Q_{radiation}$) can be expressed by Fourier's Law as per Equation (1) [39], Newton's Law of Cooling as per Equation (2) [40] and Stefan–Boltzmann's Law as per Equations (3) and (4) [41].

$$Q_{conduction} = kA\frac{dT}{dx} \tag{1}$$

where k is the thermal conductivity of the material expressed in W/mK, A is the cross-sectional area perpendicular to heat flow expressed in m^2, and $\frac{dT}{dx}$ is the temperature gradient expressed in K/m.

$$Q_{convection} = h_c A\left(T_s - T_f\right) \tag{2}$$

where h_c is the surface heat transfer coefficient, A is the surface area, T_s is the surface temperature, and T_f is the fluid temperature.

$$Q_{radiation} = \sigma A_1 \varepsilon_1 \left(T_1^{\,4} - T_2^{\,4}\right) \tag{3}$$

$$Q_{radiation} = h_r A_1 (T_1 - T_2) \tag{4}$$

where σ is the Stefan–Boltzmann constant, h_r is the coefficient of heat transfer, A_1 is the area of the first surface, ε_1 is the emissivity of the first surface, T_1 is the absolute temperature of the first surface, and T_2 is the absolute temperature of the second surface.

The ability of the roof surface to reject solar heat, as indicated by the solar reflectance index (SRI), was also taken into consideration. SRI refers to the relative steady-state temperature of a surface with respect to a standard white, which is given the SRI value of 100, and a standard black, which is given the SRI value of 0, under standard solar and ambient conditions. It is calculated as per Equation (5) [42]. As its definition and method of calculation are based on the steady-state temperatures of a standard black, which has a reflectance of 0.05 and an emittance of 0.90, and a standard white, which has a reflectance of 0.80 and an emittance of 0.90, it is possible for SRI values to be slightly negative or exceed 100.

$$SRI = \frac{\left(T_{black} - T_{surface}\right)}{\left(T_{black} - T_{white}\right)} \times 100 \tag{5}$$

where T_{black}, T_{white} and $T_{surface}$ are steady-state temperatures of the standard black, standard white and material surface, respectively, which are derived from measured values of solar reflectance and infrared emittance of the material surface according to the calculations in the Standard Practice for Calculating Solar Reflectance Index of Horizontal and Low-Sloped Opaque Surfaces (ASTM E1980-11) [43].

Adoption of high-albedo roof tiles and bulk insulation within the roof assembly were aimed toward developing an energy-efficient roof assembly. Monitoring of the energy-efficiency considered the cooling load and energy savings, which signify the level of indoor thermal comfort. A conceptual framework of the present study is outlined in Figure 1.

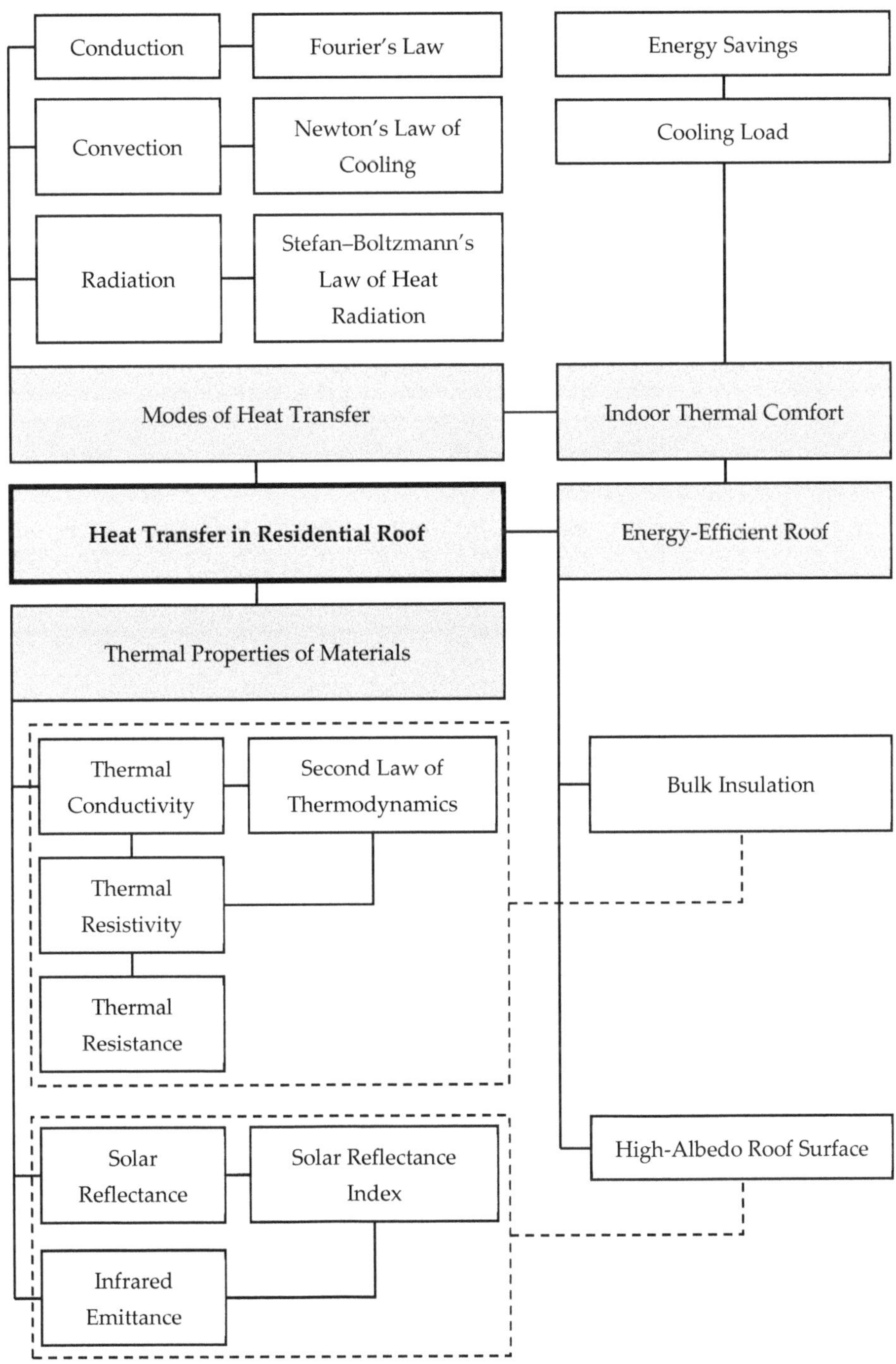

Figure 1. Conceptual framework of the present study.

BIM was selected as the methodology, as it is capable of assisting the decision-making process when it comes to sustainable design of buildings. BIM has been employed in previous studies to perform sustainable design of residential buildings that are exposed to the hot, humid climate of Malaysia. Amir et al. [44], Gardezi et al. [45] and Jamaludin et al. [46] developed building information models of pre-determined types of residential buildings. Alternatively, building information models developed in Farhan et al. [31], Halim et al. [23], Irwan et al. [24,25] and Morris et al. [26] were those of test cells that represent the conditions of a habitable space in typical urban residential buildings in Malaysia. BIM simulation data that were collected from the building information models can be validated by conducting field measurements and comparing the measured data with their counterparts from the BIM simulation based on Equation (6) as in Vangimalla et al. [47].

$$PD_{SD-FM} = \frac{SD - FM}{FM} \times 100\% \tag{6}$$

where SD and FM are the simulation and field measurement data, respectively, and PD_{SD-FM} is the percentage difference between the simulation and field measurement data.

Acceptable PD_{SD-FM} values adopted in Vangimalla et al. [47] and Leng et al. [48] are 15% and 20%, respectively, which were determined based on the 10% to 20% acceptable range recommended in Maamari et al. [49].

In the present study, two test cells, as shown in Figure 2, were constructed at the site location of 3.07° N, 101.50° E in Shah Alam, as shown in Figure 3. The test cells are identical, barring the roof tile color, where red and white roof tile colors were adopted, as shown in Figures 4 and 5, to represent conventional and high-albedo roofs, respectively. Thermocouples and data loggers were installed in the test cells to collect air and surface temperature data throughout the whole year of 2021.

Inspections were conducted prior to commencement of data collection, as well as once per week after the commencement, to ascertain the accuracy of data throughout the data collection. The inspections were conducted by comparing air and surface temperature data that were recorded using primary data loggers, which were mounted to the test cells, with those that were recorded using secondary data loggers of the same model that were minimally used. For the surface temperature data, their accuracy was further ascertained by performing supplementary comparisons between the temperatures that were measured using thermocouples, which were then recorded by the data loggers, with those that were manually measured using a thermal imaging camera.

Figure 2. Two test cells.

Figure 3. Plan view of the site location of the test cells.

Figure 4. Red (conventional) roof of one of the test cells.

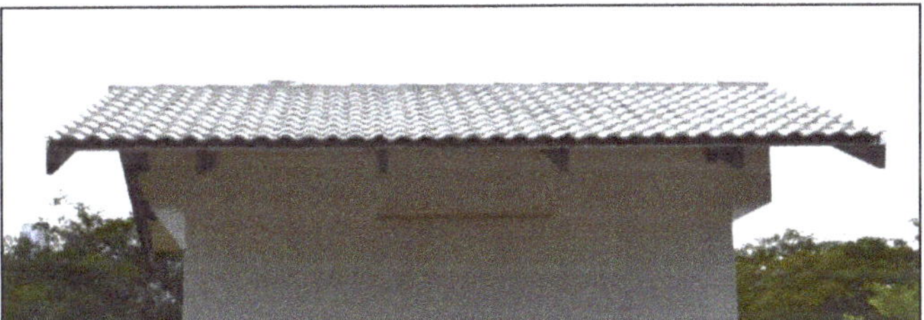

Figure 5. White (high-albedo) roof of one of the test cells.

The test cells were 4 m long, 4 m wide and 3 m high. The dimensions were selected as such to fulfil minimum size requirements as specified in UBBL 1984 [37], while minimizing the size of the test cells for feasibility of the experiment. The minimum base area, height and width of a habitable room in residential buildings were 11 m^2, 2.5 m and 2 m, respectively.

Conventional materials were employed to construct the test cells, as itemized in Table 1, inclusive of the density, thermal conductivity (k-value), specific heat and thickness of each material, with the aim of creating the conditions of a habitable space in typical urban residential buildings in Malaysia.

Table 1. Materials employed to construct the test cells.

Component	Material	Density (kg/m^3)	Specific Heat (J/kgK)	k-Value (W/mK)	Thickness (mm)
Roof	Cement Tile	1890	1000	0.836	10.0
Ceiling	Cement Board	720	1000	0.250	4.5
Window	Clear Float Glass	2800	800	0.810	6.0
Door	Solid Timber	702	2720	0.138	38.0
	Cement Plaster	1690	840	0.533	18.0
Wall	Clay Brick	1800	800	1.154	114.0
	Cement Plaster	1690	840	0.533	18.0
Floor	Reinforced Concrete	2400	1000	1.442	50.0

The test cell was modeled in IESVE as a building information model. An axonometric projection of the model is shown in Figure 6. Two-dimensional plan, front and rear, and left and right views of the model are shown in Figure 7.

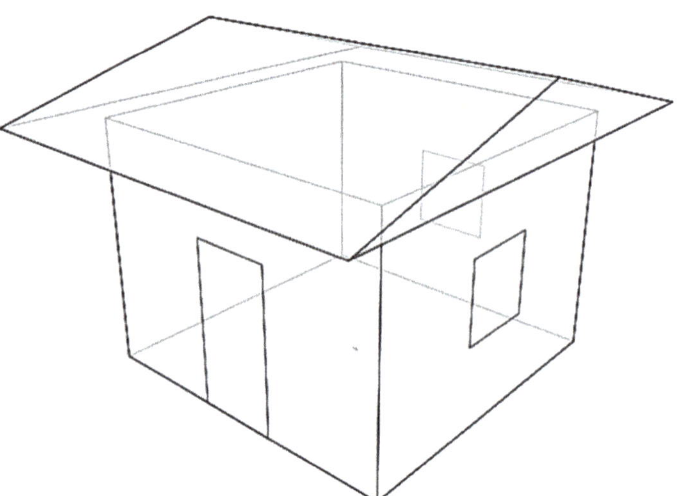

Figure 6. Axonometric projection of the building information model of the test cell.

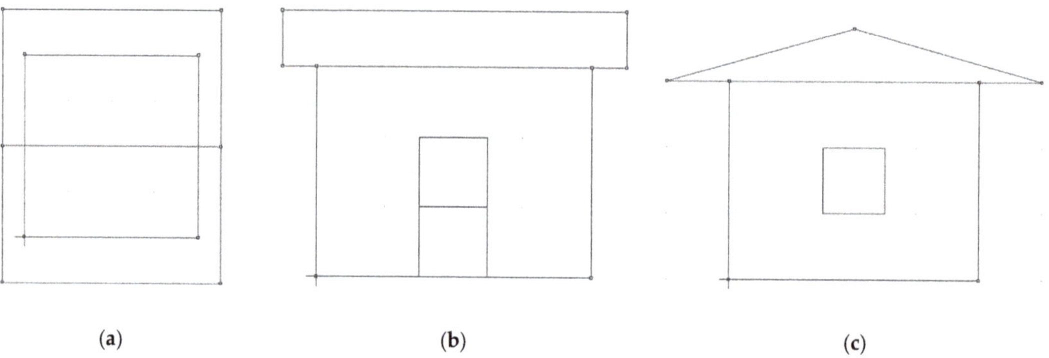

(a)   (b)   (c)

Figure 7. Two-dimensional views of the building information model of the test cell: (**a**) plan; (**b**) front and rear; (**c**) left and right.

Thermal-energy performance of the roof assembly of the building information model was evaluated by performing thermal-energy and CFD analyses using Apache and MicroFlo, respectively, in IESVE. SRI values of the roof surfaces were calculated from solar reflectance and infrared emittance values of roof surfaces of roof tile samples, which were obtained from laboratory measurements. Thermal-energy analysis was performed for a whole typical meteorological year (TMY) for the site location of the test cells at 3.07° N, 101.50° E in Shah Alam, Malaysia. Meteorological data, inclusive of solar irradiance, were generated by Meteonorm, based on data obtained from weather stations, geostationary satellites and globally calibrated aerosol climatology, as well as sophisticated interpolation models [50]. Roof-surface and attic-air temperature data generated from the thermal-energy analysis throughout the TMY were averaged to obtain annual-averaged 24-h profiles. The CFD analysis generated indoor temperature contours at peak diurnal outdoor temperature and the trough of nocturnal outdoor temperature. Configuration of the roof assembly was varied according to the roof tile color and presence of insulation to create three building information models as presented in Table 2. Then, 100-mm thick mineral wool was employed within the roof assembly as bulk insulation, as it is commonly used for building insulation in Malaysia.

Table 2. Roof assembly configurations of building information models.

Building Information Model	Roof Tile Color	Bulk Insulation
Conventional	Red	Nil
High-Albedo	White	Nil
High-Albedo + Bulk Insulation	White	Mineral Wool

The evaluation of thermal-energy performance considered the operation of a unit of a 950-W air conditioner for cooling of the indoor space with a set-point temperature of 24 °C as recommended in Malaysian Standard: Energy Efficiency and Use of Renewable Energy for Non-Residential Buildings (MS 1525:2019) [51] and also adopted in Halim et al. [23] and Irwan et al. [24,25]. Daily, weekly and monthly indoor cooling profiles were configured based on the profiles adopted in Halim et al. [23], Irwan et al. [24,25], Tang and Chin [52] and Zakaria et al. [29], which also focused on residential buildings in Malaysia. Simulation settings were configured with the assumption that no occupants and furniture are present in the indoor space, and the door and windows are closed throughout the year.

The methodology of the study is elucidated in Figures 8 and 9.

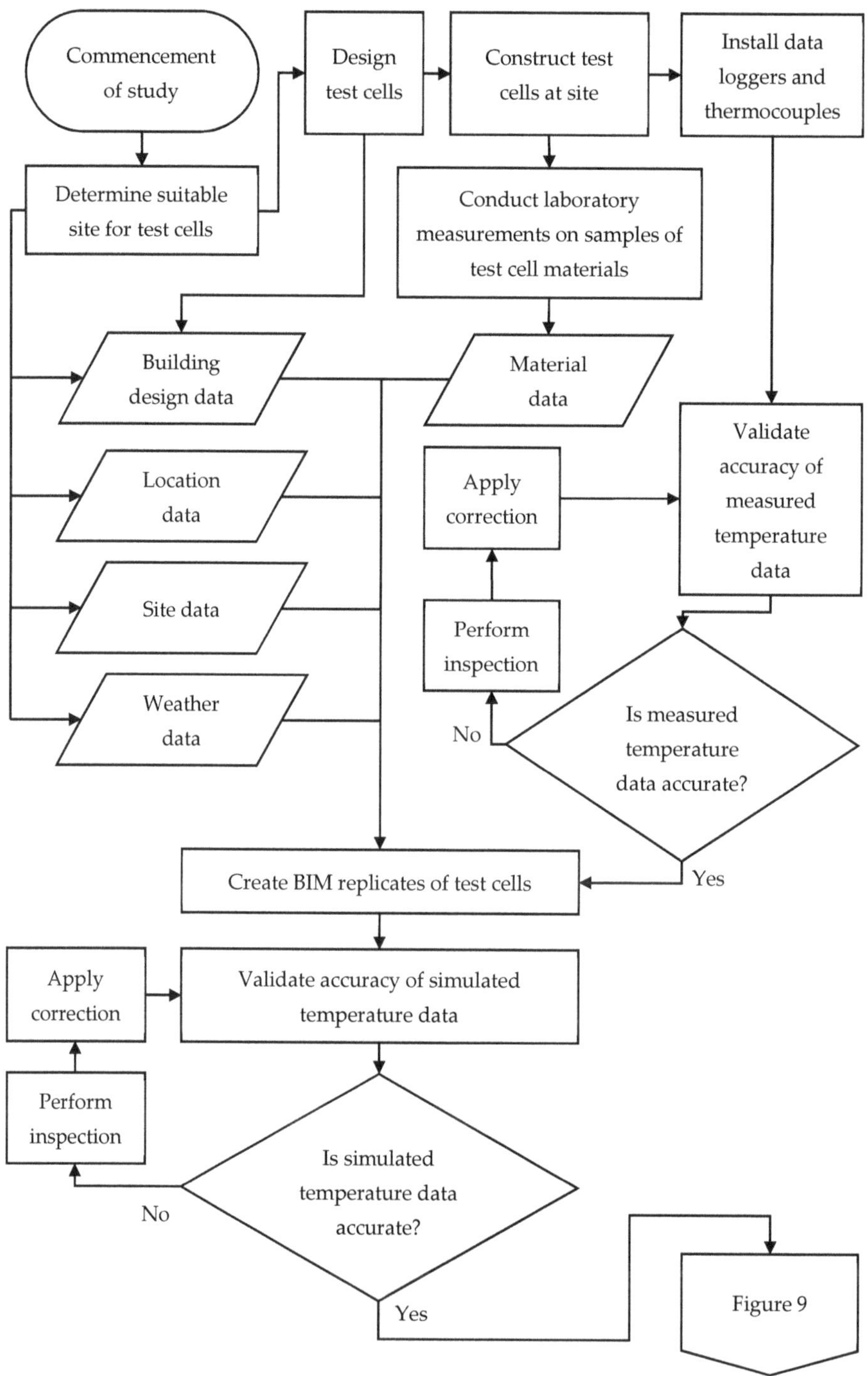

Figure 8. Methodology of the present study.

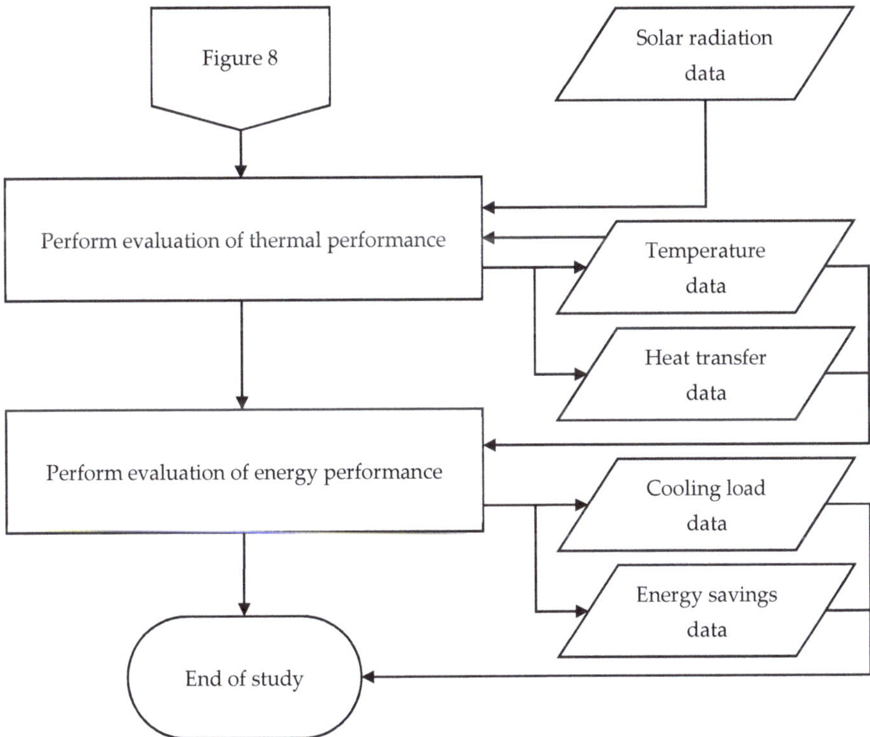

Figure 9. Methodology of the present study (continued).

3. Results and Discussion

The site location of the test cells at 3.07° N, 101.50° E in Shah Alam, Malaysia is within the tropical rainforest region as per the Köppen–Geiger climate classification [6]. Consequently, the test cells are exposed to a hot, humid climate throughout the year [1,5]. Minimum, mean and maximum annual-averaged profiles of solar irradiance and outdoor air temperature throughout the TMY of the location are presented in Figure 10. For the most part, solar irradiance throughout diurnal periods is relatively high, particularly in the afternoon, owing to the high altitude of the sun path as mentioned in Alam et al. [1] and Al-Obaidi et al. [2]. Minimum, mean and maximum solar irradiance profiles peaked at 73.36, 587.52 and 1070.25 W/m^2, respectively. The solar irradiance culminated at 13:30, which is about halfway through the diurnal period. Inversely, there is zero solar irradiance throughout the nocturnal period from 20:30 to 5:30. Accordingly, as outdoor air temperature is directly impacted by solar irradiance, the trend of the outdoor air temperature profiles trailed those of solar irradiance. Minimum, mean and maximum profiles of the outdoor air temperatures peaked at 26.9, 31.23 and 36.00 °C, respectively, with the temperatures culminating from 15:30 to 16:30.

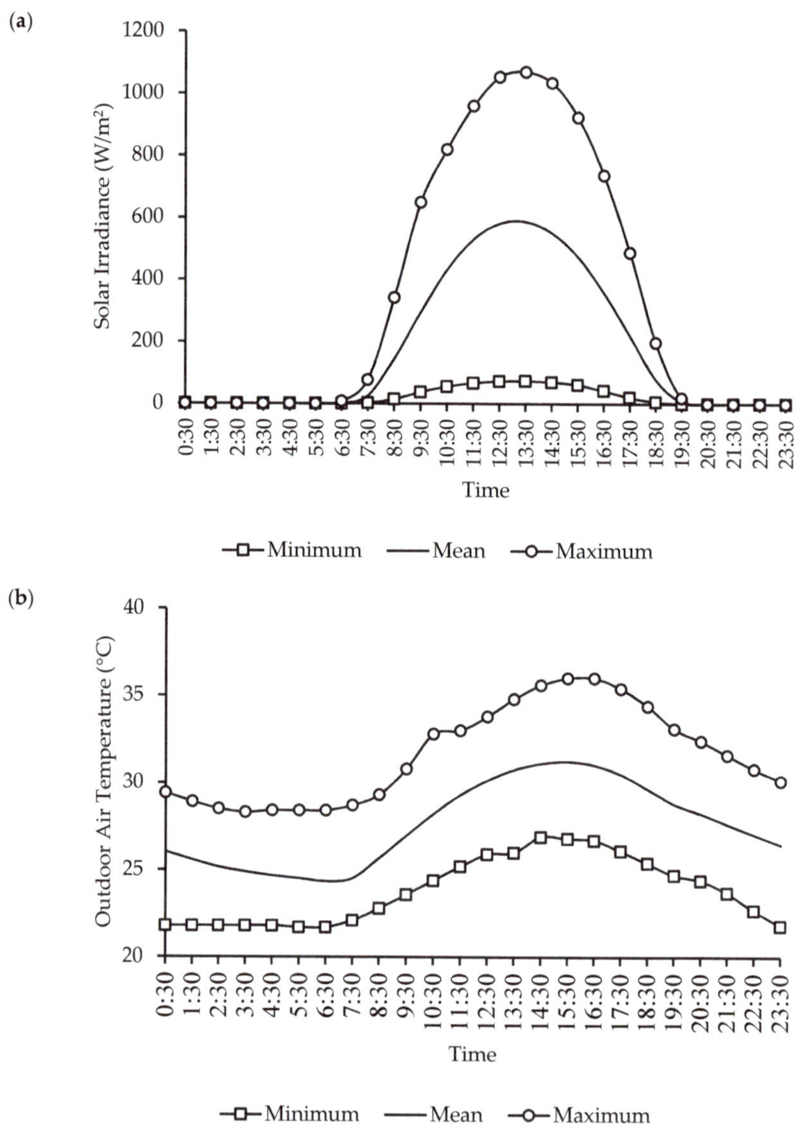

Figure 10. Annual-averaged profiles of (**a**) solar irradiance and (**b**) outdoor air temperature throughout the typical meteorological year of the site location.

Solar reflectance and infrared emittance of the roof surfaces are, as measured in the laboratory, presented in Figure 11. SRI values, as shown in Figure 12, were calculated from the solar reflectance and infrared emittance values, as per Santamouris et al. [42] and ASTM E1980-11 [43]. The change from a conventional to a high-albedo roof surface has led to an increase in the solar reflectance from 0.20 to 0.73, no change in the infrared emittance at 0.90, and an increase in the SRI from 19 to 90. Application of white paint that brought about the high-albedo roof surface can significantly reduce heat transfer through the roof assembly, as, according to Raeissi and Taheri [53], the wavelength of light is reflected by the white pigment at the roof surface and, as a consequence, less solar radiation is absorbed.

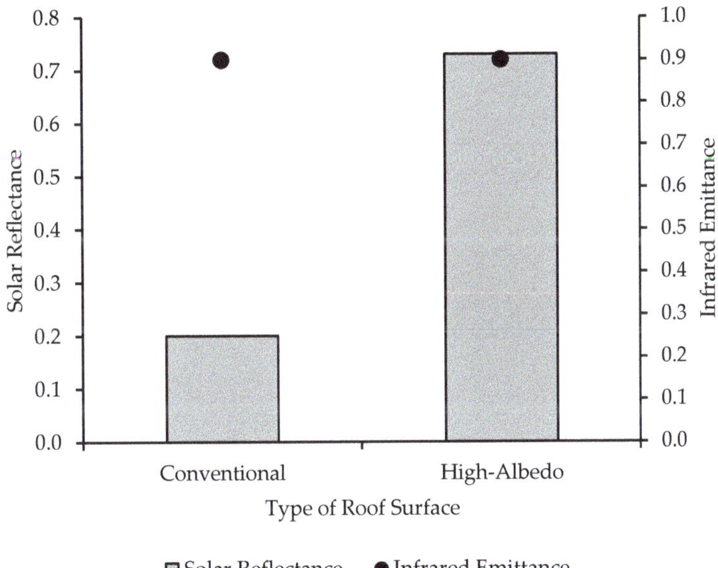

Figure 11. Solar reflectance and infrared emittance of the conventional and high-albedo roof surfaces.

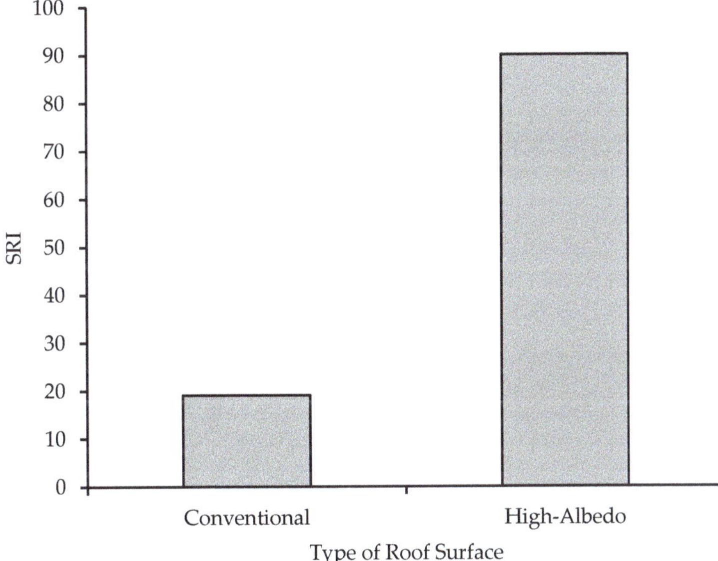

Figure 12. Solar reflectance index (SRI) of the conventional and high-albedo roof surfaces.

Roof surface temperature (T_{RS}) profiles of the non-insulated conventional and high-albedo roof assemblies are compared in Figure 13. As T_{RS} is heavily influenced by solar irradiance, waveform of the T_{RS} profiles in Figure 13 bears resemblances to that of the solar irradiance profiles in Figure 10. The change from a conventional to a high-albedo roof surface has led to the reduction in T_{RS} throughout the diurnal segment, where T_{RS} culminated at 50.50 and 35.84 °C for the conventional and high-albedo roof surfaces, respectively. The strong peak reduction of -14.79 °C in T_{RS}, as illustrated in Figure 14, which presents the change in T_{RS} (ΔT_{RS}), transpired owing to the relatively higher SRI of the high-albedo roof surface of 90, in comparison to that of the conventional roof surface of 19. The higher SRI resulted in a higher rate of reflection and accordingly a lower rate of absorption of solar radiation that is incident on the roof surface. As opposed to that of the diurnal segment, the increase in SRI did not influence T_{RS} throughout the nocturnal segment due to the absence of solar radiation throughout the nocturnal period.

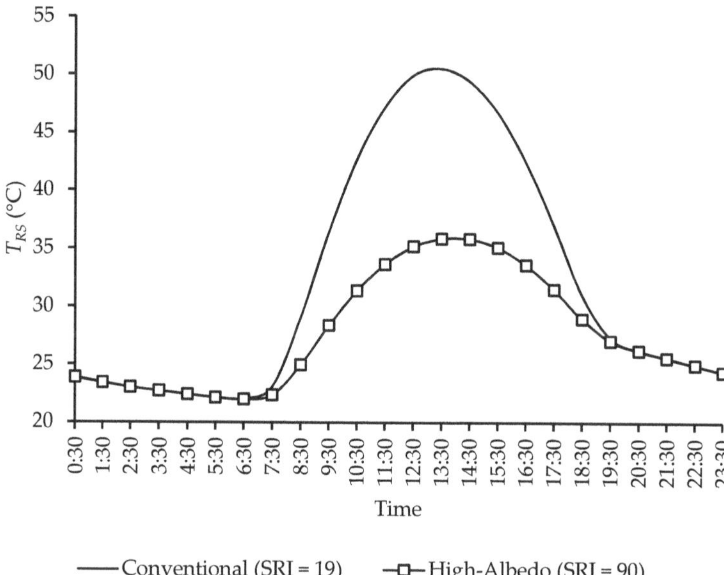

Figure 13. Roof surface temperature (T_{RS}) profiles of non-insulated roof assemblies.

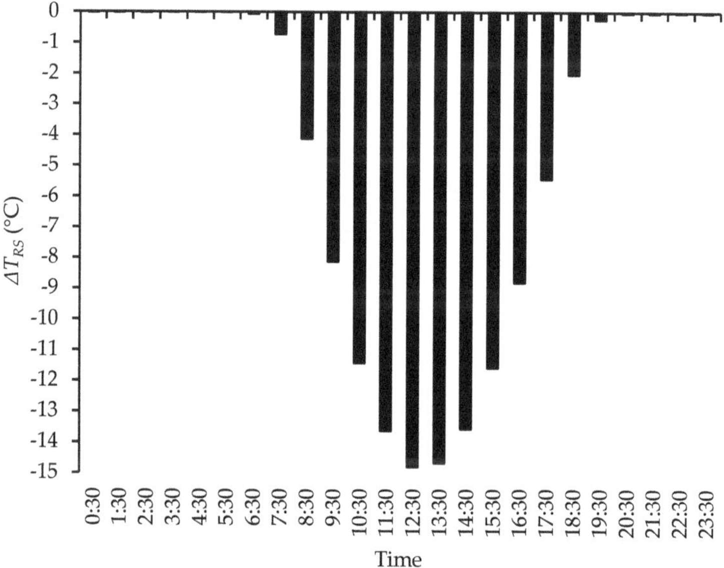

Figure 14. Change in T_{RS} (ΔT_{RS}) when SRI was increased from that of the conventional roof assembly (SRI = 19) to that of the non-insulated high-albedo roof assembly (SRI = 90).

Plot of T_{RS} versus solar irradiance for the non-insulated conventional and high-albedo roof assemblies are presented in Figure 15. Correlations between T_{RS} and solar irradiance are positive, with coefficient of determination (R^2) values of 0.9662 and 0.8233, and gradients of 0.0446 and 0.0197, for the conventional (SRI = 19) and high-albedo (SRI = 90) roof assemblies, respectively. The lower R^2 and gradient for the high-albedo roof assembly signify that the change from the conventional to the high-albedo roof surface has led to the reduction in the influence of solar irradiance on T_{RS} by virtue of the higher rate of reflection and lower rate of absorption of solar radiation on the high-albedo roof surface in comparison to that on the conventional roof surface.

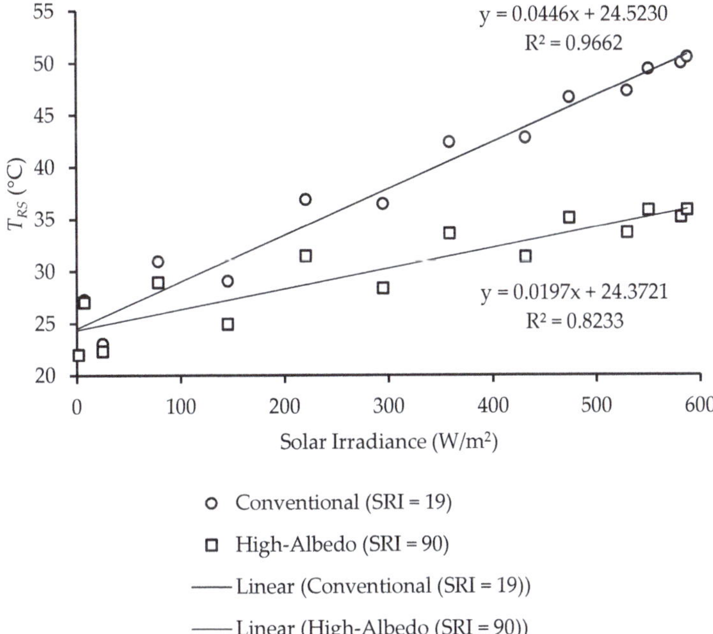

Figure 15. Plot of T_{RS} versus solar irradiance for the non-insulated roof assemblies.

T_{RS} profiles of the non-insulated and insulated high-albedo roof assemblies are compared in Figure 16. The installation of bulk insulation within the high-albedo roof assembly has led to the reduction in T_{RS} throughout the diurnal segment, where the peak T_{RS} further declined from 35.84 to 32.63 °C. The presence of insulation has led to a further peak reduction in T_{RS} of −4.06 °C as illustrated in Figure 17, as the insulation material hinders heat conduction through the roof assembly [11].

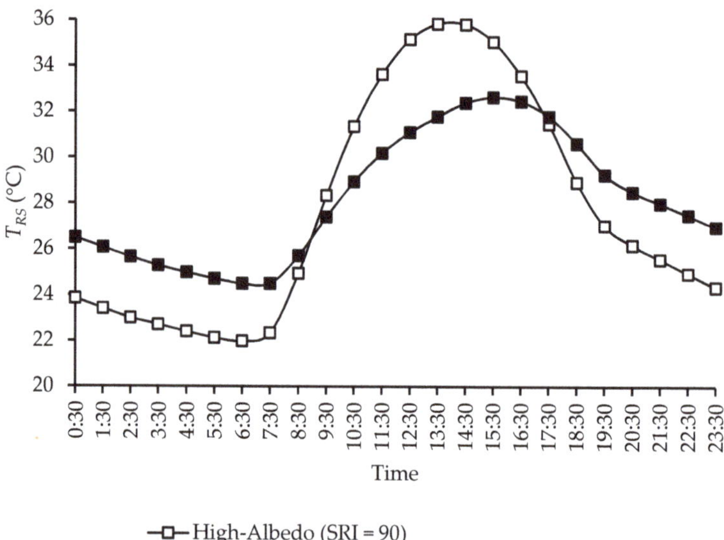

Figure 16. T_{RS} profiles of high-albedo roof assemblies.

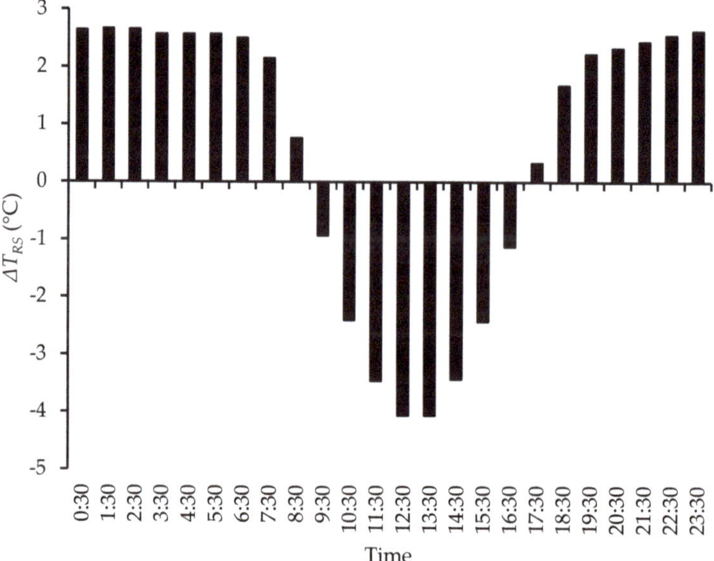

Figure 17. ΔT_{RS} when bulk insulation was added to the non-insulated high-albedo roof.

Throughout the nocturnal segment, the absence of solar radiation has caused the heat conduction transfer to invert. The transposition occurred due to the reduction in the average effective sky temperature, which resulted in the radiation of heat from the roof surface to the sky during the nocturnal period. Accordingly, heat within the indoor space and roof assembly flows toward the roof tiles and attempts to escape the building to achieve thermal equilibrium as mentioned in Farhan et al. [31] and Tang and Chin [52]. Under the circumstances, the presence of insulation within the roof assembly contributed toward hampering the heat transfer out of the building. As a consequence, T_{RS} increased throughout the nocturnal period by up to 2.64 °C, as shown in Figure 17.

Indoor temperature (T_i) contours were generated at peak diurnal outdoor temperature (T_o) as shown in Figure 18, which was on 17 March at 16:00, and the trough of nocturnal T_o as shown in Figure 19, which was on 22 September at 6:00. The contours were generated for all of the configurations of the roof assembly, which were conventional, high-albedo, and high-albedo with bulk insulation. T_i contour for the conventional roof assembly discloses that, during peak diurnal T_o, the high T_{RS}, which culminated at 45.00 °C, caused the attic air temperature (T_{AA}) to elevate to within the range from 39.55 to 43.18 °C. Then, heat transfer into the indoor space resulted in the increase in room air temperature (T_{RA}) to within the range from 34.09 to 37.73 °C. T_i contour for the high-albedo roof assembly exhibits that, resulting from the change from a conventional to a high-albedo roof surface, at peak diurnal T_o, the range of T_{RS} greatly reduced to within the range from 25.00 to 39.55 °C. The decline in T_{RS} transpired due to the adoption of the high-albedo roof surface, which reduced heat transfer into the attic space. Accordingly, T_{AA} and T_{RA} reduced to within the range from 32.27 to 34.09 °C. T_i contour for the high-albedo roof assembly with bulk insulation reveals that the presence of bulk insulation caused T_{AA} and T_{RA} to further reduce to within the range from 30.45 to 32.27 °C by hampering the heat transfer from the roof surface to the attic space.

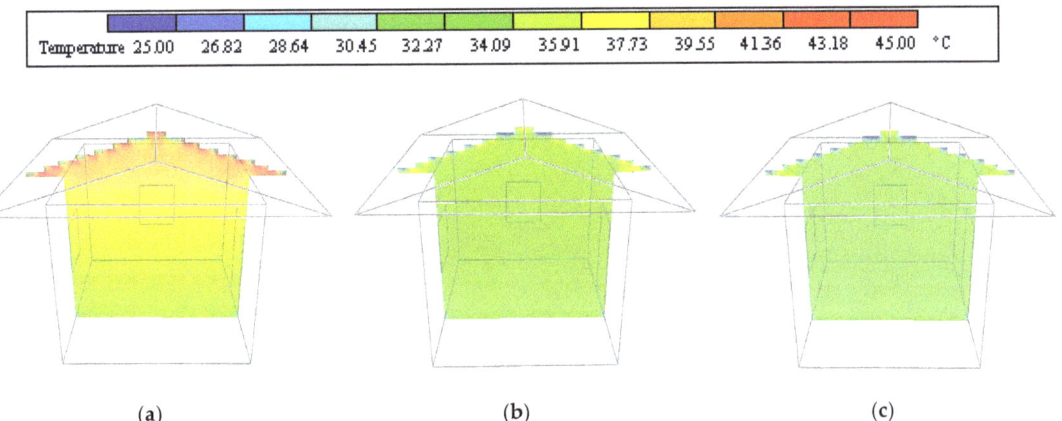

Figure 18. Indoor temperature (T_i) contours at peak diurnal outdoor temperature (T_o): (**a**) conventional, (**b**) high-albedo and (**c**) high-albedo with bulk insulation.

At the trough of nocturnal T_o, increase in the SRI of the roof surface did not influence T_{RS}, T_{AA} and T_{RA} by virtue of the absence of solar radiation, which induced heat radiation from the roof surface to the sky during the nocturnal period and is in agreement with Farhan et al. [31] and Tang and Chin [52]. T_{RS} ranged from 20.00 to 21.36 °C while T_{AA} and T_{RA} ranged from 21.82 to 22.27 °C. T_i contour for the high-albedo roof assembly with bulk insulation shows that, due to the presence of bulk insulation, T_{RS}, T_{AA} and T_{RA} increased owing to the obstruction of heat from escaping the building from the roof surface toward the sky during the nocturnal period. Consequently, heat transfer from the indoor space to the roof surface was hindered.

The change from a conventional to a high-albedo roof surface, followed by the installation of bulk insulation, has resulted in reduction of indoor annual cooling load of the building information model as presented in Figure 20. The indoor annual cooling load reduced from 2.67 MWh for the conventional roof assembly to 2.32 MWh for the high-albedo roof assembly. Installation of bulk insulation within the high-albedo roof assembly has led to further reduction of the indoor annual cooling load from 2.32 to 2.28 MWh.

Energy savings of 15.13% have been generated when the change from a conventional to a high-albedo roof surface was performed, while 17.00% have been generated when the installation of bulk insulation was performed on the high-albedo roof assembly.

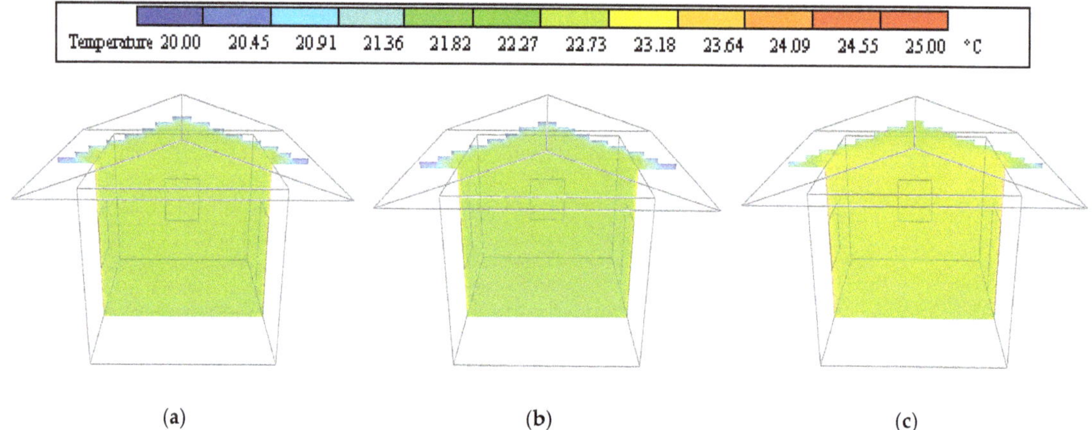

Figure 19. T_i contours at the trough of nocturnal T_o: (**a**) conventional, (**b**) high-albedo and (**c**) high-albedo with bulk insulation.

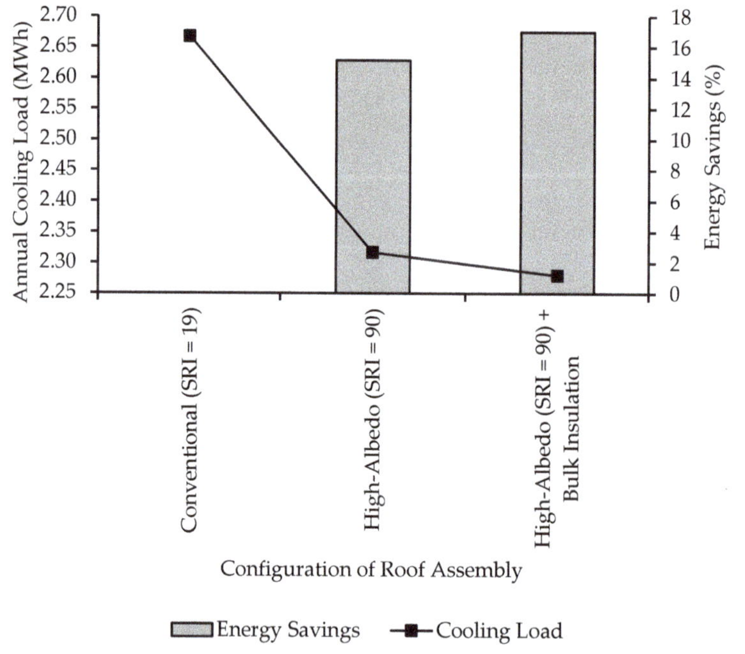

Figure 20. Annual cooling load and energy savings of the roof assembly at various configurations.

4. Conclusions

Thermal-energy performance of an urban pitched residential roof assembly, which adopted varying configurations of high-albedo roof tiles and bulk insulation under the hot, humid climate, was evaluated. Thermal-energy and CFD analyses were performed on a building information model.

Change from the conventional to the high-albedo roof surface has led to the reduction in the influence of solar irradiance on roof surface temperature due to the higher rate of reflection of solar radiation on the roof surface. However, the change did not influence the roof surface temperature throughout the nocturnal segment due to the absence of solar radiation. Installation of bulk insulation within the high-albedo roof assembly has led to further reduction in roof surface temperature throughout the diurnal segment. However, the reduction coincided with the increase in the roof surface temperature throughout the nocturnal period as heat transfers out of the building, owing to the absence of solar radiation that has caused the direction of heat conduction transfer to invert, which is hampered by the insulation material. Despite the negative impact of installing bulk insulation throughout the nocturnal period, on the whole, energy savings have been achieved, which are 15.13%, which is from 2.67 to 2.32 MWh when the change from a conventional to a high-albedo roof surface was performed, and 17.00%, which is from 2.32 to 2.28 MWh when the installation of bulk insulation was performed on the high-albedo roof assembly.

For future research, studies that consider the variation in height of the building and surrounding buildings and degree of the placement of high-albedo materials can be considered. Development of a solar-reflective coating that can further increase the solar reflectance of the roof surface and potentially eliminate dependence toward insulation is also recommended. Alternatively, engineering of novel materials that possess extremely low thermal conductivity, which can potentially be applied within the roof assembly with minuscule thicknesses, is proposed.

Author Contributions: Conceptualization, M.D.; methodology, S.A.F.; software, N.S.; validation, H.A.-M.; formal analysis, R.I.; investigation, M.D.; resources, N.S.; data curation, H.A.-M.; writing—original draft preparation, S.A.F.; writing—review and editing, M.D. and S.A.F.; visualization, R.I.; supervision, N.S.; project administration, H.A.-M.; funding acquisition, R.I. All authors have read and agreed to the published version of the manuscript.

Funding: The Article Processing Charge was self-funded by the authors. The research also received financial support from Yayasan Universiti Teknologi PETRONAS (UTP), Malaysia, via grant number 015LC0–155, which was administered by Research Management Center of UTP.

Institutional Review Board Statement: Not applicable.

Informed Consent Statement: Not applicable.

Data Availability Statement: The data presented in this study are available on request from the corresponding author, S.A.F.

Acknowledgments: The authors are thankful to the Institute of Science of Universiti Teknologi MARA Shah Alam, Malaysia, as well as Yarmouk University and Jadara University, Jordan, for the support and collaboration.

Conflicts of Interest: The authors declare no conflict of interest.

Nomenclature

Abbreviations

BIM	Building Information Modeling
CFD	Computational Fluid Dynamics
IESVE	Integrated Environmental Solutions <Virtual Environment>
k-value	Thermal conductivity
R-value	Thermal resistance
SRI	Solar Reflectance Index
TMY	Typical Meteorological Year

Notations

A	Cross-sectional area perpendicular to heat flow (for calculation of $Q_{conduction}$), or surface area (for calculation of $Q_{convection}$)
A_1	Area of the first surface
$\frac{dT}{dx}$	Temperature gradient
FM	Field measurement data
h_c	Surface heat transfer coefficient
h_r	Coefficient of heat transfer
k	Thermal conductivity
$PD_{SD\text{-}FM}$	Percentage difference between the simulation and field measurement data
$Q_{conduction}$	Heat transfer by conduction
$Q_{convection}$	Heat transfer by convection
$Q_{radiation}$	Heat transfer by radiation
SD	Simulation data
T_{AA}	Attic air temperature
T_{black}	Steady-state temperature of the standard black
T_f	Fluid temperature
T_i	Indoor temperature
T_o	Outdoor temperature
T_{RS}	Roof-top surface temperature
T_s	Surface temperature
$T_{surface}$	Steady-state temperature of the material surface
T_{white}	Steady-state temperature of the standard white
T_1	Absolute temperature of the first surface
T_2	Absolute temperature of the second surface
ΔT_{RS}	Change in roof-top surface temperature
ε_1	Emissivity of the first surface
σ	Stefan–Boltzmann constant

References

1. Alam, S.S.; Omar, N.A.; Ahmad, M.S.; Siddiquei, H.R.; Nor, S.M. Renewable energy in Malaysia: Strategies and development. *Environ. Manag. Sustain. Dev.* **2013**, *2*, 51–66. [CrossRef]
2. Al-Obaidi, K.M.; Ismail, M.; Abdul Rahman, A.M. Passive cooling techniques through reflective and radiative roofs in tropical houses in Southeast Asia: A literature review. *Front. Archit. Res.* **2014**, *3*, 283–297. [CrossRef]
3. Miranville, F.; Boyer, H.; Mara, T.; Garde, F. On the thermal behaviour of roof-mounted radiant barriers under tropical and humid climatic conditions: Modelling and empirical validation. *Energy Build.* **2003**, *35*, 997–1008. [CrossRef]
4. Al-Obaidi, K.M.; Ismail, M.; Abdul Rahman, A.M. Investigation of passive design techniques for pitched roof systems in the tropical region. *Mod. Appl. Sci.* **2014**, *8*, 182–191. [CrossRef]
5. Tang, K.H.D. Climate change in Malaysia: Trends, contributors, impacts, mitigation and adaptations. *Sci. Total Environ.* **2019**, *650*, 1858–1871. [CrossRef]
6. Beck, H.E.; Zimmermann, N.E.; McVicar, T.R.; Vergopolan, N.; Berg, A.; Wood, E.F. Present and future Köppen-Geiger climate classification maps at 1-km resolution. *Sci. Data* **2018**, *5*, 180214. [CrossRef]
7. MetMalaysia: Maklumat Iklim. Available online: https://www.met.gov.my/iklim/laporanringkasan/maklumatiklim (accessed on 11 January 2022).
8. MetMalaysia: Iklim Malaysia. Available online: https://www.met.gov.my/pendidikan/iklim/iklimmalaysia (accessed on 11 January 2022).
9. Mekhilef, S.; Safari, A.; Mustaffa, W.E.S.; Saidur, R.; Omar, R.; Younis, M.A.A. Solar energy in Malaysia: Current state and prospects. *Renew. Sustain. Energy Rev.* **2012**, *16*, 386–396. [CrossRef]
10. Vijaykumar, K.C.K.; Srinivasan, P.S.S.; Dhandapani, S. A performance of hollow clay tile (HCT) laid reinforced cement concrete (RCC) roof for tropical summer climates. *Energy Build.* **2007**, *39*, 886–892. [CrossRef]
11. Tuck, N.W.; Zaki, S.A.; Hagishima, A.; Rijal, H.B.; Yakub, F. Affordable retrofitting methods to achieve thermal comfort for a terrace house in Malaysia with a hot–humid climate. *Energy Build.* **2020**, *223*, 110072. [CrossRef]
12. Papadopoulos, A.M.; Giama, E. Environmental performance evaluation of thermal insulation materials and its impact on the building. *Build. Environ.* **2007**, *42*, 2178–2187. [CrossRef]
13. Husna, N.; Farhan, S.A.; Abdul Wahab, M.M.; Shafiq, N.; Sharif, M.T.; Abd Razak, S.N.; Ismail, F.I. Effect of granular silica aerogel as filler on tensile and flexural strengths and moduli of stone-wool-fibre-reinforced composite as rigid board roof insulation material. *IOP Conf. Ser. Earth Environ. Sci.* **2021**, *945*, 012061. [CrossRef]

14. Ismail, F.I.; Farhan, S.A.; Shafiq, N.; Husna, N.; Sharif, M.T.; Affan, S.U.; Veerasenan, A.K. Nano-porous silica-aerogel-incorporated composite materials for thermal-energy-efficient pitched roof in the tropical region. *Appl. Sci.* **2021**, *11*, 6081. [CrossRef]
15. Nuruddin, M.F.; Husna, N.; Azizli, K.A.M.; Farhan, S.A.; Wan Zainal Abidin, W.A.A. Prospect of adopting kapok fibre as roof insulation. *Appl. Mech. Mater.* **2014**, *567*, 482–487. [CrossRef]
16. Farhan, S.A.; Khamidi, M.F.; Husna, N. Thickness optimization of kapok fibre insulation below roof pitch of residential buildings in hot-humid climate with mathematical formulation. *Appl. Mech. Mater.* **2014**, *699*, 864–870. [CrossRef]
17. Omar, M.F.; Abdullah, M.A.H.; Rashid, N.A.; Abdul Rani, A.L.; Illias, N.A. The application of coconut fiber as insulation ceiling board in building construction. *IOP Conf. Ser. Mater. Sci. Eng.* **2020**, *864*, 012196. [CrossRef]
18. Al Yacouby, A.M.; Khamidi, M.F.; Teo, Y.W.; Nuruddin, M.F.; Farhan, S.A.; Sulaiman, S.A.; Razali, A.E. Housing developers and home owners awareness on implementation of building insulation in Malaysia. *WIT Trans. Ecol. Environ.* **2011**, *148*, 219–230. [CrossRef]
19. *MS 2680:2017*; Energy Efficiency and Use of Renewable Energy for Residential Buildings—Code of Practice. Malaysian Standard: Cyberjaya, Malaysia, 2017.
20. Green Building Index Assessment Criteria for Residential New Construction Version 3.0. Available online: https://www.greenbuildingindex.org/Files/Resources/GBI%20Tools/GBI%20RNC%20Residential%20Tool%20V3.0.pdf (accessed on 11 January 2022).
21. *Sel. P.U. 142/2012*; Selangor Uniform Building (Amendment) (No. 2) By-Laws 2012. State of Selangor: Putrajaya, Malaysia, 2012.
22. Farhan, S.A.; Khamidi, M.F.; Al Yacouby, A.M.; Idrus, A.; Nuruddin, M.F. Critical Review of Published Research on Building Insulation: Focus on Building Components and Climate. In Proceedings of the 2012 IEEE Business, Engineering & Industrial Applications Colloquium (BEIAC 2012), Kuala Lumpur, Malaysia, 7–8 April 2012. [CrossRef]
23. Halim, N.H.A.; Zain-Ahmed, A.; Zakaria, N.Z. Thermal and Energy Analysis of Ceiling and Pitch Insulation for Buildings in Malaysia. In Proceedings of the 3rd International Symposium & Exhibition in Sustainable Energy & Environment (ISESEE 2011), Malacca, Malaysia, 1–3 June 2011. [CrossRef]
24. Irwan, S.S.; Zain-Ahmed, A.; Zakaria, N.Z.; Ibrahim, N. Thermal and energy performance of conditioned building due to insulated sloped roof. *AIP Conf. Proc.* **2010**, *1250*, 476. [CrossRef]
25. Irwan, S.S.; Zain-Ahmed, A.; Ibrahim, N.; Zakaria, N.Z. Roof Angle for Optimum Thermal and Energy Performance of Insulated Roof. In Proceedings of the 3rd International Conference on Energy and Environment (ICEE 2009), Malacca, Malaysia, 7–8 December 2009. [CrossRef]
26. Morris, F.; Zain-Ahmed, A.; Zakaria, N.Z. Thermal Performance of Naturally Ventilated Test Building with Pitch and Ceiling Insulation. In Proceedings of the 3rd International Symposium & Exhibition in Sustainable Energy & Environment (ISESEE 2011), Malacca, Malaysia, 1–3 June 2011. [CrossRef]
27. Nuruddin, M.F.; Puad, N.H.A.; Zaidi, N.H.A. Effectiveness of aerogel roofing system on temperature reduction in Malaysian residential buildings. *J. Eng. Appl. Sci.* **2017**, *12*, 4057–4062. [CrossRef]
28. Puad, N.H.A.; Nuruddin, M.F.; Lian, J.J.; Othman, I. Roof Insulation Material from Low Density Polyethylene (LDPE), Kapok Fibre and Silica Aerogel. In *Engineering Challenges for Sustainable Future, Proceedings of the 3rd International Conference on Civil, Offshore and Environmental Engineering (ICCOEE 2016), Kuala Lumpur, Malaysia, 15–17 August 2016*; CRC Press: Boca Raton, FL, USA, 2016. [CrossRef]
29. Zakaria, N.Z.; Zain-Ahmed, A.; Ariffin, N.N.; Abdul Halim, N.H.; Morris, F. Thermal Energy Evaluation of Building with Ceiling Insulation in Warm-Humid Tropical Climate. In Proceedings of the 2011 IEEE Colloquium on Humanities, Science and Engineering (CHUSER 2011), Penang, Malaysia, 5–6 December 2011. [CrossRef]
30. Al Yacouby, A.M.; Khamidi, M.F.; Nuruddin, M.F.; Farhan, S.A.; Razali, A.E. Study on Roof Tile's Colors in Malaysia for Development of New Anti-Warming Roof Tiles with Higher Solar Reflectance Index (SRI). In *Energy and Sustainability: Exploring the Innovative Minds, Proceedings of the National Postgraduate Conference (NPC 2011), Perak, Malaysia, 19–20 September 2011*; IEEE: Piscataway, NJ, USA, 2011.
31. Farhan, S.A.; Ismail, F.I.; Kiwan, O.; Shafiq, N.; Zain-Ahmed, A.; Husna, N.; Hamid, A.I.A. Effect of roof tile colour on heat conduction transfer, roof-top surface temperature and cooling load in modern residential buildings under the tropical climate of Malaysia. *Sustainability* **2021**, *13*, 4665. [CrossRef]
32. Synnefa, A.; Santamouris, M.; Akbari, H. Estimating the effect of using cool coatings on energy loads and thermal comfort in residential buildings in various climatic conditions. *Energy Build.* **2007**, *39*, 1167–1174. [CrossRef]
33. Pisello, A.L. High-albedo roof coatings for reducing building cooling needs. In *Eco-Efficient Materials for Mitigating Building Cooling Needs—Design, Properties and Applications*; Woodhead Publishing: Cambridge, UK, 2015; pp. 243–268. [CrossRef]
34. Syuhada, A.; Maulana, M.I. Heat transfer capability of solar radiation in colored roof and influence on room thermal comfort. *AIP Conf. Proc.* **2018**, *1931*, 030054. [CrossRef]
35. Zingre, K.T.; Wan, M.P.; Wong, S.K.; Toh, W.B.T.; Lee, I.Y.L. Modelling of cool roof performance for double-skin roofs in tropical climate. *Energy* **2015**, *82*, 813–826. [CrossRef]
36. Thongkanluang, T.; Chirakanphaisarn, N.; Limsuwan, P. Preparation of NIR reflective brown pigment. *Procedia Eng.* **2012**, *32*, 895–901. [CrossRef]
37. *G.N. 5178/85*; UBBL 1984. Uniform Building By-Laws 1984. International Law Book Services: Petaling Jaya, Malaysia, 2013.

38. Sekerka, R.F. *Thermodynamics and Statistical Mechanics for Scientists and Engineers*; Elsevier Inc.: Amsterdam, The Netherlands, 2015. [CrossRef]
39. Jayamaha, L. *Energy Efficient Building Systems: Green Strategies for Operation and Maintenance*; McGraw-Hill: New York, NY, USA, 2006.
40. Ghiaasiaan, S.M. *Convective Heat and Mass Transfer*; Cambridge University Press: New York, NY, USA, 2011.
41. Howell, J.R.; Siegel, R.; Mengüç, M.P. *Thermal Radiation Heat Transfer*, 5th ed.; CRC Press: Boca Raton, FL, USA, 2011.
42. Santamouris, M.; Synnefa, A.; Karlessi, T. Using advanced cool materials in the urban built environment to mitigate heat islands and improve thermal comfort conditions. *Sol. Energy* **2011**, *85*, 3085–3102. [CrossRef]
43. *ASTM E1980-11*; Standard Practice for Calculating Solar Reflectance Index of Horizontal and Low-Sloped Opaque Surfaces. ASTM: West Conshohocken, PA, USA, 2019.
44. Amir, A.; Mohamed, M.F.; Sulaiman, M.K.A.; Yusoff, W.F.M. Assessment of indoor thermal condition of a low-cost single story detached house: A case study in Malaysia. *Alam Cipta* **2019**, *12*, 80–88.
45. Gardezi, S.S.S.; Shafiq, N.; Wan Abdullah Zawawi, N.A.; Khamidi, M.F.; Farhan, S.A. Minimization of embodied carbon footprint from housing sector of Malaysia. *Chem. Eng. Trans.* **2015**, *45*, 1927–1932. [CrossRef]
46. Jamaludin, N.; Mohammed, N.I.; Farhan, S.A. Indoor Thermal Environment in Residential Buildings at Different Micro-Climates in Malaysia. In *Recent Advances in Environmental and Biological Engineering, Proceedings of the 3rd International Conference on Sustainable Cities, Urban Sustainability and Transportation (SCUST '14), Istanbul, Turkey, 15–17 December 2014*; WSEAS Press: Athens, Greece, 2014.
47. Vangimalla, P.R.; Olbina, S.J.; Issa, R.R.; Hinze, J. Validation of Autodesk Ecotect™ Accuracy for Thermal and Daylighting Simulations. In Proceedings of the 2011 Winter Simulation Conference, Phoenix, AZ, USA, 11–14 December 2011.
48. Leng, P.C.; Ahmad, M.H.; Ossen, D.R.; Hamid, M. Investigation of Integrated Environmental Solutions-Virtual Environment Software Accuracy for Air Temperature and Relative Humidity of the Test Room Simulations. In Proceedings of the Universiti Malaysia Terengganu 11th International Annual Symposium on Sustainability Science and Management, Terengganu, Malaysia, 9–11 July 2012.
49. Maamari, F.; Andersen, M.; de Boer, J.; Carroll, W.L.; Dumortier, D.; Greenup, P. Experimental validation of simulation methods for bi-directional transmission properties at the daylighting performance level. *Energy Build.* **2006**, *38*, 878–889. [CrossRef]
50. Meteonorm: Global Climate Database. Available online: https://meteonorm.com (accessed on 11 January 2022).
51. *MS 1525:2019*; Energy Efficiency and Use of Renewable Energy for Non-Residential Buildings—Code of Practice (Third Revision). Malaysian Standard: Cyberjaya, Malaysia, 2019.
52. Tang, C.K.; Chin, N. *Building Energy Efficiency Technical Guideline for Passive Design*; Building Sector Energy Efficiency Project (BSEEP): Kuala Lumpur, Malaysia, 2013.
53. Raeissi, S.; Taheri, M. Cooling load reduction of buildings using passive roof options. *Renew. Energy* **1996**, *7*, 301–313. [CrossRef]

 *sustainability*

Article

Investigation on the Potential of Various Biomass Waste for the Synthesis of Carbon Material for Energy Storage Application

Brenda Ai-Lian Lim [1,2,*], Steven Lim [1,2,*], Yean Ling Pang [1,2], Siew Hoong Shuit [1,2], Kam Huei Wong [1,2] and Jong Boon Ooi [3]

[1] Department of Chemical Engineering, Lee Kong Chian Faculty of Engineering and Science, Universiti Tunku Abdul Rahman (UTAR), Jalan Sungai Long, Kajang 43000, Malaysia; pangyl@utar.edu.my (Y.L.P.); shuitsh@utar.edu.my (S.H.S.); wongkh@utar.edu.my (K.H.W.)

[2] Centre of Photonics and Advanced Material Research, Universiti Tunku Abdul Rahman (UTAR), Bandar Sungai Long, Kajang 43000, Malaysia

[3] Mechanical Engineering Discipline, School of Engineering, Monash University Malaysia, Jalan Lagoon Selatan, Bandar Sunway, Subang Jaya 47500, Malaysia; ooi.jongboon@monash.edu

* Correspondence: brendalim2707@1utar.my (B.A.-L.L.); stevenlim@utar.edu.my (S.L.)

Abstract: The metal–air battery (MAB) has been a promising technology to store energy, with its outstanding energy density, as well as safety features. Yet, the current material used as air cathode is costly and not easily available. This study investigated a few biomass wastes with good potential, including the oil palm empty fruit bunch and garlic peel, as well as the oil palm frond, to determine a sufficiently environmentally-safe, yet efficient, precursor to produce carbon material as an electro-catalyst for MAB. The precursors were carbonized at different temperatures (450, 600, and 700 °C) and time (30, 45, and 60 min) followed by chemical (KOH) activation to synthesize the carbon material. The synthesized materials were subsequently studied through chemical, as well as physical characterization. It was found that PF presented superior tunability that can improve electrical conductivity, due to its ability to produce amorphous carbon particles with a smaller size, consisting of hierarchical porous structure, along with a higher specific surface area of up to 777.62 m^2g^{-1}, when carbonized at 600 °C for 60 min. This paper identified that PF has the potential as a sustainable and cost-efficient alternative to carbon nanotube (CNT) as an electro-catalyst for energy storage application, such as MAB.

Keywords: metal–air battery; carbon particles; biomass waste; electro-catalyst

Citation: Lim, B.A.-L.; Lim, S.; Pang, Y.L.; Shuit, S.H.; Wong, K.H.; Ooi, J.B. Investigation on the Potential of Various Biomass Waste for the Synthesis of Carbon Material for Energy Storage Application. *Sustainability* **2022**, *14*, 2919. https://doi.org/10.3390/su14052919

Academic Editor: Thanikanti Sudhakar Babu

Received: 30 December 2021
Accepted: 9 February 2022
Published: 2 March 2022

Publisher's Note: MDPI stays neutral with regard to jurisdictional claims in published maps and institutional affiliations.

1. Introduction

Energy sources that are renewable have gradually developed and been implemented to compensate for the decreasing and non-environmentally safe sources, which are mainly fossil fuel derivatives [1]. With the advanced development in science and technology, metal–air batteries (MAB) are manufactured as devices to store energy from the unstable renewable yet green energy source that depend primarily on the surrounding nature, including the weather that affects the availability of the sun, wind, and water currents. Currently, energy storage technologies are used to store energy in many off grid devices, such as electrical automobiles, hand phones, power banks, laptops, and many other household appliances [2]. The conventional energy storage technology, which is commonly known as the lithium-ion battery (LiB), has been generally established, while being commonly used in the industry. Nonetheless, the LiB possessed its own drawbacks, which includes slow charging time, bulky in size, and safety concerns, such as electrolyte leakage and flammable components, as well as had approached its theoretical energy density and recyclability limit [3]. The first metal–air battery was created back in the year 1800, and continuous research had been carried out to improve the efficiency and affordability of the device [4]. The MAB was generally design with primary components consisting of a cathode, an anode, and a liquid or non-liquid electrolyte. It works through the electrochemical reactions that take place among the components, whereby the atmospheric oxygen is reduced, and the metal electrode is oxidized, which assembles into reversible metal-oxides [5]. An advantage to MAB is the application of a half-open system, which can facilitate the oxygen reduction reaction (ORR) that takes place at the cathode [5]. The open structure allows a substantial size and weight reduction of the battery [6]. MAB is applicable in the storage of any intermittent energy created from renewable energy sources, including hydropower, as well as geothermal power, to consistently maintain continuously energy supply. MAB has shown great potential for a device to store energy, with its high theoretical energy density (as much as 5928 Wh kg^{-1}), improved recyclability, and light and compact design, compared to LiBs [6,7]. However, the MAB technology is still lacking, in terms of economics, endurance, and efficiency, which affects the status of commercialization of the device.

Many researchers came to understand that majority of the chemical and physical properties of the air cathode material had a large impact on the capability of a MAB [5,8]. According to Ha et al., the catalytic efficiency of the battery can be controlled with the properties of the air cathode, as the charge and discharge reactions relies on the availability of active sites to store discharge products, as well as the diffusion pathway [9]. The commercial material used to fabricate the air cathode that is used for MAB includes rare metals, such as the Pt-based materials, RuO_2, and IrO_2 [10]. These rare metal materials are not economical, less stable, and are inclined to capitulate, due to poisoning from methanol, as well as carbon dioxide. There are researchers who had also disclosed that carbon-based materials contained advantageous characteristics that give them greater suitability as precursors to synthesize electro-catalyst in MABs, due to its highly porous structure [11]. Carbon nanotubes (CNTs) were utilized as an electro-catalyst material due to its elevated energy density and exceptional cyclic-life [12]. Unfortunately, CNTs are usually very expensive, due to its highly complex production method. To mitigate this issue, researchers had initiated studies to find a suitable carbon material substitute by using biomass wastes as precursors that are more economical and environmentally friendly.

Biomass wastes are typically generated from redundant plants or animal parts found in industrial, agricultural, and municipal waste. They are primarily made of carbohydrates and polysaccharides, which provides high content of carbon, oxygen, and hydrogen atoms, as well as less significant components [13]. The unique and natural composition found in biomass wastes rendered them as a competent, yet sustainable, precursor material to produce carbon material. Besides, the natural availability of other components, including silicon and nitrogen, increases the possibility for self-heteroatom doping to occur [14]. Heteroatom doping was found to be able to improve the electrochemical efficiency of the electro-catalyst material by altering its chemical and physical structures, thus improving the electrical conductivity, surface wettability, and transfer of charges of the battery [15]. Currently, a large amount of biomass wastes were being created globally, which presented a huge threat to the environment if not being disposed properly. The usage of biomass waste as precursor material to produce of carbon material can not only reduce the waste accumulation, but also acted as a substitute for higher cost products. Different biomass wastes can produce carbon materials with different properties with their unique structures, as well as elemental and lignocellulosic compositions [16]. Due to this, the study of carbon material from biomass waste will be vital to synthesize a high performing electro-catalyst for energy storage applications, such as MAB.

Lignocellulosic components found in biomass wastes are bonded by covalent and non-covalent bonding [17]. The carbonization of lignocellulosic biomass wastes produces carbon materials with hierarchical pores which is a mixture of different pore size ranging from micropores to macropores. Hierarchical porous structures provide greater specific surface area and plenty of diffusion channels that can improve the electrochemical reactions that takes place at the cathode of the battery [18]. Each lignocellulosic component, including lignin, cellulose, and hemicellulose, decomposes at different temperatures range, as each component has an individual thermal stability, due to the different chemical bondings present. Hemicellulose is predicted to thermally break down at temperature ranging from 180 to 270 °C, while cellulose is anticipated to disintegrate at 270 to 350 °C [19]. Meanwhile, lignin has a greater structural complexity, whereby it begins to disintegrate only when the temperature exceeds 300 °C.

Various studies have been done on the synthesis of carbon materials for electric storage application from different types of biomass waste with different synthesis methods [20,21]. These studies focused on individual biomass waste as a precursor, while studying the effect of different synthesis parameters, such as the types of doping agent, carbonization temperature, and carbonization method. There were other studies done on multiple biomass wastes, and these studies focused mainly on the effect of heteroatom content [22] on the synthesis methods [16,23] and synthesis parameters [24]. However, there is insufficient information available that focuses on the fundamental relationship between the pyrolysis temperature and time, as well as the composition of the lignocellulosic component in the precursor to synthesize carbon material to be used in metal air batteries. This information is vital when it comes to the selection of suitable precursors and synthesis parameters to synthesize an optimum electro-catalyst for electric storage application.

For this study, there are three individual biomass wastes containing unique properties that were selected as raw materials to synthesize carbon material as electro-catalyst, which were the oil palm empty fruit bunch (EFB) and garlic peel (GP), as well as oil palm frond (PF). These biomass wastes are easily accessible, as Malaysia is presently second to the world largest producer of palm, while garlic is one of the basic cooking ingredients in the country [25]. These biomass wastes were all obtained without any charge. The oil palm frond and garlic peel have no commercial value yet, as these wastes are currently not reused. On the other hand, the oil palm empty fruit bunch is currently reused for other applications, such as the bio-fertilizer, as well as methane production [26,27]. According to Salleh et al., the price of EFB in the year 2018 was c.a. USD 70.85 per ton, which was calculated to be c.a. USD 72.62 per ton in the year 2021, when the CPI inflation rate was taken into consideration [28,29]. EFB, PF, and GP have different elemental and lignocellulosic compositions, as well as surface textures, which can result in the production of carbon materials with different properties. Currently, there are no studies done on these precursors for electrical storage applications in MAB [30,31]. Therefore, this study aimed to identify their feasibility as a suitable raw material to synthesize a carbon material as an electro-catalyst for MAB, based on the chemical and physical characterization tests conducted.

2. Materials and Methods

2.1. Materials

The biomass wastes selected as precursor in this study includes EFB and PF, as well as GP, as shown in Figure 1. EFB, along with PF, were both obtained by car from an oil palm plantation locally found in Tangjung Tualang, Perak on the 3rd of May 2021. On the other hand, the acquisition of GP had been done from a locally based food and beverage factory located within the industrial area of Kampung Baru, Selangor on the 21st of April 2021; 37% concentrated hydrochloric acid (HCl) and 98% potassium hydroxide (KOH) were bought from Merck and Friendemann Schmidt Chemicals, respectively, to synthesize carbon material.

Figure 1. (a) Raw OPEFB; (b) raw OPF [32]; (c) raw GP.

2.2. Synthesis of Carbon Material

Figure 2 shows the process flow of the production of a carbon material for this study. The biomass waste was first washed multiple times to remove other solid impurities before being placed in the drying oven (Memmert, UNB 100) at a constant temperature of 80 °C for approximately 24 h. After that, the size of the dried precursor was then reduced by grinding and sieving to obtain the size of less than 300 μm and stored in individual sealed containers prior to usage. For the study of the carbonization temperature parameter, the carbonization process of dried precursor was accomplished by filling the crucible with 5.0 g of biomass precursor powder and inserted into a carbolite furnace (Carbolite, RHF 15/8) before setting the carbonization temperature to 450 °C, 600 °C and 700 °C with a carbonization duration of half an hour. On the other hand, the carbonization time was manipulated from 30 to 60 min whereas the carbonization temperature was set constant at 600 °C for the study of carbonization time. Subsequently, the raw material that was carbonized was left to cool overnight, before it underwent activation using the wet chemical impregnation method, whereby the carbonized precursor was infused with activating agent KOH at a 3 to 1 weight ratio (carbon to KOH), and 5 mL of water was added to dissolve the added KOH before being stirred continuously with a magnetic stirrer for 24 h. Then, the slurry was then left to dry in the same oven at a temperature of 80 °C until all the moisture content was removed. The agglomerated dried mixture was ground into fine powder form by using a pestle and mortar before it underwent heat treatment in the carbolite furnace at a temperature 600 °C for 30 min. After the treated carbon material was cooled to room temperature, the carbon material was washed with HCl solution of 1.0 mol concentration in order to rinse off the remaining inorganic salts before further rinsing with distilled water until neutral pH was achieved followed by drying it overnight in the oven. Similarly, the size of the dried agglomerated activated carbon was reduced by grounding using a pestle and mortar to remove agglomeration before being stored in individual sealed containers. Hereupon, activated carbon materials were formed and each of individual samples with unique labels EFB-x-y, GP-x-y, and PF-x-y, by which x described the carbonization temperature, while y represented the carbonization time, which were 30, 45, and 60 min.

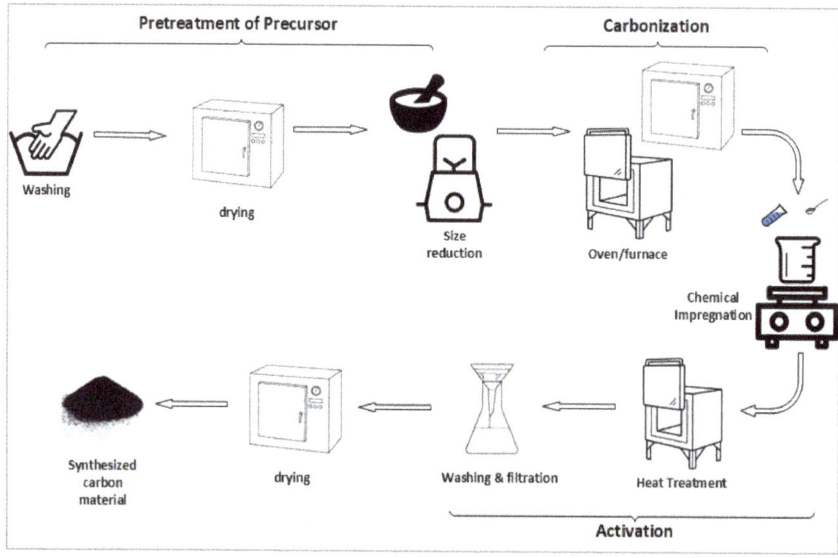

Figure 2. Carbon material synthesis process flow.

2.3. Characterization of Carbon Material

2.3.1. Lignocellulosic Component Test

The extractives were removed from the raw materials using the soxhlet extraction method with methanol as solvent for 4 h and the extractive content were calculated accordingly using Equation (1) [33]. The ash content in the raw precursors were determined in accordance with the procedure by the National Renewable Energy Laboratory [34]. The lignocellulosic component test was carried out in accordance with the Klason for lignin and Tappi T203 cm-09 test method for the determination of the cellulose and hemicellulose components with slight modification [35]. The content percentage were calculated as shown in the equations below.

Extractive content percentage:

$$Extractive \, (\%) = \frac{W_{ex,1} - W_{ex,2}}{W_{ex,1}} \times 100\% \tag{1}$$

Ash content percentage:

$$Ash \, content \, (\%) = \frac{W_{ash,2}}{W_{ash,1}} \times 100\% \tag{2}$$

Lignin content percentage:

$$Lignin \, content \, (\%) = \frac{W_{l,2}}{W_{l,1}} \times 100\% \tag{3}$$

Cellulose and Hemicellulose content percentage:

$$Hemicellulose \, content \, (\%) = \frac{6.85(V_b - V_t) \times 0.2}{V \times 1.5} \times (100 - Lignin \, content)\% \tag{4}$$

$$Cellulose \, content \, (\%) = 100\% - Lignin \, content(\%) - Hemicellulose \, content(\%) \tag{5}$$

where $W_{ex,1}$ represents the weight of the raw dried precursor before extraction while $W_{ex,2}$ indicates the dry weight of extractive-free precursor. $W_{ash,1}$ and $W_{ash,2}$ represent the weight of dried raw precursor before and after carbonization, respectively. $W_{l,1}$ and $W_{l,2}$ constitute for the weight of dried raw precursor and dried extracted lignin, respectively. V_b and V_t constitute for the volume of 0.1 N ferrous ammonium sulfate titrated into the blank solution and sample, respectively, while V indicates the volume of pulp filtrate used in the sample.

2.3.2. Chemical Characterization

The chemical properties of each of the synthesized samples were analyzed using the scanning electron microscopy-energy dispersive X-ray (SEM-EDX) (Hitachi, S-3400 N), thermogravimetric analysis (TGA) (Thermo Scientific Sorptometic, SO1990), and Fourier transform infrared spectroscopy (FTIR) (Nicolet, iS10). The SEM-EDX method was utilized to analyze the presence of elements on the sample surface. Meanwhile, TGA analysis was carried out to observe the weight changes of the raw materials, with temperatures ranging from 30 to 1000 °C at the flow rate of 10 °C min^{-1} in N$_2$ atmosphere. The weight percentage and derivative weight curves were obtained and studied for the individual raw precursors. FTIR analysis was executed to analyze and compare the presence of the chemical bonds within the wavelength spectrum, ranging from 400 to 4000 cm^{-1}, for the raw materials, along with the selected carbon samples at 700 °C and 30 min.

2.3.3. Physical Characterization

The physical properties of several selected carbon samples synthesized were being characterized through X-ray diffraction (XRD) analysis (Shidmazu, XRD-6000), the Brunauer–Emmet–Teller (BET) surface analysis (Micromeritics, 3Flex), and the field emission scanning electron microscopy (FESEM) (JEOL, JSM-6701F). The XRD analysis had been carried out, with a scan range of 10 to 60 °C, at a scan rate of 2 °C min^{-1}, in order to study and compare the crystalline characteristics of the raw precursors as well as the selected carbonized precursors. The individual specific surface area (SSA) of each of the carbon samples was verified by carrying out the BET analysis. 0.4 g of each carbon samples were inserted in each tube and left to degass at 150 °C for 6 h to remove moisture content before being purged with N_2 gas to calculate the surface area of the sample based on the adsorption as well as desorption of nitrogen molecules on the sample surface. Meanwhile, FESEM was utilized to analyze the surface morphology of each sample, synthesized with the size of the sample, porous structure, and surface texture at different magnifications of 15, 30, and 50 thousand times.

3. Results and Discussion

3.1. Chemical Properties Characterization

3.1.1. Lignocellulosic Composition Test

The results of the lignocellulosic composition test were tabulated in Table 1. Based on the result obtained, EFB had higher lignin and cellulose content than that of PF and GP. On the other hand, PF had the highest hemicellulose content, as well as lowest lignin and cellulose content, among the three precursors. The remaining dry weight of the precursors accounted for the organic solvent extractives and ash contents.

Table 1. Lignocellulosic component composition of EFB, PF, and GP.

Components (%)	EFB	PF	GP
Extractives	22.55 ± 3.30	31.10 ± 4.88	15.85 ± 0.07
Ash	6.827 ± 0.87	3.71 ± 0.18	3.84 ± 0.05
Lignin	24.67 ± 3.49	17.91 ± 2.11	19.30 ± 3.57
Hemicellulose	14.93 ± 0.20	39.51 ± 0.22	26.62 ± 0.23
Cellulose	40.69 ± 0.20	19.61 ± 0.22	37.68 ± 0.23

The specific surface area of an electro-catalyst material plays a crucial role in the electrochemical efficiency through the active sites that are present for reactions to happen [36]. Rios et al. highlighted that the biomass with high lignin content is more likely to produce carbon materials with a more macroporous structure [37]. Macropores provides diffusion channel for particles, but it also results in lower SSA, as well as the available active sites that could hinder the electrochemical performance of the carbon material. Chen et al. had separated lignin, cellulose, and hemicellulose from corn straw to form carbon material and discovered that hemicellulose provided a higher specific surface area, followed by cellulose [38]. According to Leng et al., high ash content resulted to a lower micropore structure. As the development of a porous structure is linked to the removal of volatile matters and ash, which are non-volatile inorganic residues that cause blockage of the pores and, thus, result in low micropore surfaces that account for the available active sites on the material [39]. Based on the findings above, lower lignin and ash contents, as well as higher hemicellulose content, are preferable to ensure a larger surface area of the carbon material.

3.1.2. SEM-EDX Analysis

Table 2 listed the elemental (C, N, O, S) content found on the surface of the raw biomass precursors and synthesized carbon samples at 600 °C by using SEM-EDX. By comparing the elemental content of the raw biomass precursors and the carbonized biomass precursors, the carbonization of the biomass wastes had resulted in an increase in carbon atomic percentage, as volatile matters were removed at an elevated temperature and, thus, increased the carbon purity of the samples [40]. The increase of carbon content also indicated that the raw material had broken down into carbon. Carbonization of the raw materials also caused a decrease in oxygen content, especially in garlic peel, which justified the FTIR results in the later section. The carbonization process exhibited different effects on the content of N and S heteroatoms for each precursor. The N atom content in EFB increased, while a decrease in the N content was observed for PF and GP. On the other hand, the S content in EFB and PF decreased, but increased for GP. The change in heteroatoms may be attributed to the interaction between the atoms, as well as free radicals, on the carbon surface, which occurred during the activation process [40].

Table 2. Elemental content of raw materials and synthesized materials at 600 °C with SEM-EDX.

Elements (At %)	Raw OPEFB	EFB-600-30	Raw OPF	PF-600-30	Raw GP	GP-600-30
C	54.16	80.44	53.52	79.11	51.61	86.42
N	1.29	3.07	1.46	1.35	1.63	0.82
O	44.28	16.41	44.85	19.49	46.62	12.60
S	0.26	0.08	0.18	0.05	0.14	0.16

High carbon content is an advantage as carbon possessed good conductivity, tailorable chemical properties, and superior resistance to corrosion [41]. On the other hand, the electrical resistivity of a material is subjected to the impurity present. Kolanowski et al. had reported that an excess in oxygen content might influence the storage of discharge product as it disrupts the surface conjugation of the synthesized carbon material [18]. N and S heteroatoms are common doping agents as they help to elevate the SSA of a carbon material, by means of redistributing the electron charges and alteration of spin density [42,43]. While some impurities can facilitate the electrical conductivity of the material, there are also impurities that can disrupt its conductivity performance.

3.1.3. TGA Analysis

Figure 3 present the weight percentage and derivative weight curves of raw EFB, PF and GP from room temperature to 900 °C by TGA analysis, respectively. It was observed that the derivative weight curve and weight percentage curve for all three samples displayed a small peak and a gradual slope, respectively, within room temperature to 100 °C, which represented the change in sample weight, due to the removal of moisture. Besides that, another sharper peak and sheer slope in the derivative weight curve and weight percentage curves, within the temperatures 200 to 400 °C, were observed, which aligned with the lignocellulosic component decomposition temperatures. The decomposition of lignocellulosic components had resulted in a significant drop in sample weight. The peak and slope, obtained through this characterization method, had confirmed that the lignocellulosic components decomposed within the temperature 200 to 400 °C. Nevertheless, it was observed that PF and GP exhibited a slight individual peak, at approximately 250 °C, which represented decomposition of hemicellulose [44]. This also indicated the higher hemicellulose content present in both precursors, as compared to EFB.

Above 400 °C, the changes in weight were observed to be minimal, as most lignocellulosic components had decomposed and other volatile compounds were removed. Sunphorka et al. reported that the weight change within 400 to 800 °C may be attributed to the slow lignin decomposition, due to its amorphous and highly complex structure [45]. Above 800 °C, the weight percentage curve of EFB, PF and GP exhibited minimal changes with the lowest weight percentage obtained at 6.11%, 8.59%, and 3.57%, respectively. Based on the final weight percentage, it can be concluded that PF had the highest thermal stability, as it was able to preserve more weight at an elevated temperature [46]. A precursor with a good thermal stability is preferred to prevent the loss in porous structure during the synthesis process, which will greatly impact on the availability of active sites, as well as the diffusion channel for the reaction to take place.

3.1.4. FTIR Analysis

Figure 4 displays the results obtained from the FTIR analysis carried out on the raw materials along with the selected carbon samples. It was then discovered that the O-H bond peaks within the wavenumber ranging from 3650 cm^{-1} to 3200 cm^{-1} had flattened tremendously, particularly for GP, whereby it demonstrated a drop in the natural oxygen content present in the carbonized sample, as shown in Table 2. The peak present at approximately 2900 cm^{-1} for raw GP represented the C-H bond which also flattened after carbonization [47]. Within the wavenumber ranging from 1650 cm^{-1} up to 1560 cm^{-1}, as well as 1260 cm^{-1} up to 1020 cm^{-1}, the presence of C=N double bond and C-N bond, respectively, in the samples indicated that the nitrogen was present in the raw materials which is a potential natural doping agent [48]. Many researchers had concluded that the doping of nitrogen is capable of boosting the SSA of a carbon material synthesized by redistributing the charges and various N arrangements, which enhances the interaction between the carbon and oxygen molecules [42,49]. Within the lower section of the wavenumber range (1035–1149 cm^{-1}), the amount of natural polysaccharides present in the samples were observed to have dropped after the raw materials were carbonized, which confirmed the decomposition of lignocellulosic components. However, it can be seen that the OPEFB after carbonization had a slight peak at 1035–1149 cm^{-1}, as compared to OPF and GP, which may imply that not all polysaccharides, especially lignin, were broken down for OPEFB.

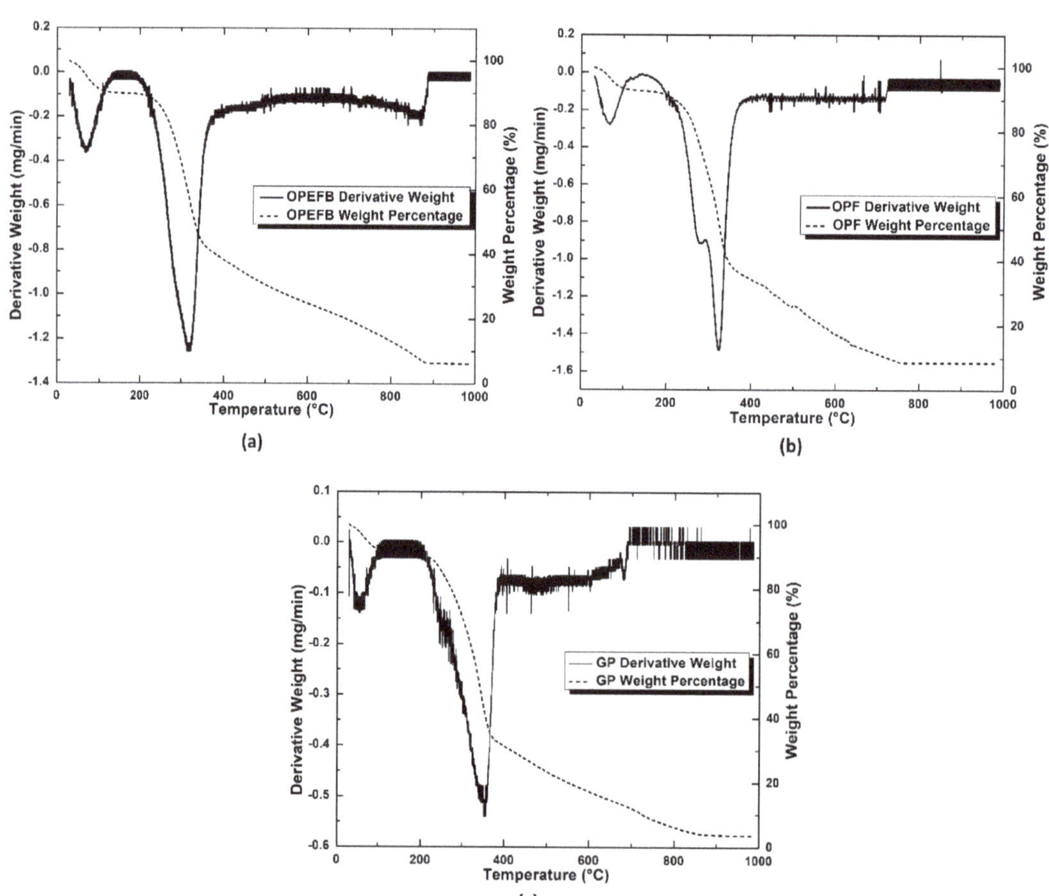

Figure 3. Weight percentage and the derivative weight curve of raw (**a**) EFB, (**b**) PF, and (**c**) GP obtained from TGA analysis.

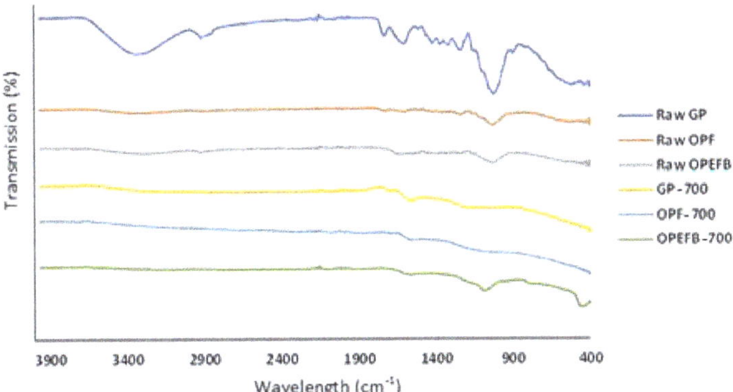

Figure 4. FTIR Spectra of Samples.

3.2. Physical Properties Characterization

3.2.1. XRD Analysis

Figure 5 depicts the XRD peaks for selected samples, which are raw EFB, raw PF, raw GP, EFB-600-30, PF-600-30, and GP-600-30, to compare the crystalline structure of each carbonized biomass precursors. All samples displayed a peak at approximately 21 to 24°, as well as a smaller peak at approximately 44°, which corresponded to the (002) and (100) standard carbon peaks [50]. It was observed that all the carbonized precursors exhibited broader peaks than the raw biomass precursors which implied that carbonization of the biomass precursors resulted in a change in the carbon structure into an amorphous structure. Another observation made was the growth in the peaks at 44 °C after carbonization of all the biomass precursors. However, it can be observed that the peaks at 44 °C for PF is slightly sharper than EFB and GP, which implied that PF may have a small degree of graphitization. According to Chakraborty et al., the extent of graphitization of a carbon material affects its electrical conductivity [51]. The degree of graphitization often increases with carbonization temperature. Bhat et al. had observed that higher the degree of graphitization caused more carbon to be hybridized into sp^2 configuration whereby electrons are more mobile which enhances the electrical conductivity of the material [52]. Hereby, it can be assumed that PF has better potential to produce graphitic carbon materials that is preferable for electrical conductivity.

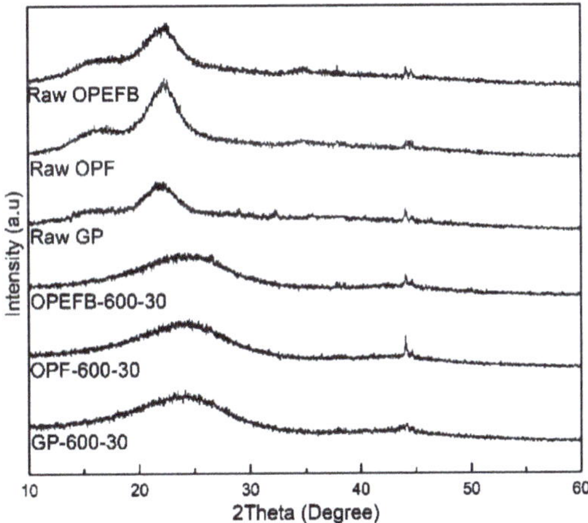

Figure 5. XRD patterns of different biomass samples.

3.2.2. BET Analysis

It was discovered from Figure 6a that increasing the carbonization temperature had caused the BET SSA of both PF and GP raw materials to increase. On the contrary, the specific surface area for EFB had increased insignificantly and then decreased even lower when the carbonization temperature was increased. PF-700-30, in turn, showed better capability in the formation of a higher specific surface area, due to its ability to obtain the greatest specific surface area of 548.26 m^2g^{-1}, when the carbonization temperature was elevated to 700 °C, in comparison with EFB-700-30 and GP-700-30 at 442.02 m^2g^{-1} and 541.01 m^2g^{-1}, respectively. This could be associated to the lignocellulosic components content in the precursors as every component have different disintegrating temperature range. The higher lignin content in EFB (24.7%), compared to PF and GP (17.9% and 19.32%, respectively) could clarify the insignificant changes in the BET SSA for EFB during the increase in carbonization temperature. The considerable rise in SSA of GP, during the rise in temperature of the carbonization process from 450 up to 600 °C could be a result of a greater content in cellulose present in GP that disintegrated at an elevated temperature in comparison to that of hemicellulose. Contrarily, PF had demonstrated a continuous hike in SSA while the temperature of the carbonization process was elevated whereby it may be associated with the greater hemicellulose content along with the decreased cellulose and lignin content in comparison to GP. These results had shown that PF do not depend on such an elevated temperature to break down and hence, it was able to obtain the highest BET SSA at 600 and 700 °C.

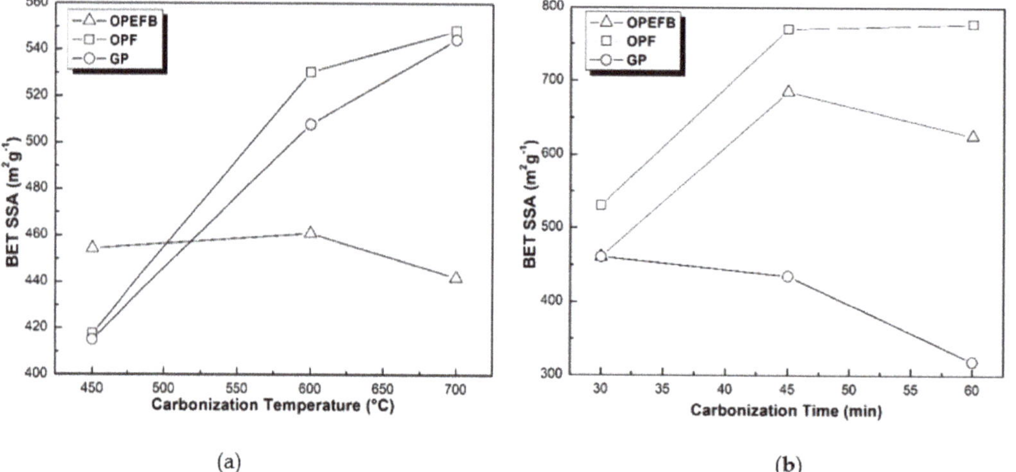

Figure 6. BET Specific Surface Area changes with (**a**) Carbonization Temperature; and (**b**) Carbonization Time.

When comparing the BET SSA of samples synthesized with different carbonization times in Figure 6b, it was observed that an increase in carbonization duration does not guarantee an increase in BET SSA. While the BET SSA of PF showed a steep increase when the carbonization time increased from 30 min to 45 min, an increase to 60 min only resulted in a slight increase in surface area. On the other hand, EFB-600-60 and GP-600-60 exhibited a lower specific surface area of 625.54 m^2g^{-1} and 319.02 m^2g^{-1} respectively compared to EFB-600-45 and GP-600-45 at 685.27 m^2g^{-1} and 435.11 m^2g^{-1}, respectively. A large specific surface area takes up a vital part in increasing the efficiency of the electro-catalyst whereby caters higher vacancy of active sites to cater to the occurrence of electrochemical reactions. Higher amount of discharge products can be stored which implied that the material exhibited high energy density [53].

Figure 7a–c shows the adsorption and desorption isotherms of the synthesized carbon samples at the carbonization temperature and time of 600 °C and 30 min. From Figure 7, all the samples exhibited type I as well as IV isotherm combination characteristics. The type I isotherm characteristics were seen within the lower relative pressure (P/P$_0$) range with high N$_2$ adsorption quantity at P/P$_0$ < 0.2. On the other hand, type IV characteristics which are represented with hysteresis loops were found within the relative pressure 0.2 < P/P$_0$ < 1.0. Both the isotherms combined together had indicated that the synthesized materials contain hierarchical porous structures with micropores, mesopores, and macropores present [54]. Hierarchical porous structure is an advantageous characteristic of an efficient electro-catalyst as mesopores improves the performance of the transfer of both electrons, as well as ions, whereas micropores are responsible for the supply of a greater specific surface area which supplies more room for the storage of discharge products [55]. Besides, macropores also aid in the transport of chemical compounds, which improves the rate of reaction that takes place in the battery [56]. The hierarchical porous structure that is present within the synthesized materials can be verified with the FESEM observation shown in Figure 8.

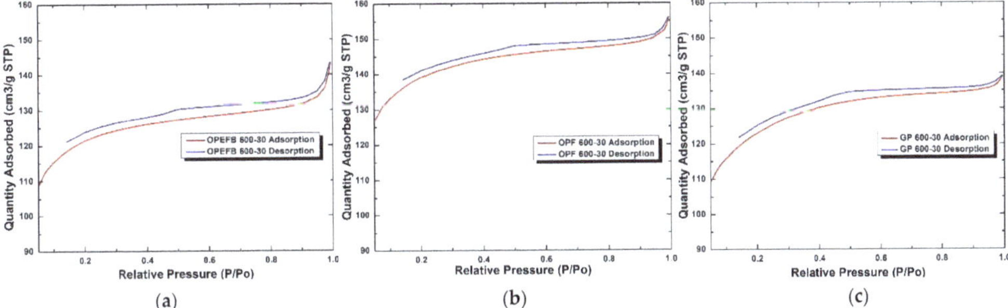

Figure 7. N$_2$ Adsorption-Desorption Isotherms of (**a**) EFB-600-30; (**b**) PF-600-30; and (**c**) GP-600-30.

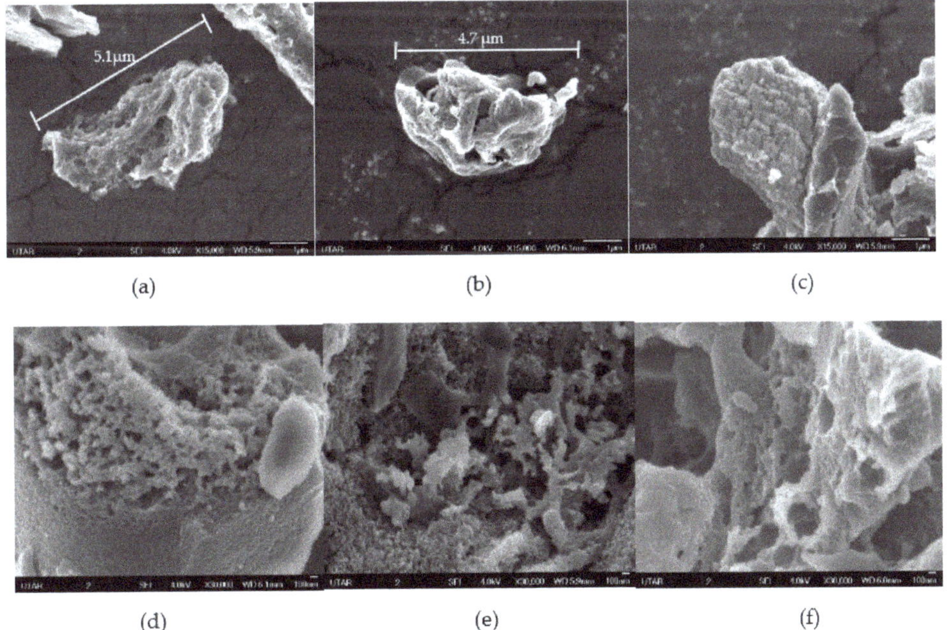

Figure 8. *Cont.*

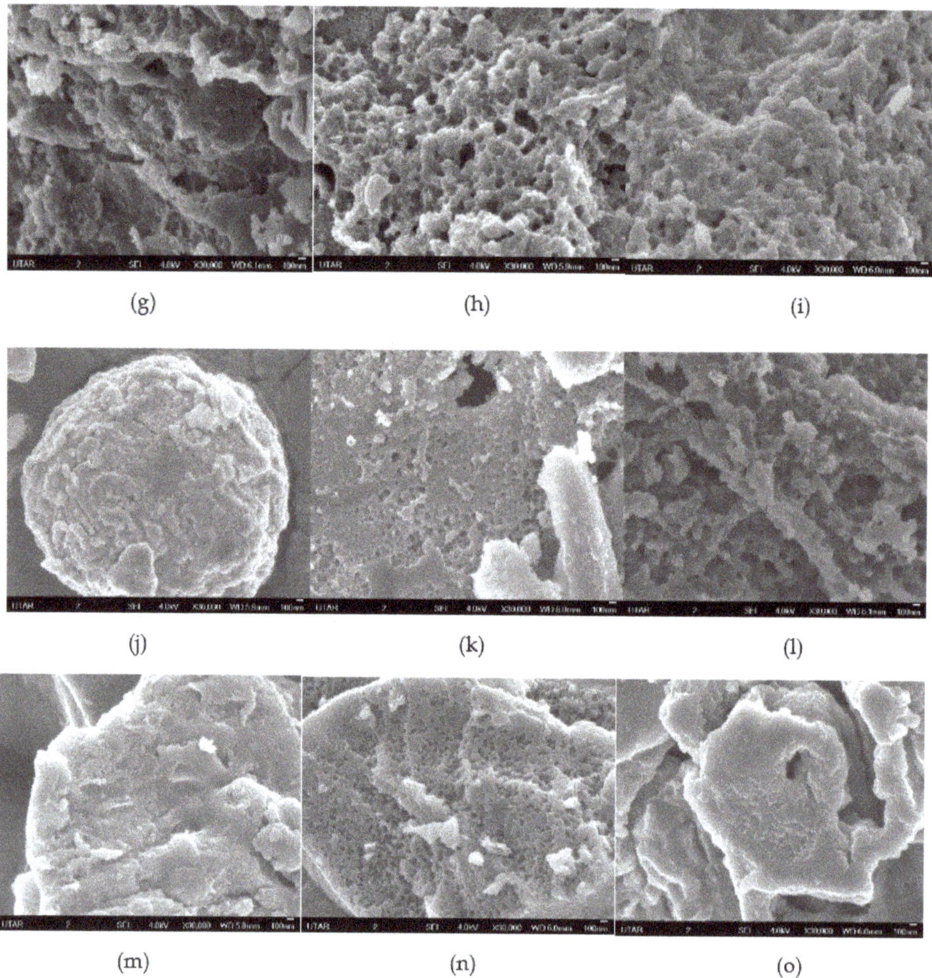

Figure 8. (a–c) EFB-450-30, PF-450-30 and GP-450-30 with 15,000 times magnification and scale of 1µm (d–o) EFB-600-30, PF-600-30, GP-600-30, EFB-700-30, PF-700-30, GP-700-30, EFB-600-45, PF-600-45, GP-600-45, EFB-600-60, PF-600-60 and GP-600-60 at 30,000 times magnification.

3.2.3. FESEM Analysis

Figure 8a–c shows EFB-450-30, PF-450-30 and GP-450 at the magnification of 15,000 times, respectively. It was observed that EFB-450-30 and PF-450-30 have achieved smaller sized particles in comparison to GP-450-30. Particles of smaller size are commonly preferred compared to bigger sized particles as it supplies more surface area with a greater number of active sites available in order for the occurrence of the electrochemical reactions. EFB-450-30 likewise had displayed more uneven surface texture compared to PF-450-30 which clarified the elevated SSA that EFB was able to achieve at carbonization temperature of 450 °C. By comparing Figure 8e,f,h, as well as Figure 8j, PF-700-30 and GP-700-30 exhibited obvious hierarchical porous structures, as well as greater structural deformity on the surface of the carbon material, in comparison to the PF and GP materials synthesized at 600 °C and 30 min. The prominent porous structure had heightened the BET SSA and hence, it was verified the elevated SSA when the carbonization temperature increased. Contrarily in

Figure 8d,g, EFB-600-30 exhibited greater porosity in comparison to EFB-700-30 which may be because of the widening of pores in the structure of the synthesized carbon from EFB when the temperature was further increased. Rios et al. found that biomasses with high lignin biomass are inclined to form macroporous structure [37].

Different precursors reacted differently to the change in carbonization time. An increase in carbonization time had resulted in producing smaller particle for EFB, as shown in Figure 8j. However in Figure 8j,m, it was observed that the surface of the OPEFB-derived carbon material particles were rougher and the porous structure reduced tremendously. Similar effect was observed for the GP-derived carbon material in Figure 8l,o whereby the pores were enlarged and less visible. By relating the surface morphology observed and the BET SSA obtained for EFB-600-45, EFB-600-60, GP-600-45 and GP-600-60, less prominent porous structure had resulted in the decrease in the SSA of the carbon material. The decrease in porosity of a carbon material could possibly be related to the thermal stability of the precursor in which the precursors were not able to withstand an elevated temperature for an extended duration. At extreme conditions, the porous structure of the material with low thermal stability collapse when the pores enlarge and the walls between pores becomes thin and weak [57,58]. According to Ornaghi et al., hemicellulose mainly attributes to the thermal stability of the lignocellulosic material, but it could not be concluded, due to the interaction between each component [59]. On the other hand, a prolonged carbonization time exhibited a positive impact on the PF-derived carbon material as its hierarchical porous structure remained intact and visible. When comparing the surface area along with the morphology of the surface of the individual material that were synthesized with respective carbonization temperatures, it can be validated that carbon materials with particles of smaller sizes as well as surfaces with immensely defective texture provided higher surface area when compared to particles with larger size and smooth texture. The hierarchical porous structure is highly desired, as the number of actives sites for reaction to take place is a function of the volume of micropores while mesopores and macropores provide channeling for ions and particles to travel which enhances the probability of occurrence of reactions [51].

Table 3 shows the comparison of the BET SSA achieved from this study with other similar studies. It can be observed that each study applied different carbonization and activation methods that produced carbon material with different SSAs despite using similar precursors. Zhang et al., had shown that a very high specific surface area of 2818 m^2g^{-1} can be achieved when the activation duration and the amount of activator added were increased [49]. From the study conducted by Bhat et al., it showed that elevated temperature itself did not necessarily result in high SSA (436.2 m^2g^{-1}) and therefore implying that activation process holds a critical part in the BET SSA elevation of a carbon material [47]. When comparing the results obtained by Bhat et al. as well as Zhang et al., the carbon material synthesized by Zhang et al. was able to achieve higher specific capacitance of 427 Fg^{-1} (0.5 Ag^{-1} current density), in comparison with Bhat et al.'s at 119.2 Fg^{-1} (0.1 Ag^{-1} current density). Despite having different current density applied, the specific capacitance of Zhang et al.'s exceeded Bhat et al.'s by several times. This indicated that the SSA of a material do affect the performance of the electro-catalyst.

According to the results obtained and observed in the previous sections, it is apparent that the change in carbonization temperature and time have different influence on the morphology of each precursor. The composition of the lignocellulosic components of each precursor is responsible for the development in porous structure as each of the component decomposes within different range of temperature in accordance with its distinctive thermal stability. For an instance, the changes in porous structure of EFB and GP were more apparent at higher carbonization temperature and longer carbonization duration, which is probably due to the complex and stretched decomposition of lignin.

By observing the characterization results in this study, it was discovered that PF possessed better potential as a raw material for the synthesis of carbon nanoparticles materials as electro-catalyst. PF had exhibited an increasing trend in BET SSA for carbonization time and temperature. In comparison with EFB and GP, PF showed better result by achieving the highest specific surface area, due to its elevated defects in the surface structure, which does not give in to the depletion in mechanical strength when the carbonation temperature and time was increased. It is also capable of achieving a smaller particle size, which implied that nano-sized particles are feasible with further experimenting, which involves extensive synthesis parameter studies, as well as electrical conductivity tests.

Table 3. SSA comparison from this study with similar studies from the literature.

Precursor	Purpose	Synthesis Condition		SSA_{BET} (m^2/g)	Remark
		Carbonization	Activation		
Garlic Peel	Supercapacitor	1000 °C, 1 h, N_2 flow 250 mL cm^{-3}, 10 °C min^{-1}	-	436.2	[47]
Garlic Stem	ZAB	HTC 180 °C, 6 h, 900 °C, 75 min, Ar flow 100 sccm	-	991.0	[60]
Garlic Peel	Supercapacitor	600 °C, 2 h, N_2 flow 5 °C min^{-1}	4:1(KOH:C) ratio, 800 °C, 2.5 h N_2 flow 5 °C min^{-1}	2818.0	[49]
Garlic Peel	MAB	700 °C, 30 min	1:3(KOH:C) ratio, 600 °C, 30 min	541.0	This study
Empty Fruit Bunch	Wastewater treatment	900 °C, 15 min, CO_2 0.1 $Lmin^{-1}$	-	345.0	[61]
Empty Fruit Bunch	Wastewater treatment	900 °C, 15 min, Steam 2.0 mL min^{-1}	-	635.6	[61]
Empty Fruit Bunch	MAB	600 °C, 30 min	1:3(KOH:C) ratio, 600 °C, 30 min	460.9	This study
Oil Palm Frond	Wastewater treatment	500 °C, 2 h, Steam 100 cm^3 min^{-1}, 10 °C min^{-1}	-	457.7	[30]
Oil Palm Frond	MAB	700 °C, 30 min	1:3(KOH:C) ratio, 600 °C, 30 min	548.26	This study

4. Conclusions

To conclude this study, three different biomass wastes, including EFB, PF, and GP, were examined, in order to choose a good raw material in the synthesis of carbon nanoparticle as an electro-catalyst for MAB. According to the obtained results, PF was determined to be the most suitable precursor, as it was capable of achieving the highest SSA at an elevated temperature of 700 °C (548.26 m^2g^{-1}) and time of 60 min (777.62 m^2g^{-1}), due to the hierarchical porous structure and smaller particle size achieved, which are advantageous characteristics for enhancing the electrochemical efficacy of an electro-catalyst for energy storage. PF is a potential alternative to replace the current costly CNT as a material for electro-catalysts, which is greener and sustainable, yet cost-efficient. A more in-depth investigation on the electrical conductivity test will be carried out to further ascertain its potential as an electro-catalyst for MAB.

Author Contributions: Conceptualization, B.A.-L.L. and S.L.; methodology, B.A.-L.L. and S.L.; validation, Y.L.P., S.H.S. and K.H.W.; formal analysis, B.A.-L.L. and S.L.; investigation, B.A.-L.L. and S.L.; resources, S.H.S. and K.H.W.; writing—original draft preparation, B.A.-L.L. and S.L.; writing—review and editing, B.A.-L.L., S.L. and J.B.O.; supervision, S.L., Y.L.P., S.H.S. and K.H.W.; funding acquisition, S.L. and Y.L.P. All authors have read and agreed to the published version of the manuscript.

Funding: This research was fully funded by Ministry of Education, Malaysia, through the Fundamental Research Grant Scheme (FRGS/1/2018/TK10/UTAR/02/1), as well as the Universiti Tunku Abdul Rahman (UTAR) Research Grant (IPSR/RMC/UTARRF/2020-C2/W01).

Institutional Review Board Statement: Not applicable.

Informed Consent Statement: Not applicable.

Data Availability Statement: The data that was presented in this paper will be available on request from the corresponding authors.

Acknowledgments: The authors would like to express their thank to the Ministry of Education (MOE), Malaysia, for providing the Fundamental Research Grant Scheme (FRGS/1/2018/TK10/UTAR/02/1), as well as Universiti Tunku Abdul Rahman (UTAR) Research Grant (IPSR/RMC/UTARRF/2020-C2/W01) for the financial support and Universiti Tunku Abdul Rahman Research Publication Scheme (6251/S06) for the scholarship to Brenda Lim Ai-Lian.

Conflicts of Interest: The participating authors had acknowledged that there is no conflict of interest.

References

1. Praene, J.P.; Fakra, D.A.H.; Benard, F.; Ayagapin, L.; Rachadi, M.N.M. Comoros's Energy Review for Promoting Renewable Energy Sources. *Renew. Energy* **2021**, *169*, 885–893. [CrossRef]
2. Liedel, C. Sustainable Battery Materials from Biomass. *ChemSusChem* **2020**, *13*, 2110–2141. [CrossRef]
3. Mori, R. Recent Developments for Aluminum–Air Batteries. *Electrochem. Energy Rev.* **2020**, *3*, 344–369. [CrossRef]
4. Chen, T.; Jin, Y.; Lv, H.; Yang, A.; Liu, M.; Chen, B.; Xie, Y.; Chen, Q. Applications of Lithium-Ion Batteries in Grid-Scale Energy Storage Systems. *Trans. Tianjin Univ.* **2020**, *26*, 208–217. [CrossRef]
5. Clark, S.; Latz, A.; Horstmann, B. A Review of Model-Based Design Tools for Metal-Air Batteries. *Batteries* **2018**, *4*, 5. [CrossRef]
6. Wang, C.; Yu, Y.; Niu, J.; Liu, Y.; Bridges, D.; Liu, X.; Pooran, J.; Zhang, Y.; Hu, A. Recent Progress of Metal–Air Batteries-A Mini Review. *Appl. Sci.* **2019**, *9*, 2787. [CrossRef]
7. Zhong, M.; Liu, M.; Li, N.; Bu, X.H. Recent Advances and Perspectives of Metal/Covalent-Organic Frameworks in Metal-Air Batteries. *J. Energy Chem.* **2021**, *63*, 113–129. [CrossRef]
8. Lu, G.; Li, Z.; Fan, W.; Wang, M.; Yang, S.; Li, J.; Chang, Z.; Sun, H.; Liang, S.; Liu, Z. Sponge-like N-Doped Carbon Materials with Co-Based Nanoparticles Derived from Biomass as Highly Efficient Electrocatalysts for the Oxygen Reduction Reaction in Alkaline Media. *RSC Adv.* **2019**, *9*, 4843–4848. [CrossRef]
9. Ha, T.A.; Pozo-Gonzalo, C.; Nairn, K.; MacFarlane, D.R.; Forsyth, M.; Howlett, P.C. An Investigation of Commercial Carbon Air Cathode Structure in Ionic Liquid Based Sodium Oxygen Batteries. *Sci. Rep.* **2020**, *10*, 7123. [CrossRef]
10. Sathiskumar, C.; Ramakrishnan, S.; Vinothkannan, M.; Kim, A.R.; Karthikeyan, S.; Yoo, D.J. Nitrogen-Doped Porous Carbon Derived from Biomass Used as Trifunctional Electrocatalyst toward Oxygen Reduction, Ox-ygen Evolution and Hydrogen Evolution Reactions. *Nanomaterials* **2020**, *10*, 76. [CrossRef] [PubMed]
11. Sekhon, S.S.; Lee, J.; Park, J.S. Biomass-Derived Bifunctional Electrocatalysts for Oxygen Reduction and Evolution Reaction: A Review. *J. Energy Chem.* **2021**, *65*, 149–172. [CrossRef]
12. Malik, S.; Marchesan, S. Growth, Properties, and Applications of Branched Carbon Nanostructures. *Nanomaterials* **2021**, *11*, 2728. [CrossRef] [PubMed]
13. Gopalakrishnan, A.; Badhulika, S. Effect of Self-Doped Heteroatoms on the Performance of Biomass-Derived Carbon for Supercapacitor Applications. *J. Power Sources* **2020**, *480*, 228830. [CrossRef]
14. He, G.; Yan, G.; Song, Y.; Wang, L. Biomass Juncus Derived Nitrogen-Doped Porous Carbon Materials for Supercapacitor and Oxygen Reduction Reaction. *Front. Chem.* **2020**, *8*, 226. [CrossRef] [PubMed]
15. Hu, W.; Xiang, R.; Lin, J.; Cheng, Y.; Lu, C. Lignocellulosic Biomass-Derived Carbon Electrodes for Flexible Supercapacitors: An Overview. *Materials* **2021**, *14*, 4571. [CrossRef]
16. Wang, N.; Li, T.; Song, Y.; Liu, J.; Wang, F. Metal-Free Nitrogen-Doped Porous Carbons Derived from Pomelo Peel Treated by Hypersaline Environments for Oxygen Reduction Reaction. *Carbon* **2018**, *130*, 692–700. [CrossRef]
17. Baruah, J.; Nath, B.K.; Sharma, R.; Kumar, S.; Deka, R.C.; Baruah, D.C.; Kalita, E. Recent Trends in the Pretreatment of Lignocellulosic Biomass for Value-Added Products. *Front. Energy Res.* **2018**, *6*, 141. [CrossRef]
18. Kolanowski, L.; Graś, M.; Bartkowiak, M.; Doczekalska, B.; Lota, G. Electrochemical Capacitors Based on Electrodes Made of Lignocellulosic Waste Materials. *Waste Biomass Valorization* **2020**, *11*, 3863–3871. [CrossRef]
19. Hendriansyah, R.; Prakoso, T.; Widiatmoko, P.; Nurdin, I.; Devianto, H. Manufacturing Carbon Material by Carbonization of Cellulosic Palm Oil Waste for Supercapacitor Material. *MATEC Web Conf.* **2018**, *156*, 03018. [CrossRef]
20. Han, J.; Lee, J.H.; Roh, K.C. Herbaceous Biomass Waste-Derived Activated Carbons for Supercapacitors. *J. Electrochem. Sci. Technol.* **2018**, *9*, 157–162. [CrossRef]
21. Elmouwahidi, A.; Bailón-García, E.; Pérez-Cadenas, A.F.; Maldonado-Hódar, F.J.; Carrasco-Marín, F. Activated Carbons from KOH and H_3PO_4-Activation of Olive Residues and Its Application as Supercapacitor Electrodes. *Electrochim. Acta* **2017**, *229*, 219–228. [CrossRef]
22. Kim, M.J.; Park, J.E.; Kim, S.; Lim, M.S.; Jin, A.; Kim, O.H.; Kim, M.J.; Lee, K.S.; Kim, J.; Kim, S.S.; et al. Biomass-Derived Air Cathode Materials: Pore-Controlled S,N-Co-Doped Carbon for Fuel Cells and Metal–Air Batteries. *ACS Catal.* **2019**, *9*, 3389–3398. [CrossRef]

23. Hao, X.; Chen, W.; Jiang, Z.; Tian, X.; Hao, X.; Maiyalagan, T.; Jiang, Z.J. Conversion of Maize Straw into Nitrogen-Doped Porous Graphitized Carbon with Ultra-High Surface Area as Excellent Oxygen Reduction Electrocatalyst for Flexible Zinc–Air Batteries. *Electrochim. Acta* **2020**, *362*, 137143. [CrossRef]

24. Zainol, M.M.; Amin, N.A.S.; Asmadi, M. Preparation and Characterization of Impregnated Magnetic Particles on Oil Palm Frond Activated Carbon for Metal Ions Removal. *Sains Malays.* **2017**, *46*, 773–782. [CrossRef]

25. Muhamad, N.A.S.; Hanoin, M.A.H.M.; Mokhtar, N.M.; Lau, W.J.; Jaafar, J. Industrial Application of Membrane Distillation Technology Using Palm Oil Mill Effluent in Malaysia. *Mater. Today Proc.* **2021**, 575. [CrossRef]

26. Nurika, I.; Shabrina, E.N.; Azizah, N.; Suhartini, S.; Bugg, T.D.H.; Barker, G.C. Application of Ligninolytic Bacteria to the Enhancement of Lignocellulose Breakdown and Methane Production from Oil Palm Empty Fruit Bunches (OPEFB). *Bioresour. Technol. Rep.* **2022**, *17*, 100951. [CrossRef]

27. Mahmud, M.S.; Chong, K.P. Formulation of Biofertilizers from Oil Palm Empty Fruit Bunches and Plant Growth-Promoting Microbes: A Comprehensive and Novel Approach towards Plant Health. *J. King Saud Univ.-Sci.* **2021**, *33*, 101647. [CrossRef]

28. Fatihah Salleh, S.; Abd Rahman, A.; Ab Rashid Tuan Abdullah, T. Potential of Deploying Empty Fruit Bunch (EFB) for Biomass Cofiring in Malaysia's Largest Coal Power Plant. In Proceedings of the 2018 IEEE International Conference on Power and Energy (PECon2018), Kuala Lumpur, Malaysia, 3–4 December 2018.

29. Department of Statistics Malaysia Malaysian CPI Inflation Calculator. Available online: https://www.dosm.gov.my/cpi_calc/ (accessed on 22 December 2021).

30. Lawal, A.A.; Hassan, M.A.; Farid, M.A.A.; Yasim-Anuar, T.A.T.; Yusoff, M.Z.M.; Zakaria, M.R.; Roslan, A.M.; Mokhtar, M.N.; Shirai, Y. Production of Biochar from Oil Palm Frond by Steam Pyrolysis for Removal of Residual Contaminants in Palm Oil Mill Effluent Final Discharge. *J. Clean. Prod.* **2020**, *265*, 121643. [CrossRef]

31. Osman, N.B.; Shamsuddin, N.; Uemura, Y. Activated Carbon of Oil Palm Empty Fruit Bunch (EFB); Core and Shaggy. *Procedia Eng.* **2016**, *148*, 758–764. [CrossRef]

32. Zeng Wei, H. Effect of Pretreatment for Synthesis of Oil Palm Frond Based Catalyst for Biodiesel Production. Ph.D. Thesis, Universiti Tunku Abdul Rahman, Kampar, Perak, Malaysia, May 2019.

33. Rodríguez-Solana, R.; Salgado, J.M.; Domínguez, J.M.; Cortés-Diéguez, S. Characterization of Fennel Extracts and Quantification of Estragole: Optimization and Comparison of Accelerated Solvent Extraction and Soxhlet Techniques. *Ind. Crops Prod.* **2014**, *52*, 528–536. [CrossRef]

34. Sluiter, A.; Ruiz, R.; Scarlata, C.; Sluiter, J.; Templeton, D. Determination of Extractives in Biomass: Laboratory Analytical Procedure (LAP). 2008. Available online: https://www.nrel.gov/docs/gen/fy08/42619.pdf (accessed on 19 December 2021).

35. Chin, D.W.K.; Lim, S.; Pang, Y.L.; Lim, C.H.; Shuit, S.H.; Lee, K.M.; Chong, C.T. Effects of Organic Solvents on the Organosolv Pretreatment of Degraded Empty Fruit Bunch for Fractionation and Lignin Removal. *Sustainability* **2021**, *13*, 6757. [CrossRef]

36. Wang, Y.J.; Fang, B.; Zhang, D.; Li, A.; Wilkinson, D.P.; Ignaszak, A.; Zhang, L.; Zhang, J. *A Review of Carbon-Composited Materials as Air-Electrode Bifunctional Electrocatalysts for Metal-Air Batteries*; Springer: Singapore, 2018; Volume 1, ISBN 0123456789.

37. Del Mar Saavedra Rios, C.; Simone, V.; Simonin, L.; Martinet, S.; Dupont, C. Biochars from Various Biomass Types as Precursors for Hard Carbon Anodes in Sodium-Ion Batteries. *Biomass Bioenergy* **2018**, *117*, 32–37. [CrossRef]

38. Chen, S.; Xia, Y.; Zhang, B.; Chen, H.; Chen, G.; Tang, S. Disassembly of Lignocellulose into Cellulose, Hemicellulose, and Lignin for Preparation of Porous Carbon Materials with Enhanced Performances. *J. Hazard. Mater.* **2021**, *408*, 124956. [CrossRef] [PubMed]

39. Leng, L.; Xiong, Q.; Yang, L.; Li, H.; Zhou, Y.; Zhang, W.; Jiang, S.; Li, H.; Huang, H. An Overview on Engineering the Surface Area and Porosity of Biochar. *Sci. Total Environ.* **2021**, *763*, 144204. [CrossRef] [PubMed]

40. Hendrawan, Y.; Sajidah, N.; Umam, C.; Fauzy, M.R.; Wibisono, Y.; Hawa, L.C. Effect of Carbonization Temperature Variations and Activator Agent Types on Activated Carbon Characteristics of Sengon Wood Waste (*Paraserianthes Falcataria* (L.) *Nielsen*). In Proceedings of the 12th International Interdisciplinary Studies Seminar—Environmental Conservation and Education for Sustainable Development, Malang, Indonesia, 14–15 November 2018; IOP Publishing: Bristol, UK, 2019; Volume 239.

41. Zhao, R.; Ni, B.; Wu, L.; Sun, P.; Chen, T. Carbon-Based Iron-Cobalt Phosphate FeCoP/C as an Effective ORR/OER/HER Triunctional Electrocatalyst. *Colloids Surf. A Physicochem. Eng. Asp.* **2022**, *635*, 128118. [CrossRef]

42. Oh, W.-D.; Lisak, G.; Webster, R.D.; Liang, Y.N.; Veksha, A.; Giannis, A.; Moo, J.G.S.; Lim, J.W.; Lim, T.T. Insights into the Thermolytic Transformation of Lignocellulosic Biomass Waste to Redox-Active Carbocatalyst: Durability of Surface Active Sites. *Appl. Catal. B Environ.* **2018**, *233*, 120–129. [CrossRef]

43. Zhao, L.; Ning, G.; Zhang, S. Green Synthesis of S-Doped Carbon Nanotubes via Gaseous Post-Treatment and Their Application as Conductive Additive in Li Ion Batteries. *Carbon* **2021**, *179*, 425–434. [CrossRef]

44. Babinszki, B.; Jakab, E.; Terjék, V.; Sebestyén, Z.; Várhegyi, G.; May, Z.; Mahakhant, A.; Attanatho, L.; Suemanotham, A.; Thanmongkhon, Y.; et al. Thermal Decomposition of Biomass Wastes Derived from Palm Oil Production. *J. Anal. Appl. Pyrolysis* **2021**, *155*, 105069. [CrossRef]

45. Sunphorka, S.; Chalermsinsuwan, B.; Piumsomboon, P. Artificial Neural Network Model for the Prediction of Kinetic Parameters of Biomass Pyrolysis from Its Constituents. *Fuel* **2017**, *193*, 142–158. [CrossRef]

46. Xu, C.A.; Qu, Z.; Meng, H.; Chen, B.; Wu, X.; Cui, X.; Wang, K.; Wu, K.; Shi, J.; Lu, M. Effect of Polydopamine-Modified Multi-Walled Carbon Nanotubes on the Thermal Stability and Conductivity of UV-Curable Polyurethane/Polysiloxane Pressure-Sensitive Adhesives. *Polymer* **2021**, *223*, 123615. [CrossRef]

47. Bhat, V.S.; Kanagavalli, P.; Sriram, G.; B, R.P.; John, N.S.; Veerapandian, M.; Kurkuri, M.; Hegde, G. Low Cost, Catalyst Free, High Performance Supercapacitors Based on Porous Nano Carbon Derived from Agriculture Waste. *J. Energy Storage* **2020**, *32*, 101829. [CrossRef]
48. Li, J.; Dou, B.; Zhang, H.; Zhang, H.; Chen, H.; Xu, Y. Thermochemical Characteristics and Non-Isothermal Kinetics of Camphor Biomass Waste. *J. Environ. Chem. Eng.* **2021**, *9*. [CrossRef]
49. Zhang, Q.; Han, K.; Li, S.; Li, M.; Li, J.; Ren, K. Synthesis of Garlic Skin-Derived 3D Hierarchical Porous Carbon for High-Performance Supercapacitors. *Nanoscale* **2018**, *10*, 2427–2437. [CrossRef] [PubMed]
50. Zhao, C.; Liu, G.; Sun, N.; Zhang, X.; Wang, G.; Zhang, Y.; Zhang, H.; Zhao, H. Biomass-Derived N-Doped Porous Carbon as Electrode Materials for Zn-Air Battery Powered Capacitive Deionization. *Chem. Eng. J.* **2018**, *334*, 1270–1280. [CrossRef]
51. Chakraborty, I.; Sathe, S.M.; Dubey, B.K.; Ghangrekar, M.M. Waste-Derived Biochar: Applications and Future Perspective in Microbial Fuel Cells. *Bioresour. Technol.* **2020**, *312*, 123587. [CrossRef]
52. Xie, Y.; Feng, C.; Guo, Y.; Hassan, A.; Li, S.; Zhang, Y.; Wang, J. Dimethylimidazole and Dicyandiamide Assisted Synthesized Rich-Defect and Highly Dispersed CuCo-N$_x$ Anchored Hollow Graphite Carbon Nanocages as Efficient Trifunctional Electrocatalyst in the Same Electrolyte. *J. Power Sources* **2022**, *517*, 230721. [CrossRef]
53. Chen, W.; Gong, Y.F.; Liu, J.H. Recent Advances in Electrocatalysts for Non-Aqueous Li–O$_2$ Batteries. *Chin. Chem. Lett.* **2017**, *28*, 709–718. [CrossRef]
54. Peng, L.; Liang, Y.; Dong, H.; Hu, H.; Zhao, X.; Cai, Y.; Xiao, Y.; Liu, Y.; Zheng, M. Super-Hierarchical Porous Carbons Derived from Mixed Biomass Wastes by a Stepwise Removal Strategy for High-Performance Supercapacitors. *J. Power Sources* **2018**, *377*, 151–160. [CrossRef]
55. Lei, W.; Deng, Y.P.; Li, G.; Cano, Z.P.; Wang, X.; Luo, D.; Liu, Y.; Wang, D.; Chen, Z. Two-Dimensional Phosphorus-Doped Carbon Nanosheets with Tunable Porosity for Oxygen Reactions in Zinc-Air Batteries. *ACS Catal.* **2018**, *8*, 2464–2472. [CrossRef]
56. Seow, Y.X.; Tan, Y.H.; Mubarak, N.M.; Kansedo, J.; Khalid, M.; Ibrahim, M.L.; Ghasemi, M. A Review on Biochar Production from Different Biomass Wastes by Recent Carbonization Technologies and Its Sustainable Applications. *J. Environ. Chem. Eng.* **2022**, *10*, 107017. [CrossRef]
57. Ahmad, M.A.; Herawan, S.G.; Yusof, A.A. Effect of Activation Time on the Pinang Frond Based Activated Carbon for Remazol Brilliant Blue R Removal. *J. Mech. Eng. Sci.* **2014**, *7*, 1085–1093. [CrossRef]
58. Fu, Y.; Shen, Y.; Zhang, Z.; Ge, X.; Chen, M. Activated Bio-Chars Derived from Rice Husk via One- and Two-Step KOH-Catalyzed Pyrolysis for Phenol Adsorption. *Sci. Total Environ.* **2019**, *646*, 1567–1577. [CrossRef] [PubMed]
59. Ornaghi, H.L.; Ornaghi, F.G.; Neves, R.M.; Monticeli, F.; Bianchi, O. Mechanisms Involved in Thermal Degradation of Lignocellulosic Fibers: A Survey Based on Chemical Composition. *Cellulose* **2020**, *27*, 4949–4961. [CrossRef]
60. Ma, Z.; Wang, K.; Qiu, Y.; Liu, X.; Cao, C.; Feng, Y.; Hu, P.A. Nitrogen and Sulfur Co-Doped Porous Carbon Derived from Bio-Waste as a Promising Electrocatalyst for Zinc-Air Battery. *Energy* **2018**, *143*, 43–55. [CrossRef]
61. Amosa, M.K.; Jami, M.S.; AlKhatib, M.F.R.; Jimat, D.N.; Muyibi, S.A. Comparative and Optimization Studies of Adsorptive Strengths of Activated Carbons Produced from Steam- and CO$_2$-Activation for BPOME Treatment. *Adv. Environ. Biol.* **2014**, *8*, 603–612.

sustainability

MDPI

Article

Enhanced Sonocatalytic Performance of Non-Metal Graphitic Carbon Nitride (g-C$_3$N$_4$)/Coconut Shell Husk Derived-Carbon Composite

Yean Ling Pang [1,2,*], Aaron Zhen Yao Koe [1], Yin Yin Chan [1], Steven Lim [1,2] and Woon Chan Chong [1,2]

[1] Department of Chemical Engineering, Lee Kong Chian Faculty of Engineering and Science, Universiti Tunku Abdul Rahman, Kajang 43000, Malaysia; aaronkoe@1utar.my (A.Z.Y.K.); 1302974@1utar.my (Y.Y.C.); stevenlim@utar.edu.my (S.L.); chongwchan@utar.edu.my (W.C.C.)

[2] Centre for Photonics and Advanced Materials Research, Universiti Tunku Abdul Rahman, Kajang 43000, Malaysia

* Correspondence: pangyl@utar.edu.my; Tel.: +60-39-086-0288; Fax: +60-39-019-8868

Abstract: This study focused on the modification of graphitic carbon nitride (g-C$_3$N$_4$) using carbon which was obtained from the pyrolysis of coconut shell husk. The sonocatalytic performance of the synthesized samples was then studied through the degradation of malachite green. In this work, pure g-C$_3$N$_4$, pure carbon and carbon/g-C$_3$N$_4$ composites (C/g-C$_3$N$_4$) at different weight percentages were prepared and characterized by using XRD, SEM-EDX, FTIR, TGA and surface analysis. The effect of carbon amount in the C/g-C$_3$N$_4$ composites on the sonocatalytic performance was studied and 10 wt% C/g-C$_3$N$_4$ showed the best catalytic activity. The optimization study was conducted by using response surface methodology (RSM) with a central composite design (CCD) model. Three experimental parameters were selected in RSM including initial dye concentration (20 to 25 ppm), initial catalyst loading (0.3 to 0.5 g/L), and solution pH (4 to 8). The model obtained was found to be significant and reliable with R^2 value (0.9862) close to unity. The degradation efficiency of malachite green was optimized at 97.11% under the conditions with initial dye concentration = 20 ppm, initial catalyst loading = 0.5 g/L, solution pH = 8 after 10 min. The reusability study revealed the high stability of 10 wt% C/g-C$_3$N$_4$ as sonocatalyst. In short, 10 wt% C/g-C$_3$N$_4$ has a high potential for industrial application since it is cost effective, reusable, sustainable, and provides good sonocatalytic performance.

Keywords: g-C$_3$N$_4$; carbon composite; coconut shell husk; characteristic; sonocatalytic degradation; malachite green

Citation: Pang, Y.L.; Koe, A.Z.Y.; Chan, Y.Y.; Lim, S.; Chong, W.C. Enhanced Sonocatalytic Performance of Non-Metal Graphitic Carbon Nitride (g-C$_3$N$_4$)/Coconut Shell Husk Derived-Carbon Composite. *Sustainability* **2022**, *14*, 3244. https://doi.org/10.3390/su14063244

Academic Editor: Farooq Sher

Received: 26 December 2021
Accepted: 25 February 2022
Published: 10 March 2022

Publisher's Note: MDPI stays neutral with regard to jurisdictional claims in published maps and institutional affiliations.

1. Introduction

Malaysia is a developing country that is currently experiencing economic growth with a rapid transition to an urban and industrialized society. This change has created various environmental problems including water pollution caused by the increase in the discharge of wastewater in Malaysia [1]. Water contamination may affect the welfare of all living things, including humans, and the economy of the country. Water pollution also impacts greatly the sustainability of water resources [2,3]. This implies that the discharge of untreated wastewater is a major challenge which must be monitored strictly in order to establish sustainable development.

With the growth in population, water demand will rise along with the discharge of municipal wastewater. However, the clean water supply for domestic purposes will be limited with the increasing number of polluted rivers in Malaysia. Therefore, water scarcity is going to be a major problem in Malaysia for coming years. The discharge of dye effluent from industry is one of the root causes of water pollution. Organic or synthetic dyes may display noticeable color in the water body even at low concentrations. The presence of dye molecules in water will block the penetration of sunlight which suppresses photosynthesis activities of aquatic plants. This will in turn cause the DO level in the river to decrease and harm the ecosystem [4]. Dyes are carcinogenic and highly toxic which will give rise to health and environmental problems. Living things that consume water from these polluted sources will have an increased risk of cancer while the toxic property of dyes will endanger human health and even animals [5].

With the world transforming towards more sustainable technology, sonocatalytic technology is in the limelight in wastewater treatment due to being environmentally friendly and highly efficient in the degradation of organic pollutants by producing enormous amounts of reactive oxygen species [6,7]. The common semiconductor used for this treatment method is titanium dioxide (TiO_2). However, TiO_2 has disadvantages of a wide band gap energy that results in unsatisfactory catalytic performance [8]. Since the main focus of treating the wastewater is to overcome water pollution, a metal-free semiconductor is proposed in this research. g-C_3N_4 is a metal-free sonocatalyst that was discovered recently [9]. With it being metal-free, g-C_3N_4 will be more cost effective than TiO_2 due to the abundance and renewability of raw materials [10]. The main drawback of g-C_3N_4 is the high recombination rate of photo-induced carriers, which can be countered by incorporating carbon materials [11].

To the best of our knowledge, there is no work reported on the modification of g-C_3N_4 using carbon materials derived from coconut shell husk. The conversion of unwanted husks into a value-added carbon material can extend the life cycle of biomass waste. Thus, this research focused on the synthesis of g-C_3N_4 incorporated by carbon material that was obtained from coconut shell husks. In this work, a sonocatalytic technique was also proposed to remove organic pollutants from wastewater using the synthesized g-C_3N_4 composites. The generation of g-C_3N_4 doped with carbon derived from coconut shell husk fulfilled the principles of green chemistry including waste prevention, designation for catalytic degradation as well as utilization of less hazardous and renewable raw materials.

2. Materials and Methods

2.1. Materials

Urea (CH_4N_2O, purity $\geq$ 99%, CAS-No. 57-13-6), sodium hydroxide pellets (NaOH, 99% purity, CAS-No. 1310-73-2), and malachite green (99% purity, CAS-No. 2437-29-8) were obtained from R&M Chemicals, Malaysia. Hydrochloric Acid (HCl, 37% purity, CAS-No. 7647-01-0) was obtained from Sigma-Aldrich, United States while hydrogen peroxide (H_2O_2, 30% purity, CAS-No. 7722-84-1) was obtained from SYSTERM, Malaysia. Deionised water was used throughout the study.

2.2. Preparation of Pure Carbon, Pure g-C_3N_4, and Carbon Composited with g-C_3N_4

To synthesis carbon material, the coconut shell husk was firstly dried for 2 days at 70 °C to remove the moisture of the husk. The dried husk was then carbonized through pyrolysis at 750 °C for 1 h. The resulting samples were crushed into powder form by using a mortar and pestle. For the preparation of pure g-C_3N_4, 5 g of urea was placed in an open crucible and heated in a carbolite furnace at 500 °C for 2 h. The resulting g-C_3N_4 sample was then ground into fine powder. These steps were repeated to synthesize C/g-C_3N_4 composites. The C/g-C_3N_4 composites at different weight percentages were prepared by adding a 0.05 g, 0.25 g, 0.5 g, 0.75 g, 1.00 g and 1.25 g of carbon in the urea to obtain 1 wt% C/g-C_3N_4, 5 wt% C/g-C_3N_4, 10 wt% C/g-C_3N_4, 15 wt% C/g-C_3N_4, 20 wt% C/g-C_3N_4, and 25 wt% C/g-C_3N_4, respectively.

2.3. Sample Characterization

Several pieces of equipment were used to characterize the synthesized samples. Characterization is an important step in studying the structure of the catalyst surface, composition of the catalyst, and chemical properties of the catalyst. The characterization of the catalyst could provide the explanation for its catalytic activity. The analysis techniques applied in this research involved X-ray diffraction analysis (XRD), scanning electron microscopy–energy dispersive X-ray (SEM-EDX), Fourier-transform infrared spectroscopy (FTIR), surface analysis and thermogravimetric analysis (TGA). XRD provided information on the crystalline structure and gave an estimation for the crystallite size. The characterization of the samples was carried out by means of Shidmazu Model XRD-6000 (Japan) using Cu-Kα radiation. The X-ray tube was set to the applied current of 25 mA and voltage of 45 kV. The prepared samples were scanned at 2θ starting from $10°$ and ending at $60°$.

SEM-EDX was conducted using the Hitachi S-3400N (Japan). The morphological information of samples could be obtained through SEM. The addition of EDX gave results on the relative abundance of elements present on the surface of the sample. The scanning voltage was set at 5.0 kV. FTIR could provide identification on the functional groups present by detecting the bonding between atoms. The characteristic peaks displayed in the graph were due to specific chemical bonding. In this work, FTIR was carried out using Nicolet Model IS10 from Thermo Scientific (Waltham, MA, USA). Surface analysis was carried out in this study to determine the surface area and porosity of samples using Micromeritics 3Flex (USA). In addition, the thermal stability of the synthesized sample was studied through TGA using Perkin Elmer STA 8000 (USA).

2.4. Sonocatalytic Degradation of Organic Dyes and Analysis of Liquid Samples

Typically, 0.3 g/L of catalyst was added to a beaker containing 100 mL of 15 ppm malachite green aqueous solution. Ultrasonic irradiation was achieved by means of a Hielscher UP400S (Germany) ultrasonic processor, which was operated at an effective output power of 80 W. The treated liquid sample was collected after 10 min of ultrasonic irradiation. The dye solution was separated from the catalyst by using a syringe filter and the residual concentration of malachite green was determined using a UV–Vis spectrophotometer (model CARY 100 from Agilent, Santa Clara, CA, USA). The measurement of maximum absorbance was performed at 617 nm.

3. Results and Discussion

3.1. Sample Characterization

3.1.1. XRD Analysis

XRD results for pure g-C_3N_4, pure carbon, and C/g-C_3N_4 composites at different weight percentages are shown in Figure 1. There were three peaks observed at 2θ = 27.5°, 37.9°, and 44.1° in the XRD pattern of pure g-C_3N_4. The existence of the dominant diffraction peak at 27.5° was due to the (002) plane which represented the interlayer stacking of the conjugated aromatic system [12]. This peak confirmed the presence of the graphitic-like layered structure of carbon nitride. In other words, this finding proved that the sample produced was indeed g-C_3N_4. Meanwhile, the peak at 44.1° represented the crystallography plane of g-C_3N_4 at (020) [13]. The XRD pattern for pure carbon showed two significant peaks at 28.3° and 40.4°. The amorphous carbon-based materials were hemicellulose, amorphous cellulose, and lignin. The peaks appearing at the diffraction angle of 28.3° and 40.4° were attributed to the (120) and (100) crystallography planes of carbon, respectively [14,15]. The results reinforced the implication that the biochar produced from coconut shell husk was indeed carbon.

It is worth noting that the characteristic peak of g-C_3N_4 at 27.5° experienced a decrement in peak intensity and shifted slightly to a lower angle with the increasing amount of carbon incorporated into the lattice of g-C_3N_4. The peak eventually disappeared when the carbon amount was increased further to 25 wt%. These changes were owing to the substitution of N atoms in the lattice of g-C_3N_4 by C atoms originating from the coconut shell husk [16]. Hence, the addition of carbon to g-C_3N_4 induced a decrement in the orderliness of the interlayer stacking structure leading to the shift of characteristic peaks and a decrement in the peak intensity [11]. At carbon weightage equal to 15 wt%, all the characteristic peaks of g-C_3N_4 were not detected from the XRD analysis. The disappearance of g-C_3N_4 peaks were due to the presence of an excessive amount of carbon in the sample.

As the amount of carbon increased beyond 15 wt%, the characteristic peaks of carbon became dominant as shown in Figure 1. The emergence of peaks at 28.3° and 40.4° in 25 wt% C/g-C$_3$N$_4$ revealed the dominating peaks of carbon over g-C$_3$N$_4$. The broad peak observed in the XRD spectra of 15 wt% C/g-C$_3$N$_4$ and 20 wt% C/g-C$_3$N$_4$ were attributed to the high amount of amorphous carbon material present. The findings were in line with the works reported by Xiao et al. [17] and Lin et al. [18] in which the characteristic peaks of g-C$_3$N$_4$ disappeared and the degree of amorphous structure was enhanced with the addition of carbon material.

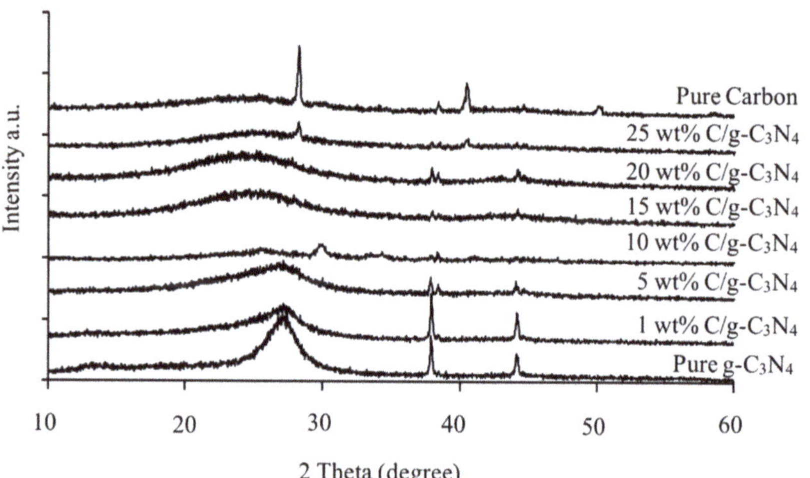

Figure 1. XRD patterns of pure g-C$_3$N$_4$, pure carbon, and C/g-C$_3$N$_4$ composites at different weight percentages.

3.1.2. SEM-EDX Analysis

Figure 2 illustrates the surface morphology of pure g-C$_3$N$_4$, pure carbon, and C/g-C$_3$N$_4$ composites at different weight percentages captured at 5000 magnification. Figure 2a shows that pure g-C$_3$N$_4$ exhibited a structure that resembled overlapping flat sheets with irregular shapes. Figure 2b reveals that 1 wt% C/g-C$_3$N$_4$ samples consisted of smoother surface with irregular shapes adhering to it. The smoother surface was attributed to the carbon material. The presence of carbon materials increased the surface area of g-C$_3$N$_4$ and hence enhanced the availability of active sites for sonocatalytic reaction. Figure 2c displays that the smoother surface became more dominant than the irregular aggregation which was observed in the pure g-C$_3$N$_4$.

Based on Figure 2d, the SEM image of 10 wt% C/g-C$_3$N$_4$ shows that fewer irregular shapes were detected, indicating the dominance of the carbon materials over g-C$_3$N$_4$. The SEM images reflect the results obtained from XRD where the characteristic peaks of g-C$_3$N$_4$ were diminished and eventually disappeared with the increasing amount of carbon. As the amount of carbon increased, the presence of the irregular shaped particles decreased as shown in Figure 2e–g which represent the composite samples with carbon amounts of 15 wt%, 20 wt%, and 25 wt%, respectively. The decrement in the irregular-shaped particles implied that there is a decrement in the amount of g-C$_3$N$_4$ attached to the carbonaceous structure. This was caused by the increased amount of carbon material which increased the surface area of the sample. The distribution of g-C$_3$N$_4$ spread out more widely with an increased specific surface area which explained the decreasing amount of g-C$_3$N$_4$ observed per unit area of carbon material. From Figure 2h, it can be seen that there were signs of cracks and residues on the carbon material with no irregular shaped particles present. The observed morphology of pure carbon might be attributed to the results of the heat treatment of coconut shell husks [19].

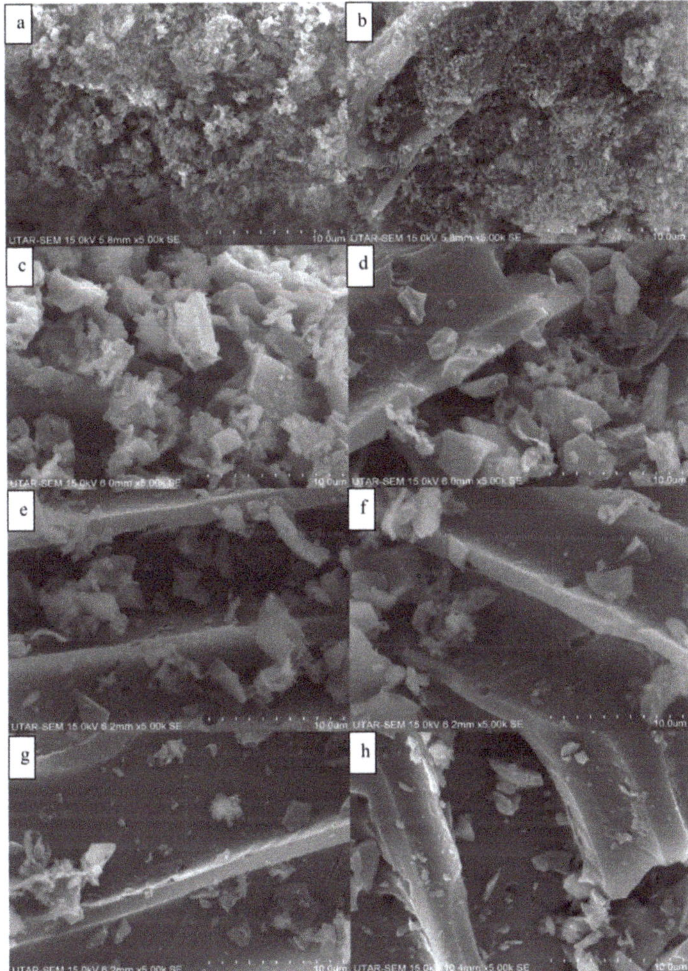

Figure 2. SEM images of (**a**) pure g-C$_3$N$_4$, (**b**) 1 wt% C/g-C$_3$N$_4$, (**c**) 5 wt% C/g-C$_3$N$_4$, (**d**) 10 wt% C/g-C$_3$N$_4$, (**e**) 15 wt% C/g-C$_3$N$_4$, (**f**) 20 wt% C/g-C$_3$N$_4$, (**g**) 25 wt% C/g-C$_3$N$_4$ and (**h**) pure carbon.

The EDX results with the composition of carbon and nitrogen elements are displayed in Table 1. Pure g-C$_3$N$_4$ exhibited the carbon-to-nitrogen ratio of 0.78 which was very close to the stoichiometric value of 0.75. The result proved that g-C$_3$N$_4$ was synthesized successfully in this study. The introduction of carbon material to the pure g-C$_3$N$_4$ caused a significant increment in carbon content. As the content of carbon materials increased beyond 5 wt%, the carbon element detected by EDX was capped at a range of 85 wt% to 90 wt%. A decrement in the amount of nitrogen was observed with the increasing amount of carbon added. The finding was in a good agreement with the SEM analysis which implied the wider distribution of g-C$_3$N$_4$ over the carbon surface at higher content of carbon materials. Hence, this resulted in a decrement in the nitrogen content per unit surface area. The observations also confirmed the successful incorporation of g-C$_3$N$_4$ into the carbon materials synthesized using coconut shell husk. Besides, the elementary analysis results showed that the pure carbon had a 2.15 wt% of nitrogen. The small fraction of nitrogen present in the sample was the residual of coconut shell husk remaining after pyrolysis [20].

Table 1. EDX results of the synthesized samples.

Samples	Carbon		Nitrogen	
	wt%	**at %**	**wt%**	**at %**
Pure g-C$_3$N$_4$	43.81	47.62	56.19	52.38
1 wt% C/g-C$_3$N$_4$	49.12	52.96	50.88	47.04
5 wt% C/g-C$_3$N$_4$	85.31	87.14	14.69	12.86
10 wt% C/g-C$_3$N$_4$	87.42	89.01	12.58	10.99
15 wt% C/g-C$_3$N$_4$	88.22	89.72	11.78	10.28
20 wt% C/g-C$_3$N$_4$	86.30	88.02	13.70	11.98
25 wt% C/g-C$_3$N$_4$	89.72	91.06	10.28	8.94
Pure carbon	97.85	98.15	2.15	1.85

3.1.3. FTIR Analysis

The results obtained from FTIR analysis for pure g-C$_3$N$_4$, pure carbon, and C/g-C$_3$N$_4$ composites are shown in Figure 3. The FTIR band for pure g-C$_3$N$_4$ exhibited a broad, weak band between 2800 cm^{-1} and 3600 cm^{-1}. The band indicated the stretching vibration modes for the N–H groups. Other than the broad peak, intense bands were observed in the range of 1200 cm^{-1} to 1700 cm^{-1}. This was due to the stretching vibration of C-N bond in the heterocyclic structure of g-C$_3$N$_4$ [21]. There were also well-defined peaks at 1239.2 cm^{-1} and 805.6 cm^{-1}. The peak observed at 1239.2 cm^{-1} represented the presence of tri-s-azine derivatives in the sample [22]. The out-of-plane bending vibration characteristic of heptazine rings was represented by the peak observed at 805.6 cm^{-1} [12].

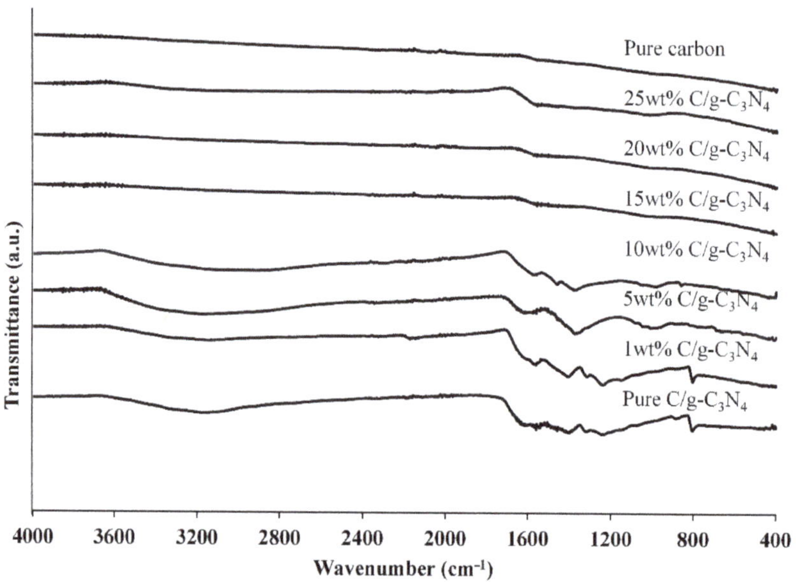

Figure 3. FTIR results of pure g-C$_3$N$_4$, pure carbon, and C/g-C$_3$N$_4$ composites at different weight percentages.

With the addition of carbon, the peaks observed in pure g-C$_3$N$_4$ became less obvious. The intensities of characteristic peaks from 1200 cm^{-1} to 1700 cm^{-1} decreased in 1 wt% C/g-C$_3$N$_4$. For 5 wt% C/g-C$_3$N$_4$ and 10 wt% C/g-C$_3$N$_4$, both samples showed that the peaks at 1239.2 cm^{-1} and 805.6 cm^{-1} disappeared. The results revealed that the addition of carbon to g-C$_3$N$_4$ would minimize the transition energy of composites [18]. The FTIR spectra of 15 wt% C/g-C$_3$N$_4$, 20 wt% C/g-C$_3$N$_4$, and 25 wt% C/g-C$_3$N$_4$ were observed to be similar to the FTIR results of pure carbon. The finding was in line with XRD and

SEM results which showed that the composition of excess carbon would dominate the characteristics of composite over g-C$_3$N$_4$.

3.1.4. Surface Analysis

Figure 4a illustrates the nitrogen adsorption-desorption of g-C$_3$N$_4$ and 10 wt% C/g-C$_3$N$_4$. Both the samples exhibited Type IV isotherms with H3 hysteresis loop indicating the mesoporous properties of the samples [23]. The findings revealed that the adsorption capacity of 10 wt% C/g-C$_3$N$_4$ was lower than pure g-C$_3$N$_4$. This was well related to the Brunauer-Emmett-Teller (BET) surface area of samples. According to the results obtained, pure g-C3N4 had a higher surface area (4.4715 m^2/g) than 10 wt% C/g-C$_3$N$_4$ (2.9647 m^2/g). The reduction of surface area after the addition of carbon could be attributed to the pore blocking of carbon by g-C$_3$N$_4$ [24]. In Figure 4b, the pore distribution plots show that both the samples were mainly consisting of mesopores with the pore size between 2 and 50 nm. Together with the results of nitrogen adsorption-desorption isotherms, the presence of mesopores were confirmed to be dominant in the synthesized samples.

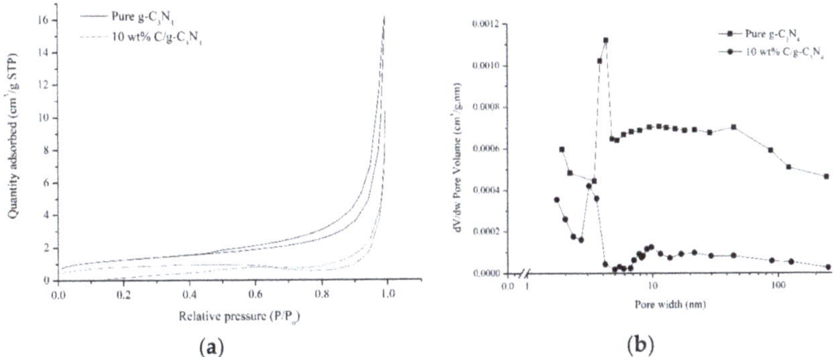

Figure 4. (a) Nitrogen adsorption-desorption isotherms and (b) pore size distribution curves of g-C$_3$N$_4$ and 10 wt% C/g-C$_3$N$_4$.

3.1.5. TGA

Figure 5 compares the thermal stability of pure g-C$_3$N$_4$, pure carbon and 10 wt% C/g-C$_3$N$_4$. All the samples experienced weight loss between 30 °C and 250 °C which was due to the removal of moisture content adsorbed on the samples [25]. The results showed that 10 wt% C/g-C$_3$N$_4$ composite decomposed faster than g-C$_3$N$_4$ at temperatures below 600 °C. This could be explained by the addition of carbon which weakened the Van der Waals force present in the π conjugated system of g-C$_3$N$_4$ [15]. However, the weight percentage of pure g-C$_3$N$_4$ declined sharply after 500 °C and the sample was decomposed nearly completely while increasing temperature up to 690 °C. In the range of temperature between 500 °C and 690 °C, the ring structure of g-C$_3$N$_4$ was destroyed to form nitrogen gas, cyanogen and cyanide [12]. As a whole, the total weight loss percentage of pure g-C$_3$N$_4$, pure carbon and 10 wt% C/g-C$_3$N$_4$ were found to be 97.68%, 92.59% and 71.58%. It is worth noting that the thermal stability of pure g-C$_3$N$_4$ was enhanced with the introduction of pure carbon synthesized using coconut shell husk.

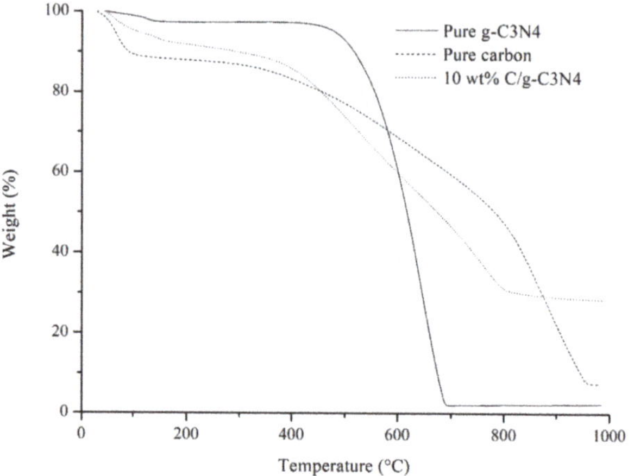

Figure 5. TGA curves of pure g-C_3N_4, pure carbon and 10 wt% C/g-C_3N_4.

3.2. Parameter Study for Sonocatalytic Degradation of Malachite Green

3.2.1. Effect of Carbon Amount Composited with g-C_3N_4

The effect of different carbon amounts composited with the g-C_3N_4 on the sonocatalytic performance was studied by comparing the degradation efficiencies of malachite green as shown in Figure 6. The findings showed that 10 wt% C/g-C_3N_4 exhibited the best degradation efficiency of 93.44%. Pure g-C_3N_4 only showed the lowest degradation efficiency of 11.20%. The 10 wt% C/g-C_3N_4 provided around eight times higher degradation efficiency of malachite green as compared to the pure g-C_3N_4. The composites with increasing amount of carbon showed a positive effect in the degradation efficiency of malachite green until maximum degradation efficiency was achieved by 10 wt% C/g-C_3N_4. The degradation efficiencies of malachite green were found to be 18.23% and 57.93% in the presence of 1 wt% C/g-C_3N_4 and 5 wt% C/g-C_3N_4 respectively. After achieving the optimum value of degradation efficiency, the further addition of carbon amount on g-C_3N_4 showed a decrement trend in the degradation efficiency of malachite green. The results reveal that 15 wt% C/g-C_3N_4, 20 wt% C/g-C_3N_4, and 25 wt% C/g-C_3N_4 exhibited 86.08%, 87.96%, and 83.28% of degradation efficiencies respectively. Meanwhile, the degradation efficiency of malachite green in the presence of pure carbon was only 67.17%.

Pure g-C_3N_4 showed a very low catalytic performance due to the rapid recombination rate of the electron-hole pairs and poor light absorption [23]. In a typical sonocatalysis process, the explosion of acoustic bubbles release energy in the form of light which is known as sonoluminescence [26]. The emission of light under ultrasound irradiation will then excite g-C_3N_4 to form electron-hole pairs leading to the generation of free radicals [27]. Hence, both light absorption and separation of charge carriers were important steps in the sonocatalytic degradation of malachite green in order to activate the sonocatalyst and generate free radicals, respectively. Although the 10 wt% C/g-C_3N_4 had a smaller area than pure g-C_3N_4, the presence of carbon in the composite could help to delay the recombination of electron-hole pairs because carbon could serve as the electron collector [11]. By allowing a higher separation efficiency of the charge carriers, the degradation efficiency of the malachite green was enhanced since more free radicals could be generated [15]. The results were in good agreement with Xiao et al. [17] where the catalytic activity of g-C_3N_4 was enhanced with the introduction of biochar. This implied that the optimum carbon content in C/g-C_3N_4 for this study was 10 wt%.

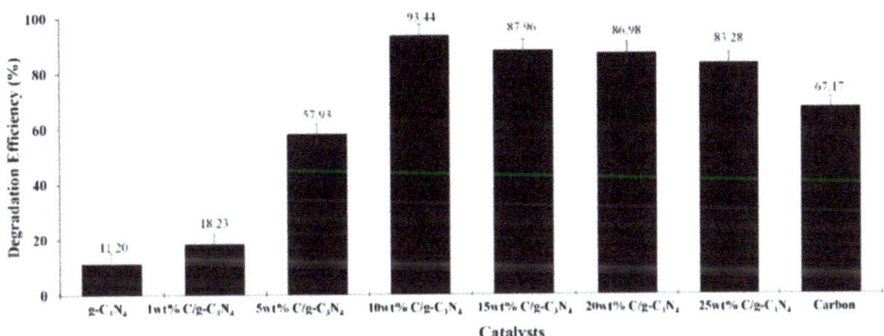

Figure 6. Degradation efficiency of carbon amount composited with g-C$_3$N$_4$ (initial malachite green concentration = 15 ppm, initial catalyst loading = 0.3 g/L, and natural solution pH for 10 min).

A decrement in degradation efficiency could be observed when the amount of carbon composited on g-C$_3$N$_4$ exceeded 10 wt%. This phenomenon occurred because the concentration of carbon composited with g-C$_3$N$_4$ was extremely high. The excessive amount of black carbon would block the light absorption of g-C$_3$N$_4$ that inhibited the sonocatalytic performance [17]. With some of the light being blocked by carbon, the degradation efficiency of malachite green dropped and remained almost constant for the other three composite samples. The results were in line with the work reported by Meng et al. [11] which proposed that the highest catalytic activity was achieved in the presence of g-C$_3$N$_4$ composited with optimum carbon content. The work also reported that the band gap energy of the composites increased with the increasing amount of biochar beyond optimum value. This would in turn reduce the formation of electron-hole pairs since higher energy was required to activate the catalyst [28]. Hence, it was speculated that moderate carbon content would improve light harvesting capacity and band gap narrowing which in turn enhanced the generation of charge carriers commanding a higher catalytic degradation efficiency of the dye.

3.2.2. Effect of H$_2$O$_2$ Dosage

The effect of H$_2$O$_2$ dosage (0 to 1.50 mmol/L) on the degradation efficiency of malachite green is shown in Figure 7. It was noted that the degradation efficiency of malachite green was the lowest with 91.03% degradation when there was no H$_2$O$_2$ added. The sonocatalytic degradation efficiency of malachite green increased to 100% with the increasing dosage of H$_2$O$_2$ until 0.5 mmol/L. The degradation efficiency of the malachite green dropped from 100% to 94.89% when H$_2$O$_2$ dosage was increased further from 0.5 mmol/L to 1.5 mmol/L.

The increment in degradation efficiency after the addition of H$_2$O$_2$ implied that H$_2$O$_2$ exhibited a positive effect in the sonocatalytic degradation of dye. This was due to the strong oxidizing power of H$_2$O$_2$ which could promote the generation of •OH radicals by self-decomposition under ultrasonic irradiation [6,29]. Besides, the presence of H$_2$O$_2$ could act as an electron trap to yield more •OH radicals and inhibit the recombination of charge carriers of the catalyst [10]. Therefore, the increased amount of •OH radicals in the solution would improve the degradation performance.

The decrement in the sonocatalytic performance was observed when excessive amounts of H$_2$O$_2$ were utilized. At high concentrations of H$_2$O$_2$, hydroperoxyl radicals (OOH•) with lower oxidation potential would be produced instead of •OH radicals [30]. The scavenging effect would reduce the amount of •OH radicals and holes present in the solution. Since the degradation efficiency of 10 wt% C/g-C$_3$N$_4$ in the absence of H$_2$O$_2$ showed an impressive result of 91.03%, it was decided that no H$_2$O$_2$ would be added for the subsequent studies.

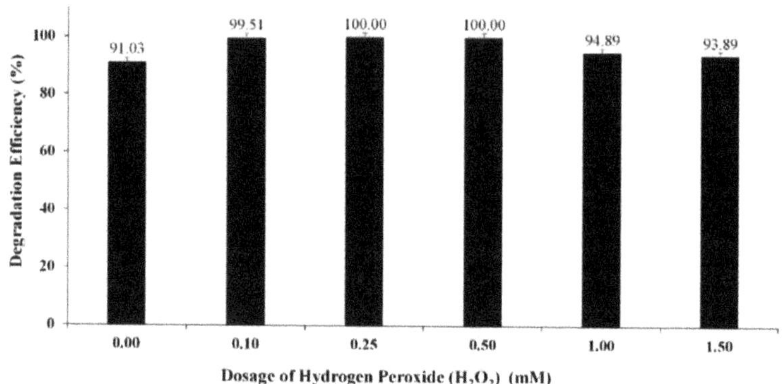

Figure 7. Degradation efficiency of malachite green with different H_2O_2 dosage by using 10 wt% C/g-C$_3$N$_4$ (initial dye concentration = 15 ppm, initial catalyst loading = 0.3 g/L, and natural solution pH for 10 min).

3.3. RSM Modelling

3.3.1. CCD Model

For this study, three factors were taken into consideration for the optimization study which were initial catalyst dosage (g/L), initial dye concentration (ppm), and solution pH. A CCD model was used to analyze the interactive effect of these three parameters on the degradation efficiency of malachite green.

The experimental and predicted degradation efficiency values are presented in Table 2. The experimental degradation efficiency of malachite green by sonocatalytic degradation in the presence of 10 wt% C/g-C$_3$N$_4$ was found to be in the range of 65.91% to 97.11%. The experimental data fitted well into the quadratic vs. 2FI model with coefficient of determination (R^2) equal to 0.9862. The modification of the model was made based on the backward elimination method. Since modelling with insignificant model terms would affect the accuracy of the model, it was decided to remove the insignificant interactive term, $X_{1 \times 2}$ which showed *p*-value (0.1354) greater than 0.05.

Table 2. Three factor CCD matrix and value of degradation efficiency.

Order No.	Point Type	Initial Dye Concentration, ppm (X_1)	Initial Catalyst Loading, g/L (X_2)	Solution pH (X_3)	Experimental Values	Predicted Values	Residuals
1	Factorial	20.00 (−1)	0.300 (−1)	4.0 (−1)	77.62	77.54	0.08
2	Factorial	25.00 (+1)	0.300 (−1)	4.0 (−1)	68.86	68.03	0.83
3	Factorial	20.00 (−1)	0.500 (+1)	4.0 (−1)	88.69	86.96	1.73
4	Factorial	25.00 (+1)	0.500 (+1)	4.0 (−1)	76.12	77.46	−1.34
5	Factorial	20.00 (−1)	0.300 (−1)	8.0 (+1)	94.23	94.42	−0.19
6	Factorial	25.00 (+1)	0.300 (-1)	8.0 (+1)	91.07	91.55	−0.48
7	Factorial	20.00 (−1)	0.500 (+1)	8.0 (+1)	97.11	96.69	0.42
8	Factorial	25.00 (+1)	0.500 (+1)	8.0 (+1)	92.21	93.82	−1.61
9	Axial	18.30 (−1.682)	0.400 (0)	6.0 (0)	91.88	93.16	−1.28
10	Axial	26.70 (+1.682)	0.400 (0)	6.0 (0)	84.23	82.75	1.48
11	Axial	22.50 (0)	0.232 (−1.682)	6.0 (0)	83.60	83.81	−0.21

Table 2. *Cont.*

Order No.	Point Type	Actual and Coded Independent Variable Levels			Degradation Efficiency, %		Residuals
		Initial Dye Concentration, ppm (X_1)	Initial Catalyst Loading, g/L (X_2)	Solution pH (X_3)	Experimental Values	Predicted Values	
12	Axial	22.50 (0)	0.568 (+1.682)	6.0 (0)	94.05	93.64	0.41
13	Axial	22.50 (0)	0.400 (0)	2.6 (−1.682)	65.91	66.75	−0.84
14	Axial	22.50 (0)	0.400 (0)	9.4 (+1.682)	95.73	94.70	1.03
15	Centre	22.50 (0)	0.400 (0)	6.0 (0)	85.64	85.69	−0.05
16	Centre	22.50 (0)	0.400 (0)	6.0 (0)	84.19	85.69	−1.50
17	Centre	22.50 (0)	0.400 (0)	6.0 (0)	86.85	85.69	1.16
18	Centre	22.50 (0)	0.400 (0)	6.0 (0)	85.23	85.69	−0.46
19	Centre	22.50 (0)	0.400 (0)	6.0 (0)	85.96	85.69	0.27
20	Centre	22.50 (0)	0.400 (0)	6.0 (0)	86.22	85.69	0.53

The dependent and independent variables could be correlated by using a second-order polynomial response equation. The empirical relationship between the response and independent variables for the sonocatalytic degradation of malachite green in the presence of 10 wt% C/g-C$_3$N$_4$ is shown in Equation (1) which is presented in terms of coded units.

$$y_{\text{pred}} = 85.69 - 3.09 \times 1 + 2.92X_2 + 8.31X_3 + 1.66X_1X_3 - 1.79X_2X_3 + 0.8023X_1{}^2 + 1.07X_2{}^2 - 1.76X_3{}^2 \tag{1}$$

where y_{pred} represents predicted degradation efficiency.

The predicted values of degradation efficiency listed in Table 2 were computed using Equation (1). The small residual values for all the runs indicated the good match between experimental and predicted values. The sign of the terms in Equation (1) implied the effect of that parameter on the degradation efficiency. A positive sign showed the synergistic effect of a particular term on the degradation efficiency of malachite green while a negative term showed its antagonistic effect on the degradation of malachite green. In this study, the terms X_2, X_3, X_1X_3, $X_1{}^2$ and $X_2{}^2$ exhibited a positive effect on the response while X_1, X_2X_3 and $X_3{}^2$ showed a negative impact on the degradation efficiency. Based on Table 3, high F value (98.10) indicated that the model constructed was significant with satisfactory *p*-value (<0.0001). In addition, adequate precision of this model was found to be 34.4670 which reflected a high ratio of signal-to-noise. Together with the insignificant lack of fit value, the model was again proved to be fitted well with the experimental data.

Table 3. ANOVA results of sonocatalytic degradation of malachite green.

Factors	Sum of Squares	Degree of Freedom	Mean Square	F Value	Probability, *p*-Value	
Modified Model	1316.11	8	164.51	98.10	<0.0001	Significant
X_1	130.74	1	130.74	77.97	<0.0001	
X_2	116.72	1	116.72	69.60	<0.0001	
X_3	942.97	1	942.97	562.31	<0.0001	
X_1X_3	22.01	1	22.01	13.13	0.0040	
X_2X_3	25.60	1	25.60	15.26	0.0024	
$X_1{}^2$	9.28	1	9.28	5.53	0.0384	
$X_2{}^2$	16.64	1	16.64	9.92	0.0092	
$X_3{}^2$	44.42	1	44.42	26.49	0.0003	
Residual	18.45	11	1.68			
Lack of fit	14.28	6	2.38	2.86	0.1345	Insignificant
Pure error	4.16	5	0.83			
Corrected Total	1334.55	19				

$R^2 = 0.9862$; adequate precision = 34.4670.

3.3.2. Effect of Experimental Variables on the Degradation Efficiency

Figure 8a shows the response surface plot of the degradation efficiency with solution pH and initial dye concentration. Runs 13 and 14 which had a degradation efficiency of 66.75% and 94.70%, respectively, indicated that the solution's pH greatly affected the degradation efficiency of malachite green. For the effect of initial dye concentration, Runs 9 and 10 that had the degradation efficiency of 93.16% and 82.75%, respectively, showed that the change in initial dye concentration did not affect the response factor significantly. By comparing Run 1 and Run 6, the interaction between the two factors being studied could be observed. Run 1 was being conducted at a lower solution pH and initial dye concentration demonstrated a degradation efficiency of 77.62%. Meanwhile, Run 6 that was performed at a higher solution pH and initial dye concentration showed a higher degradation efficiency of 91.07%. This comparison showed that solution pH was the major factor that affected the response factor while the initial dye concentration only affected the degradation efficiency by a small fraction as compared to solution pH.

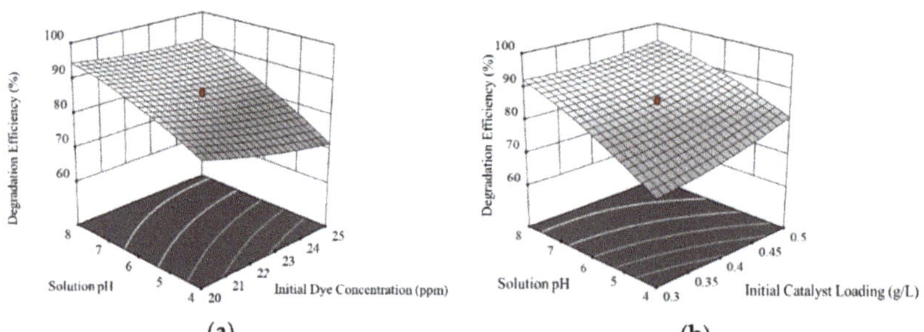

Figure 8. Three-dimensional response surface plot of degradation efficiency of malachite green against (**a**) solution pH and initial dye concentration at constant initial catalyst loading = 0.4 g/L and (**b**) solution pH and initial catalyst dosage at initial dye concentration = 22.5 ppm.

A change in solution pH could affect the electrostatic interaction between the dye molecule and catalyst surface. Solution pH variation could also alter the production of radicals that directly affected the degradation efficiency of malachite green. When solution pH increased towards the alkaline region, the degradation efficiency of malachite green increased greatly. This phenomenon could be explained by the increased production of hydroxyl radicals ($\bullet$OH). The alkaline conditions implied that more hydroxyl ions were present in the solution [31]. With more hydroxyl ions being converted to $\bullet$OH radicals, the degradation of malachite green could be improved greatly.

When the solution pH decreased toward the acidic region, the degradation efficiency of malachite green dropped drastically. This was caused by the presence of excess hydrogen ions in the dye solution leading to the deprotonation of the catalyst [32]. In other words, the hydrogen ions would compete with malachite green molecules to adsorb on the catalyst surface. Besides, excessive amount of H^+ ions would react with $\bullet$OH radicals as shown in Equation (2) leading to the decline in the amount of $\bullet$OH radicals available for degradation of the malachite green [33]. Chloride ions contributed from HCl that was used to adjust the solution pH would also scavenge $\bullet$OH radicals to form $HOCl\bullet^-$ with lower oxidation potential [34]. The scavenging effects under acidic condition are displayed in Equations (2) and (3).

$$\bullet OH + e^- + H^+ \rightarrow H_2O \tag{2}$$

$$\bullet OH + Cl^- \rightarrow HOCl\bullet^- \tag{3}$$

The scavenging effects of $\bullet$OH radicals and adsorption of hydrogen ions on the catalyst surface explained the reduction of degradation efficiency of malachite green in acidic conditions.

The initial dye concentration was another factor that would affect the degradation efficiency of malachite green. The increment in initial dye concentration would cause a slight decrement in the response factor. This trend could be explained by both physical and chemical processes. The mass transfer between the dye molecules and catalyst surface was the physical process involved in the sonocatalytic degradation of malachite green while the concentration of radical species was the chemical factor. At higher initial dye concentrations, the number of dye molecules present in the solution increased at constant catalyst dosage, indicating the number of adsorption sites was limited with increasing initial dye concentration [35]. In terms of chemical factors, the active sites of the catalyst were occupied by higher numbers of dye molecules. This would in turn prevent the absorption of energy by the catalyst and hinder the yield of charge carriers [7]. Hence, the generation of free radicals decreased at higher initial dye concentrations and was insufficient to degrade the dye molecules present in the solution [36]. In this case, the amount of free radicals became the limiting step that reduced the degradation efficiency. The increment in initial dye concentration would decrease the degradation efficiency of malachite green owing to limiting factors which were attributed to the concentration of radical species present and the availability of adsorption sites on the catalyst surface.

The interaction between solution pH and initial catalyst loading is shown in Figure 8b. The results indicated that solution pH affected the degradation efficiency more than the initial catalyst loading. This relationship was shown in their difference in F values from the ANOVA results. The initial catalyst dosage had the lowest F value among three factors which implied that it might affect the degradation efficiency insignificantly. By comparing Run 11 and Run 12, the degradation efficiency shown by these runs were 83.60% and 94.05%, respectively. The difference between Run 11 and Run 12 was only the initial catalyst dosage used. Based on the surface plot in Figure 8b, the initial catalyst dosage showed the least effect on the response factor when the solution pH was at alkaline condition. The interaction could be done by comparing four runs which were Run 1, Run 3, Run 5, and Run 7. Run 1 and Run 3 showed a degradation efficiency of 77.62% and 88.69% at a constant low solution pH when the initial catalyst loading was varied. Run 5 and Run 7 showed degradation efficiencies of malachite green of 94.23% and 97.11% at solution pH 8, respectively, with the increasing initial catalyst dosage.

The third factor that was studied in this model was the initial catalyst dosage. An increment in initial catalyst dosage would cause an increment in the response factor. This phenomenon could be explained by the increment in surface area of the catalyst to generate more pairs of charge carriers [37]. The increased surface area of the catalyst also provided more active sites for the nucleation of bubbles and the generation of free radicals [38]. This would in turn improved the degradation efficiency of malachite green. The insignificant effect of initial catalyst loading at high solution pH might be due to the presence of hydroxyl ions in the solution. With more hydroxyl ions, the amount of •OH radicals produced could be greatly increased which caused the increment in the catalyst to be redundant. In other words, the increment in catalyst loading would improve the sonocatalytic performance by generating more radicals but the effect would be insignificant once the optimum amount of radicals in the solution was exceeded.

3.3.3. Optimisation and Model Validation

In this study, RSM was used to obtain the optimum conditions for the sonocatalytic degradation of malachite green by the means of the CCD model. The solution selected from the numerical optimization showed the highest desirability of 0.986. The values of the optimum operating parameters for initial dye concentration, initial catalyst loading, and solution pH were 20 ppm, 0.5 g/L, and 8, respectively. The predicted and experimental values for the degradation efficiency of malachite green under optimum conditions were 96.69% and 97.11%, respectively. The percentage error for the predicted and actual value was only 0.43%. The low percentage error between these two values indicated that the applied model was reliable for making predictions on the response factor for real conditions.

Table 4 compares the removal efficiencies of malachite green through different techniques proposed by other works such as heterogeneous catalysis, adsorption and catalysis. Based on Table 4, it is clear that the sonocatalytic process using non-metal g-C$_3$N$_4$/coconut shell husk derived-carbon composites was a novel and potential method to remove malachite green as compared to the literature. It is interesting to note that adsorption using ZIF-8@fe/Ni which was proposed by Zhang et al. [39] could remove 99.0% of malachite green with 50 mg/L initial dye concentration. However, malachite green molecules were not eliminated from the environment but only transferred to another phase without degradation. Hence, an extra post-treatment step was needed in order to decompose malachite green. Furthermore, the results obtained in the current work were comparable to the findings reported by Pandey et al. [40] which proposed the application of g-C$_3$N$_4$/NiO as photocatalyst in the removal of malachite green. In this work, 10 wt% C/g-C3N4 also exhibited a high sonocatalytic performance in the degradation of malachite green within a relatively shorter time. From this point of view, the 10 wt% C/g-C$_3$N$_4$ composite synthesized in this research is a promising material to be applied commercially as a sonocatalyst owing to the fact that it could be categorized as a green material which involves the usage of renewable materials.

Table 4. Comparison of malachite green removal with other works.

Method	Adsorbent/Catalyst	Initial Dye Concentration (mg/L)	Adsorbent/Catalyst Dosage (g/L)	Process Time (mins)	Removal Efficiency (%)	Reference
Adsorption	ZIF-8@fe/Ni	50	1.0	30	99.0	[39]
Photocatalysis	BiVO4/g-C$_3$N$_4$	30	1.0	60	98.3	[41]
Photocatalysis	g-C$_3$N$_4$/NiO	10	0.1	12	96.0	[40]
Heterogeneous catalysis	MnFe$_2$O$_4$	50	0.6	60	~100	[42]
Sonocatalysis	10 wt% C/g-C$_3$N$_4$	20	0.5	10	97.11	Current work

3.4. Reusability Study and Economic Analysis of 10 wt% C/g-C$_3$N$_4$ Catalyst

The reusability of a catalyst was an important factor to determine the stability and practicality of using the studied catalyst for wastewater treatment processes. Although the cost of producing 10 wt% C/g-C$_3$N$_4$ was low since it was derived from biomass and urea, the reusability of the catalyst was still an important factor to reduce waste production.

Figure 9 shows the reusability study of the synthesized composite by carrying out five consecutive runs of the experiment. The degradation efficiency for the first, second, third, fourth and fifth run were 93.24%, 92.80%, 88.04%, 86.50% and 85.25%, respectively. The stability of the prepared 10 wt% C/g-C$_3$N$_4$ as sonocatalyst was confirmed owing to the insignificant decline in the degradation efficiency of malachite green (<10%) after five consecutive runs. The decrement in degradation efficiency for the reused runs could be explained by two factors. The first factor was the loss of catalyst during the centrifugation and rinsing process [43]. The loss of catalyst was caused by the hydrophilicity of the 10 wt% C/g-C$_3$N$_4$. The presence of functional groups such as carbonyl (C = O), hydroxyl (-OH) on the catalyst surface caused the catalyst to be hydrophilic. The reused sample would have a lower initial catalyst loading for each repeated run since it could be washed away easily by forming hydrogen bonds with water molecules [44]. The second factor was the partial disintegration of the catalyst caused by ultrasonic irradiation. The propagation of ultrasound waves created a mechanical force that would partially disintegrate the catalyst surface. The disintegration effect would disrupt the active sites of the catalyst and inhibit the generation of free radicals [45].

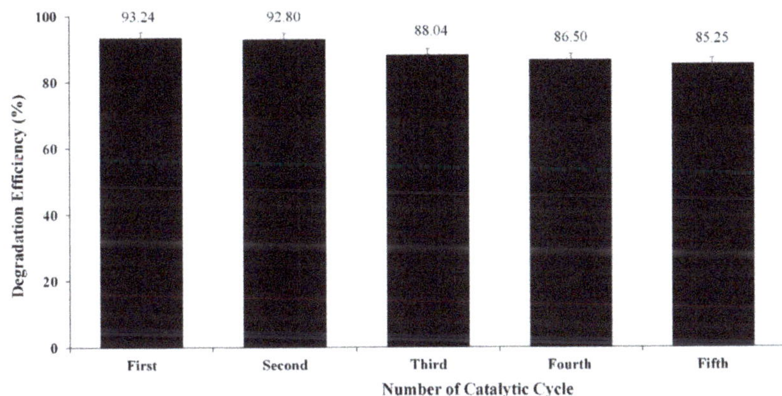

Figure 9. Degradation efficiency of malachite green with five reusability catalytic cycles (initial dye concentration = 15 ppm, initial catalyst loading = 1 g/L, and natural solution pH for 10 min).

The application of the sonocatalysis process on a large economic scale might be hampered by high costs due to a lack of proper reactor design strategies. However, Abdurahman et al. reported that it is an economically feasible decomposition treatment technology if a well-formulated cost objective function involving specific energy consumption per amount of materials is produced. Table 5 shows the cost estimation for the sonocatalytic degradation of malachite green using 10 wt% C/g-C3N4. The production cost of 10 wt% C/g-C_3N_4 was approximately MYR 1610.72/kg which was equivalent to USD 384.88/kg. The production of 10 wt% C/g-C_3N_4 was more cost effective as compared to other metal catalysts such as Ag_2O/CuO (USD 5435/kg) [46] and Bi_2WO_6/Au (USD 678/kg) [47]. In addition, the cost for sonocatalytic degradation of malachite green was estimated to be MYR 317.07/m^3 or USD 75.77/m^3. Together with the relatively low catalyst production cost and sonocatalytic degradation cost, 10 wt% C/g-C3N4 was a promising material that could be applied as sonocatalyst in the commercial wastewater treatment.

Table 5. Cost estimation for the production of 10 wt% C/g-C_3N_4 and sonocatalytic degradation of malachite green.

10 wt% C/g-C_3N_4 Production Cost			
Materials/Process	Unit Cost	Unit	Cost (MYR)
Urea	MYR 33/500g	5 g	0.33
Coconut shell husk	MYR 5/500g	2.5 g	0.03
Drying of coconut shell husk	MYR 0.234/kWh	1.4 kW; 2 days	5.24
Pyrolysis of coconut shell husk	MYR 0.234/kWh	3.5 kW; 1 h	0.82
Synthesis of 10 wt% C/g-C_3N_4	MYR 0.234/kWh	3.5 kW; 2 h	1.64
	Total cost to produce 5 g 10 wt% C/g-C_3N_4		8.05
Sonocatalytic Degradation of Malachite Green			
10 wt% C/g-C_3N_4	MYR 8.0536/5 g	0.05 g for 5 cycles	0.08
Sonocatalysis	MYR 0.234/kWh	0.4 kW; 10 min; 5 cycles	0.08
	Total cost to degrade 500 mL MG		0.16

3.5. Possible Mechanisms for Degradation of Malachite Green

Figure 10 shows the possible sonocatalytic degradation mechanism in water. There are two possible mechanisms to degrade the malachite green molecule. The first mechanism requires the presence of 10 wt% C/g-C_3N_4. The sonocatalyst would absorb the energy released from the collapse of the cavitation bubble in the form of heat and sonoluminescence [6]. Once the energy is absorbed, the electrons in the valence band (VB) of the sonocatalyst would be excited to the conduction band (CB) which would leave a positively charged hole at the VB. This would cause the formation of the electron-hole pair [31]. The electron in the CB has strong reduction potential that would reduce the oxygen present in the solution to a free radical that is known as a superoxide radical ($\bullet O_2{}^-$). The holes left in the VB would have high oxidizing potential that would be able to oxidize the water molecules and produce $\bullet OH$ [26]. The free radicals generated would degrade the dye molecule to carbon dioxide and water molecules. The second mechanism is the generation of $\bullet OH$ and $\bullet H$ radicals through water pyrolysis under ultrasonic irradiation which is known as sonolysis [35]. Both mechanisms lead to the generation of radicals which implies that malachite green molecules are degraded by the free radicals in the solution.

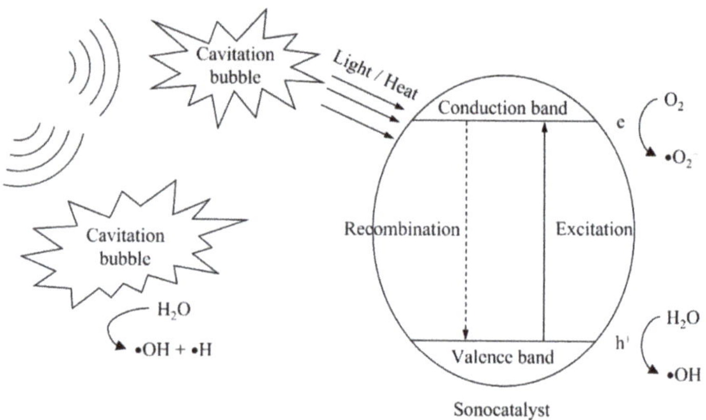

Figure 10. Mechanism of sonocatalytic dye degradation.

4. Conclusions

In this study, pure carbon was synthesized successfully using coconut shell husk. Pure carbon was added to g-C_3N_4 to form C/g-C_3N_4 composites at different weight percentages. Among the synthesized samples, 10 wt% C/g-C_3N_4 showed the best sonocatalytic degradation efficiency of malachite green with a value of 93.44%. The addition of H_2O_2 during the sonocatalytic degradation of malachite green could accelerate the degradation efficiency up to 100%. By the means of RSM, an optimization study was conducted including three experimental parameters which were initial dye concentration (20 to 25 ppm), initial catalyst loading (0.3 to 0.5 g/L), and solution pH (4 to 8). A CCD model was constructed successfully and fitted well to the experimental data with high R^2 value (0.9862). The optimum degradation efficiency of malachite green was found to be 97.11% at 20 ppm of initial dye concentration, 0.5 g/L of catalyst loading and a solution pH of 8. In the reusability test, the sample showed high stability in the sonocatalytic degradation performance even after five catalytic cycles.

Author Contributions: Data curation, S.L.; methodology, Y.L.P.; resources, Y.L.P.; visualization, S.L. and W.C.C.; writing original draft, A.Z.Y.K.; writing review and editing, Y.Y.C. All authors have read and agreed to the published version of the manuscript.

Funding: This research was funded by Ministry of Education (MOE) Malaysia that provided the Fundamental Research Grant Scheme (FRGS/1/2018/TK10/UTAR/02/2), Universiti Tunku Abdul Rahman (UTAR) Research Fund (IPSR/RMC/UTARRF/2020-C2/P01) and Kurita Water and Environment Foundation (KWEF) that provided the Overseas Research Grant (21Pmy076-25K).

Institutional Review Board Statement: Not applicable.

Informed Consent Statement: Not applicable.

Data Availability Statement: The data presented in this study are available on request from the corresponding author.

Acknowledgments: The authors would like to thank the Ministry of Education (MOE) Malaysia that provided the Fundamental Research Grant Scheme (FRGS/1/2018/TK10/UTAR/02/2), the Universiti Tunku Abdul Rahman (UTAR) Research Fund (IPSR/RMC/UTARRF/2020-C2/P01) and Kurita Water and Environment Foundation (KWEF) that provided the Overseas Research Grant (21Pmy076) for the financial support.

Conflicts of Interest: The authors declare no conflict of interest.

References

1. Razak, M.R.; Aris, A.Z.; Zakaria, N.A.C.; Wee, S.Y.; Ismail, N.A.H. Accumulation and risk assessment of heavy metals employing species sensitivity distributions in Linggi River, Negeri Sembilan, Malaysia. *Ecotoxicol. Environ. Saf.* **2021**, *211*, 111905. [CrossRef] [PubMed]
2. Joshi, P.; Sharma, O.P.; Ganguly, S.K.; Srivastava, M.; Khatri, O.P. Fruit waste-derived cellulose and graphene-based aerogels: Plausible adsorption pathways for fast and efficient removal of organic dyes. *J. Colloid Interface Sci.* **2021**, *608*, 2870–2883. [CrossRef] [PubMed]
3. Naseeb, F.; Ali, N.; Khalil, A.; Khan, A.; Asiri, A.M.; Kamal, T.; Bakhsh, E.M.; Ul-Islam, M. Photocatalytic degradation of organic dyes by U3MnO10 nanoparticles under UV and sunlight. *Inorg. Chem. Commun.* **2021**, *134*, 109075. [CrossRef]
4. Anantha, M.S.; Jayanth, V.; Olivera, S.; Anarghya, D.; Venkatesh, K.; Jayanna, B.K.; Sachin, H.P.; Muralidhara, H.B. Microwave treated Bermuda grass as a novel photocatalyst for the treatment of methylene blue dye from wastewater. *Environ. Nanotechnol. Monit. Manag.* **2021**, *15*, 100447. [CrossRef]
5. Göçenoğlu Sarıkaya, A.; Erden Kopar, E. Biosorption of Sirius Blue azo-dye by Agaricus campestris biomass: Batch and continuous column studies. *Mater. Chem. Phys.* **2022**, *276*, 125381. [CrossRef]
6. Sadeghi Rad, T.; Ansarian, Z.; Khataee, A.; Vahid, B.; Doustkhah, E. N-doped graphitic carbon as a nanoporous MOF-derived nanoarchitecture for the efficient sonocatalytic degradation process. *Sep. Purif. Technol.* **2021**, *256*, 117811. [CrossRef]
7. Yang, F.; Jiang, G.; Chang, Q.; Huang, P.; Lei, M. Fe/N-doped carbon magnetic nanocubes toward highly efficient selective decolorization of organic dyes under ultrasonic irradiation. *Chemosphere* **2021**, *283*, 131154. [CrossRef]
8. Yu, Y.; Zhu, Z.; Fan, W.; Liu, Z.; Yao, X.; Dong, H.; Li, C.; Huo, P. Making of a metal-free graphitic carbon nitride composites based on biomass carbon for efficiency enhanced tetracycline degradation activity. *J. Taiwan Inst. Chem. Eng.* **2018**, *89*, 151–161. [CrossRef]
9. Mandal, A.; Mandi, S.; Mukherjee, B. Ultrasound aided sonocatalytic degradation of Rhodamine B with graphitic carbon nitride wrapped zinc sulphide nanocatalyst. *Ceram. Int.* **2021**, *48*, 10271–10279. [CrossRef]
10. Eshaq, G.; ElMetwally, A.E. Bmim[OAc]-Cu$_2$O/g-C$_3$N$_4$ as a multi-function catalyst for sonophotocatalytic degradation of methylene blue. *Ultrason. Sonochem.* **2019**, *53*, 99–109. [CrossRef]
11. Meng, L.; Yin, W.; Wang, S.; Wu, X.; Hou, J.; Yin, W.; Feng, K.; Ok, Y.S.; Wang, X. Photocatalytic behavior of biochar-modified carbon nitride with enriched visible-light reactivity. *Chemosphere* **2020**, *239*, 124713. [CrossRef]
12. Chhabra, T.; Dhingra, S.; Nagaraja, C.M.; Krishnan, V. Influence of Lewis and Brønsted acidic sites on graphitic carbon nitride catalyst for aqueous phase conversion of biomass derived monosaccharides to 5-hydroxymethylfurfural. *Carbon N. Y.* **2021**, *183*, 984–998. [CrossRef]
13. Tyborski, T.; Merschjann, C.; Orthmann, S.; Yang, F.; Lux-Steiner, M.; Th, S. Crystal Structure of Polymeric Carbon Nitride and the Determination of Its Process-Temperature-Induced Modifications. *J. Phys. Condens. Matter* **2013**, *25*, 395402. [CrossRef]
14. Luo, S.; Li, S.; Zhang, S.; Cheng, Z.; Nguyen, T.T.; Guo, M. Visible-light-driven Z-scheme protonated g-C$_3$N$_4$/wood flour biochar/BiVO$_4$ photocatalyst with biochar as charge-transfer channel for enhanced RhB degradation and Cr(VI) reduction. *Sci. Total Environ.* **2022**, *806*, 150662. [CrossRef]
15. Karpuraranjith, M.; Chen, Y.; Ramadoss, M.; Wang, B.; Yang, H.; Rajaboopathi, S.; Yang, D. Magnetically recyclable magnetic biochar graphitic carbon nitride nanoarchitectures for highly efficient charge separation and stable photocatalytic activity under visible-light irradiation. *J. Mol. Liq.* **2021**, *326*, 115315. [CrossRef]
16. Liu, Y.; Zhao, S.; Zhang, C.; Fang, J.; Xie, L.; Zhou, Y.; Zhuo, S. Hollow tubular carbon doping graphitic carbon nitride with adjustable structure for highly enhanced photocatalytic hydrogen production. *Carbon N. Y.* **2021**, *182*, 287–296. [CrossRef]
17. Xiao, Y.; Lyu, H.; Yang, C.; Zhao, B.; Wang, L.; Tang, J. Graphitic carbon nitride/biochar composite synthesized by a facile ball-milling method for the adsorption and photocatalytic degradation of enrofloxacin. *J. Environ. Sci.* **2021**, *103*, 93–107. [CrossRef]
18. Lin, M.; Li, F.; Cheng, W.; Rong, X.; Wang, W. Facile preparation of a novel modified biochar-based supramolecular self-assembled g-C3N4 for enhanced visible light photocatalytic degradation of phenanthrene. *Chemosphere* **2021**, *288*, 132620. [CrossRef]
19. Ahmad, A.; Al-Swaidan, H.M.; Alghamdi, A.H.; Alotaibi, K.M.; Alswieleh, A.M.; Albalwi, A.N.; Bajuayfir, E. Efficient sequester of hexavalent chromium by chemically active carbon from waste valorization (Phoenix Dactylifera). *J. Anal. Appl. Pyrolysis* **2021**, *155*, 105075. [CrossRef]

20. Verma, R.; Maji, P.K.; Sarkar, S. Comprehensive investigation of the mechanism for Cr(VI) removal from contaminated water using coconut husk as a biosorbent. *J. Clean. Prod.* **2021**, *314*, 128117. [CrossRef]

21. Ragupathi, V.; Madhu babu, M.; Panigrahi, P.; Ganapathi Subramaniam, N. Scalable fabrication of graphitic-carbon nitride thin film for optoelectronic application. *Mater. Today Proc.* **2021**. [CrossRef]

22. Lu, Y.; Wang, W.; Cheng, H.; Qiu, H.; Sun, W.; Fang, X.; Zhu, J.; Zheng, Y. Bamboo-charcoal-loaded graphitic carbon nitride for photocatalytic hydrogen evolution. *Int. J. Hydrog. Energy* **2021**, *47*, 3733–3740. [CrossRef]

23. Guo, F.; Shi, C.; Sun, W.; Liu, Y.; Shi, W.; Lin, X. Pomelo biochar as an electron acceptor to modify graphitic carbon nitride for boosting visible-light-driven photocatalytic degradation of tetracycline. *Chin. J. Chem. Eng.* **2021**, *47*, 3733–3740. [CrossRef]

24. Li, K.; Huang, Z.; Zhu, S.; Luo, S.; Yan, L.; Dai, Y.; Guo, Y.; Yang, Y. Removal of Cr (VI) from water by a biochar-coupled g-C$_3$N$_4$ nanosheets composite and performance of a recycled photocatalyst in single and combined pollution systems. *Appl. Catal. B Environ.* **2019**, *243*, 386–396. [CrossRef]

25. Wu, H.; Zhang, W.; Zhang, H.; Pan, Y.; Yang, X.; Pan, Z.; Yu, X.; Wang, D. Preparation of the novel g-C$_3$N$_4$ and porous polyimide supported hydrotalcite-like compounds materials for water organic contaminants removal. *Colloids Surf. A Physicochem. Eng. Asp.* **2020**, *607*, 125517. [CrossRef]

26. Hassandoost, R.; Kotb, A.; Movafagh, Z.; Esmat, M.; Guegan, R.; Endo, S.; Jevasuwan, W.; Fukata, N.; Sugahara, Y.; Khataee, A.; et al. Nanoarchitecturing bimetallic manganese cobaltite spinels for sonocatalytic degradation of oxytetracycline. *Chem. Eng. J.* **2022**, *431*, 133851. [CrossRef]

27. Yadav, A.A.; Hunge, Y.M.; Kang, S.-W. Porous nanoplate-like tungsten trioxide/reduced graphene oxide catalyst for sonocatalytic degradation and photocatalytic hydrogen production. *Surf. Interfaces* **2021**, *24*, 101075. [CrossRef]

28. Alexpandi, R.; Abirami, G.; Balaji, M.; Jayakumar, R.; Ponraj, J.G.; Cai, Y.; Pandian, S.K.; Ravi, A.V. Sunlight-active phytol-ZnO@TiO$_2$ nanocomposite for photocatalytic water remediation and bacterial-fouling control in aquaculture: A comprehensive study on safety-level assessment. *Water Res.* **2022**, *212*, 118081. [CrossRef]

29. Lee, S.; Park, J.-W. Hematite/graphitic carbon nitride nanofilm for fenton and photocatalytic oxidation of methylene blue. *Sustainability* **2020**, *12*, 2866. [CrossRef]

30. Xin, S.; Ma, B.; Liu, G.; Ma, X.; Zhang, C.; Ma, X.; Gao, M.; Xin, Y. Enhanced heterogeneous photo-Fenton-like degradation of tetracycline over CuFeO$_2$/biochar catalyst through accelerating electron transfer under visible light. *J. Environ. Manage.* **2021**, *285*, 112093. [CrossRef]

31. Khataee, A.; Fazli, A.; Zakeri, F.; Joo, S.W. Synthesis of a high-performance Z-scheme 2D/2D WO$_3$@CoFe-LDH nanocomposite for the synchronic degradation of the mixture azo dyes by sonocatalytic ozonation process. *J. Ind. Eng. Chem.* **2020**, *89*, 301–315. [CrossRef]

32. Sadeghi Rad, T.; Khataee, A.; Arefi-Oskoui, S.; Sadeghi Rad, S.; Orooji, Y.; Gengec, E.; Kobya, M. Graphene-based ZnCr layered double hydroxide nanocomposites as bactericidal agents with high sonophotocatalytic performances for degradation of rifampicin. *Chemosphere* **2022**, *286*, 131740. [CrossRef] [PubMed]

33. Cheng, Z.; Luo, S.; Li, X.; Zhang, S.; Thang Nguyen, T.; Guo, M.; Gao, X. Ultrasound-assisted heterogeneous Fenton-like process for methylene blue removal using magnetic MnFe2O4/biochar nanocomposite. *Appl. Surf. Sci.* **2021**, *566*, 150654. [CrossRef]

34. Bampos, G.; Frontistis, Z. Sonocatalytic degradation of butylparaben in aqueous phase over Pd/C nanoparticles. *Environ. Sci. Pollut. Res. Int.* **2019**, *26*, 11905–11919. [CrossRef]

35. Karaca, S.; Önal, E.Ç.; Açışlı, Ö.; Khataee, A. Preparation of chitosan modified montmorillonite biocomposite for sonocatalysis of dyes: Parameters and degradation mechanism. *Mater. Chem. Phys.* **2021**, *260*, 124125. [CrossRef]

36. Samanta, M.; Mukherjee, M.; Ghorai, U.K.; Bose, C.; Chattopadhyay, K.K. Room temperature processed copper phthalocyanine nanorods: A potential sonophotocatalyst for textile dye removal. *Mater. Res. Bull.* **2020**, *123*, 110725. [CrossRef]

37. Karim, A.V.; Shriwastav, A. Degradation of amoxicillin with sono, photo, and sonophotocatalytic oxidation under low-frequency ultrasound and visible light. *Environ. Res.* **2021**, *200*, 111515. [CrossRef]

38. Ghalamchi, L.; Aber, S. An aminated silver orthophosphate/graphitic carbon nitride nanocomposite: An efficient visible light sonophotocatalyst. *Mater. Chem. Phys.* **2020**, *256*, 123649. [CrossRef]

39. Zhang, T.; Jin, X.; Owens, G.; Chen, Z. Remediation of malachite green in wastewater by ZIF-8@Fe/Ni nanoparticles based on adsorption and reduction. *J. Colloid Interface Sci.* **2021**, *594*, 398–408. [CrossRef]

40. Pandey, S.K.; Tripathi, M.K.; Ramanathan, V.; Mishra, P.K.; Tiwary, D. Enhanced photocatalytic efficiency of hydrothermally synthesized g-C3N4/NiO heterostructure for mineralization of malachite green dye. *J. Mater. Res. Technol.* **2021**, *11*, 970–981. [CrossRef]

41. Li, Y.; Wang, X.; Wang, X.; Xia, Y.; Zhang, A.; Shi, J.; Gao, L.; Wei, H.; Chen, W. Z-scheme BiVO$_4$/g-C$_3$N$_4$ heterojunction: An efficient, stable and heterogeneous catalyst with highly enhanced photocatalytic activity towards Malachite Green assisted by H$_2$O$_2$ under visible light. *Colloids Surf. A Physicochem. Eng. Asp.* **2021**, *618*, 126445. [CrossRef]

42. De Andrade, F.V.; De Oliveira, A.B.; Siqueira, G.O.; Lage, M.M.; De Freitas, M.R.; De Lima, G.M.; Nuncira, J. MnFe$_2$O$_4$ nanoparticulate obtained by microwave-assisted combustion: An efficient magnetic catalyst for degradation of malachite green cationic dye in aqueous medium. *J. Environ. Chem. Eng.* **2021**, *9*, 106232. [CrossRef]

43. Moradi, S.; Sobhgol, S.A.; Hayati, F.; Isari, A.A.; Kakavandi, B.; Bashardoust, P.; Anvaripour, B. Performance and reaction mechanism of MgO/ZnO/Graphene ternary nanocomposite in coupling with LED and ultrasound waves for the degradation of sulfamethoxazole and pharmaceutical wastewater. *Sep. Purif. Technol.* **2020**, *251*, 117373. [CrossRef]

44. Thue, P.S.; Lima, E.C.; Sieliechi, J.M.; Saucier, C.; Dias, S.L.P.; Vaghetti, J.C.P.; Rodembusch, F.S.; Pavan, F.A. Effects of first-row transition metals and impregnation ratios on the physicochemical properties of microwave-assisted activated carbons from wood biomass. *J. Colloid Interface Sci.* **2017**, *486*, 163–175. [CrossRef]
45. Chu, J.-H.; Kang, J.-K.; Park, S.-J.; Lee, C.-G. Application of magnetic biochar derived from food waste in heterogeneous sono-Fenton-like process for removal of organic dyes from aqueous solution. *J. Water Process Eng.* **2020**, *37*, 101455. [CrossRef]
46. Akram, N.; Guo, J.; Guo, Y.; Kou, Y.; Suleman, H.; Wang, J. Enhanced synergistic catalysis of novel Ag_2O/CuO nanosheets under visible light illumination for the photodecomposition of three dyes. *J. Environ. Chem. Eng.* **2021**, *9*, 104824. [CrossRef]
47. Tian, Y.; Ma, L.; Tian, X.; Nie, Y.; Yang, C.; Li, Y.; Lu, L.; Zhou, Z. More reactive oxygen species generation facilitated by highly dispersed bimodal gold nanoparticle on the surface of Bi_2WO_6 for enhanced photocatalytic degradation of ofloxacin in water. *Chemosphere* **2021**, *269*, 128717. [CrossRef]

MDPI
St. Alban-Anlage 66
4052 Basel
Switzerland
Tel. +41 61 683 77 34
Fax +41 61 302 89 18
www.mdpi.com

Sustainability Editorial Office
E-mail: sustainability@mdpi.com
www.mdpi.com/journal/sustainability

9 78303 6 541136